P9-DHL-742

linear
algebra
SECOND EDITION

linear algebra

SECOND EDITION

MICHAEL O'NAN

Rutgers University

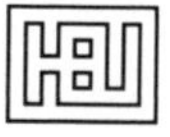

HARCOURT BRACE JOVANOVICH, INC.

NEW YORK CHICAGO SAN FRANCISCO ATLANTA

ISBN: 0-15-518560-8

Library of Congress Catalog Card Number: 76-27597

Printed in the United States of America

preface

Since the publication of *Linear Algebra*, many instructors and students have expressed an interest in more examples that deal with applications. This edition includes many such examples chosen from a variety of fields: physics, chemistry, biology, and the behavioral sciences. Generally speaking, the examples are quite simple; applications requiring lengthy explanation have been avoided.

The remaining textual revisions are of three types. First, simplifications have been made in the treatment of several topics, particularly the equivalence of linear systems and the row and column operations of determinants. Second, material that is not of general interest has been deleted, notably several complicated examples and some inessential results concerning rank. Third, the treatment of eigenvalues and canonical forms has been expanded. New exercises, many of an applied nature, have also been added.

However, the basic character of the book has not changed. The core of the book is intended for a one-semester course at the sophomore level. With some of the new material, it can also be used for a junior-level course.

In order to offset the abstractness of the subject, the spirit of the book is geometric. All concepts and examples are concretely related to their appropriate embodiments in two- and three-dimensional space. The introduction of new ideas is accompanied by detailed numerical examples. All concrete questions about the concepts encountered in the first five chapters can be solved by procedures akin to Gaussian elimination. Whenever possible, models make the phenomena of linear algebra concrete. Thus, for example, matrix products and powers and noncommuting matrices are studied in matrix algebra. Each algebraic concept is illustrated empirically. For these reasons, the discussion should be accessible to all students regardless of their field of specialization.

The general pattern of the book is to proceed from the concrete to the abstract. The material splits naturally into two parts. In the first part, we treat linear equations, column vectors, matrices, and determinants; in the second part we consider vector spaces, linear transformations, and inner products. Because of this ordering, it is possible to progress gradually from computational methods to more sophisticated concepts, thus sparing the student any traumatic plunge into the depths (or heights, according to point of view) of mathematical abstraction. This arrangement of topics has the additional advantage of providing examples that are relevant when vector spaces, linear transformations, and other abstract objects are introduced. Whenever possible, concepts are given their natural geometric interpretation—for example, vectors as directed line segments, determinants as areas and volumes, linear transformations as rotations, reflections, projections, and so on.

Chapter 1 treats simultaneous equations; in particular, their solution by Gaussian elimination is emphasized.

The second chapter begins with a study of vectors in real three-dimensional space. After defining addition and scalar multiplication geometrically, we proceed to deduce the usual operations on components, which correspond to the operations on vectors. Lines and other geometric objects are studied in this connection. Higher-dimensional spaces are introduced as spaces of column vectors. Next, matrices and the related operations of addition and multiplication are defined. Particular emphasis is given to the notion of an inverse of a matrix and its usefulness in solving systems of linear equations.

Chapter 3 deals with determinants and begins with the 2×2 case. This is done primarily to acquaint students with row and column operations, under circumstances in which the determinant is probably already familiar. For the higher-dimensional case, an inductive definition using cofactor expansions is given. More emphasis is placed on the properties of the determinant (row and column operations) than on the formal definition. The multiplicative property of determinants is proved, and their usefulness in finding inverses of matrices and solving simultaneous systems of linear equations is pointed out. In addition, the student is shown the more practical technique of inverting matrices by synthetic elimination.

Abstract vector spaces are introduced in Chapter 4. Because this may be the student's first contact with abstract mathematical objects, the approach is deliberately slow. Separate sections are devoted to the concepts of the span of a set of vectors, linear independence, subspaces, and basis. In each case, the concept is interpreted in two- and three-dimensional space, and then the principal theorems on dimension and basis are obtained.

In the fifth chapter, linear transformations are defined and explicated. Just as in the preceding chapter, the pace is unhurried; a full section is given to the range space of a linear transformation, while another treats the nullspace. We then consider the technique of calculating the rank of a matrix using row and column operations. (Since this section is not necessary for the subsequent development, it may be omitted.) Following this, the notions of inverse and isomorphism are studied. The matrix representation of a linear transformation, effected by a choice of basis, is also discussed in this chapter. The usual theorems concerning the relation of the rank and nullity of linear transformations are obtained, as are those concerning isomorphism and dimension.

The discussion of general inner products in Chapter 6 is motivated by first studying the dot and cross products in three-dimensional space. Both the geometric and algebraic definitions of these quantities are given. Later sections contain the definition of inner products in $\boldsymbol{R}^n$ *and* $\boldsymbol{C}^n$, and deal with orthogonality, orthogonal complements, orthonormal basis, and the like. Results such as the Cauchy–Schwarz inequality, Bessel's inequality, and the Gram–Schmidt orthogonalization procedure are obtained.

In the final chapter, eigenvalues are defined, and their principal properties are covered. Also, the diagonalizability of symmetric matrices is demonstrated. Next, the Jordan normal form is obtained. Lastly, a brief introduction to bilinear forms is included.

Each section contains numerous exercises; some allow the student to practice routine computational skills, while others test the student's understanding of the theory and challenge him or her to establish deeper results. Answers for selected exercises are found at the end of the book.

MICHAEL O'NAN

contents

3

DETERMINANTS 67

4

VECTOR SPACES 105

5

LINEAR TRANSFORMATIONS 163

linear
algebra
SECOND EDITION

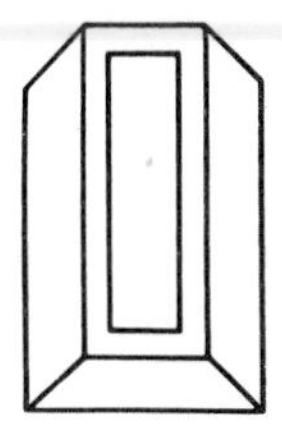

systems of linear equations

1 INTRODUCTION

Often in the course of solving problems in mathematics or its applications in the physical, biological, or social sciences, we encounter systems of simultaneous linear equations. Historically, linear algebra arose in the attempt to develop methods for solving such systems. Thus, it is appropriate to begin this book by studying systems of linear equations.

Consider, as a first example, the problem of finding all real numbers x and y that satisfy the two equations

$$x + y = 4$$
$$2x - y = 5$$

This problem has a geometric interpretation. Each equation defines a line in the Cartesian plane. Since the problem requires finding a point satisfying both equations, we must locate the point where the two lines intersect. The two lines are neither parallel nor identical, so there is a unique point of intersection. (See Figure 1-1.) In order to determine the coordinates of the point of intersection, we resort to the familiar procedure of elimination. We begin with the original system of equations.

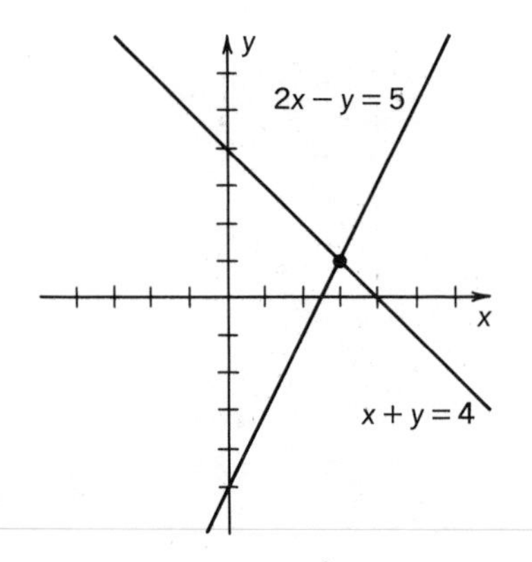

FIGURE 1-1

$$\begin{aligned} x+y&=4\\ 2x-y&=5 \end{aligned}$$

Add the first equation to the second.

$$\downarrow$$

$$\begin{aligned} x+y&=4\\ 3x\quad&=9 \end{aligned}$$

Divide the second equation by 3.

$$\downarrow$$

$$\begin{aligned} x+y&=4\\ x\quad&=3 \end{aligned}$$

Subtract the second equation from the first equation.

$$\downarrow$$

$$\begin{aligned} y&=1\\ x\quad&=3 \end{aligned}$$

Thus, we see that the point (3, 1) is the only solution to the new system of equations. On the other hand, since each step in our sequence of operations is reversible, we see that (3, 1) is indeed a solution to the original system of equations.

In the last example, we saw that the system of linear equations had a unique solution. However, there are systems of equations that possess no solutions and systems that have infinitely many solutions.

As an example of the former, we have the system

$$2x + 4y = 3$$

$$3x + 6y = 1$$

To see this, assume that a solution to the system does exist and multiply the first equation by 3 and the second by 2. We obtain

$$6x + 12y = 9$$

$$6x + 12y = 2$$

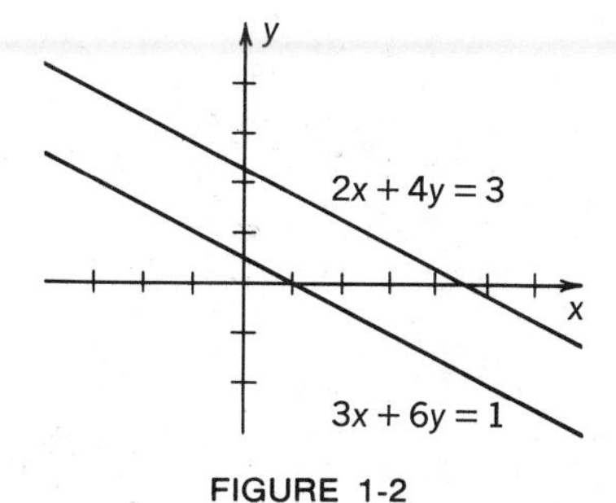

FIGURE 1-2

which means that $9 = 2$, a patent absurdity. Thus, in this case, no solution to the given system of equations exists. Geometrically, the lines determined by the equations are nonidentical parallel lines (see Figure 1-2), and it is futile to attempt to find a point of intersection.

As an example of a system having infinitely many solutions, we have the equations

$$\begin{aligned} x + 2y &= 3 \\ 3x + 6y &= 9 \end{aligned}$$

The equations in this case are formally different, but they, in fact, represent the same line. Indeed, the second equation is a multiple of the first. It is clear that the entire line is the solution set to the system.

There are many problems, both practical and amusing, whose solution may be obtained by solving a system of linear equations. We present one of the latter type.

Example 1 A man has a son and a daughter. The man is four times as old as the son, and the son is four years older than the daughter. Three years from now the man will be five times as old as the daughter. Find the ages of the man, the son, and the daughter.

We let m be the age of the man, s the age of the son, and d the age of the daughter. Since the man is four times as old as the son, $m = 4s$. Since the son is four years older than the daughter, $s = 4 + d$. Three years from now the age of the man will be $m + 3$, and that of the daughter will be $d + 3$. Thus, $m + 3 = 5(d + 3)$. Rearranging, we obtain the system of equations

$$\begin{aligned} m - 4s &= 0 \\ s - d &= 4 \\ m - 5d &= 12 \end{aligned}$$

Subtract the first equation from the third.

$$\downarrow$$

$$\begin{aligned} m - 4s &= 0 \\ s - d &= 4 \\ 4s - 5d &= 12 \end{aligned}$$

Multiply the second equation by -4 and add to the third.

$$\downarrow$$

$$\begin{aligned} m - 4s \quad &= 0 \\ s - d &= 4 \\ -d &= -4 \end{aligned}$$

Subtract the third equation from the second.

$$\downarrow$$

$$\begin{aligned} m - 4s \quad &= 0 \\ s \quad &= 8 \\ -d &= -4 \end{aligned}$$

Multiply the second equation by 4 and add to the first. This gives us

$$\downarrow$$

$$\begin{aligned} m \quad &= 32 \\ s \quad &= 8 \\ -d &= -4 \end{aligned}$$

or $m = 32$, $s = 8$, and $d = 4$.

The following shows how a system of linear equations might arise in a simple physical problem.

Example 2 Three metal balls and a meter stick are given. The mass of ball 1 and the mass of ball 2 are to be found, while that of ball 3 is 2 kilograms. The center of the meter stick is placed on a fulcrum, and the balls are suspended from the meter stick by means of string (whose weight we ignore). Two ways are found to position the balls so that the meter stick is balanced.

In the first stable position, ball 1 is 40 centimeters to the left of the fulcrum, ball 2 is 15 centimeters to the left, and ball 3 is 50 centimeters to the right. In the second position, ball 1 is 50 centimeters to the right, ball 2 is 25 centimeters to the left, and ball 3 is 25 centimeters to the right. (See Figure 1-3.) Find the mass of ball 1 and that of ball 2.

We let x and y be the masses of ball 1 and ball 2, respectively. A basic principle of physics states that when the system is in equilibrium, the sum of the moments on the left side of the fulcrum must equal the sum of the moments on the right. The moment is defined to be the mass of the object times its distance from the fulcrum. In the first balanced position, the moments of balls 1, 2, and 3 are, respectively, $40x$, $15y$, and $2 \cdot 50$. Thus, $40x + 15y = 100$. In the second case, the moments of balls 2, 3, and 1 are, respectively, $25y$, $25 \cdot 2$, and $50x$. Thus, $25y = 50 + 50x$. So, we obtain the system

$$\begin{aligned} 40x + 15y &= 100 \\ -50x + 25y &= 50 \end{aligned}$$

Solving this as we did earlier systems, we find that $x = 1$ and $y = 4$.

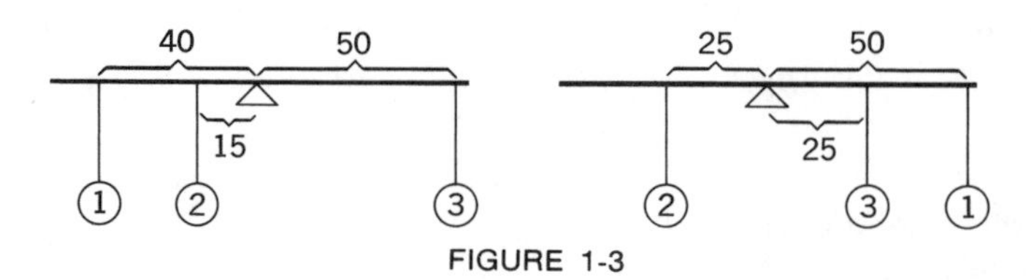

FIGURE 1-3

EXERCISES

1. Consider the following systems of linear equations in two unknowns. For each pair of equations, determine whether a solution exists and what the solution is, if it exists, and interpret geometrically.

(a) $2x + y = 7$, $x - 3y = 2$ (b) $x + y = 3$, $3x + 2y = 8$ (c) $4x - 3y = 8$, $8x - 6y = 24$

(d) $x - y = 8$, $2x + 8y = 6$ (e) $x + 2y = 3$, $2x + 4y = 6$ (f) $2x - 6y = 7$, $3x - 9y = 4$

2. Consider the system of equations

$$ax + by = 0 \qquad cx + dy = 0,$$

with a, b, c, and d real numbers.
(a) Show that the system has at least one solution.
(b) If $ad - bc \neq 0$, show that the system has only one solution. Interpret your result geometrically.

3. Solve the systems

(a) $x + y + z = 6$, $3x - 2y + 3z = 3$, $2x - y + z = 4$ (b) $x - y + z = 0$, $2x - y + z = 1$, $3x + y - 3z = 2$

4. In a certain ecosystem, there are 100 individuals of a given species. Five years later, there are 240. During that five-year period, the number of males has doubled, and the number of females has tripled. How many males and females were there originally?

5. A factory possesses two machines, M and N. M is available for 15 hours a day, and N is available for 10 hours a day. The factory produces two products, A and B. To produce one unit of A requires using M for one hour and N for one hour. To produce one unit of B requires using M for $\frac{1}{2}$ hour and N for $\frac{1}{4}$ hour. How many units of A and B should be produced daily to insure that M and N are in operation for the full time for which they are available?

6. The most general solution to the differential equation $y'' - y = 0$ is a function of the form $y(x) = Ae^{x} + Be^{-x}$. Find a solution for which $y(0) = 1$ and $y'(0) = 0$.

2 EQUIVALENCE OF LINEAR SYSTEMS

In the last section, we employed the technique of elimination to solve several systems of linear equations. Let us now abstract from these earlier examples some general principles and techniques for dealing with linear systems.

By a **linear equation**, we mean an equation of the form $a_1x_1 + a_2x_2 + \cdots + a_nx_n = b$, where $a_1, a_2, \ldots, a_n$ and b are constants and $x_1, x_2, \ldots, x_n$ are the quantities to be determined. The x_i's are called **variables.** The number a_i is called the coefficient of x_i in the equation. A **solution** to the equation is an n-tuple of real numbers $(c_1, c_2, \ldots, c_n)$ such that $a_1c_1 + a_2c_2 + \cdots + a_nc_n = b$. For example, $3x_1 - x_2 + 2x_3 - 5x_4 = 12$ is a linear equation. Its variables are x_1, x_2, x_3, and x_4. The coefficient of x_1 is 3. The coefficient of x_2 is -1, and so on. Examples of solutions are (4, 0, 0, 0) and (0, 2, 7, 0).

A **system of linear equations** is a collection of several linear equations. For example,

$$\begin{aligned} 4x_1 + 5x_2 - 3x_3 &= 1 \\ x_1 + 2x_2 + 4x_3 &= 0 \\ -x_1 + 4x_2 - 2x_3 &= 4 \end{aligned}$$

is a system of linear equations.

By a **solution to a system of linear equations** in n variables, we mean an n-tuple of numbers which satisfies each equation of the system. For example, $(3, -1, 0, 2)$ is a solution to the system

$$\begin{aligned} x_1 + x_2 - x_3 + 4x_4 &= 10 \\ -x_1 - x_2 + 2x_3 + x_4 &= 0 \\ 10x_1 + 3x_2 \qquad + x_4 &= 29 \end{aligned}$$

Suppose that we are given two systems of linear equations in n variables. We say that these systems are **equivalent** if any n-tuple of numbers which is a solution to one system is a solution to the other system as well. To state this in another way, the two systems are equivalent if they have the same solutions.

According to this definition, the systems

$$\begin{aligned} x + y &= 4 \\ 2x - y &= 5 \end{aligned} \quad \text{and} \quad \begin{aligned} 2x + 2y &= 8 \\ 4x + y &= 13 \end{aligned}$$

are equivalent, since in both cases the solution to the system is (3, 1).

It is clear that changing the order in which the equations are listed has no effect on the solutions and, hence, provides an equivalent system of equations. In addition, there are two other simple ways for changing a system of equations into an equivalent system.

(1) Suppose that a given system of equations is changed to a second system by multiplying a single equation in the first system by a nonzero number and leaving all other equations the same. Then the second system of equations is equivalent to the first system.

PROOF Let $a_1x_1 + a_2x_2 + \cdots + a_nx_n = b$ be the equation in the first system that is to be multiplied by the nonzero number, c. In the second system, this equation becomes $ca_1x_1 + ca_2x_2 + \cdots + ca_nx_n = cb$. If $(g_1, g_2, \ldots, g_n)$ is a solution to the first system, $a_1g_1 + a_2g_2 + \cdots + a_ng_n = b$. Therefore, $(ca_1)g_1 + \cdots + (ca_n)g_n = cb$. Since the remaining equations of the two systems are the same, $(g_1, g_2, \ldots, g_n)$ is a solution to the second system.

Suppose, on the other hand, that $(h_1, h_2, \ldots, h_n)$ is a solution to the second system of equations. Then, $ca_1h_1 + ca_2h_2 + \cdots + ca_nh_n = cb$. Multiplying the equation by c^{-1}, we have $a_1h_1 + a_2h_2 + \cdots + a_nh_n = b$. Again, all the remaining equations in the two systems are identical. Therefore, $(h_1, h_2, \ldots, h_n)$ is a solution to the first system. ■

For example, using this principle, we see that the systems

$$\begin{aligned} x + 5y - 2z &= 0 \\ 2x + 3y + z &= 3 \end{aligned} \quad \text{and} \quad \begin{aligned} x + 5y - 2z &= 0 \\ 6x + 9z + 3z &= 9 \end{aligned}$$

are equivalent. Using (1) twice it follows that

$$\begin{aligned} x + 3y &= 4 \\ -x + 4y &= 2 \end{aligned} \quad \text{and} \quad \begin{aligned} 2x + 6y &= 8 \\ x - 4y &= -2 \end{aligned}$$

are equivalent.

The second principle of equivalence is:

(2) Suppose that a given system of equations is changed to a second system by adding one equation of the first system to another equation of the first system, while leaving all other equations the same. Then the second system of equations is equivalent to the first system.

According to the second principle, the system of equations

$$\begin{aligned} x + y - z &= 1 \\ x + 3y - 2z &= 0 \\ x + y + 4z &= 1 \end{aligned}$$

is equivalent to the system

$$\begin{aligned} x + y - z &= 1 \\ 2x + 4y - 3z &= 1 \\ x + y + 4z &= 1 \end{aligned}$$

since the second equation in the second system was obtained by adding the first equation of the first system to the second equation of the first system, while all other equations remain the same.

PROOF Let $a_1x_1 + a_2x_2 + \cdots + a_nx_n = b$ be an equation of the first system. Let $c_1x_1 + c_2x_2 + \cdots + c_nx_n = d$ be the equation in the first system that is to be added to $a_1x_1 + a_2x_2 + \cdots + a_nx_n = b$. Thus, in the second system the equation $a_1x_1 + a_2x_2 + \cdots + a_nx_n = b$ is replaced by $(a_1 + c_1)x_1 + (a_2 + c_2)x_2 + \cdots + (a_n + c_n)x_n = b + d$, while all other equations in the two systems are the same. Now suppose $(g_1, g_2, \ldots, g_n)$ is a solution to the first system. Then, we have

$$a_1 g_1 + a_2 g_2 + \cdots + a_n g_n = b$$

$$c_1 g_1 + c_2 g_2 + \cdots + c_n g_n = d$$

Therefore, $(a_1 + c_1) g_1 + (a_2 + c_2) g_2 + \cdots + (a_n + c_n) g_n = b + d$. Since all other equations in the first system are identical with those of the second system, it follows that $(g_1, g_2, \ldots, g_n)$ is a solution to the second system.

Next, let $(h_1, h_2, \ldots, h_n)$ be a solution to the second system. Then

$$\begin{aligned} (a_1 + c_1)h_1 + (a_2 + c_2)h_2 + \cdots + (a_n + c_n)h_n &= b + d \\ c_1 h_1 + c_2 h_2 + \cdots + c_n h_n &= d \end{aligned}$$

Therefore, $a_1 h_1 + a_2 h_2 + \cdots + a_n h_n = b$. Again, all remaining equations in the two systems are the same, so $(h_1, h_2, \ldots, h_n)$ is a solution to the first system. ■

By repeated use of these two principles, we can obtain simpler systems of equations.

EXERCISES

1. Using the definition of equivalence of systems of equations, determine which of the following pairs of systems are equivalent.

(a) $\begin{aligned} 2x + y &= 1 \\ x - y &= 0 \end{aligned}$ $\quad \begin{aligned} x + 2y &= 5 \\ 3x + y &= 5 \end{aligned}$

(b) $\begin{aligned} x + y &= 0 \\ x - 2y &= 0 \end{aligned}$ $\quad \begin{aligned} 3x + 2y &= 0 \\ x - y &= 0 \end{aligned}$

(c) $\begin{aligned} x + y &= 0 \\ x - 2y &= 0 \end{aligned}$ $\quad \begin{aligned} 2x + 2y &= 0 \\ 3x + 3y &= 0 \end{aligned}$

(d) $\begin{aligned} 2x + y &= 5 \\ x - y &= -2 \end{aligned}$ $\quad \begin{aligned} 3x - y &= 0 \\ x + y &= 4 \end{aligned}$

(e) $\begin{aligned} x - y &= 2 \\ x + y &= 4 \end{aligned}$ $\quad \begin{aligned} x + 2y &= 8 \\ 3x + y &= 4 \end{aligned}$

(f) $\begin{aligned} x + 2y &= 1 \\ 3x + 6y &= 0 \end{aligned}$ $\quad \begin{aligned} x + y &= 5 \\ 2x - y &= 1 \end{aligned}$

2. Consider the two systems of equations

$$\begin{aligned} a_{11}x_1 + a_{12}x_2 + a_{13}x_3 &= b_1 \\ a_{21}x_1 + a_{22}x_2 + a_{23}x_3 &= b_2 \end{aligned} \qquad \begin{aligned} c_{11}x_1 + c_{12}x_2 + c_{13}x_3 &= 0 \\ c_{21}x_1 + c_{22}x_2 + c_{23}x_3 &= 0 \end{aligned}$$

If $b_1 \neq 0$ or $b_2 \neq 0$, show that these systems are not equivalent.

3 METHOD OF GAUSSIAN ELIMINATION

If we are given a system of m linear equations in n unknowns,

$$\begin{aligned}
a_{11}x_1 + a_{12}x_2 + a_{13}x_3 + \cdots + a_{1n}x_n &= b_1 \\
a_{21}x_1 + a_{22}x_2 + a_{23}x_3 + \cdots + a_{2n}x_n &= b_2 \\
a_{31}x_1 + a_{32}x_2 + a_{33}x_3 + \cdots + a_{3n}x_n &= b_3 \\
&\vdots \\
a_{m1}x_1 + a_{m2}x_2 + a_{m3}x_3 + \cdots + a_{mn}x_n &= b_m
\end{aligned}$$

where m and n are positive integers, a_{ij} and b_i are constants for $i = 1, 2, \ldots, m$ and $j = 1, 2, \ldots, n$, and x_j is a variable for $j = 1, 2, \ldots, n$, we can apply the following reduction procedure.

(1) Choose a variable with a nonzero coefficient. Suppose x_s is such a variable and that its coefficient in the tth equation is $a_{ts} \neq 0$. Multiply the tth equation by a_{ts}^{-1}.
(2) For each $i \neq t$ and $i = 1, 2, \ldots, m$ add $-a_{is}$ times equation t to the ith equation in the system resulting from step (1). This step eliminates x_s from all equations but the tth. x_s is now called a used variable and the tth equation is called a used equation.
(3) Next choose a new variable, x_r, with a nonzero coefficient in an unused equation. Use steps (1) and (2) to eliminate x_r from all other equations, including those that have already been used. Again the equation we have worked with in this step is called used. Step (3) is repeated until no unused equations remain, or until only equations of the form $0 = c$ are left, where c is some number.

If there are solutions to the system of equations, they will all be obtained from this procedure. If no solutions exist, a contradiction will be obtained using this method.

It is important to note that by the principles of the preceding section, the final system of equations that we obtain is equivalent to our original system.

We now present several examples illustrating this technique.

Example 1

$$\begin{aligned}
x + y \phantom{{}- 2z} &= 1 \\
6x \phantom{{}+ 3y} - 2z &= -8 \\
3y - z &= -3
\end{aligned}$$

Let y be the first variable to be used. Choosing the third equation as the equation to be used, we multiply it by $\frac{1}{3}$.

$\downarrow$

$$\begin{aligned} x + y \quad &= 1 \\ 6x \quad - 2z &= -8 \\ y - \tfrac{1}{3}z &= -1 \end{aligned}$$

Multiply the third equation by (-1) and add it to the first; multiply the third equation by 0 and add it to the second.

$\downarrow$

$$\begin{aligned} x \quad + \tfrac{1}{3}z &= 2 \\ 6x \quad - 2z &= -8 \\ y - \tfrac{1}{3}z &= -1 \end{aligned}$$

Choose the first equation as the next equation to be used, and z as next variable to be used. Multiply equation one by 3.

$\downarrow$

$$\begin{aligned} 3x \quad + z &= 6 \\ 6x \quad - 2z &= -8 \\ y - \tfrac{1}{3}z &= -1 \end{aligned}$$

Add 2 times the first equation to the second; add $\frac{1}{3}$ times the first equation to the third.

$\downarrow$

$$\begin{aligned} 3x \quad + z &= 6 \\ 12x \quad &= 4 \\ x + y \quad &= 1 \end{aligned}$$

Choose the second equation and x as equation and variable to be used; multiply the second equation by $\frac{1}{12}$.

$\downarrow$

$$\begin{aligned} 3x \quad + z &= 6 \\ x \quad &= \tfrac{1}{3} \\ x + y \quad &= 1 \end{aligned}$$

Add (-3) times equation two to equation one; add (-1) times equation two to equation three.

$\downarrow$

$$\begin{aligned} z &= 5 \\ x &= \tfrac{1}{3} \\ y &= \tfrac{2}{3} \end{aligned}$$

Thus, in this case, a unique solution is obtained.

Example 2

$$\begin{aligned} 2x + 4y &= 3 \\ 3x + 6y &= 1 \end{aligned}$$

Choose x as the first variable to be used and equation one as the first equation to be used. Multiply the first equation by $\frac{1}{2}$.

$$\downarrow$$

$$\begin{aligned} x + 2y &= \tfrac{3}{2} \\ 3x + 6y &= 1 \end{aligned}$$

Add (-3) times equation one to equation two.

$$\downarrow$$

$$\begin{aligned} x + 2y &= \tfrac{3}{2} \\ 0 &= -\tfrac{7}{2} \end{aligned}$$

The second equation in this system is, of course, absurd. Strictly speaking, it should be written as $0 \cdot x + 0 \cdot y = -\frac{7}{2}$, and this equation can never be satisfied. But what does this mean? Recall again the definition of equivalence of systems of linear equations: two systems are equivalent if they have the same solutions. In this case, the last system that we obtained had no solutions. Consequently, the first system has no solution. As in the first section, one can check that the equations represent distinct parallel lines in the plane, and, thus, the lines have no point of intersection.

Example 3

$$\begin{aligned} x_1 \qquad - x_3 + 2x_4 &= 0 \\ 2x_1 - 3x_2 \qquad - 4x_4 &= 0 \\ x_2 + 2x_3 + 3x_4 &= 0 \\ 3x_1 - 2x_2 + x_3 + x_4 &= 0 \end{aligned}$$

Use x_1 and the first equation.

$$\downarrow$$

$$\begin{aligned} x_1 \qquad - x_3 + 2x_4 &= 0 \\ -3x_2 + 2x_3 - 8x_4 &= 0 \\ x_2 + 2x_3 + 3x_4 &= 0 \\ -2x_2 + 4x_3 - 5x_4 &= 0 \end{aligned}$$

Use x_2 and the third equation.

$$\downarrow$$

$$\begin{aligned} x_1 \qquad - x_3 + 2x_4 &= 0 \\ 8x_3 + x_4 &= 0 \\ x_2 + 2x_3 + 3x_4 &= 0 \\ 8x_3 + x_4 &= 0 \end{aligned}$$

Use x_4 and the second equation.

$$\downarrow$$

$$\begin{aligned} x_1 \qquad -17x_3 \qquad &= 0 \\ 8x_3 + x_4 &= 0 \\ x_2 - 22x_3 \qquad &= 0 \\ 0 &= 0 \end{aligned}$$

The method can be used no farther since the only unused variable, x_3, does not occur with a nonzero coefficient in the unused equation. From the above system of equations we see that if we choose any value for x_3, say $x_3 = c$, and if we then let

$$\begin{aligned} x_1 &= 17x_3 = 17c \\ x_4 &= -8x_3 = -8c \\ x_2 &= 22x_3 = 22c \end{aligned}$$

we obtain a solution to the above system, and that any solution to the system is of this type for some value of c. In other words, the most general solution is the 4-tuple $(17c, 22c, c, -8c)$. Since c may vary over all the real numbers, there are infinitely many solutions to this equation.

Now that we have seen three examples of the elimination procedure, perhaps it would be appropriate to discuss the character of the results obtained in these examples.

In the first example, the equations were consistent and determined the answer exactly. In the second example, there was no solution. Here we say the system is inconsistent or "over-determined." In the third case, there were infinitely many solutions. Here the equations are consistent but "under-determined."

The first case is probably the most familiar. In an actual problem, this case would occur if the problem is well-posed, and there are enough data to determine the answer. In examples 1 and 2 of the last section, we encountered problems of this type. In both of these problems, the data were consistent and sufficient to solve the problems.

Inconsistent systems are also easy to construct: take a system of equations that has only one solution, choose another equation that this solution does not satisfy, and enlarge the system by including this equation. For example, using the three equations of example 1, we make a larger system by including the equation $3x + 6y - z = 8$. Since $x = \frac{1}{3}, y = \frac{2}{3}, z = 5$ is not a solution to this equation, the enlarged system is inconsistent.

In practice, inconsistent equations arise for a number of reasons. A mistake in the mathematical formulation of a problem might cause inconsistency. However, some problems can be solved by showing that a system is inconsistent. We present one of this type.

Example 4 Do the points $P_1 = (1, 0)$, $P_2 = (1, 1)$, $P_3 = (0, 2)$, and $P_4 = (-1, 3)$ lie on a circle?

To solve this recall that the equation of a circle is of the form: $x^2 + y^2 + Ax + By + C = 0$. We attempt to determine A, B, and C so that the four points will lie on this circle. Substituting the coordinates of these four points into the equation, we obtain four equations for A, B, and C:

$$\begin{aligned} A \qquad + C &= -1 \\ A + B + C &= -2 \\ 2B + C &= -4 \\ -A + 3B + C &= -10 \end{aligned}$$

Use A and the first equation. $\downarrow$

$$\begin{aligned} A \qquad + C &= -1 \\ B \qquad &= -1 \\ 2B + C &= -4 \\ 3B + 2C &= -11 \end{aligned}$$

Use B and the second equation. $\downarrow$

$$\begin{aligned} A \qquad + C &= -1 \\ B \qquad &= -1 \\ C &= -2 \\ 2C &= -8 \end{aligned}$$

We could continue, but already from the last two equations we have the contradiction, $C = -2$ and $C = -4$. Thus, the system of equations is inconsistent. This means we cannot choose A, B, and C so that the four points satisfy the equation $x^2 + y^2 + Ax + By + C = 0$. In other words, the four points do not lie on any circle.

If we modify the points P_i of the last example so that they all do lie on some circle, instead of a contradiction, we would find A, B, and C, and the equation of the circle.

Inconsistent systems arise from contradictory data; systems with infinitely many solutions arise from lack of data. While the solution to a system may seem unsatisfactory when it is not unique, there are some problems when such an answer provides all the information that is desired. We present an example from chemistry.

Example 5 Balance the chemical equation

$$Ca + H_3PO_4 \longrightarrow Ca_3P_2O_8 + H_2$$

Here the problem is to find integers, h, i, j, and k, so that in the equation

$$hCa + iH_3PO_4 \longrightarrow jCa_3P_2O_8 + kH_2$$

the number of atoms of each of the elements on the right side equals the number on the left. There are h atoms of calcium on the left and $3j$ on the right. Thus, $h = 3j$. Using hydrogen, $3i = 2k$. From phosphorus, $i = 2j$.

Lastly, oxygen gives $4i = 8j$. Rewriting,

$$\begin{aligned} h \quad - 3j \quad\quad &= 0 \\ 3i \quad\quad - 2k &= 0 \\ i - 2j \quad\quad &= 0 \\ 4i - 8j \quad\quad &= 0 \end{aligned}$$

Use i and the third equation.

$$\downarrow$$

$$\begin{aligned} h \quad - 3j \quad\quad &= 0 \\ 6j - 2k &= 0 \\ i - 2j \quad\quad &= 0 \\ 0 &= 0 \end{aligned}$$

At this point, it is clear that if j is arbitrary, $(3j, 2j, j, 3j)$ is the solution. Normally, one would choose $j = 1$ and the other numbers accordingly. But if j were chosen as 2 and the other numbers by the above rule, the equation would still be balanced.

Example 6

$$\begin{aligned} x_1 + x_2 + x_3 + x_4 &= 1 \\ x_1 - x_2 + x_3 - x_4 &= 1 \\ x_1 \quad\quad + x_3 \quad\quad &= 1 \end{aligned}$$

Use x_1 and the third equation.

$$\downarrow$$

$$\begin{aligned} x_2 \quad\quad + x_4 &= 0 \\ - x_2 \quad\quad - x_4 &= 0 \\ x_1 \quad\quad + x_3 \quad\quad &= 1 \end{aligned}$$

Use x_2 and the first equation.

$$\downarrow$$

$$\begin{aligned} x_2 \quad\quad + x_4 &= 0 \\ 0 &= 0 \\ x_1 \quad\quad + x_3 \quad\quad &= 1 \end{aligned}$$

Again, we cannot proceed any further since both unused variables, x_3 and x_4, have a zero coefficient in the only unused equation. This implies that if x_3 and x_4 are chosen arbitrarily, i.e., $x_3 = c$ and $x_4 = d$, and if we let

$$x_2 = -x_4 = -d$$

$$x_1 = 1 - x_3 = 1 - c$$

then $(1 - c, -d, c, d)$ is a solution to the system. Thus, in this case, we obtain a "two parameter family" of solutions. Later we shall see how to make this statement more precise.

Example 7 By applying our elimination procedure, we attempt to determine for what values of a, b, and c a solution to the following system exists.

$$\begin{aligned} 2x - y + z &= a \\ x + 2y + z &= b \\ 3x + y + 2z &= c \end{aligned}$$

Use x and the second equation.

$$\downarrow$$

$$\begin{aligned} -5y - z &= a - 2b \\ x + 2y + z &= b \\ -5y - z &= c - 3b \end{aligned}$$

Use z and the first equation.

$$\downarrow$$

$$\begin{aligned} -5y - z &= a - 2b \\ x - 3y \qquad &= a - b \\ 0 &= c - a - b \end{aligned}$$

The elimination procedure comes to a halt since in our unused equation (the third) all variables have coefficients 0. It follows from the third equation that for any solutions to exist at all, we must have $c = a + b$.

If we assume this condition is satisfied and choose y arbitrarily and let $x = 3y + (a - b)$, $z = -5y - a + 2b$, then we have a solution to the third system. By the equivalence of the systems, likewise, we have a solution to the first system.

Thus, a necessary and sufficient condition that solutions exist for the first system is that $c = a + b$.

Before ending this section, we point out that there is no relationship, in general, between the consistency of a system of equations and the number of equations and variables in the system. For example, it is perfectly possible to have a system of two equations in ten variables which is inconsistent; likewise, it is possible to have a system of ten equations in two variables which is consistent. We give an example of a type of problem that often arises in the behavioral sciences. In this system, the number of equations exceeds the number of unknowns.

Example 8 In a certain country there are two political parties, D and R. Everyone belongs to one party or the other, and every year an election is held. During the course of a year, $\frac{4}{5}$ of the members of D remain in D, but $\frac{1}{5}$ of the members switch to R. Of the members of R, $\frac{3}{5}$ remain in R, and $\frac{2}{5}$ switch to D.

Each year at polling time, in spite of all the switching, the fraction of people in D and that in R remain the same. What fraction of the electorate belongs to D?

To solve this take x to be the fraction of the people in D and y the fraction in R. (For example, $x = \frac{5}{9}$ if $\frac{5}{9}$ of the people belong to D.) Since all of the people belong to R or D, $x + y = 1$.

Next, we compute the membership of D one year later by taking into account two factors: (1) D retains $\frac{4}{5}$ of its members, (2) D acquires $\frac{2}{5}$ of the members of R.

Now D retains $\frac{4}{5}$ of its members, and its members constitute an xth part of the population. So from this source, D has a $(\frac{4}{5})x$th part of the population. Also, D acquires $\frac{2}{5}$ of the members of R, and R has a yth part of the population. So from this source, D has a $(\frac{2}{5})y$th part of the population. Thus, altogether D has $(\frac{4}{5})x + (\frac{2}{5})y$ part of the population one year later. Since this is the same fraction of the population that D had one year earlier, $(\frac{4}{5})x + (\frac{2}{5})y = x$, or $-(\frac{1}{5})x + (\frac{2}{5})y = 0$.

Performing the same analysis on R, we find that $(\frac{1}{5})x + (\frac{3}{5})y = y$, or $(\frac{1}{5})x - (\frac{2}{5})y = 0$. Thus, we have the system

$$\begin{aligned} x + \quad y &= 1 \\ -(\tfrac{1}{5})x + (\tfrac{2}{5})y &= 0 \\ (\tfrac{1}{5})x - (\tfrac{2}{5})y &= 0 \end{aligned}$$

Solving as usual, we find that $x = \frac{2}{3}$ and $y = \frac{1}{3}$.

In this problem, for simplicity we assumed that there were only two parties. Consequently, the last two equations were almost the same. However, if this type of problem had been attempted in a three-party state, there would have been four equations. These four equations would, in general, be quite different. (See exercise 14.)

Later, we shall encounter other problems of this type and develop more efficient methods for formulating them.

EXERCISES

1. Solve the following systems of linear equations, indicating if the system has no solutions and giving the general form if it has many solutions.

(a) $$\begin{aligned} x - y + z &= 5 \\ 2x + y - z &= -2 \\ 3x - y - z &= -7 \end{aligned}$$

(b) $$\begin{aligned} x_1 - x_2 + 2x_3 &= -5 \\ 3x_1 + 4x_2 + 15x_3 &= 2 \\ 2x_1 - x_2 + x_3 &= 1 \end{aligned}$$

(c) $$\begin{aligned} x + 3y + z &= 2 \\ 3x + 4y - z &= 1 \\ x - 2y - 3z &= 1 \end{aligned}$$

(d) $$\begin{aligned} 2x - 3y + 4z &= 3 \\ x - y + z &= 1 \\ x - 2y + 3z &= 2 \end{aligned}$$

(e) $$\begin{aligned} x_1 + 2x_2 + x_3 - 2x_4 &= 1 \\ 2x_1 + x_2 - x_3 + x_4 &= 0 \\ x_1 - x_2 - 2x_3 + 3x_4 &= 1 \end{aligned}$$

(f) $$\begin{aligned} x - y + z &= 9 \\ 2x + y - 3z &= 0 \\ x + 4y + z &= 4 \\ 3x + y - 5z &= -1 \end{aligned}$$

(g) $$\begin{aligned} 2x_1 + x_2 - x_3 + x_4 &= 0 \\ x_1 - x_2 + x_3 + x_4 &= 0 \\ 4x_1 - x_2 + x_3 + 3x_4 &= 0 \end{aligned}$$

(h) $$\begin{aligned} -x_1 + x_2 + x_3 &= 4 \\ x_1 - x_2 + x_3 &= 2 \\ x_1 + x_2 - x_3 &= 0 \end{aligned}$$

2. Solve the following systems if a, b, and c are arbitrary.

$$\text{(a)}\quad \begin{aligned} x_2 + x_3 &= a \\ x_1 \quad + x_3 &= b \\ x_1 + x_2 \quad &= c \end{aligned} \qquad \text{(b)}\quad \begin{aligned} 2x + 5y &= a \\ x + 3y &= b \end{aligned}$$

3. Solve to find x and y as functions of t.

$$\begin{aligned} (1-t)x + \quad ty &= 0 \\ -tx + (1+t)y &= 1 \end{aligned}$$

4. Determine a necessary and sufficient condition on a, b, and c so that the following systems of equations admit solutions.

$$\text{(a)}\quad \begin{aligned} 2x + 3y - 2z &= a \\ x - 2y + z &= b \\ x - 9y + 5z &= c \end{aligned} \qquad \text{(b)}\quad \begin{aligned} x + 4y - 2z &= a \\ 2x - 2y + 3z &= b \\ x - 6y + 5z &= c \end{aligned}$$

5. Let A, B, C, and D be arbitrary real numbers. Prove that there is a polynomial, f, of degree at most 3 such that $f(0) = A$, $f'(0) = B$, $f(1) = C$, $f'(1) = D$. (Here $f'(a)$ denotes the derivative of f at a.)

6. A family consists of a mother, a father, a son, and a daughter. The father is two years older than the mother. Two years from now the father will be three times as old as the son, and the son will be twice as old as the daughter. Three years from now, the mother will be five times as old as the daughter. What are the ages of the family members?

7. Three cups are placed on a table, and a certain number of pennies are placed in each cup. It is known how many pennies are placed in each pair of cups. Is it possible to tell how many pennies are in each individual cup?

8. Derive a formula for converting temperatures from Celsius to Fahrenheit, given that 0°C = 32°F and 100°C = 212°F. (Common sense shows that the equation is linear.)

9. Four weights and a meter stick are given. The center of the meter stick is placed on a fulcrum, and the weights are suspended from the meter stick. Three positions of equilibrium are observed as described in the following table.

	WEIGHT 1	WEIGHT 2	WEIGHT 3	WEIGHT 4
Position 1	50 (left)	20 (left)	10 (right)	30 (right)
Position 2	25 (left)	50 (left)	50 (right)	30 (right)
Position 3	20 (left)	30 (left)	30 (right)	20 (right)

What are the ratios of the weights?

10. Do the points (2, 9), (0, 12), and (−4, 18) lie on a circle?

11. Balance the chemical equation

$$KMnO_4 + H_2SO_4 + KBr \rightarrow K_2SO_4 + Br_2 + MnSO_4 + H_2O$$

12. Four chemical elements A, B, C, and D are given. It is also known that the chemical compounds ABC, ABD_3, and A_3C exist. What can be said about the valences of the chemical elements?

13. A trucking company has available three types of trucks, 1, 2, and 3. It has contracted to carry three kinds of equipment, L, M, and N. The following table shows how many units of each type of equipment each truck can carry.

EQUIPMENT	TRUCK		
	1	2	3
L	2	1	5
M	4	2	3
N	3	1	1

For example, a truck of type 2 can carry 1 unit of L, 2 of M, and 1 of N. Suppose the order specifies 20 units of L, 26 units of M, and 15 units of N. How many trucks of each type should be used to insure that all trucks are full?

14. Extending example 8, we suppose that there are three parties, A, B, and C. Suppose that each year A retains $\frac{7}{10}$ of its members, $\frac{2}{10}$ of the members of A switch to B, and $\frac{1}{10}$ switch to C. B retains $\frac{6}{10}$ of its members, $\frac{3}{10}$ switch to A, and $\frac{1}{10}$ switch to C. C retains $\frac{7}{10}$ of its members, $\frac{2}{10}$ switch to A, and $\frac{1}{10}$ switch to B. Let x, y, and z be the fraction of the electorate held by A, B, and C, respectively. As in example 8, suppose that the fraction of the electorate held by each of the parties does not change from year to year.
    (a) Show $x + y + z = 1$.
    (b) By analyzing the membership of A, show that $(\frac{7}{10})x + (\frac{3}{10})y + (\frac{2}{10})z = x$.
    (c) Using B, show that $(\frac{2}{10})x + (\frac{6}{10})y + (\frac{1}{10})z = y$.
    (d) Using C, show that $(\frac{1}{10})x + (\frac{1}{10})y + (\frac{7}{10})z = z$.
    (e) Find x, y, and z.

4 HOMOGENEOUS SYSTEMS

We call a system of equations of the form

$$\begin{aligned}
a_{11}x_1 + a_{12}x_2 + a_{13}x_3 + \cdots + a_{1n}x_n &= 0 \\
a_{21}x_1 + a_{22}x_2 + a_{23}x_3 + \cdots + a_{2n}x_n &= 0 \\
a_{31}x_1 + a_{32}x_2 + a_{33}x_3 + \cdots + a_{3n}x_n &= 0 \\
&\vdots \\
a_{m1}x_1 + a_{m2}x_2 + a_{m3}x_3 + \cdots + a_{mn}x_n &= 0
\end{aligned}$$

a **homogeneous** system. In other words, a system is homogeneous if all quantities on the right-hand side of the system vanish.

Any system of homogeneous equations always has at least one solution, namely $x_1 = 0, x_2 = 0, \ldots, x_n = 0$. This solution is often called the **trivial** solution. Any other solution is called **nontrivial**.

There is one case in which we can guarantee that a homogeneous system has a nontrivial solution. Indeed, in the above system of equations suppose that $m < n$, i.e., there are more unknowns than equations. In this case we have the following important theorem.

Theorem In a homogeneous system of linear equations in which the number of variables is greater than the number of equations, a nontrivial solution exists.

PROOF Let the system be

$$\begin{aligned} a_{11}x_1 + a_{12}x_2 + \cdots + a_{1n}x_n &= 0 \\ a_{21}x_1 + a_{22}x_2 + \cdots + a_{2n}x_n &= 0 \\ &\vdots \\ a_{m1}x_1 + a_{m2}x_2 + \cdots + a_{mn}x_n &= 0 \end{aligned}$$

where $m < n$.

If all the coefficients of x_1 in the system are zero, i.e., if $a_{11} = a_{21} = \cdots = a_{m1} = 0$, then $x_1 = 1, x_2 = x_3 = \cdots = x_n = 0$ is a nontrivial solution. So suppose some coefficient of x_1 is nonzero. By renumbering the system we may suppose $a_{11} \neq 0$. Multiply equation one by a_{11}^{-1} and eliminate x_1 from the remaining equations to obtain

$$\begin{aligned} x_1 + b_{12}x_2 + b_{13}x_3 + \cdots + b_{1n}x_n &= 0 \\ b_{22}x_2 + b_{23}x_3 + \cdots + b_{2n}x_n &= 0 \\ &\vdots \\ b_{m2}x_2 + b_{m3}x_3 + \cdots + b_{mn}x_n &= 0 \end{aligned}$$

If the coefficients of x_2 in the equations other than the first are all zero, i.e., if $b_{22} = b_{32} = \cdots = b_{m2} = 0$, we may find a nontrivial solution by setting $x_3 = x_4 = \cdots = x_n = 0$ and $x_1 = -b_{12}, x_2 = 1$.

Now suppose that after renumbering the equations $b_{22} \neq 0$. Multiply the second equation by b_{22}^{-1} and eliminate x_2 from all other equations to obtain a system of the form

$$\begin{aligned} x_1 \qquad + c_{13}x_3 + \cdots + c_{1n}x_n &= 0 \\ x_2 + c_{23}x_3 + \cdots + c_{2n}x_n &= 0 \\ c_{33}x_3 + \cdots + c_{3n}x_n &= 0 \\ &\vdots \\ c_{m3}x_3 + \cdots + c_{mn}x_n &= 0 \end{aligned}$$

By continuing this process and using the fact that the number of variables is greater than the number of equations, we eventually obtain a system of the form

$$\begin{aligned} x_1 \qquad\qquad + d_{1r}x_r + \cdots + d_{1n}x_n &= 0 \\ x_2 \qquad + d_{2r}x_r + \cdots + d_{2n}x_n &= 0 \\ &\vdots \\ x_{r-1} + d_{r-1,r}x_r + \cdots + d_{r-1,n}x_n &= 0 \\ 0 &= 0 \\ &\vdots \end{aligned}$$

where $r < n$. By choosing $x_r = 1$, $x_{r+1} = \cdots = x_n = 0$, and $x_1 = -d_{1r}$, $x_2 = -d_{2r}, \ldots, x_{r-1} = -d_{r-1,r}$, we obtain a nontrivial solution to the system. ■

Let us illustrate this proof with examples.

Example 1

$$\begin{aligned} x_1 + 3x_2 - x_3 + x_4 &= 0 \\ 3x_1 - x_2 + 2x_3 + x_4 &= 0 \\ -x_1 + 5x_2 - x_3 + 2x_4 &= 0 \end{aligned}$$

Use equation one and x_1. ↓

$$\begin{aligned} x_1 + 3x_2 - x_3 + x_4 &= 0 \\ -10x_2 + 5x_3 - 2x_4 &= 0 \\ 8x_2 - 2x_3 + 3x_4 &= 0 \end{aligned}$$

Multiply equation two by $-\frac{1}{10}$. ↓

$$\begin{aligned} x_1 + 3x_2 - x_3 + x_4 &= 0 \\ x_2 - \tfrac{1}{2}x_3 + \tfrac{1}{5}x_4 &= 0 \\ 8x_2 - 2x_3 + 3x_4 &= 0 \end{aligned}$$

Use equation two and x_2. ↓

$$\begin{aligned} x_1 \qquad + \tfrac{1}{2}x_3 + \tfrac{2}{5}x_4 &= 0 \\ x_2 - \tfrac{1}{2}x_3 + \tfrac{1}{5}x_4 &= 0 \\ 2x_3 + \tfrac{7}{5}x_4 &= 0 \end{aligned}$$

Multiply equation three by $\frac{1}{2}$. ↓

$$\begin{aligned} x_1 \qquad + \tfrac{1}{2}x_3 + \tfrac{2}{5}x_4 &= 0 \\ x_2 - \tfrac{1}{2}x_3 + \tfrac{1}{5}x_4 &= 0 \\ x_3 + \tfrac{7}{10}x_4 &= 0 \end{aligned}$$

Use equation three and x_3. ↓

$$\begin{aligned} x_1 \qquad\qquad + \tfrac{1}{20}x_4 &= 0 \\ x_2 \qquad + \tfrac{11}{20}x_4 &= 0 \\ x_3 + \tfrac{7}{10}x_4 &= 0 \end{aligned}$$

Thus, if $x_4 = c$ is an arbitrary real number, $(-\frac{1}{20}c, -\frac{11}{20}c, -\frac{7}{10}c, c)$ is the solution. If $c \neq 0$, the solution is nontrivial.

Example 2

$$\begin{aligned} 2x - 2y + z &= 0 \\ 3x - y + 2z &= 0 \end{aligned}$$

Multiply the first equation by $\frac{1}{2}$. ↓

$$\begin{aligned} x - y + \tfrac{1}{2}z &= 0 \\ 3x - y + 2z &= 0 \end{aligned}$$

Use x and the first equation.

$$\downarrow$$

$$\begin{aligned} x - y + \tfrac{1}{2}z &= 0 \\ 2y + \tfrac{1}{2}z &= 0 \end{aligned}$$

Multiply the second equation by $\frac{1}{2}$.

$$\downarrow$$

$$\begin{aligned} x - y + \tfrac{1}{2}z &= 0 \\ y + \tfrac{1}{4}z &= 0 \end{aligned}$$

Use y and the second equation.

$$\downarrow$$

$$\begin{aligned} x \quad + \tfrac{3}{4}z &= 0 \\ y + \tfrac{1}{4}z &= 0 \end{aligned}$$

Choosing $z = c \neq 0$, $x = -\frac{3}{4}c$, $y = -\frac{1}{4}c$, we have a nontrivial solution.

EXERCISES

1. Solve the following systems of equations.

(a)
$$\begin{aligned} 3x + y + z &= 0 \\ x - 2y + z &= 0 \end{aligned}$$

(b)
$$\begin{aligned} x + y - 3z &= 0 \\ 2x + 3y + z &= 0 \\ -x + 4y + 2z &= 0 \end{aligned}$$

(c)
$$\begin{aligned} 2x_1 + 2x_2 + x_3 - x_4 &= 0 \\ x_1 - x_2 + 3x_3 + x_4 &= 0 \\ 2x_1 - x_2 + x_3 + 2x_4 &= 0 \end{aligned}$$

(d)
$$\begin{aligned} 2x_1 + 2x_2 + x_3 - x_4 &= 0 \\ x_1 - x_2 + 2x_3 + x_4 &= 0 \\ 3x_1 + x_2 + 2x_3 - x_4 &= 0 \end{aligned}$$

2. Suppose we have a system of homogeneous equations in n variables. Let $(h_1, h_2, \ldots, h_n)$ be a solution and $(g_1, g_2, \ldots, g_n)$ be a solution. Show that $(h_1 + g_1, h_2 + g_2, \ldots, h_n + g_n)$ and $(ch_1, ch_2, \ldots, ch_n)$ are solutions.

3. Suppose $(h_1, h_2, h_3, \ldots, h_n)$ is a solution to the system

$$\text{(I)} \qquad \begin{aligned} a_{11}x_1 + a_{12}x_2 + a_{13}x_3 + \cdots + a_{1n}x_n &= b_1 \\ a_{21}x_1 + a_{22}x_2 + a_{23}x_3 + \cdots + a_{2n}x_n &= b_2 \\ &\vdots \\ a_{m1}x_1 + a_{m2}x_2 + a_{m3}x_3 + \cdots + a_{mn}x_n &= b_m \end{aligned}$$

and $(g_1, g_2, \ldots, g_n)$ is a solution to the same system with $(b_1, b_2, \ldots, b_m)$ replaced by zero in all cases. Show that $(h_1 + g_1, h_2 + g_2, \ldots, h_n + g_n)$ is a solution to I.

4. A system of m linear equations in n variables is given. Suppose that n is greater than m and that the system has a solution. Show that the solution is not unique.

5. Let $(x_1, y_1), \ldots, (x_5, y_5)$ be five points in the plane. Show that the five points lie on a curve having an equation of the form, $Ax^2 + Bxy + Cy^2 + Dx + Ey + F = 0$, where A, B, C, D, and E are not all 0, i.e., a conic section.

vectors and matrices

1 VECTORS

Many physical quantities such as temperature, mass, and energy can be described in terms of a single real number r and a unit. Just as in our study of systems of linear equations, we often found it necessary to consider n-tuples of numbers, so, too, other precise physical descriptions, such as the position or velocity of a body in space, require the use of several real numbers. Accordingly, in this section, we wish to study n-tuples of numbers and methods for manipulating them in greater detail.

We define a **column n-vector** to be an n-tuple of numbers written vertically as

$$\begin{bmatrix} u_1 \\ u_2 \\ \vdots \\ u_n \end{bmatrix}$$

If the numbers u_i are real, we have a real column n-vector; if they are complex, we have a complex column n-vector. The number u_i in the ith slot is called the ith **component** of the vector. For example, the following are column vectors:

$$\begin{bmatrix} 2 \\ -1 \end{bmatrix} \quad \begin{bmatrix} 3 \\ 0 \\ 5 \\ -6 \end{bmatrix} \quad \begin{bmatrix} -1 \\ 14 \\ 6 \end{bmatrix} \quad \begin{bmatrix} 1+i \\ 0 \\ 1-i \end{bmatrix}$$

The first is a real column 2-vector; the second is a real column 4-vector; the fourth is a complex column 3-vector. In the second vector, the first component is 3, the second is 0, the third is 5, and the fourth is -6.

In a similar fashion, we define a **row n-vector** to be an n-tuple of numbers written horizontally, as in $[u_1, u_2, \ldots, u_n]$. Thus, $[5, -1, 4, 3]$ is a row 4-vector.

We call the collection of all real column n-vectors $\boldsymbol{R}^n$. With this definition, $\boldsymbol{R}^2$ consists of all column vectors of the form $\begin{bmatrix} a \\ b \end{bmatrix}$, where a and b are real numbers. The collection of all complex column n-vectors is denoted by $\boldsymbol{C}^n$.

If $\boldsymbol{a} = \begin{bmatrix} \alpha_1 \\ \alpha_2 \\ \vdots \\ \alpha_n \end{bmatrix}$ and $\boldsymbol{b} = \begin{bmatrix} \beta_1 \\ \beta_2 \\ \vdots \\ \beta_n \end{bmatrix}$ are two column n-vectors, we say $\boldsymbol{a} = \boldsymbol{b}$ if and only if $\alpha_1 = \beta_1, \alpha_2 = \beta_2, \ldots, \alpha_n = \beta_n$.

In other words two vectors are equal when and only when they have the same ith components for all i.

We now wish to endow the collection of all real column n-vectors with an algebraic structure. If $\boldsymbol{a}$ and $\boldsymbol{b}$ are two column n-vectors, we define the **sum** of $\boldsymbol{a}$ and $\boldsymbol{b}$, written as $\boldsymbol{a} + \boldsymbol{b}$, as follows:

$$\text{If} \quad \boldsymbol{a} = \begin{bmatrix} \alpha_1 \\ \alpha_2 \\ \vdots \\ \alpha_n \end{bmatrix} \quad \text{and} \quad \boldsymbol{b} = \begin{bmatrix} \beta_1 \\ \beta_2 \\ \vdots \\ \beta_n \end{bmatrix}, \quad \text{then} \quad \boldsymbol{a} + \boldsymbol{b} = \begin{bmatrix} \alpha_1 + \beta_1 \\ \alpha_2 + \beta_2 \\ \vdots \\ \alpha_n + \beta_n \end{bmatrix}$$

For example,

$$\begin{bmatrix} -3 \\ 1 \\ 0 \\ 2 \end{bmatrix} + \begin{bmatrix} 6 \\ -1 \\ 0 \\ 8 \end{bmatrix} = \begin{bmatrix} 3 \\ 0 \\ 0 \\ 10 \end{bmatrix} \quad \text{and} \quad \begin{bmatrix} -6 \\ 1 \\ 0 \end{bmatrix} + \begin{bmatrix} 8 \\ 2 \\ 7 \end{bmatrix} = \begin{bmatrix} 2 \\ 3 \\ 7 \end{bmatrix}$$

Notice that addition is only defined when two vectors are of the same size, i.e., when they have the same number of components.

Generally when we are dealing with vectors we refer to numbers as **scalars**. Real numbers are real scalars. Complex numbers are complex scalars.

Another important operation is scalar multiplication. If $\boldsymbol{c}$ is a column n-vector and α is a scalar, we define the **scalar multiple** of the vector $\boldsymbol{c}$ by the scalar α, written $\alpha\boldsymbol{c}$, as follows:

$$\text{If} \quad \boldsymbol{c} = \begin{bmatrix} c_1 \\ c_2 \\ \vdots \\ c_n \end{bmatrix}, \quad \text{then} \quad \alpha\boldsymbol{c} = \begin{bmatrix} \alpha c_1 \\ \alpha c_2 \\ \vdots \\ \alpha c_n \end{bmatrix}$$

For example,

$$2\begin{bmatrix} 1 \\ 7 \end{bmatrix} = \begin{bmatrix} 2 \\ 14 \end{bmatrix}, \quad 3\begin{bmatrix} 8 \\ 0 \\ -31 \end{bmatrix} = \begin{bmatrix} 24 \\ 0 \\ -93 \end{bmatrix}, \quad \text{and} \quad (-1)\begin{bmatrix} 8 \\ \frac{1}{2} \\ \frac{1}{3} \\ 0 \end{bmatrix} = \begin{bmatrix} -8 \\ -\frac{1}{2} \\ -\frac{1}{3} \\ 0 \end{bmatrix}$$

Similar definitions can be made for addition of complex n-vectors and multiplication of complex n-vectors by complex scalars.

In terms of components, our definitions may be restated as follows: The ith component of the sum of two vectors is the sum of the ith components of each vector. The ith component of $\alpha\boldsymbol{c}$ is α times the ith component of $\boldsymbol{c}$.

The operations of addition and scalar multiplication satisfy several algebraic laws that are worthwhile to remember. In the following we suppose $\boldsymbol{a}$, $\boldsymbol{b}$, and $\boldsymbol{c}$ are n-vectors, μ, λ are scalars, and

$$\boldsymbol{a} = \begin{bmatrix} \alpha_1 \\ \alpha_2 \\ \vdots \\ \alpha_n \end{bmatrix}, \quad \boldsymbol{b} = \begin{bmatrix} \beta_1 \\ \beta_2 \\ \vdots \\ \beta_n \end{bmatrix}, \quad \text{and} \quad \boldsymbol{c} = \begin{bmatrix} \gamma_1 \\ \gamma_2 \\ \vdots \\ \gamma_n \end{bmatrix}$$

First we verify the **commutative law** for vector addition.

Proposition 1 $\boldsymbol{a} + \boldsymbol{b} = \boldsymbol{b} + \boldsymbol{a}$.

PROOF By definition of vector addition,

$$\boldsymbol{a} + \boldsymbol{b} = \begin{bmatrix} \alpha_1 + \beta_1 \\ \alpha_2 + \beta_2 \\ \vdots \\ \alpha_n + \beta_n \end{bmatrix} \quad \text{and} \quad \boldsymbol{b} + \boldsymbol{a} = \begin{bmatrix} \beta_1 + \alpha_1 \\ \beta_2 + \alpha_2 \\ \vdots \\ \beta_n + \alpha_n \end{bmatrix}$$

Since $\alpha_i + \beta_i = \beta_i + \alpha_i$ for all real numbers, $\boldsymbol{a} + \boldsymbol{b} = \boldsymbol{b} + \boldsymbol{a}$, by the definition of the equality of vectors. ■

Next we demonstrate the **associative law** for vector addition.

Proposition 2 $(\boldsymbol{a} + \boldsymbol{b}) + \boldsymbol{c} = \boldsymbol{a} + (\boldsymbol{b} + \boldsymbol{c})$.

PROOF By the definition of vector addition,

$$\boldsymbol{a} + \boldsymbol{b} = \begin{bmatrix} \alpha_1 + \beta_1 \\ \alpha_2 + \beta_2 \\ \vdots \\ \alpha_n + \beta_n \end{bmatrix} \quad \text{and} \quad \boldsymbol{b} + \boldsymbol{c} = \begin{bmatrix} \beta_1 + \gamma_1 \\ \beta_2 + \gamma_2 \\ \vdots \\ \beta_n + \gamma_n \end{bmatrix}$$

$$(\boldsymbol{a} + \boldsymbol{b}) + \boldsymbol{c} = \begin{bmatrix} (\alpha_1 + \beta_1) + \gamma_1 \\ (\alpha_2 + \beta_2) + \gamma_2 \\ \vdots \\ (\alpha_n + \beta_n) + \gamma_n \end{bmatrix} \quad \text{and} \quad \boldsymbol{a} + (\boldsymbol{b} + \boldsymbol{c}) = \begin{bmatrix} \alpha_1 + (\beta_1 + \gamma_1) \\ \alpha_2 + (\beta_2 + \gamma_2) \\ \vdots \\ \alpha_n + (\beta_n + \gamma_n) \end{bmatrix}$$

Since $(\alpha_i + \beta_i) + \gamma_i = \alpha_i + (\beta_i + \gamma_i)$ for all real numbers, we have $(\boldsymbol{a} + \boldsymbol{b}) + \boldsymbol{c} = \boldsymbol{a} + (\boldsymbol{b} + \boldsymbol{c})$, by the definition of the equality of vectors. ■

The zero vector, which we will denote by $\mathbf{0}$, is the n-vector

$$\begin{bmatrix} 0 \\ 0 \\ \vdots \\ 0 \end{bmatrix}$$

i.e., the vector all of whose components are 0. It will be clear from the context what the size of the vector denoted by $\mathbf{0}$ is intended to be.

Proposition 3 $\mathbf{0} + \boldsymbol{a} = \boldsymbol{a} + \mathbf{0} = \boldsymbol{a}$.

PROOF By definition of vector addition,

$$\mathbf{0} + \boldsymbol{a} = \begin{bmatrix} 0 + \alpha_1 \\ 0 + \alpha_2 \\ \vdots \\ 0 + \alpha_n \end{bmatrix}$$

However, we know that $0 + \alpha_i = \alpha_i$ for all real numbers α_i. So

$$\mathbf{0} + \boldsymbol{a} = \begin{bmatrix} \alpha_1 \\ \alpha_2 \\ \vdots \\ \alpha_n \end{bmatrix} = \boldsymbol{a}$$

Likewise, $\boldsymbol{a} + \mathbf{0} = \boldsymbol{a}$. ■

If $\boldsymbol{a}$ is a vector, we define the *negative* of $\boldsymbol{a}$, which we denote by $-\boldsymbol{a}$, as follows:

$$\text{If} \quad \boldsymbol{a} = \begin{bmatrix} \alpha_1 \\ \alpha_2 \\ \vdots \\ \alpha_n \end{bmatrix}, \quad \text{then} \quad -\boldsymbol{a} = \begin{bmatrix} -\alpha_1 \\ -\alpha_2 \\ \vdots \\ -\alpha_n \end{bmatrix}$$

Thus,

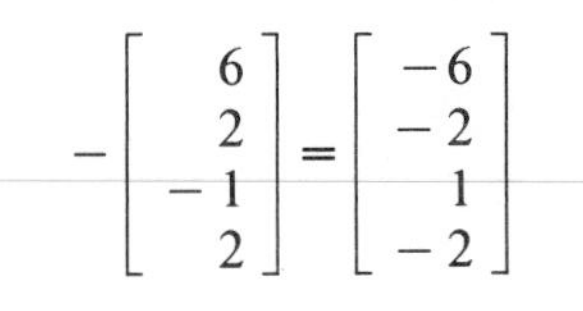

$$-\begin{bmatrix} 6 \\ 2 \\ -1 \\ 2 \end{bmatrix} = \begin{bmatrix} -6 \\ -2 \\ 1 \\ -2 \end{bmatrix}$$

Now the following proposition is immediate from the definitions above:

Proposition 4 $\boldsymbol{a} + (-\boldsymbol{a}) = (-\boldsymbol{a}) + \boldsymbol{a} = \boldsymbol{0}$.

There are also several algebraic laws satisfied by scalar multiplication.

Proposition 5 $\lambda(\boldsymbol{a} + \boldsymbol{b}) = \lambda\boldsymbol{a} + \lambda\boldsymbol{b}$.

PROOF By the definition of vector addition,

$$(\boldsymbol{a} + \boldsymbol{b}) = \begin{bmatrix} \alpha_1 + \beta_1 \\ \alpha_2 + \beta_2 \\ \vdots \\ \alpha_n + \beta_n \end{bmatrix}. \quad \text{Then,} \quad \lambda(\boldsymbol{a} + \boldsymbol{b}) = \begin{bmatrix} \lambda(\alpha_1 + \beta_1) \\ \lambda(\alpha_2 + \beta_2) \\ \vdots \\ \lambda(\alpha_n + \beta_n) \end{bmatrix}$$

by definition of scalar multiplication. Likewise,

$$\lambda\boldsymbol{a} = \begin{bmatrix} \lambda\alpha_1 \\ \lambda\alpha_2 \\ \vdots \\ \lambda\alpha_n \end{bmatrix} \quad \text{and} \quad \lambda\boldsymbol{b} = \begin{bmatrix} \lambda\beta_1 \\ \lambda\beta_2 \\ \vdots \\ \lambda\beta_n \end{bmatrix}. \quad \text{Thus,} \quad \lambda\boldsymbol{a} + \lambda\boldsymbol{b} = \begin{bmatrix} \lambda\alpha_1 + \lambda\beta_1 \\ \lambda\alpha_2 + \lambda\beta_2 \\ \vdots \\ \lambda\alpha_n + \lambda\beta_n \end{bmatrix}$$

by the definition of vector addition. Since $\lambda(\alpha_i + \beta_i) = \lambda\alpha_i + \lambda\beta_i$ for all real numbers, we have, using the definition of vector equality, $\lambda(\boldsymbol{a} + \boldsymbol{b}) = \lambda\boldsymbol{a} + \lambda\boldsymbol{b}$.
■

The proofs of these important laws are left to the reader:

$$(\lambda + \mu)\boldsymbol{a} = \lambda\boldsymbol{a} + \mu\boldsymbol{a}$$

$$\lambda(\mu\boldsymbol{a}) = (\lambda\mu)\boldsymbol{a}$$

$$1\boldsymbol{a} = \boldsymbol{a}$$

We define subtraction of vectors as follows: If $\boldsymbol{a}$ and $\boldsymbol{b}$ are vectors, $\boldsymbol{a} - \boldsymbol{b} = \boldsymbol{a} + (-\boldsymbol{b})$. In terms of components,

$$\boldsymbol{a} - \boldsymbol{b} = \boldsymbol{a} + (-\boldsymbol{b}) = \begin{bmatrix} \alpha_1 \\ \alpha_2 \\ \vdots \\ \alpha_n \end{bmatrix} + \begin{bmatrix} -\beta_1 \\ -\beta_2 \\ \vdots \\ -\beta_n \end{bmatrix} = \begin{bmatrix} \alpha_1 - \beta_1 \\ \alpha_2 - \beta_2 \\ \vdots \\ \alpha_n - \beta_n \end{bmatrix}$$

The laws we have derived are useful in so far as they enable us to perform algebraic manipulations on vectors without continually referring back to their components.

For example, to solve the equation $\boldsymbol{x} + \boldsymbol{a} = \boldsymbol{b}$ for the vector $\boldsymbol{x}$, we add $-\boldsymbol{a}$ to both sides to obtain

$$(\boldsymbol{x} + \boldsymbol{a}) + (-\boldsymbol{a}) = \boldsymbol{b} + (-\boldsymbol{a})$$

$$\boldsymbol{x} + (\boldsymbol{a} - \boldsymbol{a}) = \boldsymbol{b} - \boldsymbol{a}$$

$$\boldsymbol{x} + \boldsymbol{0} = \boldsymbol{b} - \boldsymbol{a}$$

$$\boldsymbol{x} = \boldsymbol{b} - \boldsymbol{a}$$

In this instance, we deliberately showed all steps involved in solving the equation. However, after a little practice, the solution can be written down immediately. Generally speaking, the algebraic laws exhibited above enable us to perform manipulation of vectors much as we do with numbers.

As we mentioned earlier, a vector may be used to describe an object with a single mathematical symbol when such an object cannot be described with a single number. We present a simple problem that shows how this may be advantageous.

Four urns are placed on a table, and balls are placed in each of the urns. There are 35 balls in the first urn, 18 in the second, 21 in the third, and 37 in the fourth.

An operation of shifting balls between urns is to be performed according to the following rule:

(1) Four balls are removed from the first urn. Three of these balls are placed in the second urn, and one is placed in the third urn.
(2) Three balls are removed from the second urn and placed in the third.
(3) Two balls from the third urn are moved to the fourth.
(4) Five balls are taken from the fourth urn. One is placed in the first urn and four in the second urn.

If this whole operation is repeated five times, how many balls will be in each of the urns?

We describe the system with a vector whose ith component is the number of balls in the ith urn. Thus, the original placement of balls is given by the vector:

$$\boldsymbol{n} = \begin{bmatrix} 35 \\ 18 \\ 21 \\ 37 \end{bmatrix}$$

We shall also describe each step in our sequence of operations with a vector. The first step we describe with the vector

$$\boldsymbol{s}_1 = \begin{bmatrix} -4 \\ 3 \\ 1 \\ 0 \end{bmatrix}$$

Observe that adding $\boldsymbol{s}_1$ to $\boldsymbol{n}$ produces the same effect as performing the first step in our operation, i.e., 4 units are removed from the first component, 3 are added to the second, and 1 to the third. Thus, the vector, $\boldsymbol{n} + \boldsymbol{s}_1$, is the vector representing the state of the system after the first step is performed.

We describe the remaining three steps of the operation with the vectors:

$$\boldsymbol{s}_2 = \begin{bmatrix} 0 \\ -3 \\ 3 \\ 0 \end{bmatrix}, \quad \boldsymbol{s}_3 = \begin{bmatrix} 0 \\ 0 \\ -2 \\ 2 \end{bmatrix}, \quad \boldsymbol{s}_4 = \begin{bmatrix} 1 \\ 4 \\ 0 \\ -5 \end{bmatrix}$$

Thus, in the second step, 3 units are removed from the second component and placed in the third, and so on.

After the second step is performed, the state of the system is represented by the vector, $\boldsymbol{n} + \boldsymbol{s}_1 + \boldsymbol{s}_2$. After the third and fourth steps are performed, the state is $\boldsymbol{n} + \boldsymbol{s}_1 + \boldsymbol{s}_2 + \boldsymbol{s}_3 + \boldsymbol{s}_4$. Lastly, if the whole operation is repeated five times, the state of the system is given by the vector $\boldsymbol{n} + 5(\boldsymbol{s}_1 + \boldsymbol{s}_2 + \boldsymbol{s}_3 + \boldsymbol{s}_4)$.

Since $\boldsymbol{s}_1 + \boldsymbol{s}_2 + \boldsymbol{s}_3 + \boldsymbol{s}_4 = \begin{bmatrix} -3 \\ 4 \\ 2 \\ -3 \end{bmatrix}$, the final state is given by the vector

$$\boldsymbol{n} + 5\begin{bmatrix} -3 \\ 4 \\ 2 \\ -3 \end{bmatrix} = \begin{bmatrix} 35 \\ 18 \\ 21 \\ 37 \end{bmatrix} + \begin{bmatrix} -15 \\ 20 \\ 10 \\ -15 \end{bmatrix} = \begin{bmatrix} 20 \\ 38 \\ 31 \\ 22 \end{bmatrix}$$

Of course, this result could be obtained without resorting to the use of vectors. But the reflective reader will realize that the use of vectors gives us a methodical way of tabulating the data and avoiding careless errors.

EXERCISES

1. Calculate.

(a) $\begin{bmatrix} 0 \\ 3 \\ 5 \end{bmatrix} + \begin{bmatrix} 4 \\ -1 \\ 7 \end{bmatrix}$ (b) $2\begin{bmatrix} -7 \\ 7 \\ 3 \end{bmatrix} + \begin{bmatrix} 5 \\ 4 \\ 6 \end{bmatrix}$ (c) $3\begin{bmatrix} 4 \\ 11 \\ -5 \end{bmatrix} + 2\begin{bmatrix} 5 \\ 3 \\ 0 \end{bmatrix}$

(d) $\begin{bmatrix} 13 \\ 8 \\ 9 \\ -2 \end{bmatrix} + 3\begin{bmatrix} 6 \\ 4 \\ 9 \\ -8 \end{bmatrix}$ (e) $\begin{bmatrix} -1 \\ 1 \\ 5 \\ 7 \end{bmatrix} + 3\begin{bmatrix} 6 \\ 2 \\ 8 \\ 4 \end{bmatrix} - \begin{bmatrix} 4 \\ 6 \\ 1 \\ -2 \end{bmatrix}$

2. Solve each of the following equations for $\boldsymbol{x}$.

$$\text{(a) } \boldsymbol{x} + \begin{bmatrix} 4 \\ -2 \\ 0 \\ 5 \end{bmatrix} = \begin{bmatrix} 7 \\ 8 \\ 2 \\ 3 \end{bmatrix} \qquad \text{(b) } 3\boldsymbol{x} + \begin{bmatrix} 7 \\ 0 \\ 3 \\ -3 \end{bmatrix} = \begin{bmatrix} -6 \\ 5 \\ 0 \\ 4 \end{bmatrix}$$

3. Prove the following laws:

$$(\mu + \lambda)\boldsymbol{a} = \mu\boldsymbol{a} + \lambda\boldsymbol{a}$$

$$\mu(\lambda\boldsymbol{a}) = (\mu\lambda)\boldsymbol{a}$$

$$1 \cdot \boldsymbol{a} = \boldsymbol{a}$$

4. If $\boldsymbol{a}$ and $\boldsymbol{b}$ are n-vectors and $\boldsymbol{a} + 3\boldsymbol{b} = \boldsymbol{a}$, show that $\boldsymbol{b} = 0$.

5. If $\alpha\boldsymbol{x} = \boldsymbol{0}$ and $\alpha \neq 0$, show that $\boldsymbol{x} = \boldsymbol{0}$. If $\alpha\boldsymbol{x} = \beta\boldsymbol{x}$, and $\alpha \neq \beta$, show that $\boldsymbol{x} = \boldsymbol{0}$.

6. Find two 2-vectors $\boldsymbol{a}$ and $\boldsymbol{b}$ such that the equation $\alpha\boldsymbol{a} = \boldsymbol{b}$ cannot be solved for the scalar α.

7. In the following, $\boldsymbol{x}$, $\boldsymbol{a}$, $\boldsymbol{b}$, and $\boldsymbol{c}$ are n-vectors. Solve for $\boldsymbol{x}$ in terms of $\boldsymbol{a}$, $\boldsymbol{b}$, and $\boldsymbol{c}$. $((\boldsymbol{x} - \boldsymbol{a}) + \boldsymbol{c}) = \boldsymbol{a} + \boldsymbol{b}$.

8. Solve the equations, $\boldsymbol{x} + 2\boldsymbol{y} = \boldsymbol{a}$ and $2\boldsymbol{x} + 5\boldsymbol{y} = \boldsymbol{b}$, to find the vectors, $\boldsymbol{x}$ and $\boldsymbol{y}$, in terms of the vectors, $\boldsymbol{a}$ and $\boldsymbol{b}$.

9. Show that there are exactly 2^n n-vectors, all of whose components are 0 and 1.

10. Show that all the laws for vector addition and scalar multiplication hold for complex vectors.

11. Find scalars x and y so that $\begin{bmatrix} 3 \\ 4 \end{bmatrix} = x\begin{bmatrix} 1 \\ 2 \end{bmatrix} + y\begin{bmatrix} 3 \\ 7 \end{bmatrix}$.

12. Given the three 2-vectors

$$\boldsymbol{a} = \begin{bmatrix} 2 \\ 1 \end{bmatrix} \qquad \boldsymbol{b} = \begin{bmatrix} 5 \\ 9 \end{bmatrix} \qquad \boldsymbol{c} = \begin{bmatrix} 2 \\ -3 \end{bmatrix}$$

find scalars λ, μ, and γ not all 0 such that $\lambda\boldsymbol{a} + \mu\boldsymbol{b} + \gamma\boldsymbol{c} = \boldsymbol{0}$. Prove that this is possible if $\boldsymbol{a}$, $\boldsymbol{b}$, and $\boldsymbol{c}$ are arbitrary 2-vectors.

13. Three urns are placed on a table, and a certain number of balls are placed in each urn. We represent the number of balls in each urn with a vector $\boldsymbol{v} = \begin{bmatrix} n_1 \\ n_2 \\ n_3 \end{bmatrix}$, where

n_i is the number of balls in the ith urn. Suppose initially $v = \begin{bmatrix} 23 \\ 10 \\ 6 \end{bmatrix}$. The following shifting operation is performed: Take two balls from the urn having the most balls, and place one in each of the other urns.

(a) If this operation is repeated for long enough, show that the only states of the system will be: $\begin{bmatrix} 14 \\ 13 \\ 12 \end{bmatrix}, \begin{bmatrix} 12 \\ 14 \\ 13 \end{bmatrix}, \begin{bmatrix} 13 \\ 12 \\ 14 \end{bmatrix}$.

(b) What is the state of the system after the operation is performed 100 times?

14. An economic survey is made of 30 families to determine how the families spend their money. With the ith family is associated a vector, $\boldsymbol{x}_i = \begin{bmatrix} a_1 \\ a_2 \\ a_3 \\ a_4 \end{bmatrix}$ where a_1 is the percent of the family's income spent on food, a_2 the percent spent on housing, a_3 the percent spent on clothing, and a_4 the percent spent on other items. What is the meaning of the vector $\frac{1}{30}(\boldsymbol{x}_1 + \cdots + \boldsymbol{x}_{30})$?

15. Three villages, which we label A, B, and C, have the respective populations: 500, 450, and 600. During an average year, 50 people leave A with 20 moving to B and 30 to C. During an average year, 60 people leave B with half moving to each A and C. Finally, during an average year, 80 people leave C with half going to A and half to B. Assuming that the populations are otherwise stable, find the population of A, B, and C after 10 years have elapsed.

16. If

$$\begin{aligned}
\boldsymbol{x}_1 + \boldsymbol{x}_2 + \boldsymbol{x}_3 + \cdots + \boldsymbol{x}_n &= \boldsymbol{y}_1 \\
\boldsymbol{x}_2 + \boldsymbol{x}_3 + \cdots + \boldsymbol{x}_n &= \boldsymbol{y}_2 \\
\boldsymbol{x}_3 + \cdots + \boldsymbol{x}_n &= \boldsymbol{y}_3 \\
&\vdots \\
\boldsymbol{x}_{n-1} + \boldsymbol{x}_n &= \boldsymbol{y}_{n-1} \\
\boldsymbol{x}_n &= \boldsymbol{y}_n
\end{aligned}$$

find $\boldsymbol{x}_1, \ldots, \boldsymbol{x}_n$ in terms of $\boldsymbol{y}_1, \ldots, \boldsymbol{y}_n$, where $\boldsymbol{x}_1, \ldots, \boldsymbol{x}_n$ and $\boldsymbol{y}_1, \ldots, \boldsymbol{y}_n$ are vectors.

2 GEOMETRIC INTERPRETATION OF R^2 AND R^3

Just as points in the plane may be represented as ordered pairs of real numbers, so too, points in space may be represented as triples of real numbers. To effect this representation, choose three mutually perpendicular lines that meet at a point in space. The lines are called the x axis, y axis, and z axis and the point at which they meet is called the origin. (See Figure 2-1.)

The plane formed by the y axis and z axis is called the y-z plane. (The x-y plane and the x-z plane are determined in a similar manner.) To find the x coordinate of a point P in space construct the plane through P parallel to

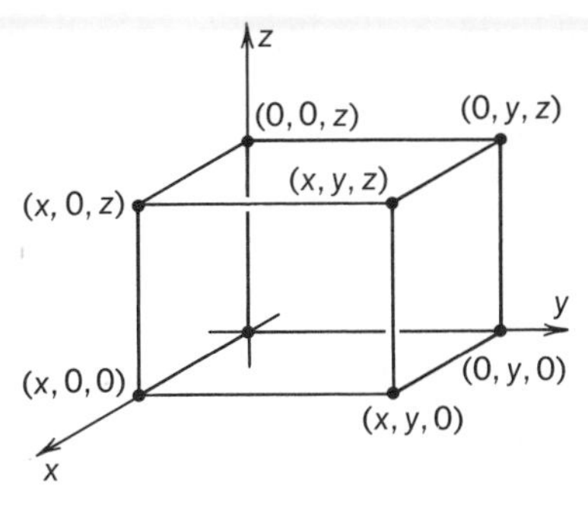

FIGURE 2-1

the y-z plane. The point at which this plane intersects the x axis is called the x coordinate of P. The y coordinate is obtained by determining where the plane through P parallel to the x-z plane intersects the y axis, and so on. Using this procedure we may associate to each point P in space a triple of real numbers (x, y, z). Also, to each triple of real numbers we may associate the point in space that has the given triple as its coordinates. The point $(0, 0, 0)$ where the three axes meet is called the origin of the coordinate system.

For example, to locate the point $(4, -3, 4)$ we go out 4 units on the x axis, -3 units on the y axis, and up 4 units on the z axis. The situation is illustrated in Figure 2-2.

With this method of representing points in mind we see that the x axis consists of the points of the form $(\alpha, 0, 0)$, where α is any real number. The x-y plane consists of the points of the form $(\alpha, \beta, 0)$. The points of the y axis, z axis, and the remaining planes can be similarly represented.

Geometrically, we define a **vector** as a **directed line segment** originating at the origin and proceeding to some point in space. Figure 2-3 shows several vectors. Vectors may be thought of as arrows beginning at the origin.

If the vector $\boldsymbol{v}$ ends at the point (x, y, z) in space, for convenience, we often write $\boldsymbol{v} = \boldsymbol{v}(x, y, z)$. Thus, $\boldsymbol{v}(4, -3, 4)$ is the directed line segment that

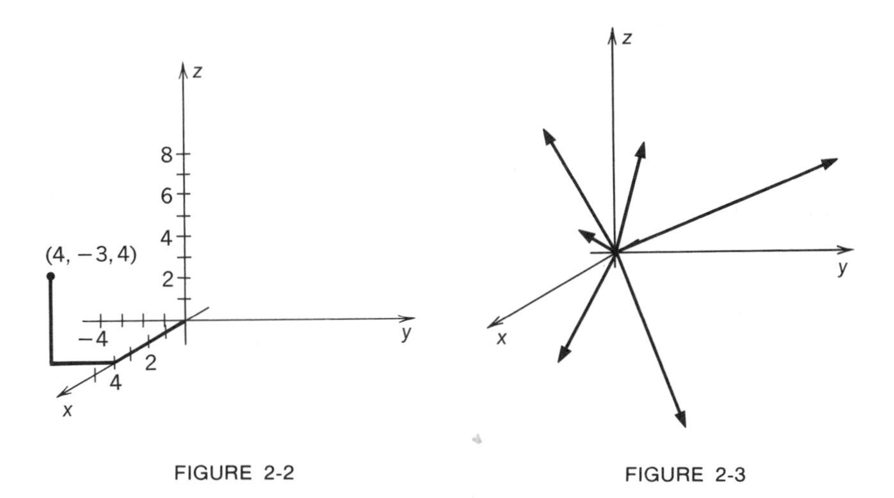

FIGURE 2-2

FIGURE 2-3

begins at the origin and ends at the point (4, − 3, 4) in space. This convention enables us to interpret geometrically those vectors that we defined in the last section as algebraic objects. Thus, we visualize the column 3-vector $\begin{bmatrix} x \\ y \\ z \end{bmatrix}$ as the geometric vector $\boldsymbol{v}(x, y, z)$.

In the last section, addition and scalar multiplication of vectors were discussed from an algebraic point of view. In this section, we discuss these operations from a geometric point of view.

Geometrically, we define vector addition as follows. In the plane formed by the vectors $\boldsymbol{v}_1$ and $\boldsymbol{v}_2$ (see Figure 2-4), form the parallelogram having $\boldsymbol{v}_1$ as one side and $\boldsymbol{v}_2$ as its adjacent side. We define $\boldsymbol{v}_1 + \boldsymbol{v}_2$ to be the directed line segment along the diagonal of the parallelogram. Let $\boldsymbol{v}_1 = \boldsymbol{v}(x, y, z)$ and $\boldsymbol{v}_2 = \boldsymbol{v}(x', y', z')$. Then, $\boldsymbol{v}_1$ and $\boldsymbol{v}_2$ correspond to the column vectors $\begin{bmatrix} x \\ y \\ z \end{bmatrix}$ and $\begin{bmatrix} x' \\ y' \\ z' \end{bmatrix}$. The sum of these two vectors is $\begin{bmatrix} x + x' \\ y + y' \\ z + z' \end{bmatrix}$, which corresponds to the vector $\boldsymbol{v}(x + x', y + y', z + z')$. We wish to show that the algebraic and geometric definitions of vector addition are consistent. To do this, we must show that $\boldsymbol{v}(x, y, z) + \boldsymbol{v}(x', y', z') = \boldsymbol{v}(x + x', y + y', z + z')$.

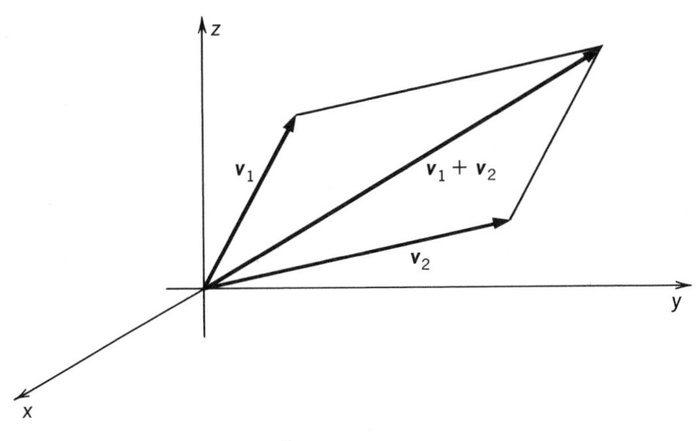

FIGURE 2-4

We prove this result in the plane, leaving the ambitious reader to formulate the proposition in three dimensional space. Thus, we wish to show that $\boldsymbol{v}(x, y) + \boldsymbol{v}(x', y') = \boldsymbol{v}(x + x', y + y')$.

In Figure 2-5 let $\boldsymbol{v}(x, y)$ be the vector ending at the point A, and $\boldsymbol{v}(x', y')$ be the vector ending at the point B. The vector $\boldsymbol{v}(x, y) + \boldsymbol{v}(x', y')$ ends at the vertex C of parallelogram $OBCA$. We wish to show that $\boldsymbol{v}(x, y) + \boldsymbol{v}(x', y') = \boldsymbol{v}(x + x', y + y')$, or in other words, that the coordinates of C are $(x + x', y + y')$.

From our figure, the reader may observe that triangle OAD is congruent to triangle CBG. Also, observe length $OD = x$, length $OE = x'$. By the

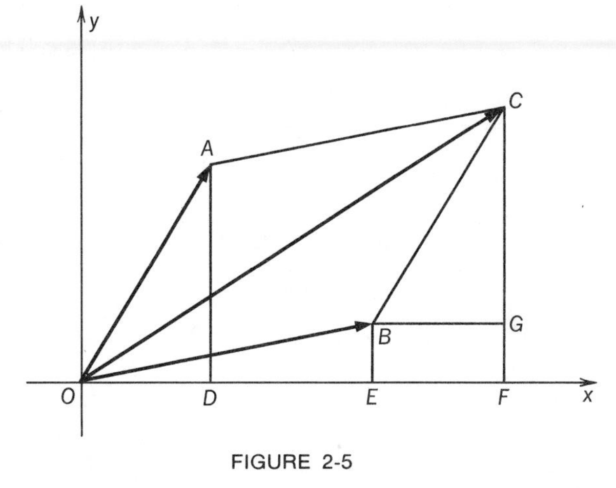

FIGURE 2-5

congruence relation, length OD = length BG, and since $BGFE$ is a rectangle, we have length OD = length EF.

But length OF = length OE + length EF, and so length $OF = x + x'$. This shows that the x coordinate of C is $x + x'$. The proof for the y coordinate is analogous. From this we see that the geometric definition of vector addition is equivalent to the algebraic definition in which we add components.

Figure 2-6 shows that we may also think of vector addition as performed by translating the directed line segment representing the vector $\boldsymbol{v}_2$ so that it begins at the vector $\boldsymbol{v}_1$. The terminal point of the resulting directed segment is the endpoint of the vector $\boldsymbol{v}_1 + \boldsymbol{v}_2$.

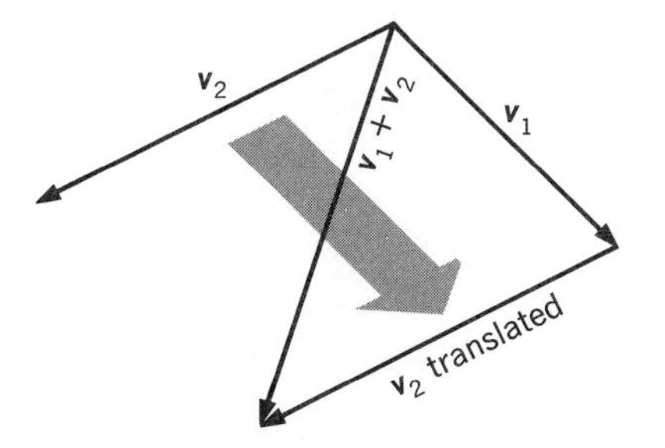

FIGURE 2-6

Scalar multiples of vectors have similar geometric interpretations. If α is a scalar and $\boldsymbol{v}$ is a vector, we may define $\alpha\boldsymbol{v}$ to be the vector that is $|\alpha|$ times as long as $\boldsymbol{v}$ and having the same direction as $\boldsymbol{v}$ if $\alpha > 0$ but the opposite direction of $\boldsymbol{v}$ if $\alpha < 0$. Figure 2-7 shows several examples.

Using an argument that depends on similar triangles, we can prove that $\alpha\boldsymbol{v}(x, y, z) = \boldsymbol{v}(\alpha x, \alpha y, \alpha z)$. Again, the geometric definition coincides with the algebraic one.

What is the vector $\boldsymbol{b} - \boldsymbol{a}$? Since $\boldsymbol{a} + (\boldsymbol{b} - \boldsymbol{a}) = \boldsymbol{b}$, it is clear that $\boldsymbol{b} - \boldsymbol{a}$ is that vector which when added to $\boldsymbol{a}$ gives $\boldsymbol{b}$. In view of this, $\boldsymbol{b} - \boldsymbol{a}$ is parallel to

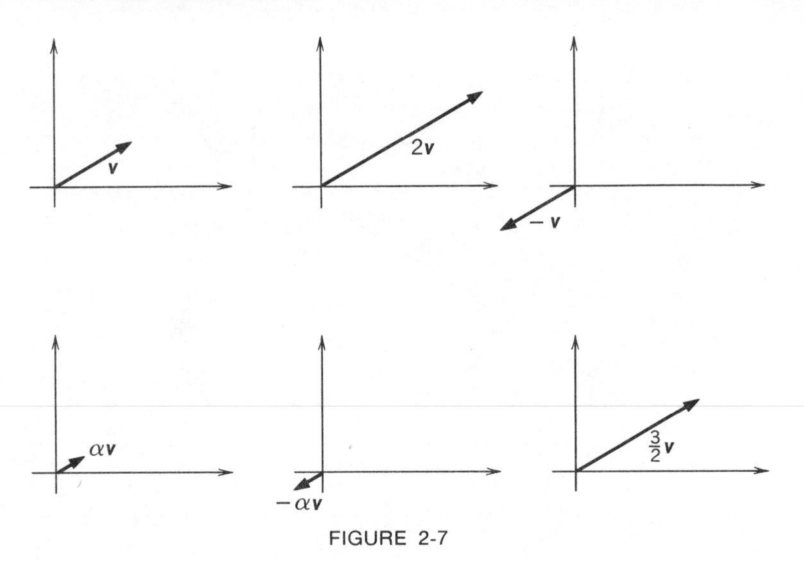

FIGURE 2-7

the directed line segment beginning at the endpoint of $\boldsymbol{a}$ and terminating at the endpoint of $\boldsymbol{b}$. (See Figure 2-8.)

Let us denote by $\boldsymbol{i}$ the vector that ends at (1, 0, 0), by $\boldsymbol{j}$ that which ends at (0, 1, 0), and by $\boldsymbol{k}$ that which ends at (0, 0, 1). Then,

$$\begin{aligned} v(x, y, z) &= v(x, 0, 0) + v(0, y, 0) + v(0, 0, z) \\ &= xv(1, 0, 0) + yv(0, 1, 0) + zv(0, 0, 1) \\ &= x\boldsymbol{i} + y\boldsymbol{j} + z\boldsymbol{k} \end{aligned}$$

Hence, we can represent any vector in three-dimensional space in terms of the unit vectors $\boldsymbol{i}$, $\boldsymbol{j}$, and $\boldsymbol{k}$.

For example, the vector that ends at (2, 3, 3) is $2\boldsymbol{i} + 3\boldsymbol{j} + 3\boldsymbol{k}$, and that ending at (0, − 1, 4) is $-\boldsymbol{j} + 4\boldsymbol{k}$. (See Figure 2-9.) The vectors $\boldsymbol{i}$, $\boldsymbol{j}$, and $\boldsymbol{k}$ are called the **standard basis** vectors for $\boldsymbol{R}^3$.

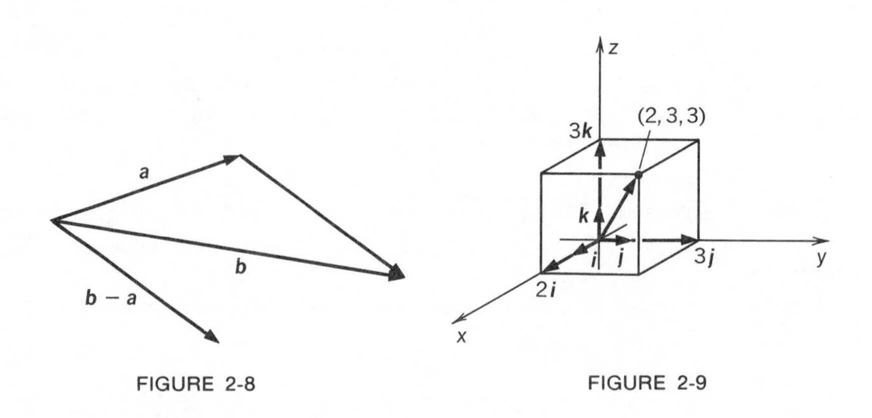

FIGURE 2-8

FIGURE 2-9

We also have

$$(x\boldsymbol{i} + y\boldsymbol{j} + z\boldsymbol{k}) + (x'\boldsymbol{i} + y'\boldsymbol{j} + z'\boldsymbol{k}) = (x + x')\boldsymbol{i} + (y + y')\boldsymbol{j} + (z + z')\boldsymbol{k}$$

and

$$\alpha(x\boldsymbol{i} + y\boldsymbol{j} + z\boldsymbol{k}) = (\alpha x)\boldsymbol{i} + (\alpha y)\boldsymbol{j} + (\alpha z)\boldsymbol{k}$$

Because of the correspondence between vectors and points we may sometimes refer to a point $\boldsymbol{a}$, under circumstances in which $\boldsymbol{a}$ has been defined to be a vector. The reader should realize that by this statement we mean the endpoint of the vector $\boldsymbol{a}$.

As an example of the use of these ideas, let us employ vectors to describe those points that lie in the interior and on the boundary of the parallelogram whose adjacent sides are the vectors $\boldsymbol{a}$ and $\boldsymbol{b}$. (See Figure 2-10.)

If P is any point in the parallelogram and we construct lines l_1 and l_2, parallel to the vectors $\boldsymbol{a}$ and $\boldsymbol{b}$, respectively, we see that l_1 intersects the side of the parallelogram determined by the vector $\boldsymbol{b}$ at some point $t\boldsymbol{b}$, where $0 \leqslant t \leqslant 1$. Likewise, l_2 intersects the side determined by the vector $\boldsymbol{a}$ at some point $s\boldsymbol{a}$, where $0 \leqslant s \leqslant 1$.

Since P is then the endpoint of the diagonal of a parallelogram having adjacent sides $s\boldsymbol{a}$ and $t\boldsymbol{b}$, and if $\boldsymbol{v}$ denotes the vector ending at P, we see that $\boldsymbol{v} = s\boldsymbol{a} + t\boldsymbol{b}$. Thus, all the points in the parallelogram are endpoints of vectors of the form $s\boldsymbol{a} + t\boldsymbol{b}$ for $0 \leqslant s \leqslant 1$ and $0 \leqslant t \leqslant 1$. By reversing our steps, we easily see that all vectors of this form end within the parallelogram.

A simple illustration of how vector operations arise in physical situations is afforded by the concept of the center of mass. Suppose that there is a system of k bodies in the plane having masses, $m_1, \ldots, m_k$, and respective position vectors, $\boldsymbol{r}_1, \ldots, \boldsymbol{r}_k$. In other words, at the endpoint of the vector $\boldsymbol{r}_k$ there is a body of mass, m_k. The center of mass of the system is defined to be the endpoint of the vector $\boldsymbol{r}_c = \dfrac{m_1\boldsymbol{r}_1 + \cdots + m_k\boldsymbol{r}_k}{m_1 + \cdots + m_k}$. A basic physical principle asserts that if the plane is suspended horizontally from the center of mass, it will remain balanced. The idea underlying this principle is, in reality, no different from that used in balancing the meter stick in example 2 of §1.1. Now, however, since the bodies are dispersed on a plane instead of a line, the position of the center of mass is given by a vector rather than a scalar.

For example, suppose that three bodies in the plane have positions (1, 0), (0, 1), and $(-1, -1)$, and respective masses, 2, 1, and 3. Then, the position vectors of the three bodies are $\boldsymbol{r}_1 = \boldsymbol{i}$, $\boldsymbol{r}_2 = \boldsymbol{j}$, and $\boldsymbol{r}_3 = -\boldsymbol{i} - \boldsymbol{j}$. Then $\boldsymbol{r}_c = \dfrac{2\boldsymbol{i} + \boldsymbol{j} + 3(-\boldsymbol{i} - \boldsymbol{j})}{2 + 1 + 3} = (\frac{1}{6})(-\boldsymbol{i} - 2\boldsymbol{j})$. Thus, the coordinates of the center of mass are $(-\frac{1}{6}, -\frac{1}{3})$. (See Figure 2-11.) If the plane is suspended horizontally from the point $(-\frac{1}{6}, -\frac{1}{3})$, it will remain balanced.

Next, we obtain the parametric equation of a line in three-dimensional space. We suppose that L is a line in space, $\boldsymbol{v}$ is a vector in the direction of L, and $\boldsymbol{a}$ is a vector whose endpoint lies on L. (See Figure 2-12.)

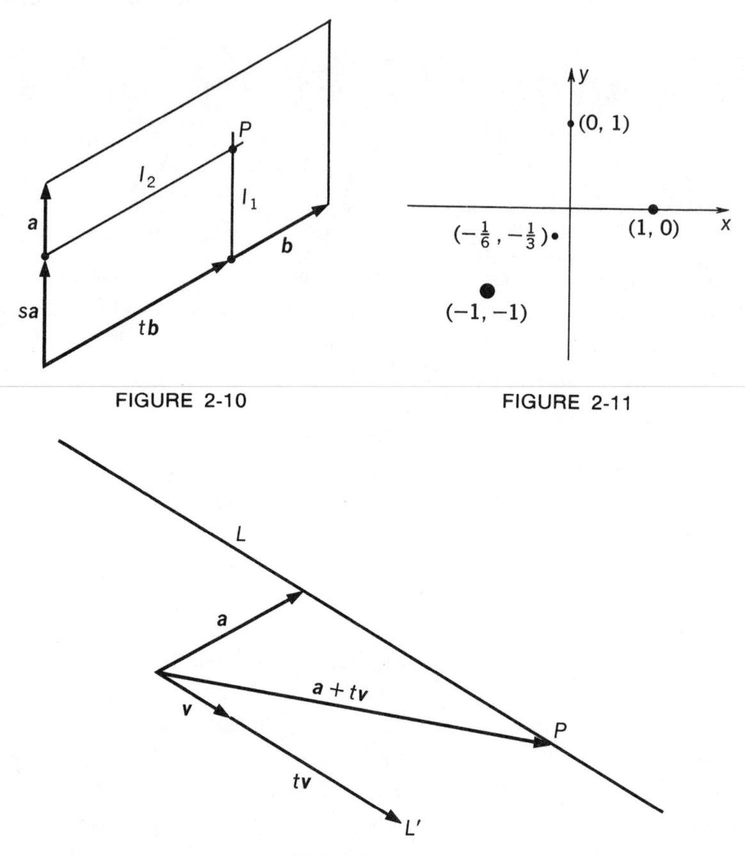

FIGURE 2-10

FIGURE 2-11

FIGURE 2-12

Let L' be the line through the origin in the direction of the vector $\boldsymbol{v}$. As t varies through all real numbers, the points of the form $t\boldsymbol{v}$ are all the scalar multiples of the vector $\boldsymbol{v}$. Thus, all points on L' are of the form $t\boldsymbol{v}$.

Now if P is a point on L, P is the endpoint of the diagonal of a parallelogram with one side $\boldsymbol{a}$ and the other lying on L'. The side of the parallelogram lying on L' is of the form $t\boldsymbol{v}$. Thus, P is the endpoint of the vector $\boldsymbol{a} + t\boldsymbol{v}$. Thus, the line may be expressed parametrically by the equation, $\boldsymbol{l}(t) = \boldsymbol{a} + t\boldsymbol{v}$.

At $t = 0$, $\boldsymbol{l}(0) = \boldsymbol{a}$. As t increases the point $\boldsymbol{l}(t)$ moves away from $\boldsymbol{a}$ in the direction of $\boldsymbol{v}$. As t decreases from $t = 0$ through negative values, $\boldsymbol{l}(t)$ moves away from $\boldsymbol{a}$ in the direction of $-\boldsymbol{v}$.

Of course, there are other parametrizations of the same line. These may be obtained by choosing a different point on the line and forming the parametric equation of the line beginning at that point and in the direction of $\boldsymbol{v}$. For example, the point $\boldsymbol{a} + \boldsymbol{v}$ is on the line $\boldsymbol{a} + t\boldsymbol{v}$, and thus $\boldsymbol{l}'(t) = \boldsymbol{a} + \boldsymbol{v} + t\boldsymbol{v}$ represents the same line.

Other parametrizations may be obtained by observing that if $\alpha \neq 0$, the vector $\alpha\boldsymbol{v}$ has the same or opposite direction as $\boldsymbol{v}$. Thus, $\boldsymbol{l}'(t) = \boldsymbol{a} + \alpha t\boldsymbol{v}$ provides us with another parametrization of $\boldsymbol{l}(t) = \boldsymbol{a} + t\boldsymbol{v}$.

Example 1 Determine the equation of the line passing through (1, 0, 0) in the direction of $\boldsymbol{j}$.

The desired line can be given parametrically as $\boldsymbol{l}(t) = \boldsymbol{i} + t\boldsymbol{j}$. (See Figure 2-13.) In terms of coordinates $x(t) = 1$, $y(t) = t$, $z(t) = 0$. In this case the line is the intersection of the planes $z = 0$ and $x = 1$.

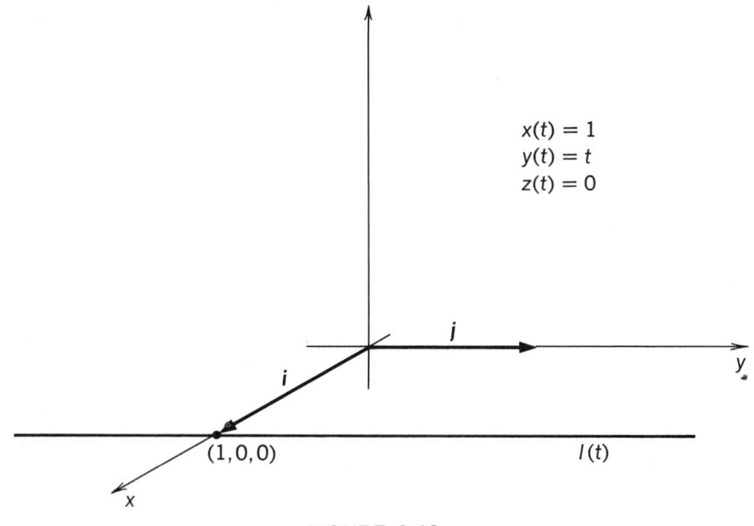

FIGURE 2-13

We may also derive the equation of a line passing through the endpoints of two given vectors $\boldsymbol{a}$ and $\boldsymbol{b}$.

In this case, since the vector $\boldsymbol{b} - \boldsymbol{a}$ is parallel to the directed line segment from $\boldsymbol{a}$ to $\boldsymbol{b}$, what we really wish to do is calculate the parametric equations of the line passing through $\boldsymbol{a}$ in the direction of $\boldsymbol{b} - \boldsymbol{a}$. (See Figure 2-14.) Thus $\boldsymbol{l}(t) = \boldsymbol{a} + t(\boldsymbol{b} - \boldsymbol{a})$, or $\boldsymbol{l}(t) = (1 - t)\boldsymbol{a} + t\boldsymbol{b}$.

As t increases from 0 to 1, $t(\boldsymbol{b} - \boldsymbol{a})$ starts as the $\boldsymbol{0}$ vector, continues in the direction of $\boldsymbol{b} - \boldsymbol{a}$, increasing in length, until at $t = 1$ it is the vector $\boldsymbol{b} - \boldsymbol{a}$. Thus, in $\boldsymbol{l}(t) = \boldsymbol{a} + t(\boldsymbol{b} - \boldsymbol{a})$, as t increases from 0 to 1, $\boldsymbol{l}(t)$ moves from the endpoint of $\boldsymbol{a}$ along the directed line segment from $\boldsymbol{a}$ to $\boldsymbol{b}$ to the endpoint of $\boldsymbol{b}$.

Example 2 Find the equation of the line passing through $(-1, 1, 0)$ and $(0, 0, 1)$. (See Figure 2-15.)

Letting $\boldsymbol{a} = -\boldsymbol{i} + \boldsymbol{j}$, $\boldsymbol{b} = \boldsymbol{k}$, we have

$$\begin{aligned} \boldsymbol{l}(t) &= (1 - t)(-\boldsymbol{i} + \boldsymbol{j}) + t\boldsymbol{k} \\ &= -(1 - t)\boldsymbol{i} + (1 - t)\,\boldsymbol{j} + t\boldsymbol{k} \end{aligned}$$

$$x(t) = -1 + t \quad y(t) = 1 - t \quad z(t) = t$$

Before continuing let us note that any vector $\boldsymbol{c}$ of the form $\boldsymbol{c} = \lambda\boldsymbol{a} + \mu\boldsymbol{b}$, where $\lambda + \mu = 1$, is on the line passing through $\boldsymbol{a}$ and $\boldsymbol{b}$. To see this, observe

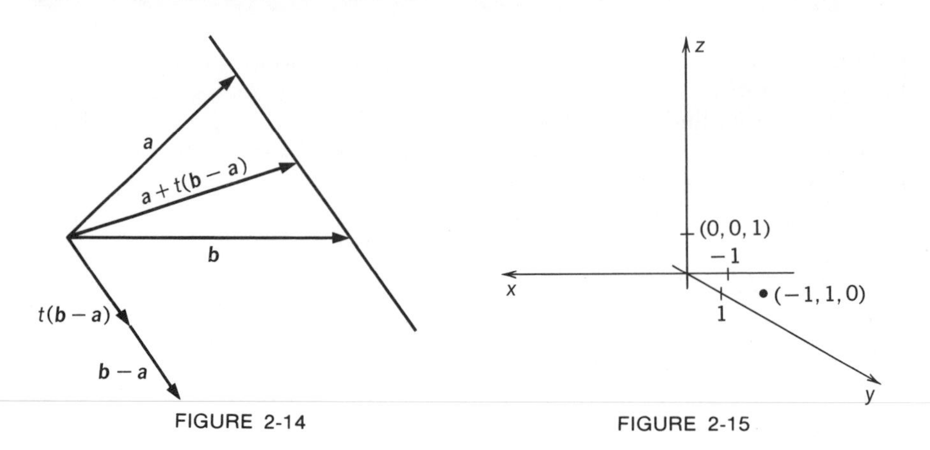

FIGURE 2-14 FIGURE 2-15

$\boldsymbol{c} = (1 - \mu)\boldsymbol{a} + \mu\boldsymbol{b} = \boldsymbol{a} + \mu(\boldsymbol{b} - \boldsymbol{a})$, and so $\boldsymbol{c}$ is on the line passing through $\boldsymbol{a}$ and $\boldsymbol{b}$.

As another example of vector methods, we prove that the diagonals of a parallelogram bisect each other.

Let the adjacent sides of the parallelogram be represented by the vectors $\boldsymbol{a}$ and $\boldsymbol{b}$, as in Figure 2-16. We first calculate the vector to the midpoint of PQ. We know $\boldsymbol{b} - \boldsymbol{a}$ is parallel to the directed segment from P to Q, and $\frac{1}{2}(\boldsymbol{b} - \boldsymbol{a})$ is therefore parallel to the directed line segment from P to the midpoint of PQ. Thus, the vector $\boldsymbol{a} + \frac{1}{2}(\boldsymbol{b} - \boldsymbol{a}) = \frac{1}{2}\boldsymbol{a} + \frac{1}{2}\boldsymbol{b}$, ends at the midpoint of PQ.

Next, we calculate the vector to the midpoint of OR. We know $\boldsymbol{a} + \boldsymbol{b}$ ends at R, thus $\frac{1}{2}(\boldsymbol{a} + \boldsymbol{b})$ ends at the midpoint of OR.

Since we know that the vector $\frac{1}{2}\boldsymbol{a} + \frac{1}{2}\boldsymbol{b}$ ends at both the midpoint of OR and the midpoint of PQ, we see that OR and PQ bisect each other.

A vector in space has both length and direction. A full discussion of length and associated concepts is given in chapter 6. At the present time, however, it is not at all difficult to give a formula for computing the length.

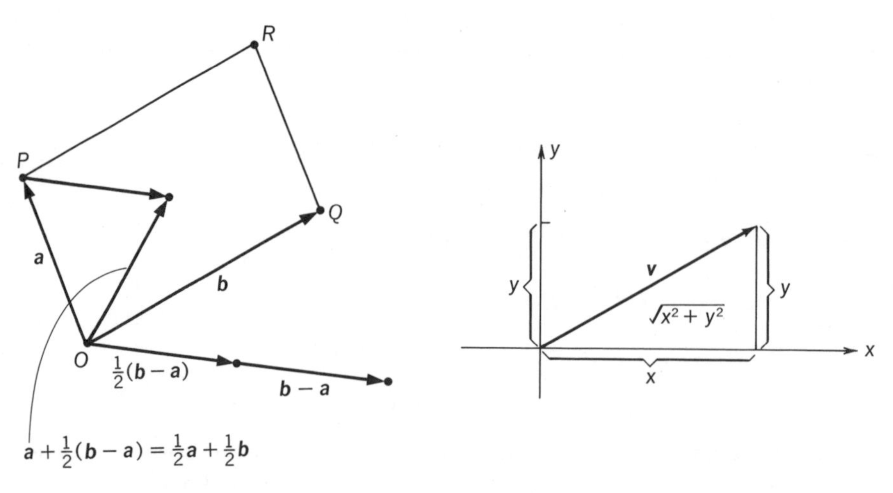

FIGURE 2-16 FIGURE 2-17

Let $v = xi + yj$ be a vector in R^2. Then using the Pythagorean theorem (see Figure 2-17), the length of v is $\sqrt{x^2 + y^2}$. For example, the length of $3i + 4j$ is $\sqrt{3^2 + 4^2} = 5$. The formula for the length of a vector in R^3 is given in excercise 6.

The geometrically defined vectors we have discussed in this section are of great usefulness in physics and engineering. Many physical quantities such as force, angular momentum, velocity, and acceleration are most naturally described as vectors of this type. For example, if a force f acts on a given body, the magnitude of the force is the length of the vector f. The direction of the force is the direction of f. When several forces act on the same body, the resultant force experienced by the body is simply the vector sum of each of the forces.

Example 3 Four men hold a tug of war. The men are equally spaced on a circle of radius 10 feet. Four ropes are tied to a metal ring which is placed in the center of the circle, and one piece of rope is given to each man. Going in counterclockwise order around the circle, the men exert forces of 100, 120, 80, and 140 pounds, respectively. What is the resultant force on the ring?

To solve this problem, take a coordinate system whose origin is at the center of the circle, with the men placed in positions (10, 0), (0, 10), (− 10, 0), and (0, − 10), as in Figure 2-18.

The man at position (10, 0) is exerting a force of 100 pounds in the direction of the positive x axis. This force may be expressed as the vector $100i$. The man at (0, 10) exerts the force $120j$. The remaining two men exert forces $-80i$ and $-140j$, respectively. The resultant force is the vector sum of these four vectors, i.e., $20i - 20j$. The magnitude of the resultant force is the length of this vector, i.e., $20\sqrt{2}$. Its direction is along the line of the vector $i - j$.

Throughout this section, vectors were regarded as directed line segments beginning at the origin. For many applications, notably in physics, the concept of a **free vector** is quite useful. A free vector is simply a directed line segment in space. Two free vectors are considered equal when they have the same direction and length. Free vectors can be manipulated just as the vectors we have considered. We only mention this idea in passing, but in other contexts it affords a more natural geometric interpretation of vectors.

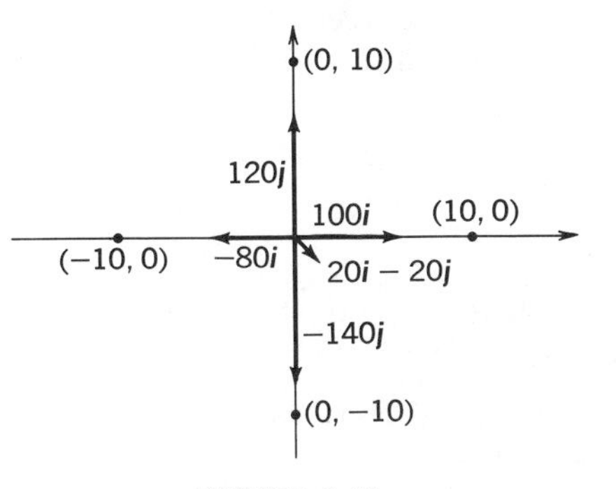

FIGURE 2-18

EXERCISES

1. Sketch the following vectors.
(a) $-2\boldsymbol{i}+\boldsymbol{j}-\boldsymbol{k}$ (b) $\boldsymbol{i}+\boldsymbol{j}+\boldsymbol{k}$ (c) $-\boldsymbol{i}+\boldsymbol{j}+\boldsymbol{k}$ (d) $2\boldsymbol{i}-\boldsymbol{j}+\boldsymbol{k}$

2. Consider the cube bounded by the six planes $x=0$, $x=1$, $y=0$, $y=1$, $z=0$, $z=1$. Find the vectors that end at its vertices.

3. Find, in parametric form, the equation of the line passing through the endpoint of $\boldsymbol{i}$ in the direction of $\boldsymbol{i}+\boldsymbol{j}+\boldsymbol{k}$. Where does this line intersect the plane $x=0$, the plane $y=2$?

4. Find the equation of the line that passes through the points $(-1, -1, 0)$ and $(2, 1, 2)$. Where does this line intersect the plane $x=0$?

5. Show that

$$x(t)=\alpha+\lambda t$$
$$y(t)=\beta+\mu t$$
$$z(t)=\gamma+\nu t$$

are the equations of a straight line in space.

6. Using the Pythagorean theorem, show that the length of the vector $x\boldsymbol{i}+y\boldsymbol{j}+z\boldsymbol{k}$ is $\sqrt{x^2+y^2+z^2}$.

7. What point on the line segment connecting $(-1, -1, 0)$ to $(1, 1, 1)$ is closest to the origin?

8. If $\boldsymbol{x}$, $\boldsymbol{y}$, and $\boldsymbol{z}$ are three noncoplanar vectors, show that $\boldsymbol{x}+(\boldsymbol{y}+\boldsymbol{z})$ is the diagonal of the parallelepiped with sides $\boldsymbol{x}$, $\boldsymbol{y}$, and $\boldsymbol{z}$. Interpret the associative law $\boldsymbol{x}+(\boldsymbol{y}+\boldsymbol{z})=(\boldsymbol{x}+\boldsymbol{y})+\boldsymbol{z}$ geometrically.

9. Interpret the laws

$$(\lambda+\mu)\boldsymbol{a}=\lambda\boldsymbol{a}+\mu\boldsymbol{a}$$
$$\lambda(\boldsymbol{a}+\boldsymbol{b})=\lambda\boldsymbol{a}+\lambda\boldsymbol{b}$$
$$\lambda(\mu\boldsymbol{a})=(\lambda\mu)\boldsymbol{a}$$

geometrically, where $\boldsymbol{a}$, $\boldsymbol{b}$ are vectors and λ, μ are scalars.

10. Show that the endpoints of the vectors $\boldsymbol{x}$, $\boldsymbol{y}$, and $\boldsymbol{z}$ are collinear if and only if there are scalars λ, μ, and ν, not all zero, such that

$$\lambda\boldsymbol{x}+\mu\boldsymbol{y}+\nu\boldsymbol{z}=0$$
$$\lambda+\mu+\nu=0$$

11. If $\boldsymbol{a}$ and $\boldsymbol{b}$ are noncollinear vectors, show that $\boldsymbol{a}+\boldsymbol{b}$, $-\boldsymbol{a}+\boldsymbol{b}$, $\boldsymbol{a}-\boldsymbol{b}$, and $-\boldsymbol{a}-\boldsymbol{b}$ are the vertices of a parallelogram.

12. Show that the figure formed by connecting the midpoints of adjacent sides of a parallelogram is a parallelogram.

13. Find the midpoint of the line segment connecting each of the following pairs of points. Sketch your results.
    (a) (1, 0, 0) and (1, 1, 1)
    (b) (1, 1, 1) and (1, 1, − 1)
    (c) (1, 1, 2) and (− 1, − 1, 0)
14. What point is on the line segment connecting (x_1, y_1, z_1) to (x_2, y_2, z_2) and $\frac{1}{3}$ of the way from (x_1, y_1, z_1) to (x_2, y_2, z_2)?
15. If $\boldsymbol{l}_1(t)$ and $\boldsymbol{l}_2(t)$ are parametric equations of two lines, show that the lines are parallel if and only if there are constants, λ_1 and λ_2, not both zero, such that $\lambda_1\boldsymbol{l}_1(t) + \lambda_2\boldsymbol{l}_2(t)$ is a constant vector.
16. Three forces act on a body. The position of the body is at the origin. One force of 20 pounds is in the direction of the positive x axis. A force of 30 pounds acts in the positive y direction, and a force of 40 pounds is in the negative x direction. Find the resultant force on the body.
17. Four bodies sit at the corners of the unit square, i.e., at (0, 0), (0, 1), (1, 1), (1, 0). Their masses are, in order, 1, 2, 3, 4. Find the center of mass of the system.

3 MATRICES

We define a **matrix** as a rectangular array of numbers, real or complex:

$$\begin{bmatrix} a_{11} & a_{12} & a_{13} & \cdots & a_{1n} \\ a_{21} & a_{22} & a_{23} & \cdots & a_{2n} \\ a_{31} & a_{32} & a_{33} & \cdots & a_{3n} \\ & & \vdots & & \\ a_{m1} & a_{m2} & a_{m3} & \cdots & a_{mn} \end{bmatrix}$$

The numbers in the array are the **elements** or **entries** of the matrix. The subscripts i and j of the element a_{ij} serve to identify the row and column where a_{ij} is located.

For example,

$$\begin{bmatrix} -3 & 0 & 13 \\ 2 & 1 & 2 \\ 1 & 3 & 0 \\ 0 & -1 & 2 \end{bmatrix}$$

is a matrix. The element a_{23} is located in the second row and third column and in this case is 2. The entry in position (3, 3) is 0.

A matrix which has m rows and n columns is said to be an $m \times n$ matrix, or an m by n matrix. When $m = n$, i.e., when the matrix is **square**, we simply say that we have a matrix of order n. For example, $\begin{bmatrix} -3 & 2 \\ 0 & 1 \\ 4 & 6 \end{bmatrix}$ is a 3×2 matrix, while $\begin{bmatrix} 3 & 7 & -1 \\ 0 & 2 & 3 \end{bmatrix}$ is a 2×3 matrix.

Matrices arise naturally in tabulating large amounts of data. For example, a mileage chart between the cities of Boston, New York, Washington, and Philadelphia can be tabulated as a matrix:

BOSTON	NEW YORK		WASHINGTON	PHILADELPHIA
Boston	0	180	410	265
New York	180	0	225	75
Washington	410	225	0	140
Philadelphia	265	75	140	0

In this case, a_{12} is the distance from Boston to New York, a_{13} the distance from Boston to Washington, and a_{42} is the distance from Philadelphia to New York. Note, that in this case, $a_{ii} = 0$, and $a_{ij} = a_{ji}$.

Often, we abbreviate the above notation to $[a_{ij}]_{(mn)}$. The subscript mn denotes the size of the matrix, and the whole symbol means the $m \times n$ matrix whose entry in position (i, j) is a_{ij}. For example, in this notation $[ij]_{(23)}$ means the 2×3 matrix which has the product ij in position (i, j), or $\begin{bmatrix} 1 & 2 & 3 \\ 2 & 4 & 6 \end{bmatrix}$.

Two matrices are equal when and only when they are of the same order and have the same entry in each position (i, j); in other words, $[a_{ij}]_{(mn)} = [b_{ij}]_{(pq)}$ if and only if $m = p$, $n = q$, and $a_{ij} = b_{ij}$ for all i and j.

If $A = [a_{ij}]_{(mn)}$ and $B = [b_{ij}]_{(mn)}$ are two $m \times n$ matrices, we define $A + B$ to be the matrix $[a_{ij} + b_{ij}]_{(mn)}$. For example,

$$\begin{bmatrix} 3 & -1 & 2 \\ 0 & 13 & 7 \\ 8 & 10 & 12 \end{bmatrix} + \begin{bmatrix} \frac{1}{2} & -3 & 1 \\ 0 & -\frac{1}{2} & 1 \\ 2 & 0 & 0 \end{bmatrix} = \begin{bmatrix} \frac{7}{2} & -4 & 3 \\ 0 & \frac{25}{2} & 8 \\ 10 & 10 & 12 \end{bmatrix}$$

$$\begin{bmatrix} 1 & 3 \\ 0 & \frac{1}{2} \\ 2 & \frac{1}{3} \end{bmatrix} + \begin{bmatrix} 0 & 0 \\ -\frac{1}{2} & 1 \\ 2 & 7 \end{bmatrix} = \begin{bmatrix} 1 & 3 \\ -\frac{1}{2} & \frac{3}{2} \\ 4 & \frac{22}{3} \end{bmatrix}$$

We define the matrix $-A = [-a_{ij}]_{(mn)}$. For example,

$$-\begin{bmatrix} 1 & 0 & 2 \\ 0 & -1 & 3 \end{bmatrix} = \begin{bmatrix} -1 & 0 & -2 \\ 0 & 1 & -3 \end{bmatrix}$$

In other words, to add matrices we add corresponding entries. To take the negative of a matrix we take the negatives of the corresponding entries.

We define the scalar multiple of the matrix A by the scalar α, written αA, by $\alpha A = [\alpha a_{ij}]_{(mn)}$. For example,

$$2\begin{bmatrix} 7 & 0 & 7 \\ -8 & 1 & -3 \end{bmatrix} = \begin{bmatrix} 14 & 0 & 14 \\ -16 & 2 & -6 \end{bmatrix}$$

We illustrate these operations with a simple example. A university is divided into two campuses. The numbers of men and women, graduate students and undergraduates, are tabulated in the matrices:

$$A_1 = \begin{matrix} \\ (\text{m.}) \\ (\text{w.}) \end{matrix} \begin{matrix} (\text{u.g.}) & (\text{g.}) \\ \left[\begin{matrix} 80 \\ 60 \end{matrix}\right. & \left.\begin{matrix} 30 \\ 20 \end{matrix}\right] \end{matrix} \qquad A_2 = \begin{matrix} \\ (\text{m.}) \\ (\text{w.}) \end{matrix} \begin{matrix} (\text{u.g.}) & (\text{g.}) \\ \left[\begin{matrix} 200 \\ 160 \end{matrix}\right. & \left.\begin{matrix} 90 \\ 70 \end{matrix}\right] \end{matrix}$$

Thus, for example, there are 80 male undergraduates on the first campus, and 70 woman graduate students on the second. The sum of the two matrices

$$A_1 + A_2 = \begin{bmatrix} 280 & 120 \\ 220 & 90 \end{bmatrix}$$

gives the number of students in each category for the whole university. Thus, there are 120 men who are also graduate students.

Addition and scalar multiplication of matrices satisfy the same laws as addition and scalar multiplication of real numbers and vectors.

Proposition 1 $(A + B) + C = A + (B + C)$.

PROOF Let

$$A = [a_{ij}]_{(mn)} \qquad B = [b_{ij}]_{(mn)} \qquad C = [c_{ij}]_{(mn)}$$

$(A + B) + C = [(a_{ij} + b_{ij}) + c_{ij}]_{(mn)}$ while $A + (B + C) = [a_{ij} + (b_{ij} + c_{ij})]_{(mn)}$. But $(a_{ij} + b_{ij}) + c_{ij} = a_{ij} + (b_{ij} + c_{ij})$ for all real numbers. Therefore, $(A + B) + C = A + (B + C)$. ■

The proofs of all other laws are similar.

Proposition 2 $A + B = B + A$.

Proposition 3 $A + (-A) = -A + A = 0$, where 0 denotes the matrix of the same order as A, all of whose entries are 0.

Proposition 4 $A + 0 = 0 + A = A$.

Proposition 5

$$(\alpha + \beta)A = \alpha A + \beta A$$

$$\alpha(A + B) = \alpha A + \alpha B$$

$$\alpha(\beta A) = (\alpha\beta)A$$

$$1A = A$$

An $n \times 1$ matrix is, according to the definition given in §2.1, just a column vector. A $1 \times n$ matrix is a row vector.

EXERCISES

1. Write out the indicated matrices as arrays.
   (a) $[i + j^2]_{(22)}$ (b) $[ij]_{(23)}$
   (c) $[i^2 + j^2]_{(23)}$ (d) $[2i + 3j]_{(33)}$

2. Calculate.
   (a) $\begin{bmatrix} 2 & 3 & 1 \\ 0 & 1 & 5 \end{bmatrix} + \begin{bmatrix} -1 & 4 & 8 \\ 5 & -3 & 4 \end{bmatrix} - \begin{bmatrix} 0 & 1 & 0 \\ 5 & 6 & 7 \end{bmatrix}$
   (b) $\begin{bmatrix} 3 & 9 \\ 1 & -2 \end{bmatrix} + 2\begin{bmatrix} 6 & 0 \\ -1 & 5 \end{bmatrix} - 3\begin{bmatrix} 2 & -4 \\ 1 & 2 \end{bmatrix}$
   (c) $\begin{bmatrix} 1 & 3 & 6 \\ 4 & 2 & 7 \\ 0 & 3 & 5 \end{bmatrix} + \begin{bmatrix} 9 & 0 & 3 \\ 0 & -1 & 3 \\ 4 & 9 & 0 \end{bmatrix}$

3. Let X, A, B, and C be $m \times n$ matrices. Solve the equation $(X + A) + (B - C) = 3X + 4A$ to find X in terms of A, B, and C.

4. Prove the laws of matrix addition and scalar multiplication (Propositions 2–5) that were not proved in the text.

5. If A, B, X, and Y are $m \times n$ matrices, solve the equations

$$2X + 3Y = A$$
$$X + 2Y = B$$

   to find X and Y in terms of A and B.

6. Show that there are 2×2 matrices X, Y, and Z, not all 0, that satisfy the equations

$$a_{11}X + a_{12}Y + a_{13}Z = 0$$
$$a_{21}X + a_{22}Y + a_{23}Z = 0$$

   with a_{ij} arbitrary real numbers.

7. How many 2×3 matrices are there all of whose entries are 0 or 1?

8. Given a 2×2 matrix A, show that there are numbers a, b, c, and d such that
   (a) $A = a\begin{bmatrix} 1 & 0 \\ 0 & 0 \end{bmatrix} + b\begin{bmatrix} 0 & 1 \\ 0 & 0 \end{bmatrix} + c\begin{bmatrix} 0 & 0 \\ 1 & 0 \end{bmatrix} + d\begin{bmatrix} 0 & 0 \\ 0 & 1 \end{bmatrix}$
   (b) $A = a\begin{bmatrix} -1 & 1 \\ 1 & 1 \end{bmatrix} + b\begin{bmatrix} 1 & -1 \\ 1 & 1 \end{bmatrix} + c\begin{bmatrix} 1 & 1 \\ -1 & 1 \end{bmatrix} + d\begin{bmatrix} 1 & 1 \\ 1 & -1 \end{bmatrix}$

9. The square below has sides of length 1 unit.

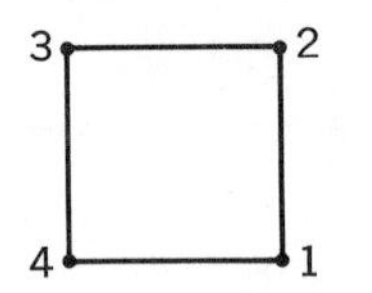

   With this notation let a_{ij} be the distance between points i and j. Construct the matrix of distances $A = [a_{ij}]_{(44)}$.

10. Four points of the plane are given with $P_1 = (1, 1)$, $P_2 = (0, 2)$, $P_3 = (-1, 0)$, and $P_4 = (0, -1)$. Construct the matrix, A, whose entry in position (i, j) is the distance from P_i to P_j.

11. A graph is a collection of points connected by certain lines. The following are examples of graphs.

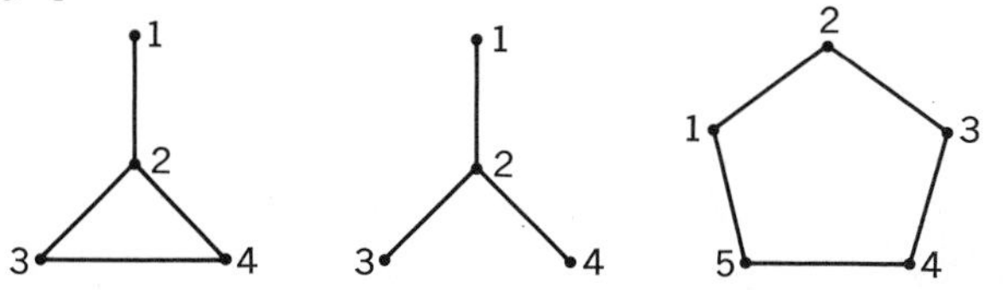

The points of the graph are often called vertices. With a graph having k vertices we associate a $k \times k$ matrix A which is called the incidence matrix of the graph. The entries a_{ij} of the matrix are given by the rule $a_{ij} = 0$ if i and j are not connected, $a_{ij} = 1$ if i and j are connected.
Thus, for the first graph the incidence matrix is

$$\begin{bmatrix} 0 & 1 & 0 & 0 \\ 1 & 0 & 1 & 1 \\ 0 & 1 & 0 & 1 \\ 0 & 1 & 1 & 0 \end{bmatrix}$$

Note, by convention, $a_{ii} = 0$.
What are the incidence matrices of the other two graphs?

12. Suppose that A, B, C, and D are four measures of length. The following is a conversion table.

$$\begin{array}{c} \\ A \\ B \\ C \\ D \end{array} \begin{array}{c} \begin{array}{cccc} A & B & C & D \end{array} \\ \begin{bmatrix} 1 & 3 & 6 & 24 \\ \frac{1}{3} & 1 & 2 & 8 \\ \frac{1}{6} & \frac{1}{2} & 1 & 4 \\ \frac{1}{24} & \frac{1}{8} & \frac{1}{4} & 1 \end{bmatrix} \end{array}$$

Thus, one unit of A is 6 units of C, one unit of B is 8 units of D, one unit of C is $\frac{1}{2}$ unit of B. Regard this table as a matrix and explain why $a_{ij}a_{jk} = a_{ik}$.

4 MATRIX MULTIPLICATION

In addition to the operations of addition and scalar multiplication defined on matrices, there is a third algebraic operation, called matrix multiplication, that is often encountered.

Let $A = [a_{ij}]_{(mn)}$ be an $m \times n$ matrix and $B = [b_{ij}]_{(np)}$ be an $n \times p$ matrix. We define an $m \times p$ matrix AB, called the product of A and B, by

$$AB = \left[\sum_{j=1}^{n} a_{ij} b_{jk} \right]_{(mp)}$$

In the 2×2 case this definition becomes more explicitly

$$\begin{bmatrix} a_{11} & a_{12} \\ a_{21} & a_{22} \end{bmatrix} \begin{bmatrix} b_{11} & b_{12} \\ b_{21} & b_{22} \end{bmatrix} = \begin{bmatrix} a_{11}b_{11} + a_{12}b_{21} & a_{11}b_{12} + a_{12}b_{22} \\ a_{21}b_{11} + a_{22}b_{21} & a_{21}b_{12} + a_{22}b_{22} \end{bmatrix}$$

Notice we go across the ith row of the first matrix and down the kth column of the second matrix to obtain the entry in position (i, k).

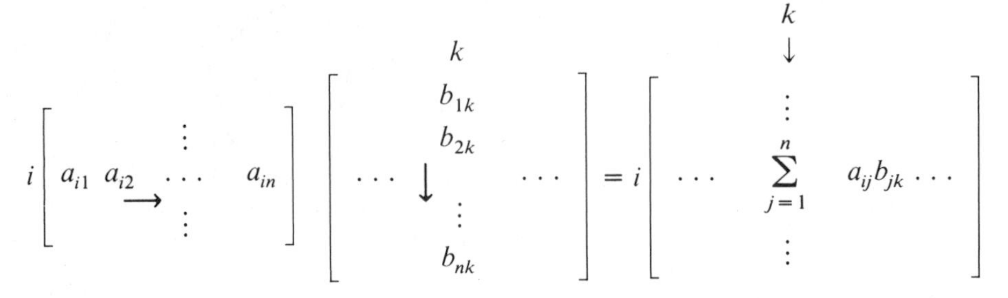

For example,

$$\begin{bmatrix} 3 & 1 & -2 \\ 6 & 3 & 4 \end{bmatrix}\begin{bmatrix} 4 & 7 \\ 3 & 0 \\ 1 & 2 \end{bmatrix} = \begin{bmatrix} 3\cdot4+1\cdot3+(-2)\cdot1 & 3\cdot7+1\cdot0+(-2)(2) \\ 6\cdot4+3\cdot3+4\cdot1 & 6\cdot7+3\cdot0+4\cdot2 \end{bmatrix}$$

$$= \begin{bmatrix} 13 & 17 \\ 37 & 50 \end{bmatrix}$$

$$\begin{bmatrix} 1 & 0 & 7 \\ 0 & 3 & 1 \\ 2 & 4 & 6 \end{bmatrix}\begin{bmatrix} -1 & 3 & 7 \\ -2 & 1 & 6 \\ 0 & 2 & 8 \end{bmatrix}$$

$$= \begin{bmatrix} 1\cdot(-1)+0\cdot(-2)+7\cdot0 & 1\cdot3+0\cdot1+7\cdot2 & 1\cdot7+0\cdot6+7\cdot8 \\ 0\cdot(-1)+3\cdot(-2)+(1)\cdot0 & 0\cdot3+3\cdot1+1\cdot2 & 0\cdot7+3\cdot6+1\cdot8 \\ 2\cdot(-1)+4\cdot(-2)+6\cdot0 & 2\cdot3+4\cdot1+6\cdot2 & 2\cdot7+4\cdot6+6\cdot8 \end{bmatrix}$$

$$= \begin{bmatrix} -1 & 17 & 63 \\ -6 & 5 & 26 \\ -10 & 22 & 86 \end{bmatrix}$$

There are two things worth remembering about the number of rows and columns of the matrices A and B in forming their matrix product. First, in order to have the matrix product AB defined, the number of columns of A must equal the number of rows of B. Second, the product AB has the same number of rows as A and the same number of columns as B.

The following example shows how matrix multiplication might arise in a biological context.

Example 1 In a certain region there are two species of carnivores C_1 and C_2 and two species of herbivores H_1 and H_2 on which the carnivores subsist. The herbivores live on three species of plants P_1, P_2, and P_3.

The following matrix describes the amount in grams of each type of plant that the average herbivore eats in a single day.

$$\begin{array}{c} \\ H_1 \\ H_2 \end{array}\begin{array}{c} \begin{array}{ccc} P_1 & P_2 & P_3 \end{array} \\ \begin{bmatrix} 200 & 350 & 150 \\ 300 & 150 & 200 \end{bmatrix} \end{array} = A$$

Thus, a typical member of H_2 consumes 150 grams of P_2 in one day.

Another matrix describes the number of herbivores the average carnivore consumes in one day.

$$\begin{array}{c} \\ C_1 \\ C_2 \end{array}\begin{array}{c} \begin{array}{cc} H_1 & H_2 \end{array} \\ \begin{bmatrix} 3 & 1 \\ 4 & 2 \end{bmatrix} \end{array} = B$$

Thus, an average member of C_2 consumes 4 members of H_1 and 2 of H_2 in a single day.

We wish to determine how many grams of each type of plant one of the carnivores consumes indirectly.

For example, how many grams of P_1 does a member of C_1 consume indirectly? First, a member of C_1 consumes 3 members of H_1. Each member of H_1 consumes 200 grams of P_1. Thus, by eating members of H_1 a member of C_1 consumes indirectly $3 \cdot 200$ grams of P_1. Also, a member of C_1 consumes 1 member of H_2, and this member of H_2 eats 300 grams of P_1. Thus, from this source a member of C_1 consumes indirectly $1 \cdot 300$ grams of P_1. Altogether, then a member of C_1 consumes $3 \cdot 200 + 1 \cdot 300$ grams of P_1 indirectly. Observe that this number is just the entry (1, 1) of the matrix product BA. Similarly, the entry in position (1, 2) is the number of grams of P_2 that a member of C_1 eats indirectly, and so on.

We compute

$$BA = \begin{bmatrix} 3 & 1 \\ 4 & 2 \end{bmatrix}\begin{bmatrix} 200 & 350 & 150 \\ 300 & 150 & 200 \end{bmatrix} = \begin{bmatrix} 900 & 1200 & 650 \\ 1400 & 1700 & 1000 \end{bmatrix}.$$

From this, we can read off the desired information. For example, each member of C_2 consumes indirectly 1000 grams of P_3.

There are several algebraic laws connected with the multiplication of matrices that we now state.

Proposition 1 If A is an $m \times n$ matrix, B is an $n \times p$ matrix, and C is a $p \times q$ matrix, then $(AB)C = A(BC)$. (Associative law of matrix multiplication.)

PROOF Let

$$A = [a_{ij}]_{(mn)} \qquad B = [b_{jk}]_{(np)} \qquad C = [c_{kl}]_{(pq)}$$

Then,

$$AB = \left[\sum_{j=1}^{n} a_{ij} b_{jk} \right]_{(mp)}$$

$$(AB)(C) = \left[\sum_{k=1}^{p} \left(\sum_{j=1}^{n} a_{ij} b_{jk} \right) c_{kl} \right]_{(mq)} = \left[\sum_{k=1}^{p} \sum_{j=1}^{n} a_{ij} b_{jk} c_{kl} \right]_{(mq)}$$

$$BC = \left[\sum_{k=1}^{p} b_{jk} c_{kl} \right]_{(nq)}$$

$$A(BC) = \left[\sum_{j=1}^{n} a_{ij} \left(\sum_{k=1}^{p} b_{jk} c_{kl} \right) \right] = \left[\sum_{j=1}^{n} \sum_{k=1}^{p} a_{ij} b_{jk} c_{kl} \right]_{(mq)}$$

Since $\sum_{k=1}^{p} \sum_{j=1}^{n} a_{ij} b_{jk} c_{kl} = \sum_{j=1}^{n} \sum_{k=1}^{p} a_{ij} b_{jk} c_{kl}$, we have $(AB)C = A(BC)$. ■

For example,

$$\left(\begin{bmatrix} 1 & 2 \\ 3 & 1 \end{bmatrix} \begin{bmatrix} -1 & 0 \\ 2 & 1 \end{bmatrix} \right) \begin{bmatrix} 1 & 6 & 3 \\ 4 & 2 & 1 \end{bmatrix} = \begin{bmatrix} 3 & 2 \\ -1 & 1 \end{bmatrix} \begin{bmatrix} 1 & 6 & 3 \\ 4 & 2 & 1 \end{bmatrix} = \begin{bmatrix} 11 & 22 & 11 \\ 3 & -4 & -2 \end{bmatrix}$$

$$\begin{bmatrix} 1 & 2 \\ 3 & 1 \end{bmatrix} \left(\begin{bmatrix} -1 & 0 \\ 2 & 1 \end{bmatrix} \begin{bmatrix} 1 & 6 & 3 \\ 4 & 2 & 1 \end{bmatrix} \right) = \begin{bmatrix} 1 & 2 \\ 3 & 1 \end{bmatrix} \begin{bmatrix} -1 & -6 & -3 \\ 6 & 14 & 7 \end{bmatrix} = \begin{bmatrix} 11 & 22 & 11 \\ 3 & -4 & -2 \end{bmatrix}$$

We can also verify the distributive laws for matrix multiplication.

Proposition 2 $A(B + C) = AB + AC$ and $(A + B)C = AC + BC$.

PROOF Let

$$A = [a_{ij}]_{(mn)} \qquad B = [b_{jk}]_{(np)} \qquad C = [c_{jk}]_{(np)}$$

Then, $B + C = [b_{jk} + c_{jk}]_{(np)}$.

$$A(B + C) = \left[\sum_{j=1}^{n} a_{ij}(b_{jk} + c_{jk}) \right]_{(mp)}$$

$$AB = \left[\sum_{j=1}^{n} a_{ij} b_{jk} \right]_{(mp)} \qquad AC = \left[\sum_{j=1}^{n} a_{ij} c_{jk} \right]_{(mp)}$$

$$AB + AC = \left[\sum_{j=1}^{n} a_{ij} b_{jk} + \sum_{j=1}^{n} a_{ij} c_{jk} \right]_{(mp)}$$

Since $\sum_{j=1}^{n} a_{ij}(b_{jk} + c_{jk}) = \sum_{j=1}^{n} a_{ij} b_{jk} + \sum_{j=1}^{n} a_{ij} c_{jk}$ we have $A(B + C) = AB + AC$. We leave the proof of the other distributive law as an exercise for the reader. ■

The last two laws show that matrix multiplication behaves to some extent like the multiplication of scalars. However, there are many ways in which the two are different.

For instance, it is **not**, in general, true that $AB = BA$. This is easy to see if we suppose that A is an $m \times n$ matrix and B an $n \times p$ matrix. In order to have BA defined we must have $m = p$. Thus, AB may be defined, but BA may not be defined. Moreover, even if both A and B are square of order n, in which case both AB and BA are defined, we still need not have $AB = BA$.

For example,

$$\begin{bmatrix} 1 & 2 \\ 0 & 3 \end{bmatrix}\begin{bmatrix} -1 & 2 \\ 1 & 3 \end{bmatrix} = \begin{bmatrix} 1 & 8 \\ 3 & 9 \end{bmatrix}$$

$$\begin{bmatrix} -1 & 2 \\ 1 & 3 \end{bmatrix}\begin{bmatrix} 1 & 2 \\ 0 & 3 \end{bmatrix} = \begin{bmatrix} -1 & 4 \\ 1 & 11 \end{bmatrix}$$

In mathematical terms we say that the multiplication of matrices is, in general, noncommutative. If it does happen that both AB and BA are defined and $AB = BA$, we say A and B commute.

The following illustrates, in an empirical way, why matrices do not always commute.

Example 2 Three beakers of water are placed on a table. Each beaker is of the same size. The amount of water in each beaker is given by the vector $v = \begin{bmatrix} x_1 \\ x_2 \\ x_3 \end{bmatrix}$. Thus, in the first beaker there are x_1 units of water, and so on.

A two step operation is performed on the two beakers.

(1) Adjust the water in the first two beakers so that the water level is the same in both beakers. Leave the third alone.
(2) Adjust the water in the second and third beakers so that the water level is the same in both. Leave the first alone.

After the first step is performed, the water levels in beakers is given by the vector $\begin{bmatrix} \frac{1}{2}(x_1 + x_2) \\ \frac{1}{2}(x_1 + x_2) \\ x_3 \end{bmatrix}$. The effect on the vector v of performing the first step of the operation can be found by multiplying v by a suitable matrix. Indeed, let $A_1 = \begin{bmatrix} \frac{1}{2} & \frac{1}{2} & 0 \\ \frac{1}{2} & \frac{1}{2} & 0 \\ 0 & 0 & 1 \end{bmatrix}$. Then, A_1v is the vector of water levels after step 1 is performed. In other words, multiplying v by A_1 affects v in the same way that step 1 affects the water in the beakers.

Step 2 can also be achieved by matrix multiplication. Let $A_2 = \begin{bmatrix} 1 & 0 & 0 \\ 0 & \frac{1}{2} & \frac{1}{2} \\ 0 & \frac{1}{2} & \frac{1}{2} \end{bmatrix}$. Then, multiplying A_2 by a vector of water levels affects the vector in exactly the same way that step 2 affects the water levels. Thus, A_1v is the vector after step 1 is performed and A_2A_1v is the vector after both steps are performed. Computing the matrix product we find that

$$A_2A_1v = \begin{bmatrix} \frac{1}{2} & \frac{1}{2} & 0 \\ \frac{1}{4} & \frac{1}{4} & \frac{1}{2} \\ \frac{1}{4} & \frac{1}{4} & \frac{1}{2} \end{bmatrix}\begin{bmatrix} x_1 \\ x_2 \\ x_3 \end{bmatrix}.$$ Thus, the effect of the whole operation on the water level vector can be obtained by multiplying by the matrix A_2A_1.

Now, however, let us suppose that the steps are performed in the opposite order. The resulting vector of water levels is then $A_1A_2v = \begin{bmatrix} \frac{1}{2} & \frac{1}{4} & \frac{1}{4} \\ \frac{1}{2} & \frac{1}{4} & \frac{1}{4} \\ 0 & \frac{1}{2} & \frac{1}{2} \end{bmatrix}\begin{bmatrix} x_1 \\ x_2 \\ x_3 \end{bmatrix}$. Observe that $A_1A_2 \neq A_2A_1$. This means that performing the steps in different order produces a different outcome. As a specific example, suppose $v = \begin{bmatrix} 4 \\ 8 \\ 4 \end{bmatrix}$, then $A_2A_1v = \begin{bmatrix} 6 \\ 5 \\ 5 \end{bmatrix}$, but $A_1A_2v = \begin{bmatrix} 5 \\ 5 \\ 6 \end{bmatrix}$. Thus, in this case noncommutativity has a palpable interpretation.

Another striking way in which matrix multiplication differs from scalar multiplication is the failure of the cancellation law: we may have $AB = AC$ with $A \neq 0$, without having $B = C$. For example,

$$\begin{bmatrix} 1 & -1 \\ 1 & -1 \end{bmatrix}\begin{bmatrix} 1 & 2 \\ 1 & 3 \end{bmatrix} = \begin{bmatrix} 0 & -1 \\ 0 & -1 \end{bmatrix} = \begin{bmatrix} 1 & -1 \\ 1 & -1 \end{bmatrix}\begin{bmatrix} 0 & 2 \\ 0 & 3 \end{bmatrix}$$

We may also have $AB = 0$ with $A \neq 0$ and $B \neq 0$. For example,

$$\begin{bmatrix} 1 & 0 \\ 0 & 0 \end{bmatrix}\begin{bmatrix} 0 & 0 \\ 0 & 1 \end{bmatrix} = \begin{bmatrix} 0 & 0 \\ 0 & 0 \end{bmatrix} \qquad \begin{bmatrix} 1 & 1 \\ 1 & 1 \end{bmatrix}\begin{bmatrix} 1 & -1 \\ -1 & 1 \end{bmatrix} = \begin{bmatrix} 0 & 0 \\ 0 & 0 \end{bmatrix}$$

We illustrate the failure of the cancellation law using the two step process of example 2. Let us suppose that the initial water levels are given by $v = \begin{bmatrix} 4 \\ 8 \\ 4 \end{bmatrix}$. Then performing step 1 and step 2 gives $A_2A_1v = \begin{bmatrix} 6 \\ 5 \\ 5 \end{bmatrix}$. Now, however, suppose that the initial distribution is given by $v' = \begin{bmatrix} 8 \\ 4 \\ 4 \end{bmatrix}$; then, also, $A_2A_1v' = \begin{bmatrix} 6 \\ 5 \\ 5 \end{bmatrix}$. Thus, with either initial distribution the final outcome is the same. Writing down matrix products this says, $A_2A_1v = A_2A_1v'$, but $v \neq v'$. In this case it is interesting to note that performing the first step has the same effect on v and v'.

EXERCISES

1. Compute the following matrix products.

(a) $\begin{bmatrix} 1 & 5 \\ 3 & 2 \end{bmatrix}\begin{bmatrix} 3 & -2 \\ 2 & 4 \end{bmatrix}$ (b) $\begin{bmatrix} 8 \\ 2 \\ 0 \end{bmatrix}[1 \quad -1 \quad 2]$ (c) $\begin{bmatrix} -3 & 2 \\ 5 & 3 \end{bmatrix}\begin{bmatrix} 1 & 1 \\ 3 & 2 \end{bmatrix}$

(d) $[3 \quad 2 \quad 1]\begin{bmatrix} 0 \\ 1 \\ 2 \end{bmatrix}$ (e) $\begin{bmatrix} 3 & 0 & 3 \\ 2 & 1 & 2 \\ 1 & 2 & 4 \end{bmatrix}\begin{bmatrix} 1 & 2 \\ 0 & -1 \\ 3 & 1 \end{bmatrix}$ (f) $\begin{bmatrix} 1 & 2 \\ 0 & 1 \end{bmatrix}\begin{bmatrix} -1 & 3 & 4 \\ 0 & 1 & 2 \end{bmatrix}$

2. Let

$$A=\begin{bmatrix}2&1\\3&2\end{bmatrix},\ B=\begin{bmatrix}3&-1\\-3&1\end{bmatrix},\ C=\begin{bmatrix}1&0\\1&1\end{bmatrix},\ I=\begin{bmatrix}1&0\\0&1\end{bmatrix},\ D=\begin{bmatrix}1&1\\1&1\end{bmatrix}.$$

Calculate.

(a) AB (b) $(AB)C$ (c) BC (d) $A(BC)$ (e) BA
(f) $AB-BA$ (g) IA (h) AI (i) IB (j) BI
(k) BD (l) DB (m) CC

3. Find $[1\quad 2\quad 0]\begin{bmatrix}1&4&3\\0&1&2\\3&2&1\end{bmatrix}\begin{bmatrix}1\\2\\-1\end{bmatrix}$.

4. Let A and B be matrices. Suppose that AB and BA are defined and $AB = BA$. Show that A and B are square matrices of the same order.

5. If B is a matrix, show that $B \cdot 0 = 0 \cdot B = 0$.

6. Let A be a 2×2 matrix and C a 4×4 matrix. Suppose that B is another matrix for which the product $(AB)C$ is defined. What is the order of B?

7. Prove the distributive law of matrix multiplication $(A + B)C = AC + BC$.

8. Find a, b, c, and d if $\begin{bmatrix}a&b\\c&d\end{bmatrix}\begin{bmatrix}2&3\\1&2\end{bmatrix}=\begin{bmatrix}1&0\\2&1\end{bmatrix}$.

9. If B is a 2×2 matrix such that $\begin{bmatrix}1&-1\\3&-3\end{bmatrix}B=0$, show that $B=\begin{bmatrix}a&b\\a&b\end{bmatrix}$ for some numbers, a and b.

10. Show that a 2×2 matrix, A, commutes with $\begin{bmatrix}1&1\\0&1\end{bmatrix}$ if and only if $A=\begin{bmatrix}a&b\\0&a\end{bmatrix}$ for some numbers, a and b.

11. Let $A=\begin{bmatrix}2&2\\3&3\end{bmatrix}$. Find a nonzero column vector $\mathbf{x}$ such that $A\mathbf{x}=\mathbf{0}$. Find a nonzero row vector $\mathbf{y}$ such that $\mathbf{y}A=\mathbf{0}$.

12. Let A and B be 2×2 matrices all of whose column sums are 1. Show that the column sums of AB are 1.

13. Let X be a 2×2 matrix such that $\begin{bmatrix}5&3\\2&1\end{bmatrix}X=0$. Show that $X=0$.

14. In a certain college, the number of graduate students, undergraduates, men and women is given by the matrix

$$A=\begin{matrix}(\text{m.})\\(\text{w.})\end{matrix}\overset{\begin{matrix}(\text{u.g.})&(\text{g.})\end{matrix}}{\begin{bmatrix}200&100\\180&80\end{bmatrix}}$$

Thus, for example, there are 100 male graduate students. The cost of tuition, room, and board for graduates and undergraduates is given by the matrix

$$B=\begin{matrix}(\text{u.g.})\\(\text{g.})\end{matrix}\overset{\begin{matrix}(\text{t.})&(\text{r.})&(\text{b.})\end{matrix}}{\begin{bmatrix}500&800&800\\800&1000&800\end{bmatrix}}$$

Compute the matrix AB and describe, in words, what the entries of AB mean.

15. A manufacturer sells three products, A, B, and C, in two markets, M and N. The number of units of each product sold in each market in a particular year is given by the matrix

$$T = \begin{matrix} \\ M \\ N \end{matrix} \begin{matrix} \begin{matrix} A & \quad B & \quad C \end{matrix} \\ \begin{bmatrix} 5000 & 2000 & 1500 \\ 2000 & 3000 & 1000 \end{bmatrix} \end{matrix}$$

The matrices

$$S_1 = \begin{bmatrix} 2.00 \\ 4.00 \\ 3.00 \end{bmatrix} \quad \text{and} \quad S_2 = \begin{bmatrix} 1.80 \\ 3.50 \\ 2.50 \end{bmatrix}$$

give, respectively, the unit sales price and unit cost price for A, B, and C. Interpret the meanings of the entries of the matrices, TS_1, TS_2, and $TS_1 - TS_2$.

16. The manufacturer described in the last problem finds that five years later the sales in M are increased by 50% and in N by 40%. Construct a 2×2 matrix, B, so that the matrix product, BT, gives the number of units of each product sold in each market at the later date.

17. Let A, B, and C be $n \times n$ matrices. If A commutes with C and B commutes with C, show that AB commutes with C.

18. Let A and B be 2×2 matrices. If A and B commute with $\begin{bmatrix} 0 & 1 \\ -1 & 0 \end{bmatrix}$, show that A commutes with B.

19. Let A be an $m \times n$ matrix and let

$$e_i = \begin{bmatrix} 0 \\ 0 \\ \vdots \\ 0 \\ 1 \\ 0 \\ \vdots \\ 0 \end{bmatrix} \left.\begin{matrix} \\ \\ \\ \\ \\ \end{matrix}\right\} i$$

denote the n-dimensional vector all of whose components but the ith are zero and whose ith component is 1. Show that Ae_i is the ith column of the matrix A.

20. Let A be an $m \times n$ matrix. If $Ax = \mathbf{0}$ for all vectors x, show that $A = 0$. [Hint: See exercise 19.]

5 SQUARE MATRICES

The square matrices of a fixed order have properties that make them particularly interesting objects of study. For example, suppose that A and B are both square of the same order. Then, all the products, AA, BB, AB, and BA, are defined. As we shall see later in this section, we can also define powers and polynomials in the matrices, A and B.

First we shall consider the identity matrix of order n. It plays the same role in the multiplication of matrices of order n as the number 1 plays in the multiplication of numbers.

We define the **Kronecker delta** δ_{ij} as follows

$$\delta_{ij} = 1 \qquad \text{if } i = j$$

$$\delta_{ij} = 0 \qquad \text{if } i \neq j$$

For example, $\delta_{23} = 0$, $\delta_{11} = 1$, $\delta_{33} = 1$, $\delta_{45} = 0$.

Using this symbol, we define the $n \times n$ matrix $I_n = [\delta_{ij}]_{(nn)}$. This matrix has only 1's on its diagonal entries, and 0's off the diagonal. For example,

$$I_2 = \begin{bmatrix} 1 & 0 \\ 0 & 1 \end{bmatrix}, \qquad I_3 = \begin{bmatrix} 1 & 0 & 0 \\ 0 & 1 & 0 \\ 0 & 0 & 1 \end{bmatrix}$$

I_n is called the **identity matrix of order *n*.**

Theorem 1 If A is an $n \times n$ matrix, $AI_n = I_n A = A$.

PROOF Let $A = [a_{ij}]_{(nn)}$. Then,

$$AI_n = \left[\sum_{j=1}^{n} a_{ij}\delta_{jk} \right]_{(nn)} = [\,a_{ij}\,]_{(nn)}$$

since in the sum $\sum_{j=1}^{n} a_{ij}\delta_{jk}$, $a_{ij}\delta_{jk} = 0$ unless $j = k$, and then $a_{ij}\delta_{kk} = a_{ij}$. Thus, $AI_n = A$. In a similar manner, $I_n A = A$. ■

For example,

$$\begin{bmatrix} 1 & 0 \\ 0 & 1 \end{bmatrix}\begin{bmatrix} a & b \\ c & d \end{bmatrix} = \begin{bmatrix} a & b \\ c & d \end{bmatrix}\begin{bmatrix} 1 & 0 \\ 0 & 1 \end{bmatrix} = \begin{bmatrix} a & b \\ c & d \end{bmatrix}$$

If α is a scalar, a matrix of the form αI_n is called a **scalar matrix**. It may be described as a matrix whose diagonal entries are all equal to α, and whose off diagonal entries are 0.

A concept of great importance when dealing with matrices is that of the inverse of a matrix. If A is an $m \times m$ matrix and there is an $m \times m$ matrix B, such that $AB = BA = I_m$, A is said to be **invertible** and B is said to be the **inverse** of A.

Having made this definition, the first natural question is whether a matrix A can have two different inverses. Theorem 2 shows that this is not possible.

Theorem 2 Let A be an $m \times m$ matrix with inverse B. If C is another matrix such that $AC = CA = I_m$, then $C = B$.

PROOF By hypothesis, we know that $AB = BA = I_m$ and $AC = CA = I_m$. It follows that $C = CI_m = C(AB) = (CA)B = I_m B = B$. ■

Thus, if a matrix has an inverse, it has only one inverse. If the matrix A has an inverse, it is denoted by A^{-1}, and A is said to be **invertible**. For example,

$$\begin{bmatrix} 2 & 3 \\ 7 & 11 \end{bmatrix}\begin{bmatrix} 11 & -3 \\ -7 & 2 \end{bmatrix} = \begin{bmatrix} 1 & 0 \\ 0 & 1 \end{bmatrix} \quad \text{and} \quad \begin{bmatrix} 11 & -3 \\ -7 & 2 \end{bmatrix}\begin{bmatrix} 2 & 3 \\ 7 & 11 \end{bmatrix} = \begin{bmatrix} 1 & 0 \\ 0 & 1 \end{bmatrix}$$

Thus, $\begin{bmatrix} 2 & 3 \\ 7 & 11 \end{bmatrix}^{-1} = \begin{bmatrix} 11 & -3 \\ -7 & 2 \end{bmatrix}$.

In our chapter on determinants, we shall find a necessary and sufficient condition that a matrix have an inverse. At this point, we simply point out several examples. First, we show how to find the inverse by solving systems of linear equations.

Example 1 Find the inverse of $\begin{bmatrix} 3 & 5 \\ 1 & 2 \end{bmatrix}$, if it exists. Call this matrix A. We wish to find B so that $AB = BA = I_2$. Let $B = \begin{bmatrix} a & b \\ c & d \end{bmatrix}$. Since $\begin{bmatrix} 3 & 5 \\ 1 & 2 \end{bmatrix}\begin{bmatrix} a & b \\ c & d \end{bmatrix} = \begin{bmatrix} 1 & 0 \\ 0 & 1 \end{bmatrix}$, we obtain the following system of linear equations for a, b, c, and d.

$$3a + 5c = 1 \qquad 3b + 5d = 0$$

$$a + 2c = 0 \qquad b + 2d = 1$$

Solving these, as in Chapter 1, we find $a = 2$, $b = -5$, $c = -1$, and $d = 3$. Substituting these numerical values into B, we can check that $AB = BA = I_2$ without difficulty. Thus, $\begin{bmatrix} 3 & 5 \\ 1 & 2 \end{bmatrix}^{-1} = \begin{bmatrix} 2 & -5 \\ -1 & 3 \end{bmatrix}$.

Next we give an example of a matrix having no inverse.

Example 2 The matrix $A = \begin{bmatrix} 0 & 1 \\ 0 & 1 \end{bmatrix}$ has no inverse. To see this, suppose $B = \begin{bmatrix} a & b \\ c & d \end{bmatrix}$ is the inverse of A. Then, $BA = I_2$. But $BA = \begin{bmatrix} a & b \\ c & d \end{bmatrix}\begin{bmatrix} 0 & 1 \\ 0 & 1 \end{bmatrix}$ $= \begin{bmatrix} 0 & a+b \\ 0 & c+d \end{bmatrix}$. Since $I_2 = \begin{bmatrix} 1 & 0 \\ 0 & 1 \end{bmatrix}$ has 1 in entry (1, 1), it is impossible to choose B so that $BA = I_2$. Thus, A has no inverse.

Matrices that do not possess inverses are said to be **noninvertible** or **singular**.

In later sections, we shall give a more systematic procedure for deciding if a matrix has an inverse and finding the inverse. We shall also see that when the inverse exists, it is a valuable tool for solving linear systems.

Powers of a square matrix A are defined in the usual way: $A^1 = A$, $A^2 = AA$, $A^3 = AA^2, \ldots, A^n = AA^{n-1}$. Because of the associative law, A^n is just A multiplied together n times, regardless of the order in which the multiplications are performed. For example, if

$$A = \begin{bmatrix} 1 & 1 \\ 0 & 1 \end{bmatrix}, \quad \text{then } A^2 = \begin{bmatrix} 1 & 2 \\ 0 & 1 \end{bmatrix}, \quad A^3 = \begin{bmatrix} 1 & 3 \\ 0 & 1 \end{bmatrix}, \quad A^4 = \begin{bmatrix} 1 & 4 \\ 0 & 1 \end{bmatrix}$$

In this case, induction will show that $A^n = \begin{bmatrix} 1 & n \\ 0 & 1 \end{bmatrix}$.

The next example illustrates how a problem in the study of animal populations might lead us to consider powers of matrices. For simplicity, we impose numerical restrictions that may be somewhat unrealistic, but the basic principle is valid in a more general context.

Example 3 We are given a species of animal that lives for at most four years. The species has two sexes. The male population is a fixed proportion of the female population. So we ignore it and study simply the female population. We are given the following data.

(1) The number of females of ages 0–1, 1–2, 2–3, and 3–4 are given, respectively, by the numbers n_1, n_2, n_3, and n_4. Thus, the number of females between ages 1 and 2 is n_2.
(2) The number of females produced in a year by an average female in each of the age groups is, respectively, $\frac{1}{10}$, $\frac{1}{2}$, $\frac{3}{4}$, and $\frac{1}{4}$. Thus, the typical female between the ages of 2 and 3 produces $\frac{3}{4}$ female babies.
(3) The probability that a given female of each age group will survive to the next age group is, respectively, $\frac{4}{5}$, $\frac{3}{4}$, $\frac{1}{3}$, and 0. Thus, a female between ages 0 and 1 has a $\frac{4}{5}$ chance of surviving for one year. Stated in another way, one year later, $(\frac{4}{5})n_1$ members of the first age group are still alive.

We wish to use these data to predict the female population after one year, two years, and so on. We represent this with a vector $\boldsymbol{p} = \begin{bmatrix} n_1 \\ n_2 \\ n_3 \\ n_4 \end{bmatrix}$. We will also consider the matrix

$$T = \begin{bmatrix} \frac{1}{10} & \frac{1}{2} & \frac{3}{4} & \frac{1}{4} \\ \frac{4}{5} & 0 & 0 & 0 \\ 0 & \frac{3}{4} & 0 & 0 \\ 0 & 0 & \frac{1}{3} & 0 \end{bmatrix}$$

It is our claim that the female population in the different age categories one year later is given by the vector

$$T\boldsymbol{p} = \begin{bmatrix} \frac{1}{10} & \frac{1}{2} & \frac{3}{4} & \frac{1}{4} \\ \frac{4}{5} & 0 & 0 & 0 \\ 0 & \frac{3}{4} & 0 & 0 \\ 0 & 0 & \frac{1}{3} & 0 \end{bmatrix} \begin{bmatrix} n_1 \\ n_2 \\ n_3 \\ n_4 \end{bmatrix} = \begin{bmatrix} \frac{1}{10}n_1 + \frac{1}{2}n_2 + \frac{3}{4}n_3 + \frac{1}{4}n_4 \\ \frac{4}{5}n_1 \\ \frac{3}{4}n_2 \\ \frac{1}{3}n_3 \end{bmatrix}$$

To see this let us first calculate the number of females between ages 0 and 1 one year later. This is, of course, simply the number of females born in a single year. Now there are n_1 females of the first age group, and each on the average produces $\frac{1}{10}$ female offspring. Thus, altogether the first age group produces $\frac{1}{10}n_1$ new females. The n_2 females of the second age group produce a total of $\frac{1}{2}n_2$ new females. Those of the last two age groups produce $\frac{3}{4}n_3$ and $\frac{1}{4}n_4$, respectively. Consequently, the number of new females is obtained by adding these four numbers together and is, indeed, the first entry of the above vector.

The other entries are even simpler. Recall that $\frac{4}{5}$ of the females of the first age group are alive one year later. Thus, after one year the number of females in the second age group is $\frac{4}{5}n_1$. The other numbers are obtained similarly.

Thus, the population vector after one year is given by the formula $T\mathbf{p}$. After two years, it is $T^2\mathbf{p}$, after three years, $T^3\mathbf{p}$, and so forth.

It is interesting to note that if the population is given initially by the vector $\mathbf{p}_0 = \begin{bmatrix} 1000 \\ 800 \\ 600 \\ 200 \end{bmatrix}$, then $T\mathbf{p}_0 = \mathbf{p}_0$. That is to say, the population is stable from year to year.

We may also consider polynomials of matrices. If $f(x) = \alpha_0 + \alpha_1 x + \cdots + \alpha_n x^n$ is a polynomial, we define $f(A) = \alpha_0 I_n + \alpha_1 A + \cdots + \alpha_n A^n$. Thus, for example, if $A = \begin{bmatrix} 1 & 1 \\ 0 & 1 \end{bmatrix}$ and $f(x) = 1 + x + 2x^2$, then

$$f(A) = \begin{bmatrix} 1 & 0 \\ 0 & 1 \end{bmatrix} + \begin{bmatrix} 1 & 1 \\ 0 & 1 \end{bmatrix} + 2\begin{bmatrix} 1 & 2 \\ 0 & 1 \end{bmatrix} = \begin{bmatrix} 4 & 5 \\ 0 & 4 \end{bmatrix}$$

If $f(x) = x^2 - 2x + 1$, then

$$f(A) = \begin{bmatrix} 1 & 2 \\ 0 & 1 \end{bmatrix} - 2\begin{bmatrix} 1 & 1 \\ 0 & 1 \end{bmatrix} + \begin{bmatrix} 1 & 0 \\ 0 & 1 \end{bmatrix} = \begin{bmatrix} 0 & 0 \\ 0 & 0 \end{bmatrix}$$

A square matrix is said to be a **diagonal** matrix if all the entries off its diagonal vanish. For example, the following are diagonal matrices:

$$\begin{bmatrix} -1 & 0 \\ 0 & 1 \end{bmatrix} \quad \begin{bmatrix} 1 & 0 & 0 \\ 0 & 5 & 0 \\ 0 & 0 & 7 \end{bmatrix} \quad \begin{bmatrix} -7 & 0 & 0 & 0 \\ 0 & 6 & 0 & 0 \\ 0 & 0 & 1 & 0 \\ 0 & 0 & 0 & 0 \end{bmatrix}$$

EXERCISES

1. If A is an $m \times n$ matrix, show that $AI_n = A$ and $I_m A = A$.

2. Verify each of the following:

(a) $\begin{bmatrix} 6 & 7 \\ 5 & 6 \end{bmatrix}^{-1} = \begin{bmatrix} 6 & -7 \\ -5 & 6 \end{bmatrix}$ (b) $\begin{bmatrix} 7 & 4 \\ 3 & 2 \end{bmatrix}^{-1} = \frac{1}{2}\begin{bmatrix} 2 & -4 \\ -3 & 7 \end{bmatrix}$

(c) $\begin{bmatrix} 3 & 6 & 1 \\ 0 & 1 & 8 \\ 2 & 4 & 1 \end{bmatrix}^{-1} = \begin{bmatrix} -31 & -2 & 47 \\ 16 & 1 & -24 \\ -2 & 0 & 3 \end{bmatrix}$

(d) $\begin{bmatrix} 2 & 3 & 4 \\ 4 & 3 & 1 \\ 1 & 2 & 4 \end{bmatrix}^{-1} = \frac{1}{5}\begin{bmatrix} -10 & 4 & 9 \\ 15 & -4 & -14 \\ -5 & 1 & 6 \end{bmatrix}$

3. Let

$$A = \begin{bmatrix} 0 & -1 \\ 1 & -1 \end{bmatrix}$$

Calculate A^2 and A^3. Show that A is invertible and $A^{-1} = A^2$.

4. Calculate $f(A)$ if

(a) $A = \begin{bmatrix} 1 & 0 \\ -1 & 1 \end{bmatrix}$ and $f(x) = 1 + 3x + x^2$

(b) $A = \begin{bmatrix} 2 & 1 \\ 1 & 3 \end{bmatrix}$ and $f(x) = x + x^2$

(c) $A = \begin{bmatrix} d_1 & 0 \\ 0 & d_2 \end{bmatrix}$ and $f(x) = \alpha_1 + \alpha_1 x + \alpha_2 x^2$

(d) $A = \begin{bmatrix} 1 & 0 & 0 \\ 0 & 0 & 1 \\ 0 & 1 & 0 \end{bmatrix}$ and $f(x) = x^3 - x^2 - x + 1$

(e) $A = \begin{bmatrix} 0 & 1 & 0 \\ 0 & 0 & 1 \\ 0 & 0 & 0 \end{bmatrix}$ and $f(x) = \alpha_0 + \alpha_1 x + \cdots + \alpha_n x^n$

5. Show by induction or otherwise that

$$\begin{bmatrix} 1 & x & y \\ 0 & 1 & x \\ 0 & 0 & 1 \end{bmatrix}^n = \begin{bmatrix} 1 & nx & ny + \dfrac{n(n-1)}{2}x^2 \\ 0 & 1 & nx \\ 0 & 0 & 1 \end{bmatrix}$$

6. If $A = \begin{bmatrix} 0 & 1 & 0 \\ 0 & 0 & 1 \\ 5 & 0 & 0 \end{bmatrix}$, show that $A^3 = 5I_3$. What is A^{-1}?

7. Find all 3×3 diagonal matrices A for which $A^2 = I_3$.

8. Define $P_t = \begin{bmatrix} \cos t & -\sin t \\ \sin t & \cos t \end{bmatrix}$, with t a real number. Show that (a) $P_t P_s = P_{t+s}$, (b) $(P_t)^n = P_{nt}$, with n an integer, and (c) $(P_t)^{-1} = P_{-t}$.

9. If A is an $n \times n$ matrix and p and q are positive integers, prove that $A^p A^q = A^{p+q}$ and $(A^p)^q = A^{pq}$.

10. Let A be an invertible $n \times n$ matrix. Suppose that B is an $n \times p$ matrix such that $AB = 0$. Show that $B = 0$.

11. By finding nonzero vectors $\boldsymbol{x}$ such that $A\boldsymbol{x} = \boldsymbol{0}$, show that the matrices in the following list are singular.

(a) $\begin{bmatrix} 1 & -2 \\ 7 & -14 \end{bmatrix}$ (b) $\begin{bmatrix} 1 & 7 & 1 \\ 0 & 2 & 1 \\ 1 & 3 & -1 \end{bmatrix}$

(Note that exercise 10 guarantees the matrices are noninvertible if $\boldsymbol{x}$ is found.)

12. Prove that

$$\begin{bmatrix} a & b \\ c & d \end{bmatrix}\begin{bmatrix} d & -b \\ -c & a \end{bmatrix} = (ad-bc)\begin{bmatrix} 1 & 0 \\ 0 & 1 \end{bmatrix} = \begin{bmatrix} d & -b \\ -c & a \end{bmatrix}\begin{bmatrix} a & b \\ c & d \end{bmatrix}$$

Show that $A = \begin{bmatrix} a & b \\ c & d \end{bmatrix}$ is invertible if and only if $ad - bc \neq 0$. Find a formula for A^{-1} if $ad - bc \neq 0$.

13. An $n \times n$ matrix is called a semimagic square if the sums of the entries on all the rows and columns are the same. For example, $\begin{bmatrix} 2 & 3 & 1 \\ 1 & 2 & 3 \\ 3 & 1 & 2 \end{bmatrix}$ is a semimagic square. Suppose that A and B are semimagic of the same size. Show that $aA + bB$ and AB are semimagic, with a and b any numbers.

14. Let A and B be two arbitrary $n \times n$ matrices and C an invertible $n \times n$ matrix. Prove that $C^{-1}(AB)C = (C^{-1}AC)(C^{-1}BC)$ and $C^{-1}(A^k)C = (C^{-1}AC)^k$, if k is a positive integer.

15. A certain product is made by two competing companies, A and B, that completely control the market. Each year A retains $\frac{1}{3}$ of its customers and $\frac{2}{3}$ of its customers switch to B. Each year B keeps $\frac{1}{2}$ of its customers and $\frac{1}{2}$ switch to A. We express this fact with the matrix $T = \begin{bmatrix} \frac{1}{3} & \frac{1}{2} \\ \frac{2}{3} & \frac{1}{2} \end{bmatrix}$. We let $v_0 = \begin{bmatrix} a \\ b \end{bmatrix}$ represent the initial distribution of the market. (For example, $a = \frac{2}{5}$ and $b = \frac{3}{5}$ of the market if A controls $\frac{2}{5}$ and B, $\frac{3}{5}$ of the market.)
    (a) Show that $v_1 = Tv_0$ is the distribution of the market one year later. Show that $v_k = T^k v_0$ is the distribution of the market after k years.
    (b) Show that if $a = \frac{3}{7}$ and $b = \frac{4}{7}$, the market is stable, i.e., does not change from year to year.
    (c) Show that if v_0 is a stable distribution, then $a = \frac{3}{7}$ and $b = \frac{4}{7}$. (If v_0 is stable, $Tv_0 = v_0$. Also, $a + b = 1$, by definition of v_0.)

16. If, in the last problem, $T = \begin{bmatrix} \frac{2}{3} & \frac{1}{4} \\ \frac{1}{3} & \frac{3}{4} \end{bmatrix}$, find the stable state.

17. The last two examples are examples of Markov processes. A basic principle of the study of these processes says that at no matter what point the process begins, it eventually approaches a stable state. We outline a proof of this fact for exercise 15.
    (a) Let $C = \begin{bmatrix} 1 & 3 \\ -1 & 4 \end{bmatrix}$. Show that $C^{-1} = \frac{1}{7}\begin{bmatrix} 4 & -3 \\ 1 & 1 \end{bmatrix}$.
    (b) Show that $C^{-1}TC = \begin{bmatrix} -\frac{1}{6} & 0 \\ 0 & 1 \end{bmatrix}$, and $C^{-1}T^kC = \begin{bmatrix} (-\frac{1}{6})^k & 0 \\ 0 & 1 \end{bmatrix}$.
    (c) Show that $T^k = \frac{1}{7}\begin{bmatrix} 3 & 3 \\ 4 & 4 \end{bmatrix} + \frac{1}{7}(-\frac{1}{6})^k\begin{bmatrix} 4 & -3 \\ -4 & 3 \end{bmatrix}$.
    (d) Show that $T^k v_0$ differs from the vector $\begin{bmatrix} \frac{3}{7} \\ \frac{4}{7} \end{bmatrix}$ by a vector both of whose components are in absolute value at most $\frac{1}{6^k}$. (Recall $a + b = 1$, and $a \geqslant 0$ and $b \geqslant 0$.)
    (e) Show that the market distribution approaches $(\frac{3}{7}, \frac{4}{7})$ as $k \to \infty$. (In fact, after five years a differs from $\frac{3}{7}$ by less than 0.00013.)

18. Three urns, called A, B, and C, are placed on the table. A certain number of balls are placed in each urn. We let $v_0 = \begin{bmatrix} n_1 \\ n_2 \\ n_3 \end{bmatrix}$ be the vector giving the number of balls in each urn. Thus, there are n_1 balls in A, n_2 in B, and n_3 in C. An operation of shifting balls from urn to urn is performed. In each step of the operation, three things are done:
    (1) $\frac{3}{5}$ of the balls are left in A, $\frac{1}{5}$ are removed to B, and $\frac{1}{5}$ are removed to C.
    (2) At the same time, $\frac{1}{4}$ of the balls in B are moved to A, $\frac{1}{2}$ of the balls of B are left in B, and $\frac{1}{4}$ of the balls of B are put in C.
    (3) Simultaneously, $\frac{1}{5}$ of the balls of C are moved to A, $\frac{1}{5}$ of the balls of C are moved to B, and $\frac{3}{5}$ of the balls of C are left in C.
    (a) Write a matrix T so that Tv_0 gives the number of balls in each urn after this operation is performed once.
    (b) Interpret the matrix T^2.
    (c) If the initial distribution of balls is 50, 40, 50, show that the operation does not affect the distribution of balls.

19. The population of a colony of bacteria is observed to obey the following rule: in any given hour the population is the sum of the populations of the three previous hours. Let p_k be the population of the colony on the kth hour, and let $v_k = \begin{bmatrix} p_{k+2} \\ p_{k+1} \\ p_k \end{bmatrix}$ be the vector describing the population on three successive hours. Find a matrix T so that $v_{k+1} = T(v_k)$.

20. Let A and B be $n \times n$ matrices. Show that $(A - B)(A + B) = A^2 - B^2$ if and only if A and B commute.

21. Let D be a 3×3 diagonal matrix and f a polynomial. If $D = \begin{bmatrix} d_1 & 0 & 0 \\ 0 & d_2 & 0 \\ 0 & 0 & d_3 \end{bmatrix}$, show that $f(D) = \begin{bmatrix} f(d_1) & 0 & 0 \\ 0 & f(d_2) & 0 \\ 0 & 0 & f(d_3) \end{bmatrix}$.

22. Show that a diagonal matrix is invertible if and only if all its diagonal entries are nonzero. What is its inverse?

23. Show that sums and products of diagonal matrices are again diagonal matrices. Show that any pair of diagonal matrices commutes.

24. If A is invertible and B is invertible, show that AB is invertible and that $(AB)^{-1} = B^{-1}A^{-1}$.

25. Show that if A is invertible, so is A^m, and $(A^m)^{-1} = (A^{-1})^m$.

26. A matrix N is said to be **nilpotent** if for some positive integer k, $N^k = 0$. Show that a nilpotent matrix is not invertible.

27. If N is a nilpotent $n \times n$ matrix, such that $N^k = 0$, show that $I_n - N$ is invertible and $(I_n - N)^{-1} = I_n + N + N^2 + \cdots + N^{k-1}$.

28. Using exercise 27, calculate the inverse of

$$\text{(a)} \begin{bmatrix} 1 & \alpha \\ 0 & 1 \end{bmatrix} \qquad \text{(b)} \begin{bmatrix} 1 & 1 & 0 \\ 0 & 1 & 0 \\ 0 & 0 & 1 \end{bmatrix}$$

$$\text{(c)} \begin{bmatrix} 1 & 1 & 0 \\ 0 & 1 & 1 \\ 0 & 0 & 1 \end{bmatrix} \qquad \text{(d)} \begin{bmatrix} 1 & 0 & 1 & 1 \\ 0 & 1 & 0 & 1 \\ 0 & 0 & 1 & 0 \\ 0 & 0 & 0 & 1 \end{bmatrix}$$

29. Let E_{ij} denote the $n \times n$ matrix that has zeros in all entries except entry (i, j), and 1 in entry (i, j).
    (a) Show that $E_{ij}E_{kl} = \delta_{jk}E_{il}$.
    (b) Show that $E_{ij}^2 = 0$ if $i \neq j$ and that $E_{ij}^2 = E_{ij}$ if $i = j$.
    (c) If $A = [a_{ij}]_{(nn)}$, show that $E_{ij}A$ has all entries zero except in the ith row which is the same as the jth row of A.
    (d) Show that AE_{ij} has all entries zero except in the jth column which is identical to the ith column of A. [Hint: If you have trouble, write out things explicitly in the 3×3 case.]

30. Show that an $n \times n$ matrix that commutes with all $n \times n$ matrices must be a scalar multiple of the identity matrix. In addition, show that any scalar multiple of the identity matrix commutes with all $n \times n$ matrices. [Hint: Use exercise 29.]

31. If E_{ij} is as defined in exercise 29 and $i \neq j$, show that $I_n + E_{ij}$ is invertible. [Hint: Use exercise 27.]

32. Let A and B be $n \times n$ matrices and let $C_1 = \alpha_1 A + \beta_1 B$ and $C_2 = \alpha_2 A + \beta_2 B$. Suppose $\alpha_1 \beta_2 - \alpha_2 \beta_1 \neq 0$. Show that C_1 and C_2 commute if and only if A and B commute.

33. If A and B are square matrices and A is invertible, show that

$$(A + B)A^{-1}(A - B) = (A - B)A^{-1}(A + B)$$

34. If A and B are commuting square matrices, show that A^m and B^n commute, where m and n are positive integers.

35. If A and B are invertible square matrices, show that the following are equivalent.
    (a) A commutes with B.
    (b) A commutes with B^{-1}.
    (c) A^{-1} commutes with B^{-1}.

6 LINEAR EQUATIONS IN MATRIX NOTATION

Suppose we have a system of linear equations

$$\begin{aligned}
a_{11}x_1 + a_{12}x_2 + a_{13}x_3 + \cdots + a_{1n}x_n &= y_1 \\
a_{21}x_1 + a_{22}x_2 + a_{23}x_3 + \cdots + a_{2n}x_n &= y_2 \\
a_{31}x_1 + a_{32}x_2 + a_{33}x_3 + \cdots + a_{3n}x_n &= y_3 \\
&\vdots \\
a_{m1}x_1 + a_{m2}x_2 + a_{m3}x_3 + \cdots + a_{mn}x_n &= y_m
\end{aligned}$$

We may express this system of equations in a more compact form using matrix notation. We define the matrix $A = [a_{ij}]_{(mn)}$ as the **matrix of coefficients**, the vector $\boldsymbol{x} = [x_j]_{(n1)}$ as the matrix of variables, and the vector $\boldsymbol{y} = [y_i]_{(m1)}$. In matrix notation the system becomes $A\boldsymbol{x} = \boldsymbol{y}$.

For example, the system

$$\begin{aligned} 3x_1 - 2x_2 + 4x_3 - x_4 &= 1 \\ x_1 + 3x_2 - x_3 + 2x_4 &= -1 \\ x_1 - x_2 + x_3 - x_4 &= 0 \end{aligned}$$

becomes

$$\begin{bmatrix} 3 & -2 & 4 & -1 \\ 1 & 3 & -1 & 2 \\ 1 & -1 & 1 & -1 \end{bmatrix} \begin{bmatrix} x_1 \\ x_2 \\ x_3 \\ x_4 \end{bmatrix} = \begin{bmatrix} 1 \\ -1 \\ 0 \end{bmatrix}$$

According to our definition of §1.4, systems of the form $A\boldsymbol{x} = \boldsymbol{0}$ are called homogeneous. We have the following theorem:

Theorem If A is an $m \times n$ matrix where $m < n$, there is an n-vector $\boldsymbol{x}$, $\boldsymbol{x} \neq \boldsymbol{0}$, such that $A\boldsymbol{x} = \boldsymbol{0}$.

PROOF If we write

$$A = \left[a_{ij}\right]_{(mn)}$$

$$\boldsymbol{x} = \left[x_j\right]_{(n1)}$$

the associated homogeneous system of linear equations has more variables than equations, and so by the theorem of §1.4, there is a nontrivial solution. ■

Thus, for example, given the matrix

$$A = \begin{bmatrix} 1 & 0 & 8 & 3 \\ 4 & 2 & -7 & 1 \\ -3 & 7 & 0 & 2 \end{bmatrix}$$

we can find some nonzero 4-vector $\boldsymbol{x}$ such that $A\boldsymbol{x} = \boldsymbol{0}$.

Suppose we are given a system of n equations in n variables, in which the matrix of coefficients is invertible. In this case we may show that the system of equations is uniquely solvable. In matrix notation the system of equations becomes $A\boldsymbol{x} = \boldsymbol{y}$.

If we let $\boldsymbol{x} = A^{-1}\boldsymbol{y}$, we have

$$A\boldsymbol{x} = A(A^{-1}\boldsymbol{y}) = (AA^{-1})\boldsymbol{y} = I_n\boldsymbol{y} = \boldsymbol{y}$$

Thus, the system is solvable. To see that the solution is unique, suppose $\boldsymbol{x}_1$ and $\boldsymbol{x}_2$ are two solutions to the system, i.e., $A\boldsymbol{x}_1 = \boldsymbol{y}$ and $A\boldsymbol{x}_2 = \boldsymbol{y}$.

Then, $A\boldsymbol{x}_1 = A\boldsymbol{x}_2$ and we can multiply both sides of the equality by A^{-1} to obtain

$$A^{-1}(A\boldsymbol{x}_1) = A^{-1}(A\boldsymbol{x}_2)$$

or

$$(A^{-1}A)\boldsymbol{x}_1 = (A^{-1}A)\boldsymbol{x}_2$$

so that $\boldsymbol{x}_1 = \boldsymbol{x}_2$.

For example,

$$\begin{bmatrix} 1 & 3 & 3 \\ 1 & 3 & 4 \\ 1 & 4 & 3 \end{bmatrix}\begin{bmatrix} 7 & -3 & -3 \\ -1 & 0 & 1 \\ -1 & 1 & 0 \end{bmatrix} = \begin{bmatrix} 7 & -3 & -3 \\ -1 & 0 & 1 \\ -1 & 1 & 0 \end{bmatrix}\begin{bmatrix} 1 & 3 & 3 \\ 1 & 3 & 4 \\ 1 & 4 & 3 \end{bmatrix} = \begin{bmatrix} 1 & 0 & 0 \\ 0 & 1 & 0 \\ 0 & 0 & 1 \end{bmatrix}$$

So

$$\begin{bmatrix} 1 & 3 & 3 \\ 1 & 3 & 4 \\ 1 & 4 & 3 \end{bmatrix}^{-1} = \begin{bmatrix} 7 & -3 & -3 \\ -1 & 0 & 1 \\ -1 & 1 & 0 \end{bmatrix}$$

Thus, when confronted with the system of equations,

$$x_1 + 3x_2 + 3x_3 = y_1$$

$$x_1 + 3x_2 + 4x_3 = y_2$$

$$x_1 + 4x_2 + 3x_3 = y_3$$

we have

$$\begin{bmatrix} x_1 \\ x_2 \\ x_3 \end{bmatrix} = \begin{bmatrix} 7 & -3 & -3 \\ -1 & 0 & 1 \\ -1 & 1 & 0 \end{bmatrix}\begin{bmatrix} y_1 \\ y_2 \\ y_3 \end{bmatrix} = \begin{bmatrix} 7y_1 & -3y_2 & -3y_3 \\ -y_1 & & +y_3 \\ -y_1 & +y_2 & \end{bmatrix}$$

The advantage of this general solution is clear. If it were necessary to solve the problem when y_1, y_2, and y_3 have many different values, the above formula would save a certain amount of work.

EXERCISES

1. Convert the following systems of linear equations to matrix form.

(a)
$$\begin{aligned} 3x_1 - 4x_2 + 5x_3 - x_4 + x_5 &= 1 \\ x_1 - x_2 + 10x_3 + x_4 + 2x_5 &= 0 \\ x_1 - x_2 + 3x_3 + x_4 + 3x_5 &= -1 \end{aligned}$$

(b)
$$\begin{aligned} x_1 - x_2 + x_3 &= 1 \\ x_1 - 2x_2 + x_3 &= 0 \\ 3x_1 - x_2 + 4x_3 &= 8 \end{aligned}$$

2. Show that

$$\begin{bmatrix} 1 & 2 & 3 & 1 \\ 1 & 3 & 3 & 2 \\ 2 & 4 & 3 & 3 \\ 1 & 1 & 1 & 1 \end{bmatrix}^{-1} = \begin{bmatrix} 1 & -2 & 1 & 0 \\ 1 & -2 & 2 & -3 \\ 0 & 1 & -1 & 1 \\ -2 & 3 & -2 & 3 \end{bmatrix}$$

and solve the system of equations

$$\begin{aligned} x_1 + 2x_2 + 3x_3 + x_4 &= y_1 \\ x_1 + 3x_2 + 3x_3 + 2x_4 &= y_2 \\ 2x_1 + 4x_2 + 3x_3 + 3x_4 &= y_3 \\ x_1 + x_2 + x_3 + x_4 &= y_4 \end{aligned}$$

3. Show that $\begin{bmatrix} 13 & 19 \\ 2 & 3 \end{bmatrix}^{-1} = \begin{bmatrix} 3 & -19 \\ -2 & 13 \end{bmatrix}$ and solve the system

$$\begin{aligned} 13x + 19y &= a \\ 2x + 3y &= b \end{aligned}$$

4. (a) If A is an $m \times n$ matrix and if there is an $n \times m$ matrix B, such that $AB = I_m$, show that the equation $Ax = y$ is always solvable.
   (b) Using the fact that

$$\begin{bmatrix} -1 & 2 & 0 \\ 8 & -17 & 1 \end{bmatrix} \begin{bmatrix} -17 & -2 \\ -8 & -1 \\ 0 & 0 \end{bmatrix} = \begin{bmatrix} 1 & 0 \\ 0 & 1 \end{bmatrix}$$

   find solutions to the equations

$$\begin{aligned} -x + 2y \quad &= \alpha \\ 8x - 17y + z &= \beta \end{aligned}$$

5. If A is an $m \times n$ matrix and $m < n$, show that there is a nonzero $n \times p$ matrix B such that $AB = 0$.

6. If A is an $m \times n$ matrix, $\boldsymbol{x}$ is an n-vector, and $\boldsymbol{y}_1$, $\boldsymbol{y}_2$ are m-vectors such that $A\boldsymbol{x} = \boldsymbol{y}_1$ and $A\boldsymbol{x} = \boldsymbol{y}_2$ can be solved, show that $A\boldsymbol{x} = \boldsymbol{y}_1 + \boldsymbol{y}_2$ can be solved.

7 THE TRANSPOSE OF A MATRIX

If A is an $m \times n$ matrix, $A = [a_{ij}]_{(mn)}$, we consider a new matrix, called the **transpose** of A, denoted by A^T, defined by $A^T = [b_{ij}]_{(nm)}$, where $b_{ij} = a_{ji}$. For example,

$$\begin{bmatrix} 1 & 0 & 7 \\ -1 & 2 & 8 \end{bmatrix}^T = \begin{bmatrix} 1 & -1 \\ 0 & 2 \\ 7 & 8 \end{bmatrix} \qquad \begin{bmatrix} 6 & -1 & 8 \\ 0 & 2 & 1 \\ 3 & 7 & 0 \end{bmatrix}^T = \begin{bmatrix} 6 & 0 & 3 \\ -1 & 2 & 7 \\ 8 & 1 & 0 \end{bmatrix}$$

If A is a complex matrix, we may define a matrix that is analogous to the transpose, called the **adjoint**, and denoted by A^*, where $A^* = [b_{ij}]_{(nm)}$ with $b_{ij} = \overline{a_{ji}}$ (the notation $\bar{a}$ denotes the complex conjugate of the number a).

Thus,

$$\begin{bmatrix} i & 0 \\ -i & 1 \\ 0 & 1+i \end{bmatrix}^* = \begin{bmatrix} -i & i & 0 \\ 0 & 1 & 1-i \end{bmatrix}$$

If A is real, then $A^* = A^T$.

The transpose operation satisfies certain laws.

$$(1)\ (A+B)^T = A^T + B^T$$

$$(2)\ (\alpha A)^T = \alpha A^T$$

$$(3)\ (AB)^T = B^T A^T$$

$$(4)\ (A^T)^T = A$$

PROOF OF (1) Let $A = [a_{ij}]_{(mn)}$ and $B = [b_{ij}]_{(mn)}$. Then, $A + B = [a_{ij} + b_{ij}]_{(mn)}$, and $(A + B)^T = [c_{ij}]_{(nm)}$, where $c_{ij} = a_{ji} + b_{ji}$. Now,

$$A^T + B^T = [\alpha_{ij}]_{(nm)} + [\beta_{ij}]_{(nm)}$$

$$= [\alpha_{ij} + \beta_{ij}]_{(nm)}$$

where $\alpha_{ij} = a_{ji}$ and $\beta_{ij} = b_{ji}$. Since $c_{ij} = \alpha_{ij} + \beta_{ij}$, we have $(A + B)^T = A^T + B^T$.

PROOF OF (3) Let $A = [a_{ij}]_{(mn)}$ and $B = [b_{ij}]_{(np)}$. So

$$(AB) = \left[\sum_{j=1}^{n} a_{ij} b_{jk}\right]_{(mp)}$$

Then, $(AB)^T = [c_{ik}]_{(pm)}$, where $c_{ik} = \Sigma_{j=1}^n a_{kj} b_{ji}$. Now, $B^T A^T = [\Sigma_{j=1}^n \beta_{ij} \alpha_{jk}]$ where $\beta_{ij} = b_{ji}$ and $\alpha_{jk} = a_{kj}$.

So $B^T A^T = [c'_{ik}]_{(pm)}$, where $c'_{ik} = \Sigma_{j=1}^n b_{ji} a_{kj} = \Sigma_{j=1}^n a_{kj} b_{ji}$. Since $c'_{ik} = c_{ik}$, we have $(AB)^T = B^T A^T$. ■

We leave the proof of (2) and (4) as exercises for the reader.

We say a matrix is **symmetric** if $A = A^T$. We say it is **Hermitian** if $A = A^*$. For example, $\begin{bmatrix} 1 & 7 & 8 \\ 7 & 0 & 4 \\ 8 & 4 & -1 \end{bmatrix}$ is symmetric, while $\begin{bmatrix} -1 & i & 4 \\ -i & 0 & -i \\ 4 & i & 2 \end{bmatrix}$ is Hermitian.

Another example of a symmetric matrix is the mileage matrix presented in the beginning of §2.3.

As an example of the use of the laws of transposition, let us show that AA^T is symmetric. Now,

$$\begin{aligned}(AA^T)^T &= (A^T)^T A^T && \text{[by (3)]}\\ &= AA^T && \text{[by (4)]}\end{aligned}$$

Since $(AA^T)^T = AA^T$, AA^T is symmetric.

EXERCISES

1. Find A^T if

(a) $A = \begin{bmatrix} 1 & -3 \\ -3 & 0 \end{bmatrix}$ (b) $A = \begin{bmatrix} 1 & 3 \\ -3 & 0 \end{bmatrix}$

(c) $A = \begin{bmatrix} 1 & 4 \\ 0 & 1 \\ 2 & -1 \\ 0 & 3 \end{bmatrix}$ (d) $A = \begin{bmatrix} 1 & -7 & 0 & 6 & 3 \\ 0 & 1 & 0 & 2 & -3 \end{bmatrix}$

2. Prove that $(\alpha A)^T = \alpha A^T$ and $(A^T)^T = A$.

3. Prove that

$$(A+B)^* = A^* + B^* \qquad (\alpha A)^* = \bar{\alpha} A^*$$

$$(AB)^* = B^* A^* \qquad (A^*)^* = A$$

[Hint: Observe $A^* = \overline{(A^T)}$; where if $B = [b_{ij}]$, $\bar{B} = [\bar{b}_{ij}]$.]

4. (a) Show that $A^T A$ is symmetric.
   (b) Show that $A + A^T$ is symmetric.

5. Prove that all symmetric matrices are square.

6. Show that A^*A, AA^*, and $A + A^*$ are Hermitian.

7. Prove that all diagonal matrices are symmetric.

8. Show that all the diagonal entries of a Hermitian matrix are real.

9. If A and B are symmetric, show that AB is symmetric if and only if A and B commute.

10. If A is invertible, show that A^T is invertible, and that $(A^T)^{-1} = (A^{-1})^T$.

11. A matrix is said to be **skew-symmetric** if $A^T = -A$. Show that every matrix can be uniquely expressed as the sum of a symmetric and a skew-symmetric matrix.

12. If A commutes with B, show that A^T commutes with B^T.

13. Find a 2×2 matrix such that $AA^T \neq A^T A$.

14. If A is an $n \times n$ matrix, and $f(x)$ is a polynomial, show that $f(A^T) = (f(A))^T$.

15. Show that every square matrix A may be expressed in the form $A = H + iK$, where H and K are Hermitian.
    (a) Show that this expression is unique.
    (b) Show that H and K commute if and only if A and A^* commute.

16. If $A = [a_{ij}]$ is an $n \times n$ matrix, we define $tr(A) = a_{11} + a_{22} + \cdots + a_{nn}$. In other words, the trace is the sum of the diagonal entries of A.
    (a) Show that $tr(A + B) = tr(A) + tr(B)$.
    (b) Show that $tr(aA) = a\,tr(A)$, if a is a scalar.
    (c) Show that $tr(AB) = tr(BA)$.

17. Show that $tr(AA^T)$ is the sum of the squares of the entries of A.

18. If A is a real $n \times n$ matrix and $AA^T = 0$, show that $A = 0$.

19. Show that the incidence matrix of a graph is symmetric. (See exercise 10 of §2.3 for the necessary definitions.)

20. Let $P_1, \ldots, P_n$ be points in the plane. Let $A = [a_{ij}]$ be the $n \times n$ matrix with a_{ij} the distance from P_i to P_j. Show that A is symmetric.

determinants

1 DETERMINANTS OF 2 × 2 MATRICES

With every $n \times n$ matrix, a scalar, called the determinant of the matrix, is associated. Later, we shall see that if a matrix has a nonzero determinant, it also possesses an inverse. Conversely, a matrix whose determinant is zero is singular. The determinant has other uses such as solving linear equations and computing the inverse of a matrix. We begin our study of determinants by considering the 2×2 case.

If $A = \begin{bmatrix} a_{11} & a_{12} \\ a_{21} & a_{22} \end{bmatrix}$ is a 2×2 matrix, then we define $\det A = a_{11}a_{22} - a_{12}a_{21}$. The quantity, $\det A$, is called the **determinant** of the matrix A. The determinant of A is often written as $\begin{vmatrix} a_{11} & a_{12} \\ a_{21} & a_{22} \end{vmatrix}$, but it is important to remember that the determinant is a scalar, not an array of numbers. Some numerical instances of 2×2 determinants are $\begin{vmatrix} 2 & 1 \\ 3 & 5 \end{vmatrix} = 2 \cdot 5 - 1 \cdot 3 = 7$ and $\begin{vmatrix} -2 & 5 \\ 2 & 3 \end{vmatrix} = -6 - 10 = -16$.

If the determinant of a 2×2 matrix, A, is nonzero, the matrix A is invertible. In fact, we have the formula

$$A^{-1} = \frac{1}{a_{11}a_{22} - a_{12}a_{21}} \begin{bmatrix} a_{22} & -a_{12} \\ -a_{21} & a_{11} \end{bmatrix}$$

To see this, check

$$A^{-1}A = \frac{1}{a_{11}a_{22} - a_{12}a_{21}} \begin{bmatrix} a_{22} & -a_{12} \\ -a_{21} & a_{11} \end{bmatrix} \begin{bmatrix} a_{11} & a_{12} \\ a_{21} & a_{22} \end{bmatrix}$$

$$= \frac{1}{a_{11}a_{22} - a_{12}a_{21}} \begin{bmatrix} a_{11}a_{22} - a_{12}a_{21} & 0 \\ 0 & a_{11}a_{22} - a_{12}a_{21} \end{bmatrix} = I_2$$

Similarly, $AA^{-1} = I_2$. Some numerical examples are:

$$\begin{bmatrix} 3 & 2 \\ 7 & 5 \end{bmatrix}^{-1} = \begin{bmatrix} 5 & -2 \\ -7 & 3 \end{bmatrix} \quad \text{and} \quad \begin{bmatrix} 5 & 7 \\ 4 & 6 \end{bmatrix}^{-1} = \tfrac{1}{2}\begin{bmatrix} 6 & -7 \\ -4 & 5 \end{bmatrix}$$

Thus, it is clear that the determinant plays an important role in computing the inverse of a 2×2 matrix.

In this chapter, we often denote a matrix A by $[A_1, A_2, \ldots, A_n]$, where A_i is the ith column of A. Thus, if

$$A = \begin{bmatrix} -1 & 1 & 3 \\ 4 & 7 & 0 \\ 8 & 3 & 6 \end{bmatrix}, \quad \text{then } A_1 = \begin{bmatrix} -1 \\ 4 \\ 8 \end{bmatrix}, \quad A_2 = \begin{bmatrix} 1 \\ 7 \\ 3 \end{bmatrix}, \quad A_3 = \begin{bmatrix} 3 \\ 0 \\ 6 \end{bmatrix}$$

In this case, we write $A = [A_1, A_2, A_3]$.

In computing determinants of $n \times n$ matrices, probably the most effective procedure is to use column operations, if n is greater than 3. In the remainder of this section, we study these column operations for 2×2 matrices. In later sections, we see that essentially the same manipulations can be performed on higher order determinants. For convenience, we label these properties as (D1)–(D7).

(D1)
$$\det[A_1 + A_1', A_2] = \det[A_1, A_2] + \det[A_1', A_2]$$
$$\det[A_1, A_2 + A_2'] = \det[A_1, A_2] + \det[A_1, A_2']$$

For example, $\begin{vmatrix} 1+4 & 7 \\ -2+3 & 6 \end{vmatrix} = \begin{vmatrix} 1 & 7 \\ -2 & 6 \end{vmatrix} + \begin{vmatrix} 4 & 7 \\ 3 & 6 \end{vmatrix}$.

To prove (D1), let

$$A_1 = \begin{bmatrix} a_{11} \\ a_{21} \end{bmatrix}, \quad A_1' = \begin{bmatrix} a_{11}' \\ a_{21}' \end{bmatrix}, \quad A_2 = \begin{bmatrix} a_{12} \\ a_{22} \end{bmatrix}$$

Then

$$\det[A_1 + A_1', A_2] = \begin{vmatrix} a_{11} + a_{11}' & a_{12} \\ a_{21} + a_{21}' & a_{22} \end{vmatrix}$$

$$= (a_{11} + a_{11}')a_{22} - (a_{21} + a_{21}')a_{12} = (a_{11}a_{22} - a_{21}a_{12}) + (a_{11}'a_{22} - a_{21}'a_{12})$$

$$= \det[A_1, A_2] + \det[A_1', A_2]$$

The proof for the other column is completely analogous. Another property somewhat similar to the first is

(D2)
$$\det[cA_1, A_2] = c \det[A_1, A_2]$$
$$\det[A_1, cA_2] = c \det[A_1, A_2]$$

For example, $\begin{vmatrix} 5c & 3 \\ 3c & -1 \end{vmatrix} = c\begin{vmatrix} 5 & 3 \\ 3 & -1 \end{vmatrix}$ and $\begin{vmatrix} 5 & 3c \\ 3 & -c \end{vmatrix} = c\begin{vmatrix} 5 & 3 \\ 3 & -1 \end{vmatrix}$. To prove (D2), we retain the same notation as before. Then

$$\det[cA_1, A_2] = \begin{vmatrix} ca_{11} & a_{12} \\ ca_{21} & a_{22} \end{vmatrix} = ca_{11}a_{22} - ca_{12}a_{21} = c \det[A_1, A_2]$$

Two additional properties are

(D3) $$\det[A, A] = 0$$
(D4) $$\det I_2 = 1$$

To prove (D3),

$$\det[A, A] = \begin{vmatrix} a & a \\ b & b \end{vmatrix} = ab - ab = 0$$

In words, (D3) says that a determinant that has two equal columns is 0. By (D4), the determinant of the identity is 1.

Using (D1), (D2), (D3), and (D4), we may formulate additional properties of the determinant.

(D5) $$\det[A_1, A_2 + cA_1] = \det[A_1, A_2]$$
$$\det[A_1 + cA_2, A_2] = \det[A_1, A_2]$$
(D6) $$\det[A_1, A_2] = -\det[A_2, A_1]$$
(D7) $$\det[A, 0] = \det[0, A] = 0$$

(D5) tells us we add a scalar multiple of one column to another without affecting the value of the determinant. (D6) says that interchanging two columns changes the sign of the determinant. (D7) means that if in a determinant one of the columns is zero, the determinant is zero.

As examples of (D5) and (D6), we have

$$\begin{vmatrix} 1 & 6+c \\ 3 & 8+3c \end{vmatrix} = \begin{vmatrix} 1 & 6 \\ 3 & 8 \end{vmatrix} \quad \text{and} \quad \begin{vmatrix} 3 & 4 \\ 1 & 2 \end{vmatrix} = -\begin{vmatrix} 4 & 3 \\ 2 & 1 \end{vmatrix}$$

To prove (D5), observe that

$$\begin{aligned} \det[A_1, A_2 + cA_1] &= \det[A_1, A_2] + \det[A_1, cA_1] && \text{[by (D1)]} \\ &= \det[A_1, A_2] + c \det[A_1, A_1] && \text{[by (D2)]} \\ &= \det[A_1, A_2] + c \cdot 0 && \text{[by (D3)]} \\ &= \det[A_1, A_2] \end{aligned}$$

To prove (D6), observe that

$$\begin{aligned} \det[A_1, A_2] &= \det[A_1 + A_2, A_2] && \text{[by (D5)]} \\ &= \det[A_1 + A_2, -A_1] && \text{[by (D5)]} \\ &= \det[A_2, -A_1] && \text{[by (D5)]} \\ &= -\det[A_2, A_1] && \text{[by (D2)]} \end{aligned}$$

Using properties (D1)–(D7), we may calculate the determinant without recourse to our original definition, as in

$$\begin{aligned}
\begin{vmatrix} 1 & -7 \\ 2 & 3 \end{vmatrix} &= \begin{vmatrix} 1 & -7+7\cdot 1 \\ 2 & 3+7\cdot 2 \end{vmatrix} && \text{[by (D5)]} \\
&= \begin{vmatrix} 1 & 0 \\ 2 & 17 \end{vmatrix} \\
&= 17\begin{vmatrix} 1 & 0 \\ 2 & 1 \end{vmatrix} && \text{[by (D2)]} \\
&= 17\begin{vmatrix} 1+(-2)\cdot 0 & 0 \\ 2+(-2)\cdot 1 & 1 \end{vmatrix} && \text{[by (D5)]} \\
&= 17\begin{vmatrix} 1 & 0 \\ 0 & 1 \end{vmatrix} = 17 && \text{[by (D4)]}
\end{aligned}$$

In the case of 2×2 matrices, it is probably simpler to calculate the determinant directly from its definition. For larger determinants, however, repeated use of the analogs of (D1)–(D7) is the more efficient procedure.

EXERCISES

1. Calculate.

(a) $\begin{vmatrix} 1 & -3 \\ 7 & 4 \end{vmatrix}$ (b) $\begin{vmatrix} 1 & 3 \\ 5 & 6 \end{vmatrix}$ (c) $\begin{vmatrix} 0 & -1 \\ 2 & 1 \end{vmatrix}$

(d) $\begin{vmatrix} 7 & 5 \\ 4 & 3 \end{vmatrix}$ (e) $\begin{vmatrix} 2 & 1 \\ 3 & 2 \end{vmatrix}$ (f) $\begin{vmatrix} 0 & 1 \\ 1 & 0 \end{vmatrix}$

2. For what values of λ, does

(a) $\begin{vmatrix} \lambda & \lambda \\ 3 & \lambda-2 \end{vmatrix} = 0$ (b) $\begin{vmatrix} 1-\lambda & 1 \\ 1 & 1+\lambda \end{vmatrix} = 0$

(c) $\begin{vmatrix} 1+\lambda & 1 \\ 2+2\lambda & 2 \end{vmatrix} = 0$

3. If $\det\begin{bmatrix} a & b \\ c & d \end{bmatrix} = 0$, show that

$$\begin{bmatrix} d & -b \\ -c & a \end{bmatrix}\begin{bmatrix} a & b \\ c & d \end{bmatrix} = \begin{bmatrix} a & b \\ c & d \end{bmatrix}\begin{bmatrix} d & -b \\ -c & a \end{bmatrix} = 0$$

Show that $\begin{bmatrix} a & b \\ c & d \end{bmatrix}$ has no inverse.

4. Find

(a) $\begin{bmatrix} 1 & 1 \\ 5 & 6 \end{bmatrix}^{-1}$ (b) $\begin{bmatrix} 7 & 4 \\ 3 & 2 \end{bmatrix}^{-1}$ (c) $\begin{bmatrix} 1 & a \\ 0 & 1 \end{bmatrix}^{-1}$

(d) $\begin{bmatrix} 1-a & a \\ -a & 1+a \end{bmatrix}^{-1}$ (e) $\begin{bmatrix} \cos\theta & \sin\theta \\ -\sin\theta & \cos\theta \end{bmatrix}^{-1}$

5. Given a function $D(A)$, defined on 2×2 matrices, and satisfying properties (D1)–(D4), show that $D(A) = \det A$. In other words, (D1)–(D4) completely characterize the determinant.

6. Show that properties (D1)–(D7) hold for rows, as well as for columns.

2 DEFINITION AND PRINCIPAL PROPERTIES OF DETERMINANTS

In this section we give an inductive definition of the determinant of an $n \times n$ matrix. In other words, knowing how the determinant is defined for 2×2 matrices, we give a definition for 3×3 matrices, and then using the definition for 3×3 matrices, we give one for 4×4 matrices. In general, using our definition for $(n-1) \times (n-1)$ matrices we give one for $n \times n$ matrices.

If A is an $n \times n$ matrix we use $\hat{A}_{ij}$ to denote the matrix obtained by deleting the ith row and jth column of the matrix A. Thus, if

$$A = \begin{bmatrix} 0 & 0 & -7 \\ 3 & 1 & 2 \\ 4 & -2 & 3 \end{bmatrix}$$

$$\hat{A}_{11} = \begin{bmatrix} 1 & 2 \\ -2 & 3 \end{bmatrix}, \quad \hat{A}_{31} = \begin{bmatrix} 0 & -7 \\ 1 & 2 \end{bmatrix}, \quad \hat{A}_{22} = \begin{bmatrix} 0 & -7 \\ 4 & 3 \end{bmatrix}$$

Assuming that we have defined the determinant for $(n-1) \times (n-1)$ matrices, we define it for an $n \times n$ matrix, $A = [a_{ij}]$, by

$$\begin{aligned} \det A &= (-1)^{1+1} a_{11} \det \hat{A}_{11} + (-1)^{1+2} a_{12} \det \hat{A}_{12} \\ &\quad + \cdots + (-1)^{1+n} a_{1n} \det \hat{A}_{1n} \\ &= \sum_{j=1}^{n} (-1)^{1+j} a_{1j} \det \hat{A}_{1j} \end{aligned}$$

For 3×3 matrices, this becomes

$$\det \begin{bmatrix} a_{11} & a_{12} & a_{13} \\ a_{21} & a_{22} & a_{23} \\ a_{31} & a_{32} & a_{33} \end{bmatrix} = a_{11} \det \begin{bmatrix} a_{22} & a_{23} \\ a_{32} & a_{33} \end{bmatrix} - a_{12} \det \begin{bmatrix} a_{21} & a_{23} \\ a_{31} & a_{33} \end{bmatrix} + a_{13} \det \begin{bmatrix} a_{21} & a_{22} \\ a_{31} & a_{32} \end{bmatrix}$$

$$= a_{11}a_{22}a_{33} - a_{11}a_{23}a_{32} + a_{12}a_{23}a_{31} - a_{12}a_{21}a_{33} + a_{13}a_{21}a_{32} - a_{13}a_{22}a_{31}$$

We often write

$$\det \begin{bmatrix} a_{11} & a_{12} & \cdots & a_{1n} \\ a_{21} & a_{22} & \cdots & a_{2n} \\ & & \vdots & \\ a_{n1} & a_{n2} & \cdots & a_{nn} \end{bmatrix} = \begin{vmatrix} a_{11} & a_{12} & \cdots & a_{1n} \\ a_{21} & a_{22} & \cdots & a_{2n} \\ & & \vdots & \\ a_{n1} & a_{n2} & \cdots & a_{nn} \end{vmatrix}_n$$

For example,

$$\begin{vmatrix} 3 & 2 & 1 \\ 1 & 4 & 2 \\ 0 & 1 & 2 \end{vmatrix} = 3\begin{vmatrix} 4 & 2 \\ 1 & 2 \end{vmatrix} - 2\begin{vmatrix} 1 & 2 \\ 0 & 2 \end{vmatrix} + 1\begin{vmatrix} 1 & 4 \\ 0 & 1 \end{vmatrix} = 3(8-2) - 2(2-0) + (1-0)$$

$$= 18 - 4 + 1 = 15$$

For 4×4 matrices our definition becomes, in more explicit form,

$$\begin{vmatrix} a_{11} & a_{12} & a_{13} & a_{14} \\ a_{21} & a_{22} & a_{23} & a_{24} \\ a_{31} & a_{32} & a_{33} & a_{34} \\ a_{41} & a_{42} & a_{43} & a_{44} \end{vmatrix} = a_{11}\begin{vmatrix} a_{22} & a_{23} & a_{24} \\ a_{32} & a_{33} & a_{34} \\ a_{42} & a_{43} & a_{44} \end{vmatrix} - a_{12}\begin{vmatrix} a_{21} & a_{23} & a_{24} \\ a_{31} & a_{33} & a_{34} \\ a_{41} & a_{43} & a_{44} \end{vmatrix}$$

$$+ a_{13}\begin{vmatrix} a_{21} & a_{22} & a_{24} \\ a_{31} & a_{32} & a_{34} \\ a_{41} & a_{42} & a_{44} \end{vmatrix} - a_{14}\begin{vmatrix} a_{21} & a_{22} & a_{23} \\ a_{31} & a_{32} & a_{33} \\ a_{41} & a_{42} & a_{43} \end{vmatrix}$$

This formula enables us to evaluate any 4×4 determinant by calculating four 3×3 determinants, and we have seen above how to compute 3×3 determinants. Similarly, we can write down a formula for 5×5 determinants. To calculate a 5×5 determinant would then be a question of computing five 4×4 determinants. The process could be continued indefinitely, but from the computational point of view it is very inefficient. Instead, we will develop analogs of properties (D1)–(D7) of §3.1. Using these properties and a procedure akin to Gaussian elimination will enable us to evaluate any determinant with much less labor than writing down formulas for determinants.

In stating and proving the analogs of (D1)–(D7) of §3.1 for higher order determinants, we confine ourselves, except in examples, to the 3×3 case. However, it should be noted at the outset that the statements we make and the methods we give are completely general. We restrict our attention to the 3×3 case primarily to avoid cumbersome notation. We shall state and prove (D1) for the 3×3 case. After this has been done, the appropriate modification for the case of 4×4 determinants is almost obvious. With a little more thought, a general inductive argument can be given that will handle all cases. With this in mind, the statement of (D1) is:

(D1)
$$\det[A_1 + A_1', A_2, A_3] = \det[A_1, A_2, A_3] + \det[A_1', A_2, A_3]$$
$$\det[A_1, A_2 + A_2', A_3] = \det[A_1, A_2, A_3] + \det[A_1, A_2', A_3]$$
$$\det[A_1, A_2, A_3 + A_3'] = \det[A_1, A_2, A_3] + \det[A_1, A_2, A_3']$$

For example,

$$\begin{vmatrix} 3 & 7+8 & 2 \\ 6 & 2+7 & -1 \\ -5 & 5+3 & 4 \end{vmatrix} = \begin{vmatrix} 3 & 7 & 2 \\ 6 & 2 & -1 \\ -5 & 5 & 4 \end{vmatrix} + \begin{vmatrix} 3 & 8 & 2 \\ 6 & 7 & -1 \\ -5 & 3 & 4 \end{vmatrix}$$

Next, we give a proof of (D1) in the 3×3 case. Note that the proof really depends only on the definition of the determinant and (D1) in the 2×2 case.

In the 3×3 case, working on the second column, (D1) becomes

$$\begin{vmatrix} a_{11} & a_{12} + a_{12}' & a_{13} \\ a_{21} & a_{22} + a_{22}' & a_{23} \\ a_{31} & a_{32} + a_{32}' & a_{33} \end{vmatrix} = \begin{vmatrix} a_{11} & a_{12} & a_{13} \\ a_{21} & a_{22} & a_{23} \\ a_{31} & a_{32} & a_{33} \end{vmatrix} + \begin{vmatrix} a_{11} & a_{12}' & a_{13} \\ a_{21} & a_{22}' & a_{23} \\ a_{31} & a_{32}' & a_{33} \end{vmatrix}$$

To prove this observe that, by our definition of the determinant,

$$\Delta = \begin{vmatrix} a_{11} & a_{12}+a'_{12} & a_{13} \\ a_{21} & a_{22}+a'_{22} & a_{23} \\ a_{31} & a_{32}+a'_{32} & a_{33} \end{vmatrix} = a_{11}\begin{vmatrix} a_{22}+a'_{22} & a_{23} \\ a_{32}+a'_{32} & a_{33} \end{vmatrix} - (a_{12}+a'_{12})\begin{vmatrix} a_{21} & a_{23} \\ a_{31} & a_{33} \end{vmatrix}$$

$$+ a_{13}\begin{vmatrix} a_{21} & a_{22}+a'_{22} \\ a_{31} & a_{32}+a'_{32} \end{vmatrix}$$

Using property (D1) in the 2×2 case,

$$\begin{vmatrix} a_{22}+a'_{22} & a_{23} \\ a_{32}+a'_{32} & a_{33} \end{vmatrix} = \begin{vmatrix} a_{22} & a_{23} \\ a_{32} & a_{33} \end{vmatrix} + \begin{vmatrix} a'_{22} & a_{23} \\ a'_{32} & a_{33} \end{vmatrix}$$

$$\begin{vmatrix} a_{21} & a_{22}+a'_{22} \\ a_{31} & a_{32}+a'_{32} \end{vmatrix} = \begin{vmatrix} a_{21} & a_{22} \\ a_{31} & a_{32} \end{vmatrix} + \begin{vmatrix} a_{21} & a'_{22} \\ a_{31} & a'_{32} \end{vmatrix}$$

Thus,

$$\Delta = a_{11}\begin{vmatrix} a_{22} & a_{23} \\ a_{32} & a_{33} \end{vmatrix} - a_{12}\begin{vmatrix} a_{21} & a_{23} \\ a_{31} & a_{33} \end{vmatrix} + a_{13}\begin{vmatrix} a_{21} & a_{22} \\ a_{31} & a_{32} \end{vmatrix}$$

$$+ a_{11}\begin{vmatrix} a'_{22} & a_{23} \\ a'_{32} & a_{33} \end{vmatrix} - a'_{12}\begin{vmatrix} a_{21} & a_{23} \\ a_{31} & a_{33} \end{vmatrix} + a_{13}\begin{vmatrix} a_{21} & a'_{22} \\ a_{31} & a'_{32} \end{vmatrix}$$

So

$$\Delta = \begin{vmatrix} a_{11} & a_{12} & a_{13} \\ a_{21} & a_{22} & a_{23} \\ a_{31} & a_{32} & a_{33} \end{vmatrix} + \begin{vmatrix} a_{11} & a'_{12} & a_{13} \\ a_{21} & a'_{22} & a_{23} \\ a_{31} & a'_{32} & a_{33} \end{vmatrix}$$

Proving (D1) for the other columns is accomplished in just the same way. Our next property is

(D2)
$$\det[cA_1, A_2, A_3] = c\det[A_1, A_2, A_3]$$
$$\det[A_1, cA_2, A_3] = c\det[A_1, A_2, A_3]$$
$$\det[A_1, A_2, cA_3] = c\det[A_1, A_2, A_3]$$

For example, $\begin{vmatrix} -1 & 4c & 8 \\ 2 & 3c & 7 \\ -7 & 8c & 1 \end{vmatrix} = c\begin{vmatrix} -1 & 4 & 8 \\ 2 & 3 & 7 \\ -7 & 8 & 1 \end{vmatrix}$.

The proof, working on the second column, is done as follows.

$$\Delta = \begin{vmatrix} a_{11} & ca_{12} & a_{13} \\ a_{21} & ca_{22} & a_{23} \\ a_{31} & ca_{32} & a_{33} \end{vmatrix} = a_{11}\begin{vmatrix} ca_{22} & a_{23} \\ ca_{32} & a_{33} \end{vmatrix} - ca_{12}\begin{vmatrix} a_{21} & a_{23} \\ a_{31} & a_{33} \end{vmatrix} + a_{13}\begin{vmatrix} a_{21} & ca_{22} \\ a_{31} & ca_{32} \end{vmatrix}$$

Using (D2) in the 2×2 case,

$$\Delta = ca_{11}\begin{vmatrix} a_{22} & a_{23} \\ a_{32} & a_{33} \end{vmatrix} - ca_{12}\begin{vmatrix} a_{21} & a_{23} \\ a_{31} & a_{33} \end{vmatrix} + ca_{13}\begin{vmatrix} a_{21} & a_{22} \\ a_{31} & a_{32} \end{vmatrix} = c\begin{vmatrix} a_{11} & a_{12} & a_{13} \\ a_{21} & a_{22} & a_{23} \\ a_{31} & a_{32} & a_{33} \end{vmatrix}$$

The next property is

(D3) $\det I_n = 1$, where, as usual, I_n is the identity of order n.

In the 3×3 case, we have

$$\begin{vmatrix} 1 & 0 & 0 \\ 0 & 1 & 0 \\ 0 & 0 & 1 \end{vmatrix} = 1\begin{vmatrix} 1 & 0 \\ 0 & 1 \end{vmatrix} - 0\begin{vmatrix} 0 & 0 \\ 0 & 1 \end{vmatrix} + 0\begin{vmatrix} 0 & 1 \\ 0 & 0 \end{vmatrix} = 1$$

Larger values of n are just as easy.
Our next property is easily stated in words:

(D4) The determinant of a matrix having two equal columns is 0.

In the 3×3 case, this would be written as:

$$\det[A, A, B] = \det[A, B, A] = \det[B, A, A] = 0$$

For example, $\begin{vmatrix} 1 & 1 & -1 \\ 2 & 2 & 17 \\ 4 & 4 & 2 \end{vmatrix} = 0.$

To prove (D4), consider

$$\Delta = \begin{vmatrix} a_1 & b_1 & b_1 \\ a_2 & b_2 & b_2 \\ a_3 & b_3 & b_3 \end{vmatrix} = a_1\begin{vmatrix} b_2 & b_2 \\ b_3 & b_3 \end{vmatrix} - b_1\begin{vmatrix} a_2 & b_2 \\ a_3 & b_3 \end{vmatrix} + b_1\begin{vmatrix} a_2 & b_2 \\ a_3 & b_3 \end{vmatrix}$$

By (D4) in the 2×2 case, the first term vanishes, while the last two cancel each other.

In a similar manner,

$$\Delta = \begin{vmatrix} b_1 & a_1 & b_1 \\ b_2 & a_2 & b_2 \\ b_3 & a_3 & b_3 \end{vmatrix} = b_1\begin{vmatrix} a_2 & b_2 \\ a_3 & b_3 \end{vmatrix} - a_1\begin{vmatrix} b_2 & b_2 \\ b_3 & b_3 \end{vmatrix} + b_1\begin{vmatrix} b_2 & a_2 \\ b_3 & a_3 \end{vmatrix}$$

By (D6) in the 2×2 case,

$$\begin{vmatrix} a_2 & b_2 \\ a_3 & b_3 \end{vmatrix} = -\begin{vmatrix} b_2 & a_2 \\ b_3 & a_3 \end{vmatrix}$$

Thus, the first and third terms cancel each other, while the second is zero by (D4) in the 2×2 case.

It is a fact, which we shall not prove, that any scalar function, defined on all $n \times n$ matrices, which satisfies (D1)–(D4) is equal to the determinant. It follows from this that using (D1)–(D4), we can compute the value of any determinant. Generally speaking, this is far more efficient than resorting to the original definition.

Actually, the property that is most useful in practice for evaluating determinants is probably the following:

(D5) The value of a determinant is unchanged if some scalar multiple of one column is added to another.

For example, $\begin{vmatrix} 0 & 4 & 1 \\ 1 & -3 & 7 \\ 2 & 1 & 3 \end{vmatrix} = \begin{vmatrix} 0 & 4+1c & 1 \\ 1 & -3+7c & 7 \\ 2 & 1+3c & 3 \end{vmatrix}$.

We can prove (D5) using (D1)–(D4). For example, suppose that in a 3×3 determinant c times the third row is added to the first. We must show that the value remains unchanged, or $\det[A_1 + cA_3, A_2, A_3] = \det[A_1, A_2, A_3]$.

By (D1), $\det[A_1 + cA_3, A_2, A_3] = \det[A_1, A_2, A_3] + \det[cA_3, A_2, A_3]$. By (D2), $\det[cA_3, A_2, A_3] = c \det[A_3, A_2, A_3]$. By (D4), $\det[A_3, A_2, A_3] = 0$. Thus, $\det[A_1 + cA_3, A_2, A_3] = \det[A_1, A_2, A_3]$. Proofs for other cases are essentially the same.

The next property is also easily stated:

(D6) Interchanging two columns of a determinant changes the sign of the determinant.

For example, $\begin{vmatrix} 1 & 4 & 5 \\ 2 & 3 & 1 \\ 2 & 5 & 6 \end{vmatrix} = -\begin{vmatrix} 1 & 5 & 4 \\ 2 & 1 & 3 \\ 2 & 6 & 5 \end{vmatrix}$.

The proof of (D6) is virtually the same as the one given in §3.1 and will be omitted. Lastly, we have:

(D7) If all entries of some column are 0, then the determinant is 0.

Indeed, $\det[A_1, 0, A_3] = \det[A_1, 0 \cdot 0, A_3] = 0 \cdot [\det A_1, 0, A_3]$, by (D2), and the last term is 0.

By using these properties repeatedly we can evaluate any determinant. Of these, probably the most useful is (D5), as we will see in the following examples.

Example 1 Evaluate the following determinant.

$$\begin{vmatrix} 4 & 1 & 3 \\ 2 & 0 & 7 \\ -1 & 3 & 2 \end{vmatrix} = \begin{vmatrix} 4+(-4)1 & 1 & 3 \\ 2+(-4)0 & 0 & 7 \\ -1+(-4)3 & 3 & 2 \end{vmatrix} = \begin{vmatrix} 0 & 1 & 3 \\ 2 & 0 & 7 \\ -13 & 3 & 2 \end{vmatrix}$$

by (D5)

$$= \begin{vmatrix} 0 & 1 & 3+(-3)1 \\ 2 & 0 & 7+(-3)0 \\ -13 & 3 & 2+(-3)3 \end{vmatrix} = \begin{vmatrix} 0 & 1 & 0 \\ 2 & 0 & 7 \\ -13 & 3 & -7 \end{vmatrix} = (-1)\begin{vmatrix} 2 & 7 \\ -13 & -7 \end{vmatrix}$$

$$= 14 - 91 = -77$$

The philosophy used in this example is quite simple and relates to Gaussian elimination. We pick a column having a nonzero entry in the top row. (In this example, we chose the second column.) By adding suitable multiples of this column to all of the other columns, we obtain zeroes in all other entries of the top row. Because of (D5), the value of the determinant remains the same. Next we apply the definition of the determinant. In the example, the problem was then reduced to one of computing a 2×2 determinant, which was easy enough. In general, however, we obtain a determinant of smaller size upon which the whole process is then repeated.

In the next example, we employ this procedure to evaluate a 4×4 determinant.

Example 2 Evaluate the following determinant.

$$\Delta = \begin{vmatrix} 1 & 2 & -3 & 1 \\ -3 & 0 & 3 & 0 \\ 4 & 1 & 0 & 0 \\ 1 & 2 & 2 & 1 \end{vmatrix}$$

$$= \begin{vmatrix} 1 & 2+(-2)(1) & -3+(3)(1) & 1+(-1)(1) \\ -3 & 0+(-2)(-3) & 3+(3)(-3) & 0+(-1)(-3) \\ 4 & 1+(-2)(4) & 0+(3)(4) & 0+(-1)(4) \\ 1 & 2+(-2)(1) & 2+(3)(1) & 1+(-1)(1) \end{vmatrix} \quad [\text{by (D5)}]$$

$$= \begin{vmatrix} 1 & 0 & 0 & 0 \\ -3 & 6 & -6 & 3 \\ 4 & -7 & 12 & -4 \\ 1 & 0 & 5 & 0 \end{vmatrix} = \begin{vmatrix} 6 & -6 & 3 \\ -7 & 12 & -4 \\ 0 & 5 & 0 \end{vmatrix}$$

$$= \begin{vmatrix} 6+(-2)(3) & -6+(2)(3) & 3 \\ -7+(-2)(-4) & 12+(2)(-4) & -4 \\ 0+(-2)(0) & 5+(2)(0) & 0 \end{vmatrix} \quad [\text{by (D5)}]$$

$$= \begin{vmatrix} 0 & 0 & 3 \\ 1 & 4 & -4 \\ 0 & 5 & 0 \end{vmatrix} = 3\begin{vmatrix} 1 & 4 \\ 0 & 5 \end{vmatrix} = 15$$

Example 3 Show that

$$\begin{vmatrix} a_{11} & a_{12} & a_{13} & \cdots & a_{1n} \\ 0 & a_{22} & a_{23} & \cdots & a_{2n} \\ 0 & 0 & a_{33} & \cdots & a_{3n} \\ 0 & 0 & 0 & \cdots & a_{4n} \\ & & \vdots & & \\ 0 & 0 & 0 & \cdots & a_{nn} \end{vmatrix} = a_{11}a_{22}a_{33} \cdots a_{nn}$$

In words, if a matrix is such that all entries below the diagonal are zero (such a matrix is said to be **upper triangular**), its determinant is the product of its diagonal entries.

If $a_{11} = 0$, then the first column is 0. By (D7), the determinant is 0. Since also, $a_{11}a_{22} \cdots a_{nn} = 0$, equality holds, and we are finished. So, suppose that $a_{11} \neq 0$. Then by subtracting suitable multiples of the first column from the other columns, we find that the determinant equals

$$\begin{vmatrix} a_{11} & 0 & 0 & \cdots & 0 \\ 0 & a_{22} & a_{23} & \cdots & a_{2n} \\ 0 & 0 & a_{33} & \cdots & a_{3n} \\ 0 & 0 & 0 & \cdots & a_{4n} \\ & & \vdots & & \\ 0 & 0 & 0 & \cdots & a_{nn} \end{vmatrix}$$

By definition of the determinant, this equals

$$a_{11}\begin{vmatrix} a_{22} & a_{23} & \cdots & a_{2n} \\ 0 & a_{33} & \cdots & a_{3n} \\ 0 & 0 & \cdots & a_{4n} \\ & \vdots & & \\ 0 & 0 & & a_{nn} \end{vmatrix}$$

Now, by induction, or by repeating the above argument, the last determinant is equal to $a_{22}a_{33} \cdots a_{nn}$. Thus, the original determinant is $a_{11}a_{22} \cdots a_{nn}$.

There are a number of examples of problems in which two determinants can be proved equal without evaluating them. For instance,

Example 4 Prove

$$D = \begin{vmatrix} na_1 + b_1 & nb_1 + c_1 & nc_1 + a_1 \\ na_2 + b_2 & nb_2 + c_2 & nc_2 + a_2 \\ na_3 + b_3 & nb_3 + c_3 & nc_3 + a_3 \end{vmatrix} = (1 + n^3)\begin{vmatrix} a_1 & b_1 & c_1 \\ a_2 & b_2 & c_2 \\ a_3 & b_3 & c_3 \end{vmatrix}$$

Rewrite as

$$
\begin{aligned}
D &= \det[nA + B,\ nB + C,\ nC + A] \\
&= \det[nA + B,\ nB + C - n(nA + B),\ nC + A] = \det[nA + B,\ C - n^2A,\ nC + A] \\
&= \det[nA + B,\ C - n^2A,\ nC + A - n(C - n^2A)] = \det[nA + B,\ C - n^2A,\ (1 + n^3)A \\
&= (1 + n^3)\det[nA + B,\ C - n^2A,\ A] = (1 + n^3)\det[nA + B - nA,\ C - n^2A + n^2A, \\
&= (1 + n^3)\det[B,\ C,\ A] = -(1 + n^3)\det[B,\ A,\ C] = (1 + n^3)\det[A,\ B,\ C]
\end{aligned}
$$

EXERCISES

1. Calculate.

(a) $\begin{vmatrix} 2 & 5 & 1 \\ 1 & 0 & -1 \\ 4 & 2 & 0 \end{vmatrix}$ (b) $\begin{vmatrix} 1 & 2 & 3 \\ 4 & 5 & 6 \\ 7 & 8 & 9 \end{vmatrix}$ (c) $\begin{vmatrix} 4 & 2 & 3 \\ 2 & 1 & 1 \\ 3 & 2 & 3 \end{vmatrix}$ (d) $\begin{vmatrix} -1 & 3 & 2 \\ 4 & 0 & 1 \\ 6 & 9 & 8 \end{vmatrix}$

(e) $\begin{vmatrix} 2 & 4 & 3 & 1 \\ 0 & 5 & 6 & 3 \\ -1 & 2 & 4 & -2 \\ 7 & 0 & 1 & 3 \end{vmatrix}$ (f) $\begin{vmatrix} 1 & 1 & 2 & 3 \\ 2 & 3 & 5 & 6 \\ 3 & 3 & 7 & 9 \\ 3 & 5 & 6 & 10 \end{vmatrix}$ (g) $\begin{vmatrix} 0 & 1 & 1 & 1 \\ 1 & 0 & 1 & 1 \\ 1 & 1 & 0 & 1 \\ 1 & 1 & 1 & 0 \end{vmatrix}$

2. Show that

$$\begin{vmatrix} a_1 & b_1 & c_1 \\ a_2 & b_2 & c_2 \\ a_3 & b_3 & c_3 \end{vmatrix} = \begin{vmatrix} c_1 & -b_1 & a_1 \\ c_2 & -b_2 & a_2 \\ c_3 & -b_3 & a_3 \end{vmatrix} = \begin{vmatrix} b_1 & c_1 & a_1 \\ b_2 & c_2 & a_2 \\ b_3 & c_3 & a_3 \end{vmatrix}$$

3. Show that

$$\begin{vmatrix} a+b & 1 & b+1 \\ b+c & 1 & c+1 \\ c+d & 1 & d+1 \end{vmatrix} = \begin{vmatrix} a & 1 & b \\ b & 1 & c \\ c & 1 & d \end{vmatrix}$$

4. Let A be a square matrix. Suppose that one column of A is a scalar multiple of another column. Show that $\det A = 0$.

5. Prove that

$$\begin{vmatrix} a_1 & b_1 + xa_1 & c_1 + yb_1 + za_1 \\ a_2 & b_2 + xa_2 & c_2 + yb_2 + za_2 \\ a_3 & b_3 + xa_3 & c_3 + yb_3 + za_3 \end{vmatrix} = \begin{vmatrix} a_1 & b_1 & c_1 \\ a_2 & b_2 & c_2 \\ a_3 & b_3 & c_3 \end{vmatrix}$$

6. Show that

$$\begin{vmatrix} a_{11} & a_{12} \\ a_{21} & a_{22} \end{vmatrix} = \begin{vmatrix} 1 & \alpha & \beta & \gamma \\ 0 & 1 & \delta & \epsilon \\ 0 & 0 & a_{11} & a_{12} \\ 0 & 0 & a_{21} & a_{22} \end{vmatrix}$$

7. Find

$$\begin{vmatrix} a_{11} & 0 & 0 & \cdots & 0 \\ a_{21} & a_{22} & 0 & \cdots & 0 \\ a_{31} & a_{32} & a_{33} & \cdots & 0 \\ & & \vdots & & \\ a_{n1} & a_{n2} & a_{n3} & \cdots & a_{nn} \end{vmatrix}$$

8. Find all values of x for which

(a) $\begin{vmatrix} x-1 & 1 & 1 \\ 0 & x-4 & 1 \\ 0 & 0 & x-2 \end{vmatrix} = 0$ (b) $\begin{vmatrix} 1 & x & x \\ x & 1 & x \\ x & x & 1 \end{vmatrix} = 0$ (c) $\begin{vmatrix} 1 & x & x^2 \\ 1 & 2 & 4 \\ 1 & 3 & 9 \end{vmatrix} = 0$

9. Let A be a square matrix of order n. Show that $\det \alpha A = \alpha^n \det A$.

10. Show that $\det[X_1 + X_2, X_2 + X_3, X_3 + X_1] = 2\det[X_1, X_2, X_3]$.

11. Show that $\det[X_1 + X_2, X_2 + X_3, X_3 + X_4, X_4 + X_1] = 0$.

12. Show that

$$\begin{vmatrix} x & 0 & 0 & \cdots & 0 & -1 \\ 0 & x & 0 & \cdots & 0 & 0 \\ 0 & 0 & x & \cdots & 0 & 0 \\ & & & \vdots & & \\ 0 & 0 & 0 & \cdots & x & 0 \\ -1 & 0 & 0 & \cdots & 0 & x \end{vmatrix}_n = x^{n-2}(x^2 - 1)$$

13. Let A be a matrix such that each column of A has exactly one nonvanishing entry and that nonvanishing entry is 1. Show that $\det A = +1, -1$, or 0. Exhibit an example of each of the three cases.

14. Show that

$$\begin{aligned} \det[X_1, X_2, X_3] &= -\det[X_2, X_1, X_3] = \det[X_2, X_3, X_1] \\ &= -\det[X_3, X_2, X_1] = \det[X_3, X_1, X_2] \\ &= -\det[X_1, X_3, X_2] \end{aligned}$$

15. Without evaluating, show that

$$\begin{vmatrix} \alpha^2 & (\alpha+1)^2 & (\alpha+2)^2 & (\alpha+3)^2 \\ \beta^2 & (\beta+1)^2 & (\beta+2)^2 & (\beta+3)^2 \\ \gamma^2 & (\gamma+1)^2 & (\gamma+2)^2 & (\gamma+3)^2 \\ \delta^2 & (\delta+1)^2 & (\delta+2)^2 & (\delta+3)^2 \end{vmatrix} = 0$$

16. Show that

$$\begin{vmatrix} 1 & -(x_1+x_2) & x_1x_2 & 0 \\ 0 & 1 & -(x_1+x_2) & x_1x_2 \\ 1 & -(y_1+y_2) & y_1y_2 & 0 \\ 0 & 1 & -(y_1+y_2) & y_1y_2 \end{vmatrix} = (x_1-y_1)(x_1-y_2)(x_2-y_1)(x_2-y_2)$$

17. Using exercise 16, show that the polynomials

$$a_0x^2 + a_1x + a_2 = 0, \qquad a_0 \neq 0$$

$$b_0x^2 + b_1x + b_2 = 0, \qquad b_0 \neq 0$$

have a common root if and only if $\begin{vmatrix} a_0 & a_1 & a_2 & 0 \\ 0 & a_0 & a_1 & a_2 \\ b_0 & b_1 & b_2 & 0 \\ 0 & b_0 & b_1 & b_2 \end{vmatrix} = 0$

18. Show that

$$\begin{vmatrix} x & 0 & 0 & 0 & \cdots & 0 & \alpha_0 \\ -1 & x & 0 & 0 & \cdots & 0 & \alpha_1 \\ 0 & -1 & x & 0 & \cdots & 0 & \alpha_2 \\ 0 & 0 & -1 & x & \cdots & 0 & \alpha_3 \\ & & & \vdots & & & \\ 0 & 0 & 0 & 0 & \cdots & -1 & \alpha_{n-1} + x \end{vmatrix} = x^n + \alpha_{n-1}x^{n-1} + \cdots + \alpha_0$$

19. Show that

$$\begin{vmatrix} 1 & 1 & 1 & 1 & 1 & \cdots & 1 & 1 \\ 1 & a_{22} & 1 & 1 & 1 & \cdots & 1 & 1 \\ 1 & a_{32} & a_{33} & 1 & 1 & \cdots & 1 & 1 \\ 1 & a_{42} & a_{43} & a_{44} & 1 & \cdots & 1 & 1 \\ 1 & a_{52} & a_{53} & a_{54} & a_{55} & \cdots & 1 & 1 \\ & & & \vdots & & & & \\ 1 & a_{n2} & a_{n3} & a_{n4} & a_{n5} & \cdots & a_{n,n-1} & a_{nn} \end{vmatrix} = (a_{22} - 1)(a_{33} - 1) \cdots (a_{nn} - 1)$$

20. Find all roots x, to the following equation.

$$\begin{vmatrix} 1 & x_1 & x_2 & x_3 & x_4 \\ 1 & x & x_2 & x_3 & x_4 \\ 1 & x_1 & x & x_3 & x_4 \\ 1 & x_1 & x_2 & x & x_4 \\ 1 & x_1 & x_2 & x_3 & x \end{vmatrix} = 0$$

3 A MULTIPLICATIVE PROPERTY OF DETERMINANTS

In this section, we show that if A and B are $n \times n$ matrices, then $\det AB = \det A \det B$. In other words, the determinant of the product of A and B is the product of their determinants. First, we note one consequence of this fact.

Corollary If A is invertible, then $\det A \neq 0$.

PROOF If A is invertible, then $AA^{-1} = I$. Thus, $\det AA^{-1} = \det I = 1$. Since $\det AA^{-1} = \det A \det A^{-1}$, it follows that $\det A \det A^{-1} = 1$. Therefore, $\det A \neq 0$. ■

Example 1 Since

$$\begin{vmatrix} 1 & 3 & 2 \\ 3 & 2 & -1 \\ -1 & 4 & 5 \end{vmatrix} = \begin{vmatrix} 1 & 3-3(1) & 2-2(1) \\ 3 & 2-3(3) & -1-2(3) \\ -1 & 4-3(-1) & 5-2(-1) \end{vmatrix} = \begin{vmatrix} -7 & -7 \\ 7 & 7 \end{vmatrix} = 0,$$

$\begin{bmatrix} 1 & 3 & 2 \\ 3 & 2 & -1 \\ -1 & 4 & 5 \end{bmatrix}$ is not an invertible matrix.

Theorem $\det A \det B = \det AB$.

PROOF We give a proof only in the 2×2 case. The proof we give generalizes in a straightforward way, but the notation is fairly cumbersome.

We let $A = [A_1, A_2]$ and $B = [b_{ij}]$. Then we observe that $AB = [b_{11}A_1 + b_{21}A_2, b_{12}A_1 + b_{22}A_2]$. To see this, let $A_1 = \begin{bmatrix} a_{11} \\ a_{21} \end{bmatrix}$, $A_2 = \begin{bmatrix} a_{12} \\ a_{22} \end{bmatrix}$. Then compute the matrix product $AB = \begin{bmatrix} a_{11}b_{11} + a_{12}b_{21} & a_{11}b_{12} + a_{12}b_{22} \\ a_{21}b_{11} + a_{22}b_{21} & a_{21}b_{12} + a_{22}b_{22} \end{bmatrix}$. Observe that the first column of AB is simply the column vector $b_{11}A_1 + b_{21}A_2$. Likewise, the second column is $b_{12}A_1 + b_{22}A_2$.

Thus,

$$\begin{aligned} \det AB &= \det[b_{11}A_1 + b_{21}A_2, b_{12}A_1 + b_{22}A_2] \\ &= \sum_{\substack{1 \leq i \leq 2 \\ 1 \leq j \leq 2}} b_{i1}b_{j2} \det[A_i, A_j] \end{aligned}$$

The last step was obtained by repeated use of (D1) and (D2) of §3.1. The summation extends over all pairs of integers (i, j): (1, 1), (1, 2), (2, 1), (2, 2).

If in some term of the above sum, $i = j$, then $A_i = A_j$. Thus, $\det[A_i, A_j] = 0$, since both columns of the determinant are equal. So, in the sum, we need only consider the possibilities $(i, j) = (1, 2)$ or $(2, 1)$. Writing this down explicitly, $\det AB = b_{11}b_{22} \det[A_1, A_2] + b_{21}b_{12} \det[A_2, A_1]$. Since $\det[A_1, A_2] = \det A$ and $\det[A_2, A_1] = -\det A$, it follows that $\det AB = (b_{11}b_{22} - b_{21}b_{12}) \det A = \det B \det A$. ■

Example 2 Here we give a numerical example. $\begin{vmatrix} 1 & 0 & 2 \\ 0 & 1 & 0 \\ 2 & 0 & 1 \end{vmatrix} =$

$$\begin{vmatrix} 1 & 0 & 0 \\ 0 & 1 & 0 \\ 2 & 0 & -3 \end{vmatrix} = -3 \quad \text{and} \quad \begin{vmatrix} 1 & 2 & 0 \\ 2 & 1 & 2 \\ 0 & 2 & 1 \end{vmatrix} = \begin{vmatrix} 1 & 0 & 0 \\ 2 & -3 & 2 \\ 0 & 2 & 1 \end{vmatrix} = -3 - 4 = -7$$

Also, $\begin{bmatrix} 1 & 0 & 2 \\ 0 & 1 & 0 \\ 2 & 0 & 1 \end{bmatrix}\begin{bmatrix} 1 & 2 & 0 \\ 2 & 1 & 2 \\ 0 & 2 & 1 \end{bmatrix} = \begin{bmatrix} 1 & 6 & 2 \\ 2 & 1 & 2 \\ 2 & 6 & 1 \end{bmatrix}$. Finally, we compute the determinant of the product: $\begin{vmatrix} 1 & 6 & 2 \\ 2 & 1 & 2 \\ 2 & 6 & 1 \end{vmatrix} = \begin{vmatrix} 1 & 0 & 0 \\ 2 & -11 & -2 \\ 2 & -6 & -3 \end{vmatrix} = 21.$

EXERCISES

(In the following, all matrices are assumed to be square.)

1. Show that $\det A^n = (\det A)^n$.
2. If A is invertible, show that $\det A^{-1} = (\det A)^{-1}$.
3. If A is a real $n \times n$ matrix and $A^k = I_n$ for some odd integer k, show that $\det A = 1$.
4. If $A^2 = I_n$, show that $\det A = \pm 1$.
5. If C is invertible, show that $\det CXC^{-1} = \det X$, for any matrix X of the same order.
6. Show that the following matrices are not invertible.

   (a) $\begin{bmatrix} 1 & -1 & 1 \\ 3 & 2 & 8 \\ 0 & 4 & 4 \end{bmatrix}$ (b) $\begin{bmatrix} 2 & 0 & 1 \\ 1 & 1 & 1 \\ 3 & 1 & 2 \end{bmatrix}$ (c) $\begin{bmatrix} 6 & 8 \\ 3 & 4 \end{bmatrix}$ (d) $\begin{bmatrix} 1 & 2 & 3 \\ 4 & 5 & 6 \\ 7 & 8 & 9 \end{bmatrix}$

7. Show that there is no real matrix B such that

   (a) $B^2 = \begin{bmatrix} 0 & 1 \\ 1 & 0 \end{bmatrix}$ (b) $\begin{bmatrix} 2 & 3 \\ 1 & 2 \end{bmatrix} B \begin{bmatrix} 3 & 2 \\ 2 & 1 \end{bmatrix} B = \begin{bmatrix} 1 & 0 \\ 0 & 1 \end{bmatrix}$

8. If A is a nilpotent matrix (that is, $A^n = 0$ for some positive integer n), show that $\det A = 0$.
9. By multiplying $\begin{bmatrix} a & b \\ -b & a \end{bmatrix}$ and $\begin{bmatrix} c & d \\ -d & c \end{bmatrix}$ and taking determinants, show that $(a^2 + b^2)(c^2 + d^2) = (ac - bd)^2 + (ad + bc)^2$.
10. Carry out the argument in the text that $\det A \det B = \det AB$ explicitly in the 3×3 case.
11. Let $Q = \begin{bmatrix} b & c & 0 \\ a & 0 & c \\ 0 & a & b \end{bmatrix}$. By calculating $\det Q$, $\det Q^T$, and $\det(QQ^T)$, show that

$$\begin{vmatrix} b^2 + c^2 & ab & ac \\ ab & a^2 + c^2 & bc \\ ac & bc & a^2 + b^2 \end{vmatrix} = 4a^2b^2c^2$$

12. Two matrices A and B are said to anticommute if $AB = -BA$. If A and B are anticommuting 3×3 matrices, show that one or the other is noninvertible.

13. Show that

$$\begin{bmatrix} a_{11} & a_{12} \\ a_{21} & a_{22} \\ a_{31} & a_{32} \end{bmatrix} \begin{bmatrix} b_{11} & b_{12} & b_{13} \\ b_{21} & b_{22} & b_{23} \end{bmatrix} = \begin{bmatrix} a_{11} & a_{12} & 0 \\ a_{21} & a_{22} & 0 \\ a_{31} & a_{32} & 0 \end{bmatrix} \begin{bmatrix} b_{11} & b_{12} & b_{13} \\ b_{21} & b_{22} & b_{23} \\ 0 & 0 & 0 \end{bmatrix}$$

14. Suppose that a 3×3 matrix, C, can be expressed as a product of a 3×2 matrix and a 2×3 matrix. Show that $\det C = 0$.

15. By computing the product

$$\begin{bmatrix} \sin x_1 & \cos x_1 \\ \sin x_2 & \cos x_2 \\ \sin x_3 & \cos x_3 \end{bmatrix} \begin{bmatrix} \cos x_1 & \cos x_2 & \cos x_3 \\ \sin x_1 & \sin x_2 & \sin x_3 \end{bmatrix}, \quad \text{show that}$$

$$\begin{vmatrix} \sin 2x_1 & \sin(x_1 + x_2) & \sin(x_1 + x_3) \\ \sin(x_1 + x_2) & \sin 2x_2 & \sin(x_2 + x_3) \\ \sin(x_1 + x_3) & \sin(x_2 + x_3) & \sin 2x_3 \end{vmatrix} = 0$$

4 ROW OPERATIONS AND COFACTOR EXPANSIONS

In §3.2, we saw that there are certain operations on the columns of a determinant that change the value of the determinant in a specified way. In this section, we see that the obvious analogs of these operations can be performed on rows with a similar effect on the value of the determinant. The seven column operations were labeled (D1)–(D7). Their analogs for rows will be labeled as (D′1)–(D′7). The correct formulation of a given row operation can be obtained by replacing "column" by "row" throughout the definition of the column operation. Alternatively, we can simply write down the correct identity and transpose all of the determinants involved. So that there can be no misunderstanding, we state a few examples.

A particular formulation of (D′1) in the 3×3 case reads:

$$\begin{vmatrix} a_1 & a_2 & a_3 \\ b_1 + c_1 & b_2 + c_2 & b_3 + c_3 \\ d_1 & d_2 & d_3 \end{vmatrix} = \begin{vmatrix} a_1 & a_2 & a_3 \\ b_1 & b_2 & b_3 \\ d_1 & d_2 & d_3 \end{vmatrix} + \begin{vmatrix} a_1 & a_2 & a_3 \\ c_1 & c_2 & c_3 \\ d_1 & d_2 & d_3 \end{vmatrix}$$

Of course, working with the first or third row, there are two other ways of formulating (D′1) in the 3×3 case.

An instance of (D′2) would appear as:

$$\begin{vmatrix} a_1 & a_2 & a_3 \\ db_1 & db_2 & db_3 \\ c_1 & c_2 & c_3 \end{vmatrix} = d \begin{vmatrix} a_1 & a_2 & a_3 \\ b_1 & b_2 & b_3 \\ c_1 & c_2 & c_3 \end{vmatrix}$$

In this case, of course, (D′3) tells nothing new. As with columns, (D′4) can be easily stated: A determinant having two equal rows is zero.

In the 2×2 case, the validity of the row operations can be demonstrated just as in §3.1. Alternatively, observe that $\det A = \det A^T$, and substitute A^T for A throughout the appropriate identity. Later in this section, we shall prove that $\det A = \det A^T$ for all $n \times n$ matrices. For the 2×2 case, however, this is particularly easy. We have:

$$\begin{vmatrix} a_{11} & a_{12} \\ a_{21} & a_{22} \end{vmatrix} = a_{11}a_{22} - a_{12}a_{21} \quad \text{and} \quad \begin{vmatrix} a_{11} & a_{21} \\ a_{12} & a_{22} \end{vmatrix} = a_{11}a_{22} - a_{21}a_{12}$$

The proofs that these row properties are valid are very much in the same spirit as the proofs for column properties. Accordingly, we will give two examples and leave the rest to the interested reader.

First, we prove (D′1) stated above. Let D be the determinant on the left. Then, by the definition of the determinant,

$$D = a_1\begin{vmatrix} b_2 + c_2 & b_3 + c_3 \\ d_2 & d_3 \end{vmatrix} - a_2\begin{vmatrix} b_1 + c_1 & b_3 + c_3 \\ d_1 & d_3 \end{vmatrix} + a_3\begin{vmatrix} b_1 + c_1 & b_2 + c_2 \\ d_1 & d_2 \end{vmatrix}$$

Now using (D′1) in the 2×2 case, it follows that

$$D = a_1\begin{vmatrix} b_2 & b_3 \\ d_2 & d_3 \end{vmatrix} + a_1\begin{vmatrix} c_2 & c_3 \\ d_2 & d_3 \end{vmatrix} - a_2\begin{vmatrix} b_1 & b_3 \\ d_1 & d_3 \end{vmatrix} - a_2\begin{vmatrix} c_1 & c_3 \\ d_1 & d_3 \end{vmatrix} + a_3\begin{vmatrix} b_1 & b_2 \\ d_1 & d_2 \end{vmatrix} + a_3\begin{vmatrix} c_1 & c_2 \\ d_1 & d_2 \end{vmatrix}$$

Now observe that the sum of the first, third, and fifth terms in this expression is the determinant $\begin{vmatrix} a_1 & a_2 & a_3 \\ b_1 & b_2 & b_3 \\ d_1 & d_2 & d_3 \end{vmatrix}$. Likewise, the sum of the second, fourth, and sixth terms is the determinant $\begin{vmatrix} a_1 & a_2 & a_3 \\ c_1 & c_2 & c_3 \\ d_1 & d_2 & d_3 \end{vmatrix}$. Thus, D is the sum of these two determinants, which is what we desired to prove.

Next we prove a particular instance of (D′4) in the 3×3 case.

Suppose that D is a determinant in which the first and third rows are equal. We wish to show that $D = 0$. Let $D = \begin{vmatrix} a_1 & a_2 & a_3 \\ b_1 & b_2 & b_3 \\ a_1 & a_2 & a_3 \end{vmatrix}$. Now if $a_1 = a_2 = a_3 = 0$, then by the definition of the determinant, $D = 0$. So we may suppose that some a_i is nonzero. First suppose that $a_1 \neq 0$. Then by (D2), $D = a_1\begin{vmatrix} 1 & a_2 & a_3 \\ b_1/a_1 & b_2 & b_3 \\ 1 & a_2 & a_3 \end{vmatrix}$. Now subtract a_2 times the first column from the second and a_3 times the first column from the third. It follows that $D = a_1\begin{vmatrix} 1 & 0 & 0 \\ b_1/a_1 & c_1 & c_2 \\ 1 & 0 & 0 \end{vmatrix}$, where c_2 and c_3 are terms we do not bother to compute. Then, applying the definition of the determinant, $D = a_1\begin{vmatrix} c_2 & c_3 \\ 0 & 0 \end{vmatrix} =$ 0. If instead $a_2 \neq 0$, then interchanging the first and second columns changes the sign of the determinant and produces a determinant of the same type with the entry in position (1, 1) nonzero. By what we have shown, the second and, hence, the first determinants are 0. The same argument works if $a_3 \neq 0$.

Probably the reader will not find it difficult to modify this argument for the 4×4 case, and so on. Note also that if the second and third rows are equal, it is much easier to show that the determinant is 0.

Now the properties (D′5), (D′6), and (D′7) can be proved from (D′1)–(D′4) just as the analogous properties were deduced for columns. These properties may be stated as:

(D′5) The value of a determinant is unchanged if some scalar multiple of one row is added to another.
(D′6) Interchanging two rows changes the sign of a determinant.
(D′7) The determinant of a matrix having a row of zeroes is 0.

As with columns, (D′5) is probably the most useful in evaluating determinants.

There are some examples in which it is quicker to compute with rows than columns.

Example 1 Compute the determinant

$$D = \begin{vmatrix} 5 & 4 & 1 \\ 3 & 4 & 1 \\ 2 & -1 & 2 \end{vmatrix}$$

In this example, we notice that the first and second rows are nearly the same. When we subtract the second row from the first,

$$D = \begin{vmatrix} 2 & 0 & 0 \\ 3 & 4 & 1 \\ 2 & -1 & 2 \end{vmatrix} = 2\begin{vmatrix} 4 & 1 \\ -1 & 2 \end{vmatrix} = 18$$

If $A = [a_{ij}]_{(nn)}$ is a matrix, we define its (i, j)th **cofactor**, written A_{ij}, by $A_{ij} = (-1)^{i+j} \det \hat{A}_{ij}$, where $\hat{A}_{ij}$ is the $(n-1) \times (n-1)$ matrix obtained by deleting the ith row and jth column of A.

For example, if $A = \begin{bmatrix} 1 & 2 & 1 \\ 0 & 1 & 2 \\ -3 & 4 & 1 \end{bmatrix}$, then

$$A_{11} = (-1)^{1+1}\begin{vmatrix} 1 & 2 \\ 4 & 1 \end{vmatrix} = -7, \quad A_{13} = (-1)^{1+3}\begin{vmatrix} 0 & 1 \\ -3 & 4 \end{vmatrix} = 3$$

$$A_{32} = (-1)^{3+2}\begin{vmatrix} 1 & 1 \\ 0 & 2 \end{vmatrix} = -2, \quad A_{22} = (-1)^{2+2}\begin{vmatrix} 1 & 1 \\ -3 & 1 \end{vmatrix} = 4$$

Using this notation, we may rewrite our original definition for the determinant as

$$\det A = \sum_{j=1}^{n} a_{1j} A_{1j}$$

In words, this says that the determinant of A is the sum of the entries in the first row times their corresponding cofactors. The cofactor corresponding to a_{1j} is the determinant of the matrix obtained by deleting the first row and jth column times the "fudge factor" $(-1)^{1+j}$. We intend to extend this formula to

$$\det A = \sum_{j=1}^{n} a_{ij} A_{ij}$$

That is, the determinant of A is the sum of the elements in the ith row times their corresponding cofactors.

This formula is called the **cofactor expansion** along the ith row and it allows us to evaluate a determinant in a variety of ways. For example, in the 3×3 case

$$\begin{vmatrix} a_{11} & a_{12} & a_{13} \\ a_{21} & a_{22} & a_{23} \\ a_{31} & a_{32} & a_{33} \end{vmatrix} = -a_{21}\begin{vmatrix} a_{12} & a_{13} \\ a_{32} & a_{33} \end{vmatrix} + a_{22}\begin{vmatrix} a_{11} & a_{13} \\ a_{31} & a_{33} \end{vmatrix} - a_{23}\begin{vmatrix} a_{11} & a_{12} \\ a_{31} & a_{32} \end{vmatrix}$$

or,

$$\begin{vmatrix} a_{11} & a_{12} & a_{13} \\ a_{21} & a_{22} & a_{23} \\ a_{31} & a_{32} & a_{33} \end{vmatrix} = a_{31}\begin{vmatrix} a_{12} & a_{13} \\ a_{22} & a_{23} \end{vmatrix} - a_{32}\begin{vmatrix} a_{11} & a_{13} \\ a_{21} & a_{23} \end{vmatrix} + a_{33}\begin{vmatrix} a_{11} & a_{12} \\ a_{21} & a_{22} \end{vmatrix}$$

A numerical example (going along the second row) is

$$\begin{vmatrix} 0 & 1 & 4 \\ 3 & 2 & 1 \\ 1 & 2 & 3 \end{vmatrix} = -3\begin{vmatrix} 1 & 4 \\ 2 & 3 \end{vmatrix} + 2\begin{vmatrix} 0 & 4 \\ 1 & 3 \end{vmatrix} - 1\begin{vmatrix} 0 & 1 \\ 1 & 2 \end{vmatrix}$$

$$= (-3)(-5) + (2)(-4) - (1)(-1) = 8$$

We now prove that the cofactor expansion along the ith row gives the correct value of the determinant.

Theorem 1 $\det A = \sum_{j=1}^{n} a_{ij} A_{ij}$.

PROOF Let $A = [a_{ij}]$. We obtain a matrix B from A as follows: Interchange the ith row of A with the $(i-1)$st row of A, then with the $(i-2)$nd row of A, and so on. The final effect of this will be to place the ith row of A on the first row of B, and shift the 1st, . . . , $(i-1)$st row of A to the 2nd, . . . , ith row of B. Since B was obtained from A by interchanging rows $i-1$ times, $\det A = (-1)^{i-1}\det B$. Now the matrix B_{1j} is the same as the matrix A_{ij}. Since

$$\det B = \sum_{j=1}^{n} (-1)^{1+j} b_{1j} \det B_{1j} = \sum_{j=1}^{n} (-1)^{1+j} a_{ij} \det A_{ij},$$

it follows that $\det A = \sum_{j=1}^{n} (-1)^{i+j} a_{ij} \det A_{ij} = \sum_{j=1}^{n} a_{ij}\, A_{ij}$. ■

For some determinants, the cofactor expansion along a different row from the first may give the quickest answer.

Example 2 $\begin{vmatrix} 1 & 3 & -2 \\ 0 & 3 & 0 \\ 4 & 5 & 2 \end{vmatrix} = 3\begin{vmatrix} 1 & -2 \\ 4 & 2 \end{vmatrix} = 30.$

There is also a cofactor expansion along columns.

Theorem 2 $\det A = \sum_{i=1}^{n} a_{ij}\, A_{ij}$.

For example, in the 4×4 case this might be written as

$$\Delta = \begin{vmatrix} a_{11} & a_{12} & a_{13} & a_{14} \\ a_{21} & a_{22} & a_{23} & a_{24} \\ a_{31} & a_{32} & a_{33} & a_{34} \\ a_{41} & a_{42} & a_{43} & a_{44} \end{vmatrix} = a_{11}\begin{vmatrix} a_{22} & a_{23} & a_{24} \\ a_{32} & a_{33} & a_{34} \\ a_{42} & a_{43} & a_{44} \end{vmatrix} - a_{21}\begin{vmatrix} a_{12} & a_{13} & a_{14} \\ a_{32} & a_{33} & a_{34} \\ a_{42} & a_{43} & a_{44} \end{vmatrix}$$

$$+ a_{31}\begin{vmatrix} a_{12} & a_{13} & a_{14} \\ a_{22} & a_{23} & a_{24} \\ a_{42} & a_{43} & a_{44} \end{vmatrix} - a_{41}\begin{vmatrix} a_{12} & a_{13} & a_{14} \\ a_{22} & a_{23} & a_{24} \\ a_{32} & a_{33} & a_{34} \end{vmatrix}$$

PROOF We confine our proof to the 4×4 case, leaving the reader to handle the details in the general case. By (D1),

$$\Delta = \begin{vmatrix} a_{11} & a_{12} & a_{13} & a_{14} \\ 0 & a_{22} & a_{23} & a_{24} \\ 0 & a_{32} & a_{33} & a_{34} \\ 0 & a_{42} & a_{43} & a_{44} \end{vmatrix} + \begin{vmatrix} 0 & a_{12} & a_{13} & a_{14} \\ a_{21} & a_{22} & a_{23} & a_{24} \\ 0 & a_{32} & a_{33} & a_{34} \\ 0 & a_{42} & a_{43} & a_{44} \end{vmatrix} + \begin{vmatrix} 0 & a_{12} & a_{13} & a_{14} \\ 0 & a_{22} & a_{23} & a_{24} \\ a_{31} & a_{32} & a_{33} & a_{34} \\ 0 & a_{42} & a_{43} & a_{44} \end{vmatrix}$$

$$+ \begin{vmatrix} 0 & a_{12} & a_{13} & a_{14} \\ 0 & a_{22} & a_{23} & a_{24} \\ 0 & a_{32} & a_{33} & a_{34} \\ a_{41} & a_{42} & a_{43} & a_{44} \end{vmatrix}$$

By using (D2), we may factor a_{11}, a_{21}, a_{31}, and a_{41} from the first columns of each of the above determinants. This leaves the first columns as, respectively,

$$\begin{bmatrix} 1 \\ 0 \\ 0 \\ 0 \end{bmatrix}, \quad \begin{bmatrix} 0 \\ 1 \\ 0 \\ 0 \end{bmatrix}, \quad \begin{bmatrix} 0 \\ 0 \\ 1 \\ 0 \end{bmatrix}, \quad \begin{bmatrix} 0 \\ 0 \\ 0 \\ 1 \end{bmatrix}$$

Then, by adding a suitable multiple of the first column to each of the remaining columns, we obtain

$$\Delta = a_{11}\begin{vmatrix} 1 & 0 & 0 & 0 \\ 0 & a_{22} & a_{23} & a_{24} \\ 0 & a_{32} & a_{33} & a_{34} \\ 0 & a_{42} & a_{43} & a_{44} \end{vmatrix} + a_{21}\begin{vmatrix} 0 & a_{12} & a_{13} & a_{14} \\ 1 & 0 & 0 & 0 \\ 0 & a_{32} & a_{33} & a_{34} \\ 0 & a_{42} & a_{43} & a_{44} \end{vmatrix} + a_{31}\begin{vmatrix} 0 & a_{12} & a_{13} & a_{14} \\ 0 & a_{22} & a_{23} & a_{24} \\ 1 & 0 & 0 & 0 \\ 0 & a_{42} & a_{43} & a_{44} \end{vmatrix}$$

$$+ a_{41}\begin{vmatrix} 0 & a_{12} & a_{13} & a_{14} \\ 0 & a_{22} & a_{23} & a_{24} \\ 0 & a_{32} & a_{33} & a_{34} \\ 1 & 0 & 0 & 0 \end{vmatrix}$$

Interchanging rows, we obtain

$$\Delta = a_{11} \det \hat{A}_{11} + (-1)a_{21} \det \hat{A}_{21} + a_{31} \det \hat{A}_{31} + (-1)a_{41} \det \hat{A}_{41}$$

$$= a_{11}A_{11} + a_{21}A_{21} + a_{31}A_{31} + a_{41}A_{41} \quad \blacksquare$$

Using row and column cofactor expansions, we may prove the following theorem.

Theorem 3 $\det A = \det A^T$.

PROOF The argument will be by induction on n. In the 2×2 case, we have

$$\det\begin{bmatrix} a & b \\ c & d \end{bmatrix} = \det\begin{bmatrix} a & c \\ b & d \end{bmatrix}$$

If $A = [a_{ij}]_{(nn)}$, let $B = A^T = [b_{ij}]_{(nn)}$, where $b_{ij} = a_{ji}$. Note that $\hat{B}_{ij} = \hat{A}_{ji}^T$. By definition,

$$\det B = \sum_{j=1}^{n} b_{1j}(-1)^{1+j} \det \hat{B}_{1j}$$

$$= \sum_{j=1}^{n} a_{j1}(-1)^{1+j} \det \hat{A}_{j1}^T$$

$$= \sum_{j=1}^{n} a_{j1}(-1)^{1+j} \det \hat{A}_{j1}$$

The last step depended on our induction hypothesis for $(n-1) \times (n-1)$ matrices. Thus,

$$\det B = \sum_{j=1}^{n} a_{j1}A_{j1} = \det A$$

by the column cofactor expansion of A. ■

Thus, for example,

$$\begin{vmatrix} 1 & a & b \\ d & 1 & c \\ e & f & 1 \end{vmatrix} = \begin{vmatrix} 1 & d & e \\ a & 1 & f \\ b & c & 1 \end{vmatrix}$$

EXERCISES

1. Calculate.

(a) $\begin{vmatrix} 1 & 6 & 7 \\ 1 & 4 & 3 \\ 1 & -1 & 2 \end{vmatrix}$ (b) $\begin{vmatrix} 0 & 2 & 3 \\ 1 & 4 & 7 \\ 0 & 3 & 1 \end{vmatrix}$ (c) $\begin{vmatrix} 3 & 4 & 2 \\ 0 & 3 & 0 \\ 4 & 1 & 2 \end{vmatrix}$ (d) $\begin{vmatrix} 3 & 1 & 4 \\ 2 & 5 & 7 \\ -1 & -2 & 3 \end{vmatrix}$

(e) $\begin{vmatrix} 4 & 3 & -1 & 2 \\ 0 & 1 & 2 & 3 \\ 1 & 0 & 4 & 1 \\ 2 & 0 & 3 & 0 \end{vmatrix}$ (f) $\begin{vmatrix} 1 & -3 & 5 & 7 \\ 2 & 4 & 0 & 1 \\ 7 & 0 & 1 & 6 \\ 4 & 1 & -1 & 0 \end{vmatrix}$ (g) $\begin{vmatrix} 1 & 3 & 0 & 2 \\ 1 & 5 & 0 & 4 \\ 0 & 2 & -1 & -3 \\ 1 & 6 & 9 & -2 \end{vmatrix}$

2. Prove that

$$\begin{vmatrix} 1 & 6 & 11 & 16 & 21 \\ 2 & 7 & 12 & 17 & 22 \\ 3 & 8 & 13 & 18 & 23 \\ 4 & 9 & 14 & 19 & 24 \\ 5 & 10 & 15 & 20 & 25 \end{vmatrix} = 0$$

[Hint: First subtract row 4 from row 5. Then subtract row 3 from row 4.]

3. If n is odd and A is an $n \times n$ matrix such that $A^T = -A$, show that $\det A = 0$.

4. If Q is a real matrix, show that $\det QQ^T \geqslant 0$.

5. Let $A = [a_{ij}]$ be an $n \times n$ matrix. Suppose that there are numbers c and d such that $a_{ij} = ci + dj$ for all i and j. (For example, if $n = 2$, $A = \begin{bmatrix} c+d & c+2d \\ 2c+d & 2c+2d \end{bmatrix}$.) If $n > 2$, show that $\det A = 0$.

6. If one row of a matrix is a scalar multiple of another row, show that the determinant of the matrix is zero.

7. If A is a symmetric matrix, show that

$$\det[A + B] = \det[A + B^T]$$

8. Show that

$$\begin{vmatrix} a & b & 0 & 0 \\ c & d & 0 & 0 \\ 0 & 0 & e & f \\ 0 & 0 & g & h \end{vmatrix} = \begin{vmatrix} a & b \\ c & d \end{vmatrix}\begin{vmatrix} e & f \\ g & h \end{vmatrix}$$

9. Let A be an $n \times n$ matrix. Suppose that all diagonal entries of A are the same number, a. Suppose all the off diagonal entries are the same number b. Show that $\det A = (a + (n-1)b)(a-b)^{n-1}$. [Hint: Add columns $2, \ldots, n$ to column 1. Factor out $a + (n-1)b$ from column 1. Then subtract b times column 1 from columns $2, \ldots, n$.]

10. Let $A = [a_{ij}]$ be an $n \times n$ matrix. Consider the new matrix, B, obtained from A by replacing each entry a_{ij} by $-a_{ij}$ if $i+j$ is odd, and by $+a_{ij}$ if $i+j$ is even. In other words, $B = [(-1)^{i+j}a_{ij}]$. Show that $\det B = \det A$.

11. Let Q be an $n \times n$ matrix. If $QQ^T = I_n$, show that $\det Q = \pm 1$.

5 INVERSE OF A MATRIX

Associated with a given matrix $A = [a_{ij}]$ there is a useful matrix called the **classical adjoint** of A, denoted by $\mathcal{Q}$, and defined by $\mathcal{Q} = [A_{ij}]^T$. In other words, the classical adjoint of A is the transpose of the matrix of cofactors of A.

For example, if $A = \begin{bmatrix} a & b \\ c & d \end{bmatrix}$, the matrix of cofactors is $\begin{bmatrix} d & -c \\ -b & a \end{bmatrix}$. The classical adjoint is $\begin{bmatrix} d & -b \\ -c & a \end{bmatrix}$. If $A = \begin{bmatrix} -1 & 0 & 3 \\ 7 & 1 & -1 \\ 2 & 3 & 0 \end{bmatrix}$, the cofactor matrix is $\begin{bmatrix} 3 & -2 & 19 \\ 9 & -6 & 3 \\ -3 & 20 & -1 \end{bmatrix}$, and the classical adjoint is

$$\begin{bmatrix} 3 & 9 & -3 \\ -2 & -6 & 20 \\ 19 & 3 & -1 \end{bmatrix}$$

In the 2×2 case, note that

$$A\mathcal{Q} = \begin{bmatrix} a & b \\ c & d \end{bmatrix}\begin{bmatrix} d & -b \\ -c & a \end{bmatrix} = \begin{bmatrix} ad-bc & 0 \\ 0 & ad-bc \end{bmatrix}$$

$\mathcal{Q}A = (\det A)I_2$ is demonstrated similarly.

We intend to generalize this result for arbitrary $n \times n$ matrices, but first we need the following lemma.

Lemma The sum of the products of the elements of one line of A by the cofactors of the corresponding elements of a different parallel line of A is always zero.

PROOF In this case, we have essentially the determinant of a matrix with two equal parallel lines, which is 0 by (D4) or (D′4). We write this down explicitly in the 3×3 case.

Let $A = \begin{bmatrix} a_{11} & a_{12} & a_{13} \\ a_{21} & a_{22} & a_{23} \\ a_{31} & a_{32} & a_{33} \end{bmatrix}$. We shall show that the sum of the products of the elements of the first row by the corresponding cofactors of the elements of the third row is 0. In other words, we will show that $a_{11}A_{31} + a_{12}A_{32} + a_{13}A_{33} = 0$. We substitute for A_{31}, A_{32}, and A_{33}, their definitions in terms of entries of A. Doing this, we find that

$$a_{11}\begin{vmatrix} a_{12} & a_{13} \\ a_{22} & a_{23} \end{vmatrix} - a_{12}\begin{vmatrix} a_{11} & a_{13} \\ a_{21} & a_{23} \end{vmatrix} + a_{13}\begin{vmatrix} a_{11} & a_{12} \\ a_{21} & a_{22} \end{vmatrix} = a_{11}A_{31} + a_{12}A_{32} + a_{13}A_{33}$$

But this last expression is simply $\begin{vmatrix} a_{11} & a_{12} & a_{13} \\ a_{11} & a_{12} & a_{13} \\ a_{21} & a_{22} & a_{23} \end{vmatrix} = 0$. ■

Theorem 1 Let A be an $n \times n$ matrix. Then

$$A\mathcal{C} = \mathcal{C}A = (\det A)I_n$$

PROOF Consider

$$A\mathcal{C} = \begin{bmatrix} a_{11} & a_{12} & \cdots & a_{1n} \\ a_{21} & a_{22} & \cdots & a_{2n} \\ & & \vdots & \\ a_{n1} & a_{n2} & \cdots & a_{nn} \end{bmatrix} \begin{bmatrix} A_{11} & A_{21} & \cdots & A_{n1} \\ A_{12} & A_{22} & \cdots & A_{n2} \\ & & \vdots & \\ A_{1n} & A_{2n} & \cdots & A_{nn} \end{bmatrix}$$

$$= \begin{bmatrix} a_{11}A_{11} + a_{12}A_{12} + \cdots + a_{1n}A_{1n} & a_{11}A_{21} + a_{12}A_{22} + \cdots + a_{1n}A_{2n} & \cdots & a_{11}A_{n1} + a_{12}A_{n2} + \cdots + a_{1n}A_{nn} \\ a_{21}A_{11} + a_{22}A_{12} + \cdots + a_{2n}A_{1n} & a_{21}A_{21} + a_{22}A_{22} + \cdots + a_{2n}A_{2n} & \cdots & a_{21}A_{n1} + a_{22}A_{n2} + \cdots + a_{2n}A_{nn} \\ \vdots & \vdots & & \\ a_{n1}A_{11} + a_{n2}A_{12} + \cdots + a_{nn}A_{1n} & a_{n1}A_{21} + a_{n2}A_{22} + \cdots + a_{nn}A_{2n} & \cdots & a_{n1}A_{n1} + a_{n2}A_{n2} + \cdots + a_{nn}A_{nn} \end{bmatrix}$$

By our lemma all off diagonal terms are zero, while each of the diagonal terms is just $\det A$. So $A\mathcal{C} = (\det A)I_n$. A similar proof shows that $\mathcal{C}A = (\det A)I_n$. ■

Theorem 2 If $\det A \neq 0$, then A^{-1} exists, and $A^{-1} = (\det A)^{-1}\mathcal{C}$.

PROOF Choosing $B = (\det A)^{-1}\mathcal{A}$, we have $AB = BA = I_n$. Thus, A^{-1} exists and equals B. ■

Thus the determinant gives an accurate indication of whether or not a given matrix is invertible. It is invertible if and only if its determinant is nonzero. The determinant also yields a means, not necessarily the most efficient, for calculating this inverse.

Example 1 If $A = \begin{bmatrix} a & b \\ c & d \end{bmatrix}$ and $ad - bc \neq 0$, then

$$A^{-1} = \frac{1}{ad - bc}\begin{bmatrix} d & -b \\ -c & a \end{bmatrix}$$

Example 2 If $A = \begin{bmatrix} -1 & 0 & 3 \\ 7 & 1 & -1 \\ 2 & 3 & 0 \end{bmatrix}$, then $\det A = 54$, and also $\mathcal{A} = \begin{bmatrix} 3 & 9 & -3 \\ -2 & -6 & 20 \\ 19 & 3 & -1 \end{bmatrix}$. Therefore, $A^{-1} = \frac{1}{54}\begin{bmatrix} 3 & 9 & -3 \\ -2 & -6 & 20 \\ 19 & 3 & -1 \end{bmatrix}$.

Example 3 If $A = \begin{bmatrix} 1 & x & 0 \\ -x & 1 & x \\ 0 & -x & 1 \end{bmatrix}$, then $\det A = 1 \cdot (1 + x^2) + (-x)(-x)$ $= 1 + 2x^2$. Thus, if x is real, A is invertible. The matrix of cofactors of A is given by $\begin{bmatrix} 1+x^2 & x & x^2 \\ -x & 1 & x \\ x^2 & -x & 1+x^2 \end{bmatrix}$. Thus,

$$A^{-1} = \frac{1}{1 + 2x^2}\begin{bmatrix} 1+x^2 & -x & x^2 \\ x & 1 & -x \\ x^2 & x & 1+x^2 \end{bmatrix}$$

EXERCISES

(In the following, all matrices are assumed to be square.)

1. Find the inverses of the following matrices.

(a) $\begin{bmatrix} 1 & -1 & 2 \\ 1 & 2 & 0 \\ 4 & 1 & 3 \end{bmatrix}$ (b) $\begin{bmatrix} 1 & 3 & 2 \\ 2 & 1 & 3 \\ 3 & 2 & 1 \end{bmatrix}$ (c) $\begin{bmatrix} 3 & 5 & 7 \\ 1 & 2 & 3 \\ 2 & 3 & 5 \end{bmatrix}$ (d) $\begin{bmatrix} 2 & 1 & 1 \\ 1 & 2 & 1 \\ 1 & 1 & 2 \end{bmatrix}$

(e) $\begin{bmatrix} 1 & 2 & 3 & 1 \\ 1 & 3 & 3 & 2 \\ 2 & 4 & 3 & 3 \\ 1 & 1 & 1 & 1 \end{bmatrix}$ (f) $\begin{bmatrix} 1 & 1 & 0 & 0 \\ 1 & 2 & 3 & 0 \\ 0 & 3 & 4 & 1 \\ 0 & 0 & 1 & 1 \end{bmatrix}$ (g) $\begin{bmatrix} 1 & 2 & 3 & 4 \\ 2 & 3 & 4 & 1 \\ 3 & 4 & 1 & 2 \\ 4 & 1 & 2 & 3 \end{bmatrix}$

2. If $A = G_1 G_2 \ldots G_n$ is invertible, show that $G_1, G_2, \ldots, G_n$ are invertible.

3. Show that

$$\begin{bmatrix} \cos\theta & -\sin\theta \\ -\sin\theta & -\cos\theta \end{bmatrix}^{-1} = \begin{bmatrix} \cos\theta & -\sin\theta \\ -\sin\theta & -\cos\theta \end{bmatrix}$$

4. If

$$Q = \begin{bmatrix} \alpha_1 & -\alpha_2 & -\alpha_3 & -\alpha_4 \\ \alpha_2 & \alpha_1 & -\alpha_4 & \alpha_3 \\ \alpha_3 & \alpha_4 & \alpha_1 & -\alpha_2 \\ \alpha_4 & -\alpha_3 & \alpha_2 & \alpha_1 \end{bmatrix}$$

show that

$$Q^{-1} = \frac{1}{(\alpha_1^2 + \alpha_2^2 + \alpha_3^2 + \alpha_4^2)} Q^T$$

5. If $\det A \neq 0$ and $AB = AC$, show that $B = C$.

6. If $AB = I_n$, show that A^{-1} exists and equals B.

7. If AB is a nonzero multiple of the identity matrix, show that $AB = BA$.

8. If A and B are nonzero matrices and $AB = 0$, show that $\det A = 0$ and $\det B = 0$.

9. If $P^2 = P$ and $P \neq I_n$, show that $\det P = 0$.

10. Suppose $P^2 = P$. If $\lambda \neq 1$, prove $I_n - \lambda P$ is invertible and

$$(I_n - \lambda P)^{-1} = I_n + (\lambda/(1-\lambda))P$$

11. Let A be a symmetric matrix with cofactors A_{ij}. Show that $A_{ij} = A_{ji}$.

12. Show that an upper triangular matrix is invertible if and only if all its diagonal entries are nonzero.

13. For the following matrices, determine for which values of x the matrices are invertible and find the inverse.

(a) $\begin{bmatrix} 1 & 0 & x \\ 0 & 1 & 0 \\ -x & 0 & 1 \end{bmatrix}$ (b) $\begin{bmatrix} x & -1 & 0 \\ -1 & x & -1 \\ 0 & -1 & x \end{bmatrix}$ (c) $\begin{bmatrix} 0 & x & 1 \\ 1 & 0 & x \\ x & 1 & 0 \end{bmatrix}$

14. Find 2×2 matrices, X and Y, such that

$$\begin{bmatrix} 3 & 1 \\ 2 & 1 \end{bmatrix} X + \begin{bmatrix} 3 & 4 \\ 2 & 3 \end{bmatrix} Y = \begin{bmatrix} 1 & 1 \\ 1 & 1 \end{bmatrix}$$

$$\begin{bmatrix} 3 & 1 \\ 2 & 1 \end{bmatrix} X - \begin{bmatrix} 3 & 4 \\ 2 & 3 \end{bmatrix} Y = \begin{bmatrix} 1 & 0 \\ 0 & 1 \end{bmatrix}$$

15. If a, b, and c are real, show that $\begin{bmatrix} 1 & a & b \\ -a & 1 & c \\ -b & -c & 1 \end{bmatrix}$ is invertible and calculate its inverse.

16. Let A and B be invertible symmetric matrices for which $AB = BA$. Show that AB, $A^{-1}B$, AB^{-1}, and $A^{-1}B^{-1}$ are symmetric.

17. If an upper triangular matrix has an inverse, show that the inverse is upper triangular.

18. Suppose that A and B are $n \times n$ matrices and $AB = I_n$. Show that A is invertible and $B = A^{-1}$.

19. Let A be an $n \times n$ matrix, all of whose entries are integers. Show that the following are equivalent.
(a) $\det A = \pm 1$
(b) All entries of A^{-1} are integers.

20. In each of the following, find a 2×2 matrix, X, that satisfies the given equations.
(a) $\begin{bmatrix} 2 & 3 \\ 1 & 2 \end{bmatrix} X \begin{bmatrix} 3 & 4 \\ 2 & 3 \end{bmatrix} = \begin{bmatrix} 1 & 2 \\ 2 & 1 \end{bmatrix}$ (b) $\begin{bmatrix} 0 & 1 \\ 1 & 0 \end{bmatrix} X \begin{bmatrix} 1 & 1 \\ 0 & 1 \end{bmatrix} = \begin{bmatrix} 2 & 1 \\ 3 & 2 \end{bmatrix}$

21. Let U be a matrix with $\det U = 1$. Suppose that U_{ij} is the cofactor in position (i, j). Show that $u_{ij} = U_{ij}$ if and only if $UU^T = I_n$.

22. Let A_1, A_2, and A_3 be invertible matrices. Let X and Y be two other matrices, and suppose all of these are $n \times n$ matrices. If $Y = A_1XA_2XA_3$, show that Y is invertible if and only if X is invertible.

23. How many invertible 2×2 matrices are there having only entries 0 and 1?

6 CRAMER'S RULE

In this section, we employ the notion of an inverse of a matrix to present an explicit formula for the solution of a system of n linear equations in n unknowns, providing that the determinant of the coefficient matrix is nonzero. This fact is of both historical and general interest, but in practice it is not very useful. For real problems, the procedure of Gaussian elimination is more effective for solving systems of linear equations.

Let the system be

$$\begin{aligned} a_{11}x_1 + a_{12}x_2 + \cdots + a_{1n}x_n &= b_1 \\ a_{21}x_1 + a_{22}x_2 + \cdots + a_{2n}x_n &= b_2 \\ &\vdots \\ a_{n1}x_1 + a_{n2}x_2 + \cdots + a_{nn}x_n &= b_n \end{aligned}$$

We let $A = [a_{ij}]_{(nn)}$ be the coefficient matrix of the system, $\boldsymbol{b} = [b_i]_{(n1)}$, and $\boldsymbol{x} = [x_i]_{(n1)}$, the vector of variables. Then, in matrix notation the system becomes

$$A\boldsymbol{x} = \boldsymbol{b}$$

If $\det A \neq 0$, we know A^{-1} exists. Letting $\boldsymbol{x} = A^{-1}\boldsymbol{b}$, we see

$$\begin{aligned} A(A^{-1}\boldsymbol{b}) &= (AA^{-1})\boldsymbol{b} \\ &= I_n\boldsymbol{b} \\ &= \boldsymbol{b} \end{aligned}$$

Thus, the system of equations is solvable, with solution $\boldsymbol{x} = A^{-1}\boldsymbol{b}$.

The solution is also unique. For if $\boldsymbol{x}_1$ and $\boldsymbol{x}_2$ are two solutions to $A\boldsymbol{x} = \boldsymbol{b}$, then

$$A\boldsymbol{x}_1 = \boldsymbol{b} = A\boldsymbol{x}_2, \quad \text{so} \quad A\boldsymbol{x}_1 = A\boldsymbol{x}_2$$

and

$$A^{-1}(A\boldsymbol{x}_1) = A^{-1}(A\boldsymbol{x}_2)$$
$$\boldsymbol{x}_1 = \boldsymbol{x}_2$$

Using the explicit formula for the inverse of a matrix obtained in §3.5, we see

$$\boldsymbol{x} = A^{-1}\boldsymbol{b} = \frac{1}{\det A}\begin{bmatrix} A_{11} & A_{21} & \cdots & A_{n1} \\ A_{12} & A_{22} & \cdots & A_{n2} \\ & & \vdots & \\ A_{1n} & A_{2n} & \cdots & A_{nn} \end{bmatrix}\begin{bmatrix} b_1 \\ b_2 \\ \\ b_n \end{bmatrix}$$

$$\begin{bmatrix} x_1 \\ x_2 \\ \vdots \\ x_n \end{bmatrix} = \frac{1}{\det A}\begin{bmatrix} b_1A_{11} + b_2A_{21} + \cdots + b_nA_{n1} \\ b_1A_{12} + b_2A_{22} + \cdots + b_nA_{n2} \\ \vdots \\ b_1A_{1n} + b_2A_{2n} + \cdots + b_nA_{nn} \end{bmatrix}$$

Writing $A = [A_1, A_2, \ldots, A_n]$, and using cofactor expansions, we may write the solutions

$$x_1 = \frac{\det[\boldsymbol{b}, A_2, A_3, \ldots, A_n]}{\det[A_1, A_2, A_3, \ldots, A_n]} = \frac{b_1A_{11} + b_2A_{21} + \cdots + b_nA_n}{\det A}$$

$$x_2 = \frac{\det[A_1, \boldsymbol{b}, A_3, \ldots, A_n]}{\det[A_1, A_2, A_3, \ldots, A_n]} = \frac{b_1A_{12} + b_2A_{22} + \cdots + b_nA_{n2}}{\det A}$$

$$\vdots$$

$$x_n = \frac{\det[A_1, A_2, \ldots, \boldsymbol{b}]}{\det[A_1, A_2, \ldots, A_n]} = \frac{b_1A_{1n} + b_2A_{2n} + \cdots + b_nA_{nn}}{\det A}$$

This method of solution for linear equations is called **Cramer's Rule**. In the 3×3 case we have, for example

$$x_1 = \frac{\begin{vmatrix} b_1 & a_{12} & a_{13} \\ b_2 & a_{22} & a_{23} \\ b_3 & a_{32} & a_{33} \end{vmatrix}}{\begin{vmatrix} a_{11} & a_{12} & a_{13} \\ a_{21} & a_{22} & a_{23} \\ a_{31} & a_{32} & a_{33} \end{vmatrix}} \qquad x_2 = \frac{\begin{vmatrix} a_{11} & b_1 & a_{13} \\ a_{21} & b_2 & a_{23} \\ a_{31} & b_3 & a_{33} \end{vmatrix}}{\begin{vmatrix} a_{11} & a_{12} & a_{13} \\ a_{21} & a_{22} & a_{23} \\ a_{31} & a_{32} & a_{33} \end{vmatrix}} \qquad x_3 = \frac{\begin{vmatrix} a_{11} & a_{12} & b_1 \\ a_{21} & a_{22} & b_2 \\ a_{31} & a_{32} & b_3 \end{vmatrix}}{\begin{vmatrix} a_{11} & a_{12} & a_{13} \\ a_{21} & a_{22} & a_{23} \\ a_{31} & a_{32} & a_{33} \end{vmatrix}}$$

In particular, if the quantity on the right-hand side of the system of equations vanishes, $\boldsymbol{b} = \mathbf{0}$, or in other words, if we have a system of homogeneous equations, we see that if the determinant of the coefficient matrix is nonzero, the only solution to $A\boldsymbol{x} = \mathbf{0}$ is $\boldsymbol{x} = \mathbf{0}$.

In practice, using Cramer's rule involves inverting the matrix of coefficients using the formula of the last section. We present several examples.

Example 1 Solve the system of equations

$$\begin{aligned} x_1 + x_2 + x_3 &= y_1 \\ x_1 + x_2 - x_3 &= y_2 \\ x_1 - x_2 - x_3 &= y_3 \end{aligned}$$

We invert the matrix of coefficients $A = \begin{bmatrix} 1 & 1 & 1 \\ 1 & 1 & -1 \\ 1 & -1 & -1 \end{bmatrix}$. First,

$$\det A = \begin{vmatrix} 1 & 1 & 1 \\ 1 & 1 & -1 \\ 1 & -1 & -1 \end{vmatrix} = \begin{vmatrix} 1 & 0 & 0 \\ 1 & 0 & -2 \\ 1 & -2 & -2 \end{vmatrix} = -4$$

Next,

$$A_{11} = \begin{vmatrix} 1 & -1 \\ -1 & -1 \end{vmatrix} = -2, \quad A_{12} = (-1)\begin{vmatrix} 1 & -1 \\ 1 & -1 \end{vmatrix} = 0, \quad A_{13} = \begin{vmatrix} 1 & 1 \\ 1 & -1 \end{vmatrix} = -2$$

$$A_{21} = (-1)\begin{vmatrix} 1 & 1 \\ -1 & -1 \end{vmatrix} = 0, \quad A_{22} = \begin{vmatrix} 1 & 1 \\ 1 & -1 \end{vmatrix} = -2, \quad A_{23} = (-1)\begin{vmatrix} 1 & 1 \\ 1 & -1 \end{vmatrix} = 2$$

and so on. Therefore, we obtain that the matrix of cofactors is $\begin{bmatrix} -2 & 0 & -2 \\ 0 & -2 & 2 \\ -2 & 2 & 0 \end{bmatrix}$, and $A^{-1} = \frac{1}{2}\begin{bmatrix} 1 & 0 & 1 \\ 0 & 1 & -1 \\ 1 & -1 & 0 \end{bmatrix}$. It follows that

$$\begin{bmatrix} x_1 \\ x_2 \\ x_3 \end{bmatrix} = \frac{1}{2}\begin{bmatrix} 1 & 0 & 1 \\ 0 & 1 & -1 \\ 1 & -1 & 0 \end{bmatrix}\begin{bmatrix} y_1 \\ y_2 \\ y_3 \end{bmatrix} = \frac{1}{2}\begin{bmatrix} y_1 + y_3 \\ y_2 - y_3 \\ y_1 - y_2 \end{bmatrix}.$$

Example 2 We will derive the law of cosines.

Consider a triangle with sides a, b, and c, and opposite angles α, β, and γ. (See Figure 3-1.) Then, using trigonometric definitions,

$$\begin{aligned} c(\cos\beta) + b(\cos\gamma) &= a \\ c(\cos\alpha) \qquad\qquad + a(\cos\gamma) &= b \\ b(\cos\alpha) + a(\cos\beta) \qquad\qquad &= c \end{aligned}$$

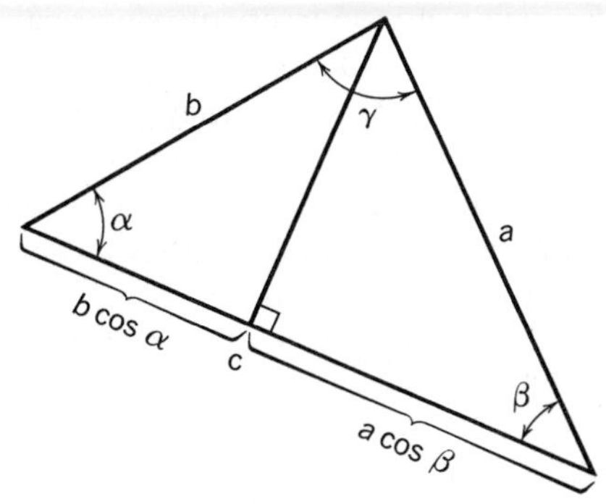

FIGURE 3-1

We wish to solve for $\cos \alpha$, $\cos \beta$, and $\cos \gamma$. That is, we wish to invert the matrix

$$A = \begin{bmatrix} 0 & c & b \\ c & 0 & a \\ b & a & 0 \end{bmatrix}$$

Its determinant is $2abc$. Hence, A is invertible if $a \neq 0$, $b \neq 0$, and $c \neq 0$, which certainly hold for a triangle.

In this case,

$$A^{-1} = \frac{1}{2abc} \begin{bmatrix} -a^2 & ab & ac \\ ab & -b^2 & bc \\ ac & bc & -c^2 \end{bmatrix}$$

So

$$\cos \alpha = \frac{-a^3 + ab^2 + ac^2}{2abc} = \frac{b^2 + c^2 - a^2}{2bc}$$

$\cos \beta$ and $\cos \gamma$ are given by similar expressions.

Example 3 The following is a diagram of a simple electrical circuit.

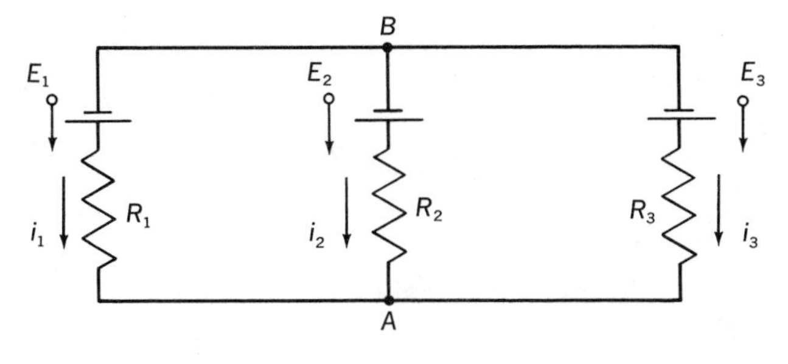

E_1, E_2, and E_3 represent sources of electrical power, for example, batteries or generators. R_1, R_2, and R_3 are resistors. They function to convert electrical energy into heat. In practice, they might be electrical heaters or ovens. The quantities, i_1, i_2, and i_3, represent the electrical current in each branch of the system. The E's are measured in volts and the R's in ohms. The i's are measured in amperes and can be negative (when the current is going in a direction opposite to the pointing arrow).

When the E's and R's are given, it is possible to compute the i's by means of Kirchhoff's laws. These are:

(1) The algebraic sum of all currents meeting at a junction is 0. (In other words, all of the current that flows into a junction must flow out.)
(2) In any closed circuit of the network, the algebraic sum of the E's (sources of potential difference) must equal the algebraic sum of the Ri terms (sources of potential loss).

Note that the currents, i_1, i_2, i_3, all flow into the junction A of the circuit. By the first law, it follows that $i_1 + i_2 + i_3 = 0$. Note that applying this law at the junction B simply gives the same equation.

Going around the first loop in a clockwise fashion, the algebraic sum of the E's is $E_2 - E_1$. The algebraic sum of the Ri terms is $R_2i_2 - R_1i_1$. Thus, by the second law $-R_1i_1 + R_2i_2 = E_2 - E_1$. From the second loop, we obtain $-R_2i_2 + R_3i_3 = E_3 - E_2$. This gives the system

$$\begin{aligned} i_1 + i_2 + i_3 &= 0 \\ -R_1i_1 + R_2i_2 \qquad &= E_2 - E_1 \\ -R_2i_2 + R_3i_3 &= E_3 - E_2 \end{aligned}$$

which we rewrite as

$$\begin{bmatrix} 1 & 1 & 1 \\ -R_1 & R_2 & 0 \\ 0 & -R_2 & R_3 \end{bmatrix} \begin{bmatrix} i_1 \\ i_2 \\ i_3 \end{bmatrix} = \begin{bmatrix} 0 \\ E_2 - E_1 \\ E_3 - E_2 \end{bmatrix}$$

By computing, we find that

$$\begin{bmatrix} 1 & 1 & 1 \\ -R_1 & R_2 & 0 \\ 0 & -R_2 & R_3 \end{bmatrix}^{-1} = \frac{1}{R_1R_2 + R_1R_3 + R_2R_3} \begin{bmatrix} R_2R_3 & -R_2 - R_3 & -R_2 \\ R_1R_3 & R_3 & -R_1 \\ R_1R_2 & R_2 & R_1 + R_2 \end{bmatrix}$$

This enables us to express the amount of current in each branch in terms of the R's and E's.

EXERCISES

1. Solve the following systems.

(a)
$$\begin{aligned} 2x + y - z &= 0 \\ x - y + 3z &= 1 \\ 2x + 2y + z &= 7 \end{aligned}$$

(b)
$$\begin{aligned} 2x + y + 2z &= 0 \\ 3x - 2y + z &= 1 \\ -x + 2y + 2z &= -7 \end{aligned}$$

(c)
$$\begin{aligned} 2x + 8y + z &= 10 \\ -x + 3y + 2z &= -2 \\ 4x + 4y - 5z &= 4 \end{aligned}$$

(d)
$$\begin{aligned} x_1 + 2x_2 - 3x_3 + x_4 &= -9 \\ 2x_1 + 3x_2 - x_3 + x_4 &= 3 \\ -x_1 + x_2 + 2x_3 - x_4 &= 0 \\ 3x_1 + 4x_2 + x_3 + 4x_4 &= 3 \end{aligned}$$

2. By calculating the inverse of the coefficient matrix, solve the following systems.

(a)
$$\begin{aligned} 2x - 3y + z &= a \\ x + 2y + 3z &= b \\ 3x - y + 2z &= c \end{aligned}$$

(b)
$$\begin{aligned} x + 2y + 4z &= a \\ -x + 3y - 2z &= b \\ 2x - y + z &= c \end{aligned}$$

(c)
$$\begin{aligned} 2x_1 + x_2 + 2x_3 - 3x_4 &= y_1 \\ 3x_1 + 2x_2 + 3x_3 - 5x_4 &= y_2 \\ 2x_1 + 2x_2 + x_3 - x_4 &= y_3 \\ 5x_1 + 5x_2 + 2x_3 - 2x_4 &= y_4 \end{aligned}$$

(d)
$$\begin{aligned} 6x_1 + x_2 + 4x_3 - 3x_4 &= y_1 \\ 2x_1 - x_2 &= y_2 \\ x_1 + x_2 + x_3 &= y_3 \\ -3x_1 - x_2 - 2x_3 + x_4 &= y_4 \end{aligned}$$

3. Consider the system

$$ax + by = \alpha + \beta t$$
$$cx + dy = \gamma + \delta t$$

where $\begin{vmatrix} a & b \\ c & d \end{vmatrix} \neq 0$, $\beta^2 + \delta^2 \neq 0$, and t is a parameter. Show that the set of solutions, as t varies, is a line. Show that the line is in the direction of the vector $\begin{bmatrix} a & b \\ c & d \end{bmatrix}^{-1} \begin{bmatrix} \beta \\ \delta \end{bmatrix}$.

4. Three particles leave the origin traveling in a straight line at a constant velocity. Their positions, as a function of time, are given by $\boldsymbol{r}_1(t) = (\boldsymbol{i} + 2\boldsymbol{j})t$, $\boldsymbol{r}_2(t) = (\boldsymbol{i} + \boldsymbol{j})t$, $\boldsymbol{r}_3(t) = (3\boldsymbol{i} + \boldsymbol{j})t$. Let $\boldsymbol{s}(t)$ be the point that is equidistant from the three particles at the time t. Find $\boldsymbol{s}(t)$. [Hint: The equation of a circle with center $(-\frac{1}{2}A, -\frac{1}{2}B)$ is of the form $x^2 + y^2 + Ax + By + C = 0$. Determine A, B, and C at a time t so that the endpoints of the three vectors lie on the circle.]

5. Three bodies having masses m_1, m_2, and m_3 are placed at the respective points $(-1, 1)$, $(-1, -1)$, and $(2, 1)$. Suppose that the center of gravity of the configuration is at $(0, 0)$, and that the sum of the masses is 1. Find m_1, m_2, and m_3.

6. In the electrical network

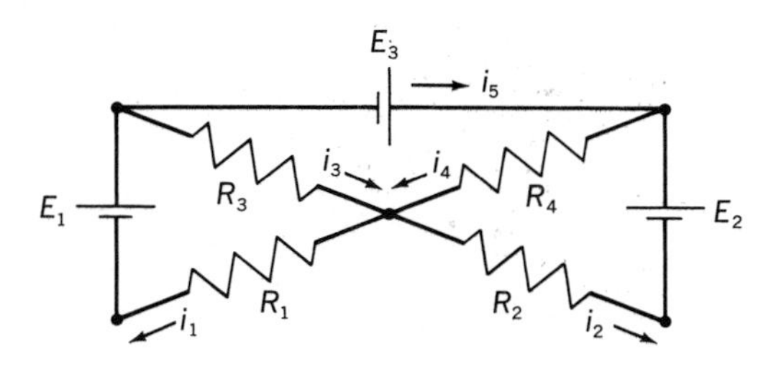

applications of Kirchhoff's laws give the equations:

$$\begin{array}{rrrrrl} i_1 & & -i_3 & & -i_5 & =0 \\ & i_2 & & -i_4 & +i_5 & =0 \\ R_1 i_1 & & +R_3 i_3 & & & =E_1 \\ & R_2 i_2 & & +R_4 i_4 & & =E_2 \\ & & -R_3 i_3 & +R_4 i_4 & & =E_3 \end{array}$$

Show that

$$\begin{bmatrix} R_1+R_3 & 0 & R_1 \\ 0 & R_2+R_4 & -R_2 \\ -R_3 & R_4 & 0 \end{bmatrix}\begin{bmatrix} i_3 \\ i_4 \\ i_5 \end{bmatrix}=\begin{bmatrix} E_1 \\ E_2 \\ E_3 \end{bmatrix}$$

and invert the matrix.

7. Let $A=[a_{ij}]$, $B=[b_{ij}]$, $C=[c_{ij}]$, $D=[d_{ij}]$ be 2×2 matrices. Let $\boldsymbol{r}$ and $\boldsymbol{s}$ be 2-vectors. Show that the equations

$$A\boldsymbol{x}+B\boldsymbol{y}=\boldsymbol{r}$$

$$C\boldsymbol{x}+D\boldsymbol{y}=\boldsymbol{s}$$

can always be solved for 2-vectors $\boldsymbol{x}$ and $\boldsymbol{y}$ if

$$\begin{vmatrix} a_{11} & a_{12} & b_{11} & b_{12} \\ a_{21} & a_{22} & b_{21} & b_{22} \\ c_{11} & c_{12} & d_{11} & d_{12} \\ c_{21} & c_{22} & d_{21} & d_{22} \end{vmatrix}\neq 0$$

7 SYNTHETIC ELIMINATION

The formula described in §3.5 for calculating the inverse of a matrix is a very inefficient procedure from the computational point of view. It is useful mainly in that it provides us with a proof of the fact that A^{-1} exists if $\det A\neq 0$. A more practical method of matrix inversion is **synthetic elimination**, which we now describe.

Given the matrix $A=[a_{ij}]_{(nn)}$, consider the system of equations

$$\begin{aligned} a_{11}x_1+a_{12}x_2+\cdots+a_{1n}x_n&=y_1 \\ a_{21}x_1+a_{22}x_2+\cdots+a_{2n}x_n&=y_2 \\ &\vdots \\ a_{n1}x_1+a_{n2}x_2+\cdots+a_{nn}x_n&=y_n \end{aligned}$$

where we regard $x_1, x_2, \ldots, x_n$ and $y_1, y_2, \ldots, y_n$ as variables. If we solve this system by Gaussian elimination, we only perform the operations of multiplying equations by nonzero constants and adding scalar multiples of one equation to another. Thus, if $\det A\neq 0$, the system is solvable and the solutions are of the form

$$\begin{aligned}
x_1 &= b_{11}y_1 + b_{12}y_2 + \cdots + b_{1n}y_n \\
x_2 &= b_{21}y_1 + b_{22}y_2 + \cdots + b_{2n}y_n \\
&\vdots \\
x_n &= b_{n1}y_1 + b_{n2}y_2 + \cdots + b_{nn}y_n
\end{aligned}$$

We let $B = [b_{ij}]_{(nn)}$. We shall show that B is the inverse of A.

Observe that if $\boldsymbol{x}$ is any vector, we have $A\boldsymbol{x} = \boldsymbol{y}$ and $B\boldsymbol{y} = \boldsymbol{x}$. It follows that $BA\boldsymbol{x} = \boldsymbol{x}$, or $(BA - I_n)\boldsymbol{x} = \boldsymbol{0}$, for all vectors, $\boldsymbol{x}$. Now we will prove a lemma that will guarantee that $BA - I_n = 0$.

Lemma If C is an $m \times n$ matrix, and $C\boldsymbol{x} = 0$ for all n-vectors $\boldsymbol{x}$, then $C = 0$.

PROOF Let $C = [c_{ij}]_{(nn)}$; then

$$\begin{bmatrix} c_{11} & c_{12} & \cdots & c_{1n} \\ c_{21} & c_{22} & \cdots & c_{2n} \\ & & \vdots & \\ c_{m1} & c_{m2} & \cdots & c_{mn} \end{bmatrix} \begin{bmatrix} 1 \\ 0 \\ \vdots \\ 0 \end{bmatrix} = \begin{bmatrix} c_{11} \\ c_{21} \\ \vdots \\ c_{m1} \end{bmatrix}$$

By hypothesis,

$$C \begin{bmatrix} 1 \\ 0 \\ \vdots \\ 0 \end{bmatrix} = 0$$

Thus, $c_{11} = c_{21} = \cdots = c_{m1} = 0$. In a similar manner, by multiplying C by the n-vectors

$$\begin{bmatrix} 0 \\ 1 \\ 0 \\ 0 \\ \vdots \\ 0 \end{bmatrix}, \begin{bmatrix} 0 \\ 0 \\ 1 \\ 0 \\ \vdots \\ 0 \end{bmatrix}, \ldots, \begin{bmatrix} 0 \\ 0 \\ 0 \\ \vdots \\ 0 \\ 1 \end{bmatrix}$$

successively, we find all other columns are 0. ■

Thus, we see that $AB = I_n$. A similar argument shows that $BA = I_n$. Thus, $B = A^{-1}$.

Example 1 In the last section, we inverted the matrix $A = \begin{bmatrix} 1 & 1 & 1 \\ 1 & 1 & -1 \\ 1 & -1 & -1 \end{bmatrix}$ by the somewhat tedious procedure of computing all the cofactors. Now, we will invert it by synthetic elimination. We consider the system of equations

$$\begin{aligned} x_1 + x_2 + x_3 &= y_1 \\ x_1 + x_2 - x_3 &= y_2 \\ x_1 - x_2 - x_3 &= y_3 \end{aligned}$$

Use the first equation and x_1.

$$\downarrow$$

$$\begin{aligned} x_1 + x_2 + x_3 &= y_1 \\ -2x_3 &= y_2 - y_1 \\ -2x_2 - 2x_3 &= y_3 - y_1 \end{aligned}$$

Use the second equation and x_3.

$$\downarrow$$

$$\begin{aligned} x_1 + x_2 &= \tfrac{1}{2}y_1 + \tfrac{1}{2}y_2 \\ x_3 &= \tfrac{1}{2}y_1 - \tfrac{1}{2}y_2 \\ -2x_2 &= y_3 - y_2 \end{aligned}$$

Use the third equation and x_2.

$$\downarrow$$

$$\begin{aligned} x_1 &= \tfrac{1}{2}y_1 + \tfrac{1}{2}y_3 \\ x_3 &= \tfrac{1}{2}y_1 - \tfrac{1}{2}y_2 \\ x_2 &= \tfrac{1}{2}y_2 - \tfrac{1}{2}y_3 \end{aligned}$$

Thus, $A^{-1} = \frac{1}{2}\begin{bmatrix} 1 & 0 & 1 \\ 0 & 1 & -1 \\ 1 & -1 & 0 \end{bmatrix}$.

This process is even more efficient for larger matrices.

Example 2 Invert

$$A = \begin{bmatrix} 1 & 2 & 3 & 1 \\ 1 & 3 & 3 & 2 \\ 2 & 4 & 3 & 3 \\ 1 & 1 & 1 & 1 \end{bmatrix}$$

$$\begin{aligned} x_1 + 2x_2 + 3x_3 + x_4 &= y_1 \\ x_1 + 3x_2 + 3x_3 + 2x_4 &= y_2 \\ 2x_1 + 4x_2 + 3x_3 + 3x_4 &= y_3 \\ x_1 + x_2 + x_3 + x_4 &= y_4 \end{aligned}$$

Use x_1 and equation 1.

$$\downarrow$$

$$\begin{aligned} x_1 + 2x_2 + 3x_3 + x_4 &= y_1 \\ x_2 + \quad + x_4 &= -y_1 + y_2 \\ -3x_3 + x_4 &= -2y_1 + y_3 \\ -x_2 - 2x_3 &= -y_1 + y_4 \end{aligned}$$

Use x_2 and equation 2.

$$\downarrow$$

$$\begin{array}{rrrrrl} x_1 & & +3x_3 & -x_4 & = & 3y_1-2y_2 \\ & x_2 & & +x_4 & = & -y_1+y_2 \\ & & -3x_3 & +x_4 & = & -2y_1+y_3 \\ & & -2x_3 & +x_4 & = & -2y_1+y_2+y_4 \end{array}$$

Use x_4 and equation 3.

$$\downarrow$$

$$\begin{array}{rrrrrl} x_1 & & & & = & y_1-2y_2+y_3 \\ & x_2 & +3x_3 & & = & y_1+y_2-y_3 \\ & & -3x_3 & +x_4 & = & -2y_1+y_3 \\ & & x_3 & & = & y_2-y_3+y_4 \end{array}$$

Use x_3 and equation 4.

$$\downarrow$$

$$\begin{array}{rrrrrl} x_1 & & & & = & y_1-2y_2+y_3 \\ & x_2 & & & = & y_1-2y_2+2y_3-3y_4 \\ & & & x_4 & = & -2y_1+3y_2-2y_3+3y_4 \\ & & x_3 & & = & y_2-y_3+y_4 \end{array}$$

Thus, $A^{-1} = \begin{bmatrix} 1 & -2 & 1 & 0 \\ 1 & -2 & 2 & -3 \\ 0 & 1 & -1 & 1 \\ -2 & 3 & -2 & 3 \end{bmatrix}$.

EXERCISES

1. Calculate the inverses of the following matrices.

(a) $\begin{bmatrix} 1 & 2 & 3 \\ 2 & 1 & 4 \\ 1 & 3 & 2 \end{bmatrix}$ (b) $\begin{bmatrix} -2 & 2 & 3 \\ 4 & 3 & -6 \\ 1 & -1 & 2 \end{bmatrix}$ (c) $\begin{bmatrix} 1 & 3 & 3 \\ 1 & 4 & 3 \\ 1 & 3 & 4 \end{bmatrix}$ (d) $\begin{bmatrix} 1 & 1 & 1 \\ 1 & 3 & 1 \\ -1 & 1 & 1 \end{bmatrix}$

(e) $\begin{bmatrix} -1 & 0 & -1 & 3 \\ -1 & -1 & -1 & 1 \\ 1 & 3 & 3 & 3 \\ -4 & 0 & -5 & 14 \end{bmatrix}$ (f) $\begin{bmatrix} 2 & 3 & 2 & 4 \\ 4 & 6 & 5 & 5 \\ 3 & 5 & 2 & 14 \\ 2 & 2 & -3 & 14 \end{bmatrix}$ (g) $\begin{bmatrix} 1 & 1 & 1 & 1 \\ 1 & 1 & -1 & -1 \\ 1 & -1 & 1 & -1 \\ 1 & -1 & -1 & 1 \end{bmatrix}$

2. For what values of x is $xI - A$ invertible? What is its inverse if A is one of the following matrices?

(a) $\begin{bmatrix} 2 & 0 \\ 0 & 3 \end{bmatrix}$ (b) $\begin{bmatrix} 1 & 1 \\ 1 & 1 \end{bmatrix}$ (c) $\begin{bmatrix} 1 & 0 & 0 \\ 0 & 0 & 1 \\ 0 & 1 & 0 \end{bmatrix}$ (d) $\begin{bmatrix} 0 & 1 & 2 \\ 0 & 1 & 3 \\ 0 & 0 & -1 \end{bmatrix}$

3. Using synthetic elimination, calculate the inverse of

$$\begin{bmatrix} 1 & a_2 & a_3 & \cdots & a_n \\ \lambda_2 & 1+\lambda_2 a_2 & \lambda_2 a_3 & \cdots & \lambda_2 a_n \\ \lambda_3 & \lambda_3 a_2 & 1+\lambda_3 a_3 & \cdots & \lambda_3 a_n \\ & & \vdots & & \\ \lambda_{n-1} & \lambda_{n-1} a_2 & \lambda_{n-1} a_3 & \cdots & \lambda_{n-1} a_n \\ \lambda_n & \lambda_n a_2 & \lambda_n a_3 & \cdots & 1+\lambda_n a_n \end{bmatrix}$$

4. Let A be an $n \times n$ matrix having exactly one nonzero entry in each row and column. Show that A is invertible and that its inverse is a matrix of the same form.

5. Invert the matrix of example 3 of §2.5. If in that example, $n_1 = 1300$, $n_2 = 1000$, $n_3 = 600$, $n_4 = 300$, how many members were in the various age groups one year earlier?

6. Find the inverse of the $n \times n$ matrices

(a) $$\begin{bmatrix} 0 & 0 & 0 & \dots & 0 & -1 \\ 1 & 0 & 0 & \dots & 0 & -1 \\ 0 & 1 & 0 & \dots & 0 & -1 \\ 0 & 0 & 1 & \dots & 0 & -1 \\ & & & \vdots & & \\ 0 & 0 & 0 & \dots & 0 & -1 \\ 0 & 0 & 0 & \dots & 1 & -1 \end{bmatrix}$$ (b) $$\begin{bmatrix} 0 & 0 & 0 & \dots & 0 & 1 \\ 1 & 0 & 0 & \dots & 0 & 0 \\ 0 & 1 & 0 & \dots & 0 & 0 \\ 0 & 0 & 1 & \dots & 0 & 0 \\ & & & \vdots & & \\ 0 & 0 & 0 & \dots & 0 & 0 \\ 0 & 0 & 0 & \dots & 1 & 0 \end{bmatrix}$$

7. Find the inverse of the $n \times n$ matrix

$$\begin{bmatrix} 1 & 1 & 1 & 1 & \dots & 1 & 1 \\ -1 & 1 & 1 & 1 & \dots & 1 & 1 \\ 1 & -1 & 1 & 1 & \dots & 1 & 1 \\ 1 & 1 & -1 & 1 & \dots & 1 & 1 \\ & & & \vdots & & & \\ 1 & 1 & 1 & 1 & \dots & -1 & 1 \end{bmatrix}$$

8. If $a \neq 0$, find the inverse of

$$\begin{bmatrix} a & b & c & d \\ 0 & a & b & c \\ 0 & 0 & a & b \\ 0 & 0 & 0 & a \end{bmatrix}$$

vector spaces

1 DEFINITION OF A VECTOR SPACE

In an earlier chapter, we gave rules for addition and scalar multiplication of vectors and matrices. We saw that these operations obeyed certain laws that enabled us to perform algebraic manipulations without continually referring back to the original definitions. In this chapter, we study abstract entities called vector spaces for which the operations of addition and scalar multiplication are also defined: these operations satisfy the same rules of addition and scalar multiplication as the vectors in chapter 2.

We shall study these objects abstractly, using only the basic postulates to develop the theory. By adopting such an approach, we simplify and clarify the proofs of many theorems, as well as achieve a considerable extension of the range of applicability of our results.

Throughout this chapter, then, we deal with sets of objects called vectors merely because they fit into our abstract framework. Column vectors, polynomials in a variable x, matrices, and functions on an interval may all be called vectors, since after certain definitions have been given, we see that each is a member of a set that can be designated as a vector space.

A **vector space** is a set V, consisting of objects called vectors, with two operations defined: addition and scalar multiplication. Addition of vectors means that given two vectors $\boldsymbol{x}$ and $\boldsymbol{y}$ in V, there is a rule determining a vector $\boldsymbol{x} + \boldsymbol{y}$ also in V, and this vector is called the **sum** of $\boldsymbol{x}$ and $\boldsymbol{y}$. By scalar multiplication we mean a rule that assigns to each vector $\boldsymbol{x}$ in V and each real scalar α a vector $\alpha\boldsymbol{x}$, also in V. This vector is called the **scalar multiple** of the vector $\boldsymbol{x}$ by the scalar α.

For example, if the set V is the collection of all polynomials with real coefficients, we could consider the sum $f + g$ of two polynomials, f and g, to be the ordinary sum. We could consider scalar multiplication to be the ordinary product of a number α with a polynomial f, the result being αf. To show that V is a vector space, we must verify that the operations satisfy certain axioms which we list below. Of course, it is possible to define addition and scalar multiplication by different rules, but if we want the set of polynomials to constitute a vector space, the new operations must still satisfy all of the axioms for a vector space.

Let V be a set for which addition and scalar multiplication are defined and let $\boldsymbol{x}$, $\boldsymbol{y}$, and $\boldsymbol{z}$ belong to V; α and β are real numbers. The axioms of a vector space are

(V1) $\boldsymbol{x} + \boldsymbol{y} = \boldsymbol{y} + \boldsymbol{x}$ (Commutative law of vector addition.)
(V2) $(\boldsymbol{x} + \boldsymbol{y}) + \boldsymbol{z} = \boldsymbol{x} + (\boldsymbol{y} + \boldsymbol{z})$ (Associative law of vector addition.)
(V3) There is an element in V, denoted by $\mathbf{0}$, such that $\mathbf{0} + \boldsymbol{x} = \boldsymbol{x} + \mathbf{0} = \boldsymbol{x}$.
(V4) For each $\boldsymbol{x}$ in V, there is an element $-\boldsymbol{x}$ in V, such that $\boldsymbol{x} + (-\boldsymbol{x}) = (-\boldsymbol{x}) + \boldsymbol{x} = \mathbf{0}$.
(V5) $(\alpha + \beta)\boldsymbol{x} = \alpha\boldsymbol{x} + \beta\boldsymbol{x}$
(V6) $\alpha(\boldsymbol{x} + \boldsymbol{y}) = \alpha\boldsymbol{x} + \alpha\boldsymbol{y}$
(V7) $(\alpha\beta)\boldsymbol{x} = \alpha(\beta\boldsymbol{x})$
(V8) $1 \cdot \boldsymbol{x} = \boldsymbol{x}$

A set V whose operations satisfy the above list of requirements is said to be a **real vector space** or a **vector space over the reals**. The vector $\mathbf{0}$, whose existence is asserted in V, is called the **zero vector**. The vector $-\boldsymbol{x}$ is called the **negative** of the vector $\boldsymbol{x}$.

Complex vector spaces are defined in a similar way. We merely postulate that the scalar multiple of the vector $\boldsymbol{x}$ and the scalar α be defined for each $\boldsymbol{x}$ in V and each complex scalar α.

The importance of the vector space concept stems largely from the extensive list of examples of objects satisfying the vector space axioms.

Example 1 Let $\boldsymbol{R}^n$ be the space of column n-vectors with addition and scalar multiplication as defined in chapter 2.

$$\begin{bmatrix} \alpha_1 \\ \alpha_2 \\ \vdots \\ \alpha_n \end{bmatrix} + \begin{bmatrix} \beta_1 \\ \beta_2 \\ \vdots \\ \beta_n \end{bmatrix} = \begin{bmatrix} \alpha_1 + \beta_1 \\ \alpha_2 + \beta_2 \\ \vdots \\ \alpha_n + \beta_n \end{bmatrix}$$

$$\mu \begin{bmatrix} \alpha_1 \\ \alpha_2 \\ \vdots \\ \alpha_n \end{bmatrix} = \begin{bmatrix} \mu\alpha_1 \\ \mu\alpha_2 \\ \vdots \\ \mu\alpha_n \end{bmatrix}$$

At this point the reader may find it a profitable exercise to repeat the proofs offered in chapter 2 showing that the addition and scalar multiplication defined above satisfy the requisite vector space axioms.

In many ways, $\boldsymbol{R}^n$ is the model real vector space. The axioms of a vector space were formulated by singling out the most important properties of addition and scalar multiplication of column vectors. These properties are used to develop in a more general setting theorems analogous to those that hold in $\boldsymbol{R}^n$. Moreover, in a sense that will later be made more precise, real vector spaces that are not "too big" are, in a natural algebraic way, equivalent to $\boldsymbol{R}^n$ for some integer n.

Example 2 The space of complex column n-vectors is a complex vector space, denoted by $\boldsymbol{C}^n$. Addition is defined by

$$\begin{bmatrix} \alpha_1 \\ \alpha_2 \\ \vdots \\ \alpha_n \end{bmatrix} + \begin{bmatrix} \beta_1 \\ \beta_2 \\ \vdots \\ \beta_n \end{bmatrix} = \begin{bmatrix} \alpha_1 + \beta_1 \\ \alpha_2 + \beta_2 \\ \vdots \\ \alpha_n + \beta_n \end{bmatrix}$$

where $\alpha_1, \alpha_2, \ldots, \alpha_n, \beta_1, \beta_2, \ldots, \beta_n$ are complex numbers, and scalar multiplication is defined by

$$\mu \begin{bmatrix} \alpha_1 \\ \alpha_2 \\ \vdots \\ \alpha_n \end{bmatrix} = \begin{bmatrix} \mu\alpha_1 \\ \mu\alpha_2 \\ \vdots \\ \mu\alpha_n \end{bmatrix}$$

with μ a complex scalar.

In the same way that $\boldsymbol{R}^n$ is the model real vector space, $\boldsymbol{C}^n$ is the model complex vector space.

Example 3 Let M_{mn} denote the collection of $m \times n$ matrices with real entries. The addition and scalar multiplication are as defined in §2.3:

$$\begin{bmatrix} a_{11} & a_{12} & \cdots & a_{1n} \\ a_{21} & a_{22} & \cdots & a_{2n} \\ & & \vdots & \\ a_{m1} & a_{m2} & \cdots & a_{mn} \end{bmatrix} + \begin{bmatrix} b_{11} & b_{12} & \cdots & b_{1n} \\ b_{21} & b_{22} & \cdots & b_{2n} \\ & & \vdots & \\ b_{m1} & b_{m2} & \cdots & b_{mn} \end{bmatrix}$$

$$= \begin{bmatrix} a_{11} + b_{11} & a_{12} + b_{12} & \cdots & a_{1n} + b_{1n} \\ a_{21} + b_{21} & a_{22} + b_{22} & \cdots & a_{2n} + b_{2n} \\ & & \vdots & \\ a_{m1} + b_{m1} & a_{m2} + b_{m2} & \cdots & a_{mn} + b_{mn} \end{bmatrix}$$

$$\mu \begin{bmatrix} a_{11} & a_{12} & \cdots & a_{1n} \\ a_{21} & a_{22} & \cdots & a_{2n} \\ & & \vdots & \\ a_{m1} & a_{m2} & \cdots & a_{mn} \end{bmatrix} = \begin{bmatrix} \mu a_{11} & \mu a_{12} & \cdots & \mu a_{1n} \\ \mu a_{21} & \mu a_{22} & \cdots & \mu a_{2n} \\ & & \vdots & \\ \mu a_{m1} & \mu a_{m2} & \cdots & \mu a_{mn} \end{bmatrix}$$

In order to show that M_{mn} is a vector space, it is necessary to check that the vector space axioms (V1)–(V8) are satisfied. All of these laws are stated and several proved in §2.3. Since column n-vectors are just $n \times 1$ matrices, example 1 is a special case of this example.

Example 4 Let P_n denote the collection of all polynomials with real coefficients, of degree less than or equal to n. If f and g belong to P_n, we add them in the ordinary way: Let

$$f = a_0 + a_1 x + \cdots + a_n x^n = \sum_{k=0}^{n} a_k x^k \quad \text{and} \quad g = b_0 + b_1 x + \cdots + b_n x^n = \sum_{k=0}^{n} b_k x^k$$

then

$$\begin{aligned} f + g &= a_0 + b_0 + (a_1 + b_1)x + \cdots + (a_n + b_n)x^n \\ &= \sum_{k=0}^{n} (a_k + b_k) x^k \end{aligned}$$

Scalar multiplication is, likewise, defined in the usual way:

$$\begin{aligned} \alpha f &= \alpha a_0 + (\alpha a_1)x + \cdots + (\alpha a_n)x^n \\ &= \sum_{k=0}^{n} (\alpha a_k) x^k \end{aligned}$$

To verify the vector space axioms explicitly in this case, we let

$$f = \sum_{k=0}^{n} a_k x^k, \qquad g = \sum_{k=0}^{n} b_k x^k, \qquad h = \sum_{k=0}^{n} c_k x^k$$

and α, β be real scalars.

We now check that the laws (V1)–(V8) are satisfied. To verify (V1) we must show that $f + g = g + f$. For (V2) we must verify that $(f + g) + h = f + (g + h)$, and so on. Of course, anyone with a little mathematical experience will realize immediately that these rules are satisfied. Nevertheless, we will check them all explicitly.

(V1)
$$f + g = \sum_{k=0}^{n} (a_k + b_k) x^k$$

$$g + f = \sum_{k=0}^{n} (b_k + a_k) x^k$$

Since a_k and b_k are real, $a_k + b_k = b_k + a_k$, and so we obtain $f + g = g + f$.

(V2)
$$(f + g) + h = \sum_{k=0}^{n} ((a_k + b_k) + c_k)x^k$$

$$f + (g + h) = \sum_{k=0}^{n} (a_k + (b_k + c_k))x^k$$

a_k, b_k, and c_k being real, we have $(a_k + b_k) + c_k = a_k + (b_k + c_k)$, and so $(f + g) + h = f + (g + h)$.

(V3) Let 0 be the zero polynomial, i.e., the polynomial all of whose coefficients are zeros. Then $f + 0 = 0 + f = f$.

(V4) If

$$f = \sum_{k=0}^{n} a_k x^k, \quad \text{let } -f = \sum_{k=0}^{n} (-a_k)x^k$$

then

$$f + (-f) = \sum_{k=0}^{n} (a_k + (-a_k))x^k = \sum_{k=0}^{n} 0 \cdot x^k = 0$$

(V5) By definition,

$$(\alpha + \beta)f = \sum_{k=0}^{n} (\alpha + \beta)a_k x^k$$

while

$$\alpha f = \sum_{k=0}^{n} \alpha a_k x^k, \qquad \beta f = \sum_{k=0}^{n} \beta a_k x^k$$

and so

$$\alpha f + \beta f = \sum_{k=0}^{n} (\alpha a_k + \beta a_k)x^k$$

For real numbers, we know $(\alpha + \beta)a_k = \alpha a_k + \beta a_k$, and so $(\alpha + \beta)f = \alpha f + \beta f$.

(V6) We have $f + g = \sum_{k=0}^{n}(a_k + b_k)x^k$, and so $\alpha(f + g) = \sum_{k=0}^{n}(\alpha(a_k + b_k))x^k$. Also $\alpha f + \alpha g = \sum_{k=0}^{n}(\alpha a_k + \alpha b_k)x^k$. Since $\alpha(a_k + b_k) = \alpha a_k + \alpha b_k$, we have $\alpha(f + g) = \alpha f + \alpha g$.

(V7) $\beta f = \sum_{k=0}^{n}(\beta a_k)x^k$, by definition, and so $\alpha(\beta f) = \sum_{k=0}^{n}(\alpha(\beta a_k))x^k$, while $(\alpha\beta)f = \sum_{k=0}^{n}((\alpha\beta)a_k)x^k$. Now, $\alpha(\beta a_k) = (\alpha\beta)a_k$ holds for real numbers, and so $\alpha(\beta f) = (\alpha\beta)f$.

(V8) $1 \cdot f = \sum_{k=0}^{n}(1 \cdot a_k)x^k = \sum_{k=0}^{n} a_k x^k = f$.

Thus, the polynomials of degree less than or equal to n, having real coefficients, with addition and scalar multiplication as defined above, form a real vector space. Because of this, all theorems proved about vector spaces in general hold for the vector space of polynomials.

In an analogous manner, it is possible to show that the polynomials of degree less than or equal to n with complex coefficients form a complex vector space under the usual addition and scalar multiplication.

In the examples above the addition and multiplication are in some sense "natural." This will also be the case in most of the examples we encounter. However, it is possible to create vector spaces in which the validity of the axioms is not so transparent. For the principle underlying example 5, see exercise 7 at the end of this section.

Example 5 On the ordered pairs of real numbers (x, y) we define an operation of addition, which we denote by $\oplus$ to distinguish it from ordinary addition, by

$$(x, y) \oplus (x', y') = (x + x' + 1, y + y' + 1)$$

We define scalar multiplication by

$$\alpha * (x, y) = (\alpha x + \alpha - 1, \alpha y + \alpha - 1)$$

We now verify (V1)–(V8).

(V1)
$$(x, y) \oplus (x', y') = (x + x' + 1, y + y' + 1)$$
$$(x', y') \oplus (x, y) = (x' + x + 1, y' + y + 1)$$

Since $x + x' + 1 = x' + x + 1$ and $y + y' + 1 = y' + y + 1$, we see that $(x, y) \oplus (x', y') = (x', y') \oplus (x, y)$.

(V2)
$$\begin{aligned}((x, y) \oplus (x', y')) \oplus (x'', y'') &= (x + x' + 1, y + y' + 1) \oplus (x'', y'') \\ &= (x + x' + x'' + 2, y + y' + y'' + 2)\end{aligned}$$
$$\begin{aligned}(x, y) \oplus ((x', y') \oplus (x'', y'')) &= (x, y) \oplus (x' + x'' + 1, y' + y'' + 1) \\ &= (x + x' + x'' + 2, y + y' + y'' + 2)\end{aligned}$$

So

$$((x, y) \oplus (x', y')) \oplus (x'', y'') = (x, y) \oplus ((x', y') \oplus (x'', y''))$$

(V3) What is the zero element of our space? Observe that

$$\begin{aligned}(x, y) \oplus (-1, -1) &= (x + (-1) + 1, y + (-1) + 1) \\ &= (x, y)\end{aligned}$$

Hence, we see that $(-1, -1)$ plays the role of the zero element.

(V4) Given (x, y), we let its negative be $(-x-2, -y-2)$. Then

$$(x, y) \oplus (-x-2, -y-2) = (x + (-x-2) + 1, y + (-y-2) + 1)$$
$$= (-1, -1)$$

which is the zero element.

(V5) $(\alpha + \beta) * (x, y)$

$$= ((\alpha + \beta)x + (\alpha + \beta) - 1, (\alpha + \beta)y + (\alpha + \beta) - 1)$$

(by definition of *)

Also, by definition,

$$\alpha * (x, y) = (\alpha x + \alpha - 1, \alpha y + \alpha - 1)$$
$$\beta * (x, y) = (\beta x + \beta - 1, \beta y + \beta - 1)$$

By definition of $\oplus$,

$(\alpha * (x, y)) \oplus (\beta * (x, y))$

$$= ((\alpha x + \alpha - 1) + (\beta x + \beta - 1) + 1, (\alpha y + \alpha - 1) + (\beta y + \beta - 1) + 1)$$
$$= ((\alpha + \beta)x + (\alpha + \beta) - 1, (\alpha + \beta)y + (\alpha + \beta) - 1)$$
$$= (\alpha + \beta) * (x, y)$$

(V6) $\alpha * ((x, y) \oplus (x', y')) = \alpha * (x + x' + 1, y + y' + 1)$

$$= (\alpha(x + x' + 1) + \alpha - 1, \alpha(y + y' + 1) + \alpha - 1)$$

while

$\alpha * (x, y) \oplus \alpha * (x', y')$

$$= (\alpha x + \alpha - 1, \alpha y + \alpha - 1) \oplus (\alpha x' + \alpha - 1, \alpha y' + \alpha - 1)$$
$$= (\alpha x + \alpha - 1 + \alpha x' + \alpha - 1 + 1, \alpha y + \alpha - 1 + \alpha y' + \alpha - 1 + 1)$$
$$= \alpha * ((x, y) \oplus (x', y'))$$

(V7) $\alpha * (\beta * (x, y)) = \alpha * (\beta x + \beta - 1, \beta y + \beta - 1)$

$$= (\alpha\beta x + \alpha\beta - \alpha + \alpha - 1, \alpha\beta y + \alpha\beta - \alpha + \alpha - 1)$$
$$= (\alpha\beta x + \alpha\beta - 1, \alpha\beta y + \alpha\beta - 1)$$
$$= (\alpha\beta) * (x, y)$$

(V8) $1 * (x, y) = (x + 1 - 1, y + 1 - 1) = (x, y)$

Thus, the collection of ordered pairs of real numbers with operation of addition $\oplus$ and scalar multiplication $*$ forms a vector space over the real numbers.

EXERCISES

1. The following is a list of sets with operations of addition and scalar multiplication defined on them. For each set, show that the set, together with its indicated operations, forms a vector space over the reals.
   (a) The real matrices of the form

$$\begin{bmatrix} a & -b \\ b & a \end{bmatrix}$$

   with

$$\begin{bmatrix} a & -b \\ b & a \end{bmatrix} + \begin{bmatrix} c & -d \\ d & c \end{bmatrix} = \begin{bmatrix} a+c & -(b+d) \\ b+d & a+c \end{bmatrix}$$

$$\alpha \begin{bmatrix} a & -b \\ b & a \end{bmatrix} = \begin{bmatrix} \alpha a & -\alpha b \\ \alpha b & \alpha a \end{bmatrix}$$

   (b) The real matrices of the form

$$\begin{bmatrix} a & b \\ 0 & c \end{bmatrix}$$

   with the usual addition and scalar multiplication.
   (c) The set of triples of real numbers (x, y, z), such that $z = x + y$, with $(x, y, z) + (x', y', z') = (x + x', y + y', z + z')$ and $\alpha(x, y, z) = (\alpha x, \alpha y, \alpha z)$.
   (d) The even polynomials of degree less than or equal to n, a positive integer, with usual addition and scalar multiplication for polynomials.
   (e) The odd polynomials of degree less than or equal to n, a positive integer, with usual addition and scalar multiplication for polynomials.
   (f) The polynomials f, of degree less than or equal to n, such that $f(1) = 0$. Addition and scalar multiplication are defined in the usual manner.
   (g) Ordered pairs of real numbers (x, y) with

$$(x, y) \oplus (x', y') = (x + x' + 1, y + y')$$

$$\alpha * (x, y) = (\alpha x + \alpha - 1, \alpha y)$$

   (h) Differentiable functions on the interval $(0, 1)$ with

$$(f + g)(x) = f(x) + g(x) \qquad (\alpha f)(x) = \alpha(f(x))$$

   and the usual addition and scalar multiplication for functions.

2. On the triples of real numbers (x, y, z) define addition by $(x, y, z) + (x', y', z') = (x + x', y + y', z + z')$, and scalar multiplication by $\alpha(x, y, z) = (0, 0, 0)$. Show that all axioms for a vector space are satisfied except (V8).

3. On the ordered pairs of real numbers (x, y) define addition by $(x, y) + (x', y') = (x + x', y + y')$ and scalar multiplication by $\alpha(x, y) = (\alpha^2 x, \alpha^2 y)$. Show that all axioms for a vector space are satisfied except (V5).

4. On the ordered pairs of real numbers define addition by $(x, y) + (x', y') = (x + x', y + y')$ and scalar multiplication by $\alpha(x, y) = (3\alpha x, 3\alpha y)$. Show that all axioms for a vector space are satisfied except (V7) and (V8).

5. Let X be a set. Consider the family $\mathfrak{F}$ of all functions from X into the real numbers. Define

$$(f+g)(x) = f(x) + g(x) \qquad (\alpha f)(x) = \alpha(f(x))$$

Show that with these indicated operations the given family of functions forms a vector space.

6. Let U be a complex vector space. Show that U is a real vector space if we use the same operation of addition while the scalar product is $\alpha \boldsymbol{x}$, for $\boldsymbol{x}$ in U and α a real number. (This definition makes sense, since the real numbers are contained in the complex numbers.)

7. Let V be real vector space, with addition denoted by + and scalar multiplication by $\cdot$. Let $\boldsymbol{t}$ be a fixed vector in V. Define a new addition on V by $\boldsymbol{x} \oplus \boldsymbol{y} = \boldsymbol{x} + \boldsymbol{y} + \boldsymbol{t}$ and scalar multiplication by $\alpha * \boldsymbol{x} = \alpha \boldsymbol{x} + (\alpha - 1)\boldsymbol{t}$. Show that V with $\oplus$ and * as its operations is a vector space.

8. On the ordered pairs of real numbers (x, y) define $\oplus$ addition by $(x, y) \oplus (x', y') = ((x^3 + (x')^3)^{1/3}, (y^3 + (y')^3)^{1/3})$ and scalar multiplication by $\alpha(x, y) = (\alpha x, \alpha y)$. Show that with the indicated operations the ordered pairs of real numbers form a vector space.

2 ADDITIONAL PROPERTIES OF VECTOR SPACES

Given any vector space, we can use the axioms (V1)–(V8) to obtain additional rules for algebraic manipulation of vectors.

Theorem 1 Let V be a vector space and $\boldsymbol{x}$ and $\boldsymbol{y}$ be vectors in V. Then there is one and only one $\boldsymbol{u}$ in V, such that $\boldsymbol{x} + \boldsymbol{u} = \boldsymbol{y}$.

PROOF First, we must show there is one such $\boldsymbol{u}$. To accomplish this, let $\boldsymbol{u} = (-\boldsymbol{x}) + \boldsymbol{y}$. Then

$$\begin{aligned} \boldsymbol{x} + \boldsymbol{u} &= \boldsymbol{x} + ((-\boldsymbol{x}) + \boldsymbol{y}) \\ &= (\boldsymbol{x} + (-\boldsymbol{x})) + \boldsymbol{y} && \text{[by (V2)]} \\ &= \boldsymbol{0} + \boldsymbol{y} && \text{[by (V4)]} \\ &= \boldsymbol{y} && \text{[by (V3)]} \end{aligned}$$

Next we show that there is only one such $\boldsymbol{u}$. Suppose $\boldsymbol{u}_1$ and $\boldsymbol{u}_2$ are vectors in V, such that

$$\boldsymbol{x} + \boldsymbol{u}_1 = \boldsymbol{y} \quad \text{and} \quad \boldsymbol{x} + \boldsymbol{u}_2 = \boldsymbol{y}$$

Then $\boldsymbol{x} + \boldsymbol{u}_1 = \boldsymbol{x} + \boldsymbol{u}_2$.

Thus,

$$\begin{aligned}
(-x) + (x + u_1) &= (-x) + (x + u_2) \\
((-x) + x) + u_1 &= ((-x) + x) + u_2 && \text{[by (V2)]} \\
\mathbf{0} + u_1 &= \mathbf{0} + u_2 && \text{[by (V4)]} \\
u_1 &= u_2 && \text{[by (V3)]} \blacksquare
\end{aligned}$$

It is customary to denote the vector $y + (-x)$ considered above simply by $y - x$. $y - x$ is called the vector obtained from y by subtracting x.

As another example of results that can be obtained from the vector space axioms, we have the following theorem.

Theorem 2 Let V be a vector space, x be a vector in V, and α be a scalar. Then

(i) $\alpha \cdot \mathbf{0} = \mathbf{0}$

(ii) $0 \cdot x = \mathbf{0}$

(iii) $\alpha x = \mathbf{0}$ implies that either $\alpha = 0$ or $x = \mathbf{0}$.

PROOF To prove (i), observe that

$$\mathbf{0} + \mathbf{0} = \mathbf{0} \qquad \text{[by (V3)]}$$

Thus,

$$\begin{aligned}
\alpha(\mathbf{0} + \mathbf{0}) &= \alpha \cdot \mathbf{0} \\
\alpha \cdot \mathbf{0} + \alpha \cdot \mathbf{0} &= \alpha \cdot \mathbf{0} && \text{[by (V6)]}
\end{aligned}$$

We know by Theorem 1 that there is only one vector u, such that $\alpha \cdot \mathbf{0} + u = \alpha \cdot \mathbf{0}$.

By (V3), one such vector is $\mathbf{0}$. Thus, $\alpha \cdot \mathbf{0} = \mathbf{0}$. Alternatively, by adding $-\alpha \cdot \mathbf{0}$ to the equation $\alpha \cdot \mathbf{0} + \alpha \cdot \mathbf{0} = \alpha \cdot \mathbf{0}$, we obtain

$$\begin{aligned}
((\alpha \cdot \mathbf{0}) + (\alpha \cdot \mathbf{0})) + (-\alpha \cdot \mathbf{0}) &= (-\alpha \cdot \mathbf{0}) + (\alpha \cdot \mathbf{0}) \\
&= \mathbf{0} && \text{[by (V4)]} \\
(\alpha \cdot \mathbf{0}) + (\alpha \cdot \mathbf{0} + (-\alpha \cdot \mathbf{0})) &= \mathbf{0} && \text{[by (V2)]} \\
(\alpha \cdot \mathbf{0}) + \mathbf{0} &= \mathbf{0} && \text{[by (V4)]} \\
\alpha \cdot \mathbf{0} &= \mathbf{0} && \text{[by (V3)]}
\end{aligned}$$

To prove (ii) observe that $0 \cdot x = (0 + 0)x$, since $0 = 0 + 0$. Thus, $0 \cdot x = 0 \cdot x + 0 \cdot x$ [by (V5)]. As before, by adding $-0 \cdot x$ to both sides of the foregoing equation, we obtain $0 \cdot x = \mathbf{0}$.

To prove (iii) suppose that $\alpha x = \mathbf{0}$. If $\alpha \neq 0$, we may multiply both sides of the above equation by α^{-1} to obtain

$$\begin{aligned} \alpha^{-1}(\alpha \boldsymbol{x}) &= \alpha^{-1} \cdot \mathbf{0} = \mathbf{0} && \text{[by (i)]} \\ (\alpha^{-1}\alpha)\boldsymbol{x} &= \mathbf{0} && \text{[by (V7)]} \\ 1 \cdot \boldsymbol{x} &= \mathbf{0} \\ \boldsymbol{x} &= \mathbf{0} && \text{[by (V8)]} \end{aligned}$$

Thus, if $\alpha \neq 0$, we have $\boldsymbol{x} = \mathbf{0}$. So we must have either $\alpha = 0$ or $\boldsymbol{x} = \mathbf{0}$. ■

The following equates the scalar multiple $(-1)\boldsymbol{x}$ of the vector $\boldsymbol{x}$ with its negative $-\boldsymbol{x}$.

Theorem 3 If V is a vector space and $\boldsymbol{x}$ is a vector in V, then

$$(-1)\boldsymbol{x} = -\boldsymbol{x}$$

PROOF Since $1 + (-1) = 0$, we have

$$(1 + (-1))\boldsymbol{x} = 0 \cdot \boldsymbol{x} = \mathbf{0} \qquad \text{[by (ii)]}$$

Thus, by (V5),

$$(1)\boldsymbol{x} + (-1)\boldsymbol{x} = \mathbf{0}$$

Or using (V8),

$$\boldsymbol{x} + (-1)\boldsymbol{x} = \mathbf{0}$$

If we add $-\boldsymbol{x}$ to both sides of this equation, we obtain $(-1)\boldsymbol{x} = -\boldsymbol{x}$. ■

Generally speaking, the vector space axioms enable us to perform algebraic operations with abstract vectors in much the same way we do with column vectors. Keeping this in mind will make it unnecessary to refer back to the vector space axioms (VI) – (V8).

EXERCISES

1. Let V be a vector space. Suppose $\boldsymbol{x}$ and $\boldsymbol{e}$ are members of V, and suppose $\boldsymbol{x} + \boldsymbol{e} = \boldsymbol{x}$. Show that $\boldsymbol{e} = \mathbf{0}$.
2. Prove by induction that if V is a vector space, $\boldsymbol{x} \in V$, and $\alpha_1, \alpha_2, \ldots, \alpha_n$ are scalars, then $(\alpha_1 + \alpha_2 + \cdots + \alpha_n)\boldsymbol{x} = \alpha_1\boldsymbol{x} + \alpha_2\boldsymbol{x} + \cdots + \alpha_n\boldsymbol{x}$.
3. If V is a vector space, $\boldsymbol{x}_1, \boldsymbol{x}_2, \ldots, \boldsymbol{x}_n$ belong to V, and α is a scalar, show by induction that $\alpha(\boldsymbol{x}_1 + \boldsymbol{x}_2 + \cdots + \boldsymbol{x}_n) = \alpha\boldsymbol{x}_1 + \alpha\boldsymbol{x}_2 + \cdots + \alpha\boldsymbol{x}_n$.

4. If V is a real vector space, $\boldsymbol{x}$ belongs to V, and $\boldsymbol{x} + \boldsymbol{x} = \boldsymbol{0}$, show that $\boldsymbol{x} = \boldsymbol{0}$.

5. If V is a vector space, $\boldsymbol{x}_1, \boldsymbol{x}_2, \boldsymbol{y}_1, \boldsymbol{y}_2$ belong to V,

$$a\boldsymbol{x}_1 + b\boldsymbol{x}_2 = \boldsymbol{y}_1$$
$$c\boldsymbol{x}_1 + d\boldsymbol{x}_2 = \boldsymbol{y}_2$$

and

$$\begin{vmatrix} a & b \\ c & d \end{vmatrix} \neq 0$$

find $\boldsymbol{x}_1$ and $\boldsymbol{x}_2$ in terms of $\boldsymbol{y}_1$ and $\boldsymbol{y}_2$.

6. If V is a vector space, $\boldsymbol{x}$ belongs to V, and α and β are scalars, show that if $\alpha\boldsymbol{x} = \beta\boldsymbol{x}$ and $\alpha \neq \beta$, then $\boldsymbol{x} = \boldsymbol{0}$.

7. If V is a vector space, $\boldsymbol{x}_1, \boldsymbol{x}_2, \boldsymbol{x}_3, \boldsymbol{y}_1, \boldsymbol{y}_2, \boldsymbol{y}_3$ are vectors in V, and

$$\begin{aligned} \boldsymbol{x}_1 + \boldsymbol{x}_2 + \boldsymbol{x}_3 &= \boldsymbol{y}_1 \\ -2\boldsymbol{x}_1 + \boldsymbol{x}_2 - 2\boldsymbol{x}_3 &= \boldsymbol{y}_2 \\ -\boldsymbol{x}_1 + 2\boldsymbol{x}_2 - \boldsymbol{x}_3 &= \boldsymbol{y}_3 \end{aligned}$$

show that $\boldsymbol{y}_1 + \boldsymbol{y}_2 - \boldsymbol{y}_3 = \boldsymbol{0}$.

8. If V is a vector space, $\boldsymbol{x}_1, \boldsymbol{x}_2, \ldots, \boldsymbol{x}_n, \boldsymbol{y}_1, \boldsymbol{y}_2, \boldsymbol{y}_3, \ldots, \boldsymbol{y}_n$ belong to V, and

$$\begin{aligned} \boldsymbol{x}_1 + \boldsymbol{x}_2 + \cdots + \boldsymbol{x}_n &= \boldsymbol{y}_1 \\ \boldsymbol{x}_2 + \cdots + \boldsymbol{x}_n &= \boldsymbol{y}_2 \\ &\vdots \\ \boldsymbol{x}_n &= \boldsymbol{y}_n \end{aligned}$$

find $\boldsymbol{x}_1, \boldsymbol{x}_2, \ldots, \boldsymbol{x}_n$ in terms of $\boldsymbol{y}_1, \boldsymbol{y}_2, \ldots, \boldsymbol{y}_n$.

9. Find a vector space V that has two vectors $\boldsymbol{x}$ and $\boldsymbol{y}$ neither of which is a scalar multiple of the other.

10. Show that any set V with operations of addition and scalar multiplication that satisfies (V2)–(V8) must also satisfy (V1). [Hint: Calculate $(1 + 1)(\boldsymbol{x} + \boldsymbol{y})$ two ways by using (V5) and (V6).]

3 SUBSPACES

If V is a vector space over the reals (or complexes), there are certain subsets of V, called subspaces, that are again vector spaces under the same algebraic operations. The purpose of this section is to study these objects.

Definition If V is a vector space and H is a nonempty subset of V having the properties

(i) Whenever $\boldsymbol{x}$ and $\boldsymbol{y}$ belong to H, then $\boldsymbol{x} + \boldsymbol{y}$ belongs to H.
(ii) If $\boldsymbol{x}$ belongs to H and α is a scalar, then $\alpha\boldsymbol{x}$ belongs to H.

H is said to be a **subspace** of the vector space V.

In other terms, a subspace of V is a subset closed under the algebraic operations of addition and scalar multiplication.

As an example, in $\boldsymbol{R}^3$, let L be the set of vectors lying on some line passing through the origin. (See Figure 4-1.) From our geometric formulation of the process of scalar multiplication, it is clear that all vectors in L are scalar multiples of a single nonzero vector in L, for definiteness, say $\boldsymbol{u}$. If $\boldsymbol{x}$ and $\boldsymbol{y}$ belong to L, we have $\boldsymbol{x} = \alpha\boldsymbol{u}$ and $\boldsymbol{y} = \beta\boldsymbol{u}$ for suitable scalars α and β. Then, $\boldsymbol{x} + \boldsymbol{y} = (\alpha + \beta)\boldsymbol{u}$. Thus, $\boldsymbol{x} + \boldsymbol{y}$ being a scalar multiple of $\boldsymbol{u}$ must necessarily belong to L. Moreover, since $\lambda\boldsymbol{x} = \lambda(\alpha\boldsymbol{u}) = (\lambda\alpha)\boldsymbol{u}$, we also see that if $\boldsymbol{x}$ belongs to L and λ is a scalar, $\lambda\boldsymbol{x}$ belongs to L. Thus, having verified conditions (i) and (ii) in the definition of a subspace above, we see that L is a subspace of $\boldsymbol{R}^3$.

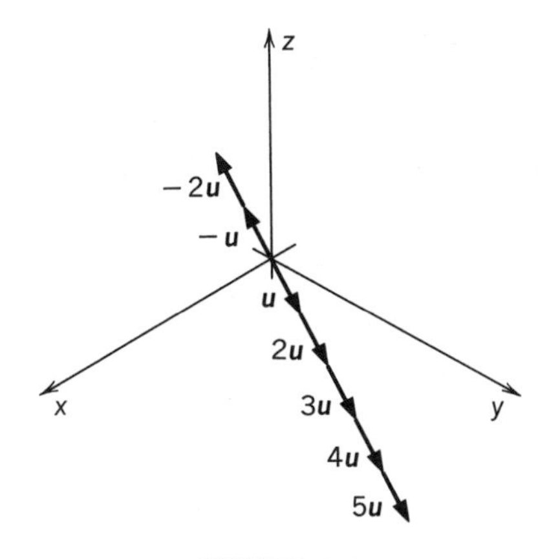

FIGURE 4-1

As another example of a subspace of $\boldsymbol{R}^3$, let H be those vectors of $\boldsymbol{R}^3$ that lie in the xy plane. In other words, H consists of those vectors of the form $\begin{bmatrix} x \\ y \\ 0 \end{bmatrix}$. If $\begin{bmatrix} x_1 \\ y_1 \\ 0 \end{bmatrix}$ and $\begin{bmatrix} x_2 \\ y_2 \\ 0 \end{bmatrix}$ are two such vectors, then their sum $\begin{bmatrix} x_1 + x_2 \\ y_1 + y_2 \\ 0 \end{bmatrix}$ still lies in H. Also if $\begin{bmatrix} x \\ y \\ 0 \end{bmatrix}$ lies in H and α is a scalar, then $\alpha\begin{bmatrix} x \\ y \\ 0 \end{bmatrix} = \begin{bmatrix} \alpha x \\ \alpha y \\ 0 \end{bmatrix}$ and the vector also lies in H. It follows that H is a subspace of $\boldsymbol{R}^3$.

The fact that H is a subspace can also be interpreted geometrically. If $\boldsymbol{a}$ and $\boldsymbol{b}$ are vectors in the xy plane, $\boldsymbol{a}$ and $\boldsymbol{b}$ may be thought of as directed line segments that begin at the origin and lie in the xy plane. Then the vector $\boldsymbol{a} + \boldsymbol{b}$ may be thought of as the diagonal of a parallelogram having adjacent sides $\boldsymbol{a}$ and $\boldsymbol{b}$. Clearly, this vector lies in the xy plane as well; see Figure 4-2.

Similarly, scalar multiples of vectors in the xy plane still lie in the xy plane.

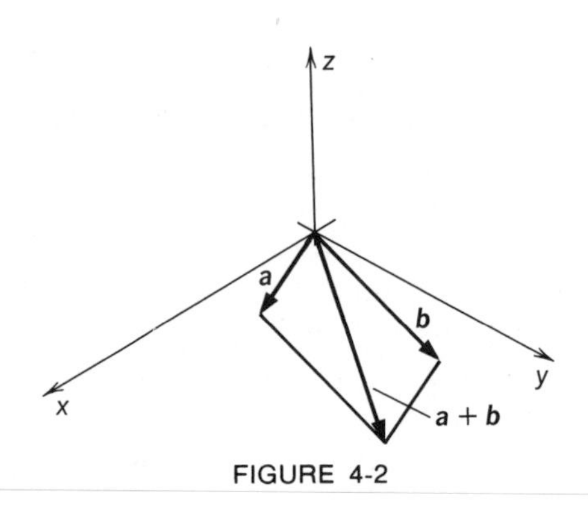

FIGURE 4-2

Any vector space V has at least two subspaces. One subspace, called the **zero subspace**, consists solely of the zero vector. It is clear that this set is closed under addition and scalar multiplication. Another subspace of the vector space V is the subspace consisting of all vectors in V. A subspace that is not the zero subspace or the whole space is said to be proper.

We next prove several results about subspaces.

Proposition 1 If H is a subspace of a vector space V, then $\mathbf{0}$ belongs to H.

PROOF Let $\boldsymbol{x}$ be an element of H. Then, since scalar multiples of vectors in H belong to H, $0\boldsymbol{x} = \mathbf{0}$ belongs to H. ■

Proposition 2 If H is a subspace of a vector space V, and $\boldsymbol{x}$ belongs to H, then $-\boldsymbol{x}$ belongs to H.

PROOF If $\boldsymbol{x}$ belongs to H, since H is closed under scalar multiplication, $(-1)\boldsymbol{x} = -\boldsymbol{x}$ belongs to H. ■

Theorem If H is a subspace of a vector space V, H is a vector space under the operations of addition and scalar multiplication defined on V.

In this case, there is not really very much to prove. H is a set. Addition and scalar multiplication are defined in H just as in V. Because of condition (i) in the definition of a subspace, the sum of two elements of H is still an element of H. By (ii), a scalar multiple of an element of H is still in H. Thus, H has a rule of addition and scalar multiplication that give elements of H. It now remains to check that H satisfies the axioms (V1)–(V8). (V1) and (V2) hold in H, because they hold in V. By Proposition 1, $\mathbf{0}$ lies in H. It follows that (V3) holds in H. Using Proposition 2, we conclude that (V4) holds in H. (V5)–(V8) are satisfied in H, since they are satisfied in V.

This theorem enables us to construct many new examples of vector spaces without going through a tedious verification of the vector space axioms (V1)–(V8).

Example 1 Let L be the subset of $\boldsymbol{R}^n$ consisting of those vectors whose first component is 0. Since the sum of two vectors with first component 0 again has first component 0, L is closed under addition. A scalar multiple of a vector with first component 0 likewise has first component 0. Thus, scalar multiples of vectors in L are still in L. Therefore, L is a subspace.

By the theorem above, we can conclude that the n-tuples belonging to L form a real vector space.

Example 2 Let A be a fixed $m \times n$ matrix with real entries. Let $N = \{\boldsymbol{x} | \boldsymbol{x} \in \boldsymbol{R}^n \text{ and } A\boldsymbol{x} = \boldsymbol{0}\}$. We claim N is a subspace of $\boldsymbol{R}^n$.

First, suppose $\boldsymbol{x}_1 \in N$ and $\boldsymbol{x}_2 \in N$; then

$$\begin{aligned} A(\boldsymbol{x}_1 + \boldsymbol{x}_2) &= A\boldsymbol{x}_1 + A\boldsymbol{x}_2 \\ &= \boldsymbol{0} + \boldsymbol{0} \\ &= \boldsymbol{0} \end{aligned}$$

Thus, $\boldsymbol{x}_1 + \boldsymbol{x}_2 \in N$. Next if $\boldsymbol{x} \in N$ and α is a scalar,

$$\begin{aligned} A(\alpha\boldsymbol{x}) &= \alpha A(\boldsymbol{x}) \\ &= \alpha\boldsymbol{0} \\ &= \boldsymbol{0} \end{aligned}$$

Thus, $\alpha\boldsymbol{x} \in N$.

Having verified conditions (i) and (ii) in the definition of subspace, we see that N is a subspace of $\boldsymbol{R}^n$. Our preceding theorem guarantees that with the proper definitions of addition and scalar multiplication, N is a vector space in its own right.

A more concrete formulation of example 2 may be found by considering solutions to the system of homogeneous equations

$$\begin{aligned} 3x_1 + 4x_2 - x_3 + 2x_4 &= 0 \\ x_1 - x_2 + 3x_3 + x_4 &= 0 \\ 4x_1 + 3x_2 - x_3 + 2x_4 &= 0 \end{aligned}$$

If

$$\begin{bmatrix} x_1 \\ x_2 \\ x_3 \\ x_4 \end{bmatrix} \quad \text{and} \quad \begin{bmatrix} x_1' \\ x_2' \\ x_3' \\ x_4' \end{bmatrix}$$

are two such solutions, we define their sum by

$$\begin{bmatrix} x_1 \\ x_2 \\ x_3 \\ x_4 \end{bmatrix} + \begin{bmatrix} x_1' \\ x_2' \\ x_3' \\ x_4' \end{bmatrix} = \begin{bmatrix} x_1 + x_1' \\ x_2 + x_2' \\ x_3 + x_3' \\ x_4 + x_4' \end{bmatrix}$$

and a scalar product by

$$\alpha \begin{bmatrix} x_1 \\ x_2 \\ x_3 \\ x_4 \end{bmatrix} = \begin{bmatrix} \alpha x_1 \\ \alpha x_2 \\ \alpha x_3 \\ \alpha x_4 \end{bmatrix}$$

Converting the system of equations to matrix notation and applying example 2, we see that the set of solutions to the system of equations together with the above addition and scalar multiplication forms a vector space over the reals. Of course, it is possible to reach the same conclusion by explicit verification of the vector space axioms (V1)–(V8).

Example 3 In the space P_n, the set of polynomials of degree less than or equal to n, with real coefficients, the subset

$$H = \left\{ f | f \in P_n \text{ and } \int_0^1 f(x)\, dx = 0 \right\}$$

is a subspace. First, suppose $f \in H$ and $g \in H$, then

$$\begin{aligned} \int_0^1 (f(x) + g(x))\, dx &= \int_0^1 f(x)\, dx + \int_0^1 g(x)\, dx \\ &= 0 + 0 \\ &= 0 \end{aligned}$$

So $f + g \in H$. Next suppose $f \in H$ and α is a scalar, then

$$\begin{aligned} \int_0^1 (\alpha f(x))\, dx &= \alpha \int_0^1 f(x)\, dx \\ &= \alpha \cdot 0 \\ &= 0 \end{aligned}$$

So, $\alpha f \in H$.

Having verified conditions (i) and (ii) in the definition of subspace, we conclude that H is a subspace of P_n. Then from our theorem we conclude that H, with the proper definitions of addition and scalar multiplication, is itself a vector space.

Now that we have seen several examples of subsets of vector spaces that are subspaces, it might be profitable to examine some subsets of $\boldsymbol{R}^2$ that are not subspaces.

Example 4 In $\boldsymbol{R}^2$, let

$$H = \left\{ \begin{bmatrix} \alpha \\ \beta \end{bmatrix} \middle| \begin{bmatrix} \alpha \\ \beta \end{bmatrix} \in \boldsymbol{R}^2 \text{ and } \alpha \geqslant 0 \right\}$$

Geometrically this set corresponds to the right half of the plane. (See Figure 4-3(a).) If $\boldsymbol{x}_1 \in H$ and $\boldsymbol{x}_2 \in H$, then

$$\boldsymbol{x}_1 = \begin{bmatrix} \alpha_1 \\ \beta_1 \end{bmatrix} \quad \text{and} \quad \boldsymbol{x}_2 = \begin{bmatrix} \alpha_2 \\ \beta_2 \end{bmatrix}$$

where $\alpha_1 \geqslant 0$ and $\alpha_2 \geqslant 0$.

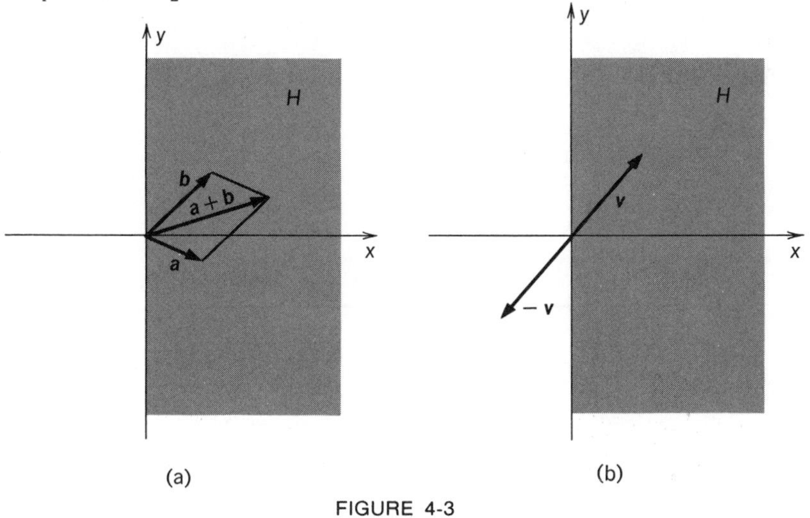

FIGURE 4-3

Thus,

$$\boldsymbol{x}_1 + \boldsymbol{x}_2 = \begin{bmatrix} \alpha_1 + \alpha_2 \\ \beta_1 + \beta_2 \end{bmatrix}$$

where $\alpha_1 + \alpha_2 \geqslant 0$, and so $\boldsymbol{x}_1 + \boldsymbol{x}_2 \in H$. Thus, H is additively closed.

However, since

$$\begin{bmatrix} 1 \\ 0 \end{bmatrix} \in H$$

and

$$(-1)\begin{bmatrix} 1 \\ 0 \end{bmatrix} = \begin{bmatrix} -1 \\ 0 \end{bmatrix}$$

is not in H, scalar multiples of vectors in H are not in H. (See Figure 4-3(b).)

Thus, H is not a subspace of $\boldsymbol{R}^2$. In this case, it is interesting to note that if $\boldsymbol{x} \in H$ and $\alpha \geqslant 0$, $\alpha\boldsymbol{x}$ also belongs to H.

Example 5 In $\boldsymbol{R}^2$, let

$$H = \left\{ \begin{bmatrix} \alpha \\ \beta \end{bmatrix} \middle| \begin{bmatrix} \alpha \\ \beta \end{bmatrix} \in \boldsymbol{R}^2 \quad \text{and} \quad \alpha = 0 \text{ or } \beta = 0 \right\}$$

Geometrically this set consists of all points on the two coordinate axes. (See Figure 4-4.) Scalar multiples of vectors in H still belong to H, but H is not additively closed. For we have

$$\begin{bmatrix} 1 \\ 0 \end{bmatrix} \in H \quad \text{and} \quad \begin{bmatrix} 0 \\ 1 \end{bmatrix} \in H$$

but

$$\begin{bmatrix} 1 \\ 0 \end{bmatrix} + \begin{bmatrix} 0 \\ 1 \end{bmatrix} = \begin{bmatrix} 1 \\ 1 \end{bmatrix}$$

does not belong to H.

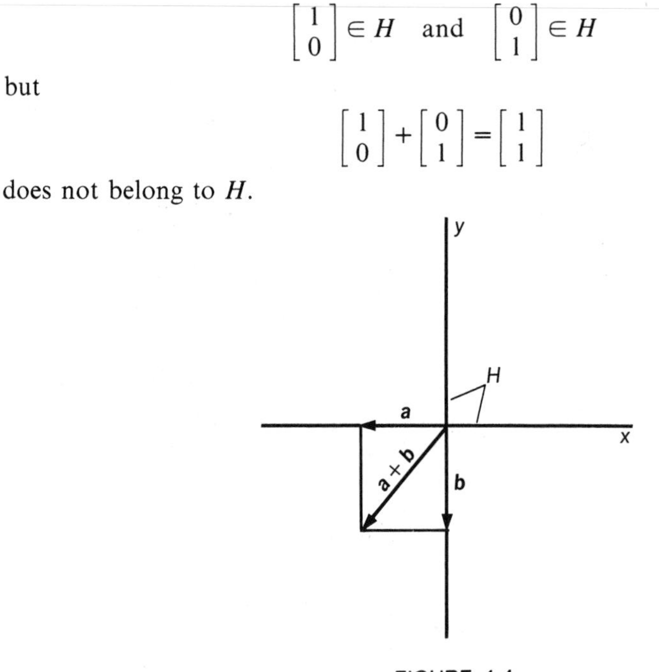

FIGURE 4-4

EXERCISES

1. Which of the following subsets of $\boldsymbol{R}^2$ are subspaces?

(a) $\{(x, y) | x = 3y\}$
(b) $\{(x, y) | x^2 = y^2\}$
(c) $\{(x, y) | x + y = 1\}$
(d) $\{(x, y) | x = y^3\}$
(e) $\{(x, y) | y = 0\}$
(f) $\{(x, y) | x \geqslant 0, y \geqslant 0\}$
(g) $\{(x, y) | x = y\}$

2. Which of the following subsets of $\boldsymbol{R}^3$ are subspaces?

(a) $\{(x, y, z) | x + 3y = 0 \text{ and } y + z = 0\}$
(b) $\{(x, y, z) | x = 0 \text{ or } y = 0\}$
(c) $\{(x, y, z) | x + y + z = 0\}$

(d) $\{(x, y, z) | x \geqslant 0, y \geqslant 0, z \geqslant 0\}$
(e) $\{(x, y, z) | x^2 + y^2 \leqslant 1\}$
(f) $\{(x, y, z) | z = 2x + 2y\}$
(g) $\{(x, y, z) | z = 0 \text{ and } x^2 + y^2 \leqslant 1\}$

3. Let H_i be the subset of $\boldsymbol{R}^n$ consisting of those vectors whose ith component is 0. Show that H_i is a subspace of $\boldsymbol{R}^n$.

4. Let P_n be the polynomials of degree less than or equal to n. Let H_e be a collection of the even polynomials in P_n and H_o be a collection of the odd polynomials in P_n. Show that H_e and H_o are subspaces of P_n.

5. Show that the following subsets of the space of 2×2 matrices M_{22} are subspaces.
   (a) All matrices of the form

$$\begin{bmatrix} a & b \\ b & a \end{bmatrix} \qquad a, b \text{ real numbers}$$

   (b) All matrices of the form

$$\begin{bmatrix} a & b \\ -b & a \end{bmatrix}, \qquad a, b \text{ real numbers}$$

   (c) All matrices of the form

$$\begin{bmatrix} a & b \\ 0 & c \end{bmatrix}, \qquad a, b, c \text{ real numbers}$$

   (d) All matrices of the form

$$\begin{bmatrix} a & c \\ c & b \end{bmatrix}, \qquad a, b, c \text{ real numbers}$$

   (e) All matrices of the form

$$\begin{bmatrix} 0 & a \\ 0 & b \end{bmatrix}, \qquad a, b \text{ real numbers}$$

   (f) All matrices of the form

$$\begin{bmatrix} a & c \\ -c & b \end{bmatrix}, \qquad a, b, c \text{ real numbers}$$

   (g) All matrices of the form

$$\begin{bmatrix} a & 0 \\ 0 & b \end{bmatrix}, \qquad a, b \text{ real numbers}$$

6. Let P denote the collection of all polynomials with real coefficients.
   (a) Show that P is a vector space over the reals.
   (b) Show that the following subsets of P are subspaces.
      (1) $\{f | \text{degree of } f \leqslant n, \text{ where } n \text{ is a fixed positive integer}\}$
      (2) $\{f | f(\alpha) = 0, \text{ where } \alpha \text{ is a fixed number}\}$
      (3) $\{f | f(\alpha) = f(\beta), \text{ where } \alpha \text{ and } \beta \text{ are fixed numbers}\}$
      (4) $\{f | f'(\alpha) = 0, \text{ where } \alpha \text{ is a fixed number}\}$
      (5) $\{f | f \text{ is divisible by } (x - 1)\}$
      (6) $\{f | f(0) = f'(0) = f''(0) = 0\}$
      (7) $\{f | f(0) - \int_0^1 f(x)\, dx = 0\}$

7. In the space of $n \times n$ matrices with real coefficients $A = [a_{ij}]_{(nn)}$, we say that a matrix is upper triangular if $a_{ij} = 0$ for $i > j$. Show that the upper triangular matrices form a subspace of the $n \times n$ matrices.

8. Show that the symmetric matrices form a subspace of the space of $n \times n$ matrices. Do the Hermitian matrices form a subspace of the $n \times n$ matrices with complex entries?

9. Show that the diagonal matrices form a subspace of the space of $n \times n$ matrices.

10. Let V be a vector space over the reals and H be a subset of V. Show that (a), (b), and (c) are equivalent.
    (a) H is a subspace.
    (b) If $\boldsymbol{x}, \boldsymbol{y} \in H$, then $\boldsymbol{x} + \boldsymbol{y} \in H$.
    If $\boldsymbol{x} \in H$, then $-\boldsymbol{x} \in H$.
    If $\boldsymbol{x} \in H$ and $\alpha \geqslant 0$ is a real scalar then $\alpha\boldsymbol{x} \in H$.
    (c) If $\boldsymbol{x}, \boldsymbol{y} \in H$ and β is a scalar then $\boldsymbol{x} + \beta\boldsymbol{y} \in H$.

11. Let M_{nn} denote the vector space of $n \times n$ matrices with real entries.
    (a) Show that the subset of invertible matrices is not a subspace of M_{nn}.
    (b) Show that the subset of noninvertible matrices is closed under scalar multiplication, i.e., a scalar multiple of a noninvertible matrix is noninvertible, but that the subset of noninvertible matrices is not a subspace of M_{nn}.

12. Let B be a fixed matrix in M_{nn}. Show that the following subsets of M_{nn} are susbspaces.

    (a) $\{A | A \in M_{nn} \text{ and } AB = BA\}$ (b) $\{A | A \in M_{nn} \text{ and } AB + BA = 0\}$
    (c) $\{A | A \in M_{nn} \text{ and } AB = 0\}$ (d) $\{A | A \in M_{nn} \text{ and } BA = 0\}$

    Give examples to show that (c) and (d) do not necessarily represent the same subspace.

13. Let $a_1, a_2, a_3, \ldots, a_n$ be real numbers. Show that the vectors

$$\begin{bmatrix} x_1 \\ x_2 \\ \vdots \\ x_n \end{bmatrix}$$

    in $\boldsymbol{R}^n$, such that $a_1x_1 + a_2x_2 + \cdots + a_nx_n = 0$ form a subspace of $\boldsymbol{R}^n$. Why is this a special case of example 2 in the text?

14. Let V be a vector space and let H and K be subspaces of V. Show that the set $H \cap K = \{\boldsymbol{x} | \boldsymbol{x} \in H \text{ and } \boldsymbol{x} \in K\}$ is a subspace of V.

15. Let V be a vector space and let H and K be subspaces of V. Show that the set of $H + K = \{\boldsymbol{x} | \boldsymbol{x} = \boldsymbol{h} + \boldsymbol{k}, \text{ where } \boldsymbol{h} \in H \text{ and } \boldsymbol{k} \in K\}$ is a subspace of V.

16. Prove by induction: If H is a subspace of a vector space V, and $\boldsymbol{x}_1, \boldsymbol{x}_2, \ldots, \boldsymbol{x}_n$ belong to H, while $\alpha_1, \alpha_2, \ldots, \alpha_n$ are scalars, then $\alpha_1\boldsymbol{x}_1 + \alpha_2\boldsymbol{x}_2 + \cdots + \alpha_n\boldsymbol{x}_n$ belongs to H.

4 SPAN

Let V be a vector space, $\boldsymbol{x}_1, \boldsymbol{x}_2, \boldsymbol{x}_3, \ldots, \boldsymbol{x}_n$ be vectors in V, and $\alpha_1, \alpha_2, \ldots, \alpha_n$ be scalars. If a vector $\boldsymbol{y}$ in V can be written in the form

$\boldsymbol{y} = \alpha_1\boldsymbol{x}_1 + \alpha_2\boldsymbol{x}_2 + \cdots + \alpha_n\boldsymbol{x}_n$, then $\boldsymbol{y}$ is said to be a **linear combination** of $\boldsymbol{x}_1, \boldsymbol{x}_2, \ldots, \boldsymbol{x}_n$. For example, in P_2, the polynomials of degree less than or equal to 2, any element can be expressed as a linear combination of 1, x, and x^2. For we have, $p(x) = \alpha_0 + \alpha_1 x + \alpha_2 x^2$. In $\boldsymbol{R}^3$, if a vector $\boldsymbol{v}$ ends at the point (x, y, z), we have $\boldsymbol{v} = x\boldsymbol{i} + y\boldsymbol{j} + z\boldsymbol{k}$, and so $\boldsymbol{v}$ is a linear combination of $\boldsymbol{i}$, $\boldsymbol{j}$, and $\boldsymbol{k}$.

Using the idea of linear combination, we may characterize subspaces as follows: A subset H of a vector space V is a subspace if and only if whenever $\boldsymbol{x}_1, \boldsymbol{x}_2, \ldots, \boldsymbol{x}_n$ belongs to H, any linear combination of $\boldsymbol{x}_1, \boldsymbol{x}_2, \ldots, \boldsymbol{x}_n$ also belongs to H.

There is a natural way to generate subspaces of a vector space V using linear combinations. If S is a subset of a vector space V, we define the **span** of S, written $\text{sp}(S)$, to be the set of vectors, each of which can be written as a linear combination of vectors in S.

For example, in $\boldsymbol{R}^3$, the span of the set

$$\left\{\begin{bmatrix}1\\0\\0\end{bmatrix}, \begin{bmatrix}0\\1\\0\end{bmatrix}\right\}$$

consists of all vectors of the form

$$\alpha\begin{bmatrix}1\\0\\0\end{bmatrix} + \beta\begin{bmatrix}0\\1\\0\end{bmatrix} = \begin{bmatrix}\alpha\\\beta\\0\end{bmatrix}$$

In this case the span is precisely that subset of $\boldsymbol{R}^3$ consisting of vectors whose third component is 0. If we consider the span of the set

$$\left\{\begin{bmatrix}1\\0\\0\end{bmatrix}, \begin{bmatrix}0\\1\\0\end{bmatrix}, \begin{bmatrix}1\\1\\0\end{bmatrix}\right\}$$

we find it consists of precisely the same collection of vectors. Indeed, any linear combination of the above three vectors has third component 0. It is clear, on the other hand, that any vector with third component 0 can be expressed as a linear combination of just the first two vectors in the given set and, thus, necessarily as a linear combination of all three vectors.

To elucidate the geometric meaning of the concept of span, we characterize the span of two noncollinear vectors $\boldsymbol{a}$ and $\boldsymbol{b}$ in $\boldsymbol{R}^3$. We will show that the span of $\boldsymbol{a}$ and $\boldsymbol{b}$ consists of those vectors that lie in the plane that passes the origin and is determined by the vectors $\boldsymbol{a}$ and $\boldsymbol{b}$. (See Figure 4-5.)

To begin with we note that, by the geometric meaning of vector addition and scalar multiplication, any linear combination of $\boldsymbol{a}$ and $\boldsymbol{b}$ must lie in the plane determined by $\boldsymbol{a}$ and $\boldsymbol{b}$. For what is the vector $\alpha\boldsymbol{a} + \beta\boldsymbol{b}$? It is clear that $\alpha\boldsymbol{a}$ and $\beta\boldsymbol{b}$, each lying on the line through the origin determined by the vectors $\boldsymbol{a}$ and $\boldsymbol{b}$, respectively, each lie in the plane through the origin determined by the vectors $\boldsymbol{a}$ and $\boldsymbol{b}$. Since $\alpha\boldsymbol{a} + \beta\boldsymbol{b}$ lies in any plane in which $\alpha\boldsymbol{a}$ and $\beta\boldsymbol{b}$ both lie, $\alpha\boldsymbol{a} + \beta\boldsymbol{b}$ lies in the plane determined by the vectors $\boldsymbol{a}$ and $\boldsymbol{b}$.

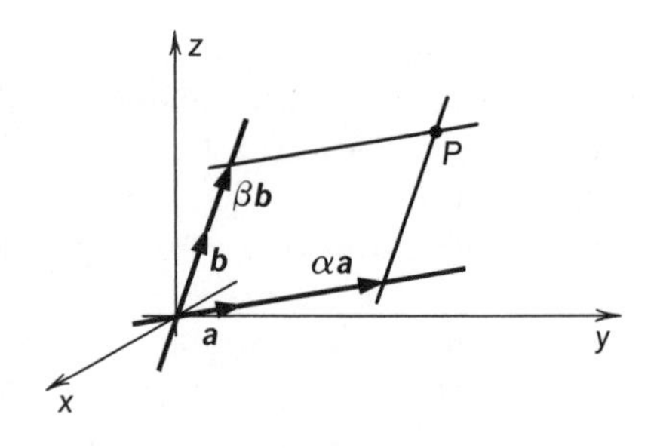

FIGURE 4-5

On the other hand let P be a point in the plane determined by the noncollinear vectors $\boldsymbol{a}$ and $\boldsymbol{b}$. In this plane, construct a line l_a, passing through the point P and parallel to the directed line segment associated with the vector $\boldsymbol{a}$. Since $\boldsymbol{a}$ and $\boldsymbol{b}$ are noncollinear, l_a is not parallel to the line generated by $\boldsymbol{b}$ and therefore intersects it in some point, say the endpoint of the vector $\beta\boldsymbol{b}$. Likewise, if we construct a line l_b through the point P and parallel to the vector $\boldsymbol{b}$, we see that it intersects the line generated by the vector $\boldsymbol{a}$ at the endpoint of some vector $\alpha\boldsymbol{a}$. From this construction P is necessarily the endpoint of the diagonal of a parallelogram having adjacent sides represented by the vectors $\alpha\boldsymbol{a}$ and $\beta\boldsymbol{b}$. Thus, if $\boldsymbol{v}$ is the vector that ends at P, $\boldsymbol{v} = \alpha\boldsymbol{a} + \beta\boldsymbol{b}$. Hence, any vector lying in the plane determined by $\boldsymbol{a}$ and $\boldsymbol{b}$ is a linear combination of $\boldsymbol{a}$ and $\boldsymbol{b}$.

The reader may have, perhaps, noted the similarity between this procedure and that for constructing Cartesian coordinates in the plane. From the foregoing examples we perceive an important property of sp(S).

Theorem If V is a vector space and S is a subset of V, then sp(S) is a subspace of V.

PROOF Suppose $\boldsymbol{x}$ and $\boldsymbol{y}$ belong to sp(S). Then

$$\boldsymbol{x} = \alpha_1\boldsymbol{x}_1 + \alpha_2\boldsymbol{x}_2 + \cdots + \alpha_n\boldsymbol{x}_n$$
$$\boldsymbol{y} = \beta_1\boldsymbol{y}_1 + \beta_2\boldsymbol{y}_2 + \cdots + \beta_m\boldsymbol{y}_m$$

where $\alpha_1, \alpha_2, \ldots, \alpha_n, \beta_1, \beta_2, \ldots, \beta_m$ are scalars and $\boldsymbol{x}_1, \boldsymbol{x}_2, \ldots, \boldsymbol{x}_n, \boldsymbol{y}_1, \boldsymbol{y}_2, \ldots, \boldsymbol{y}_m$ are vectors in S. Then

$$\boldsymbol{x} + \boldsymbol{y} = \alpha_1\boldsymbol{x}_1 + \cdots + \alpha_n\boldsymbol{x}_n + \beta_1\boldsymbol{y}_1 + \cdots + \beta_m\boldsymbol{y}_m$$

Thus, $\boldsymbol{x} + \boldsymbol{y}$ is a linear combination of vectors in S and so $\boldsymbol{x} + \boldsymbol{y}$ belongs to sp(S). Moreover, if α is a scalar, we have

$$\alpha\boldsymbol{x} = \alpha(\alpha_1\boldsymbol{x}_1 + \cdots + \alpha_n\boldsymbol{x}_n) = (\alpha\alpha_1)\boldsymbol{x}_1 + \cdots + (\alpha\alpha_n)\boldsymbol{x}_n$$

Thus $\alpha\boldsymbol{x}$, being a linear combination of vectors in S, necessarily belongs to sp(S). ■

If we apply this theorem to a previous example, we see that the totality of vectors lying in some plane through the origin constitutes a subspace of $\boldsymbol{R}^3$.

As another example, let P denote the space of all polynomials in the variable x with real coefficients. The span of the polynomials $1, x, x^2, \ldots, x^n$ in P is precisely P_n, the polynomials of degree less than or equal to n, which is, of course, a subspace of P.

If H is a subspace of a vector space V and $\boldsymbol{x}_1, \boldsymbol{x}_2, \ldots, \boldsymbol{x}_n$ is a collection of vectors in V, such that $\text{sp}(\boldsymbol{x}_1, \boldsymbol{x}_2, \ldots, \boldsymbol{x}_n) = H$, we say $\boldsymbol{x}_1, \boldsymbol{x}_2, \ldots, \boldsymbol{x}_n$ span H. Stated otherwise, the linear combinations of $\boldsymbol{x}_1, \boldsymbol{x}_2, \ldots, \boldsymbol{x}_n$ fill out the subspace H.

It is important to note that a given collection S of vectors that spans a subspace H of a vector space V may contain redundancies in the sense that it may be possible to delete certain members of S and obtain a subset with the same span. Consider, for example, the set of vectors $\{\boldsymbol{i}, \boldsymbol{k}, \boldsymbol{i} + \boldsymbol{k}\}$ in $\boldsymbol{R}^3$. (See Figure 4-6.) These vectors span the xz plane in $\boldsymbol{R}^3$. If, however, we delete any member of the set, we obtain a pair of noncollinear vectors in the same plane. By the preceding example, this pair of vectors again spans the xz plane.

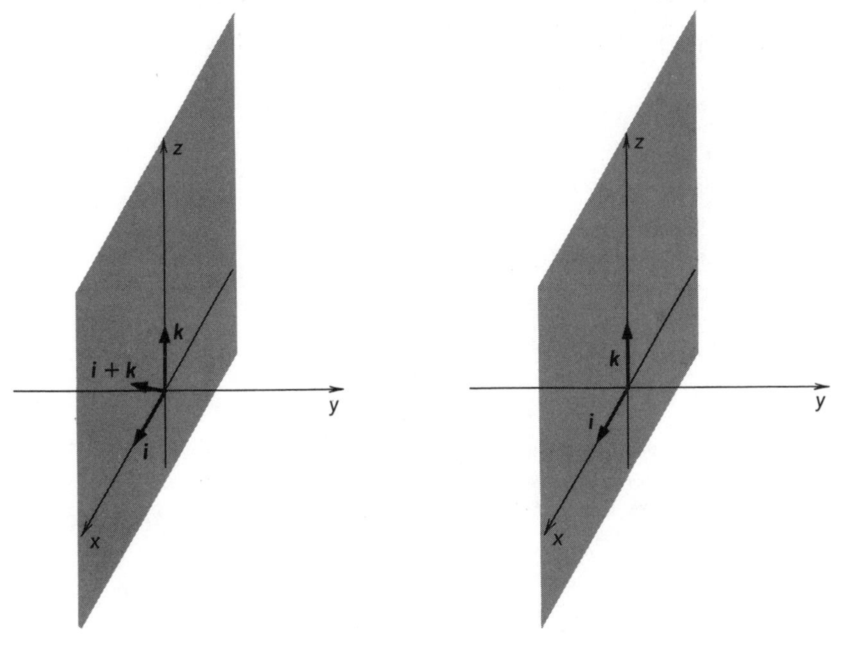

FIGURE 4-6

In the next section we investigate a condition on a given set of vectors $\{\boldsymbol{x}_1, \boldsymbol{x}_2, \ldots, \boldsymbol{x}_n\}$ that insures that these repetitions do not occur.

EXERCISES

1. Show that the following sets of vectors span the indicated vector spaces.

(a) $\begin{bmatrix} 1 \\ 0 \\ 0 \end{bmatrix}, \begin{bmatrix} 0 \\ 1 \\ 0 \end{bmatrix}, \begin{bmatrix} 0 \\ 0 \\ 1 \end{bmatrix}$; $\boldsymbol{R}^3$

(b) $\begin{bmatrix} 1 \\ 0 \\ 0 \end{bmatrix}, \begin{bmatrix} 1 \\ 1 \\ 0 \end{bmatrix}, \begin{bmatrix} 2 \\ 1 \\ 1 \end{bmatrix}$; $\boldsymbol{R}^3$

(c) $\begin{bmatrix} 1 \\ -1 \end{bmatrix}, \begin{bmatrix} 1 \\ 1 \end{bmatrix}, \begin{bmatrix} 2 \\ 1 \end{bmatrix}$; $\boldsymbol{R}^2$

(d) $\begin{bmatrix} 1 \\ 1 \end{bmatrix}, \begin{bmatrix} -1 \\ 2 \end{bmatrix}$; $\boldsymbol{R}^2$

(e) $1, x, x^2$; P_2

(f) $1, (1+x), (1+x)^2$; P_2

(g) $\begin{bmatrix} 1 & 0 \\ 0 & 0 \end{bmatrix}, \begin{bmatrix} 0 & 0 \\ 0 & 1 \end{bmatrix}, \begin{bmatrix} 0 & 1 \\ 1 & 0 \end{bmatrix}$

space of 2×2 symmetric matrices.

(h) $\begin{bmatrix} 0 & 1 \\ -1 & 0 \end{bmatrix}$

space of 2×2 skew-symmetric matrices.

(i) $1+x, 1+2x, 1+x^2, 1+2x^2$; P_2

(j) $\begin{bmatrix} 1 & 0 & 0 \\ 0 & 1 & 0 \\ 0 & 0 & 1 \end{bmatrix}, \begin{bmatrix} 1 & 0 & 0 \\ 0 & -1 & 0 \\ 0 & 0 & 0 \end{bmatrix}, \begin{bmatrix} 1 & 0 & 0 \\ 0 & 1 & 0 \\ 0 & 0 & -1 \end{bmatrix}$

space of 3×3 diagonal matrices.

2. Find two vectors that span the indicated planes in $\boldsymbol{R}^3$.
   (a) The plane $x = 0$.
   (b) The plane $x = y$.
   (c) The plane $y = z$.

3. Find n vectors that span $\boldsymbol{R}^n$.

4. Find four elements that span the space of 2×2 matrices.

5. Let $\{\boldsymbol{x}_1, \boldsymbol{x}_2, \ldots, \boldsymbol{x}_n\}$ be a set of vectors in a vector space V. If $\boldsymbol{y}_1, \boldsymbol{y}_2, \ldots, \boldsymbol{y}_n$ belong to $\text{sp}(\boldsymbol{x}_1, \boldsymbol{x}_2, \ldots, \boldsymbol{x}_n)$ and $\boldsymbol{y}_1, \boldsymbol{y}_2, \ldots, \boldsymbol{y}_n$ span V, show that $\{\boldsymbol{x}_1, \ldots, \boldsymbol{x}_n\}$ also span V.

6. Let V be a vector space and $\boldsymbol{x}_1, \boldsymbol{x}_2, \ldots, \boldsymbol{x}_n$ be vectors in V. Show that $\text{sp}(\boldsymbol{x}_1, \boldsymbol{x}_2, \ldots, \boldsymbol{x}_n)$ is the smallest subspace of V containing $\boldsymbol{x}_1, \boldsymbol{x}_2, \ldots, \boldsymbol{x}_n$. In other words, show that if H is a subspace of V containing $\boldsymbol{x}_1, \boldsymbol{x}_2, \ldots, \boldsymbol{x}_n$, then H also contains $\text{sp}(\boldsymbol{x}_1, \boldsymbol{x}_2, \ldots, \boldsymbol{x}_n)$.

7. Let V be a vector space and $\boldsymbol{x}_1, \boldsymbol{x}_2, \boldsymbol{y}_1, \boldsymbol{y}_2$ be vectors in V. If

$$\boldsymbol{x}_1 = \alpha_1 \boldsymbol{y}_1 + \alpha_2 \boldsymbol{y}_2 \quad \boldsymbol{x}_2 = \beta_1 \boldsymbol{y}_1 + \beta_2 \boldsymbol{y}_2$$

where

$$\begin{vmatrix} \alpha_1 & \alpha_2 \\ \beta_1 & \beta_2 \end{vmatrix} \neq 0$$

show that $\text{sp}(\boldsymbol{x}_1, \boldsymbol{x}_2) = \text{sp}(\boldsymbol{y}_1, \boldsymbol{y}_2)$.

8. If $A \subset B$ are subsets of a vector space V, show that $\text{sp}(A) \subset \text{sp}(B)$.

9. If $\boldsymbol{y}$ belongs to $\text{sp}(\boldsymbol{x}_1, \boldsymbol{x}_2, \ldots, \boldsymbol{x}_n, \boldsymbol{z})$ but $\boldsymbol{y}$ does not belong to $\text{sp}(\boldsymbol{x}_1, \boldsymbol{x}_2, \ldots, \boldsymbol{x}_n)$, show that $\boldsymbol{z}$ belongs to $\text{sp}(\boldsymbol{x}_1, \boldsymbol{x}_2, \ldots, \boldsymbol{x}_n, \boldsymbol{y})$.

10. Show that

$$\text{sp}(\boldsymbol{x}_1, \boldsymbol{x}_2, \ldots, \boldsymbol{x}_n, \boldsymbol{y}) = \text{sp}(\boldsymbol{x}_1, \boldsymbol{x}_2, \ldots, \boldsymbol{x}_n)$$

if and only if $\boldsymbol{y}$ is a linear combination of $\boldsymbol{x}_1, \boldsymbol{x}_2, \ldots, \boldsymbol{x}_n$.

11. Show that the invertible 2×2 matrices span the space of 2×2 matrices. Show that the noninvertible 2×2 matrices span the space of 2×2 matrices.

12. Show that the matrices

$$\begin{bmatrix} 1 & 1 \\ 0 & 1 \end{bmatrix}, \quad \begin{bmatrix} -1 & 1 \\ 0 & -1 \end{bmatrix}, \quad \begin{bmatrix} 0 & 1 \\ 0 & 0 \end{bmatrix}$$

do not span the space of 2×2 matrices.

13. Show that the vector space P of all polynomials cannot be spanned by a finite number of elements.

14. Show that the matrices of the form $AB - BA$ do not span the space of $n \times n$ matrices.

15. Is it possible to span the space of $n \times n$ matrices using the powers of a single matrix A, i.e., $I_n, A, A^2, \ldots, A^n, \ldots$?

5 LINEAR INDEPENDENCE

Closely related to the notion of span discussed in the preceding section is the concept of linear independence.

A collection $\{\boldsymbol{x}_1, \boldsymbol{x}_2, \ldots, \boldsymbol{x}_n\}$ of vectors in a vector space is said to be **linearly independent** if whenever

$$\alpha_1\boldsymbol{x}_1 + \alpha_2\boldsymbol{x}_2 + \cdots + \alpha_n\boldsymbol{x}_n = \boldsymbol{0}$$

we must have $\alpha_1 = \alpha_2 = \cdots = \alpha_n = 0$. In other words, the **only** linear combination of $\boldsymbol{x}_1, \boldsymbol{x}_2, \ldots, \boldsymbol{x}_n$ that vanishes is the obvious one:

$$0 \cdot \boldsymbol{x}_1 + 0 \cdot \boldsymbol{x}_2 + \cdots + 0 \cdot \boldsymbol{x}_n = \boldsymbol{0}$$

For example, in $\boldsymbol{R}^2$, the vectors

$$\begin{bmatrix} 1 \\ 0 \end{bmatrix} \quad \text{and} \quad \begin{bmatrix} 0 \\ 1 \end{bmatrix}$$

are linearly independent. For suppose

$$\alpha\begin{bmatrix} 1 \\ 0 \end{bmatrix} + \beta\begin{bmatrix} 0 \\ 1 \end{bmatrix} = \begin{bmatrix} 0 \\ 0 \end{bmatrix}$$

Then

$$\begin{bmatrix} \alpha \\ \beta \end{bmatrix} = \begin{bmatrix} 0 \\ 0 \end{bmatrix}$$

and so $\alpha = 0$ and $\beta = 0$. We have shown, therefore, that if

$$\alpha\begin{bmatrix} 1 \\ 0 \end{bmatrix} + \beta\begin{bmatrix} 0 \\ 1 \end{bmatrix} = \begin{bmatrix} 0 \\ 0 \end{bmatrix}$$

then $\alpha = 0$ and $\beta = 0$. Thus, the vectors

$$\begin{bmatrix} 1 \\ 0 \end{bmatrix} \quad \text{and} \quad \begin{bmatrix} 0 \\ 1 \end{bmatrix}$$

are linearly independent.

As another example, in P_2, the polynomials with real coefficients of degree less than or equal to 2, the vectors 1, $1 + x$, $(1 + x)^2$ are independent. For suppose,

$$\alpha \cdot 1 + \beta(1 + x) + \gamma(1 + x)^2 = 0$$

Then

$$\alpha \cdot 1 + \beta \cdot 1 + \beta x + \gamma \cdot 1 + 2\gamma x + \gamma x^2 = 0$$

or

$$(\alpha + \beta + \gamma) + (\beta + 2\gamma)x + \gamma x^2 = 0$$

Since a polynomial is zero only if all its coefficients are zero, we must have

$$\begin{aligned} \alpha + \beta + \gamma &= 0 \\ \beta + 2\gamma &= 0 \\ \gamma &= 0 \end{aligned}$$

This immediately implies $\alpha = \beta = \gamma = 0$. Having shown that $\alpha \cdot 1 + \beta(1 + x) + \gamma(1 + x)^2 = 0$ implies $\alpha = \beta = \gamma = 0$, we may conclude that $\{1, 1 + x, (1 + x)^2\}$ is a linearly independent set of vectors.

If a set of vectors is not linearly independent, we say it is **linearly dependent**. Thus, if $\{\boldsymbol{x}_1, \boldsymbol{x}_2, \ldots, \boldsymbol{x}_n\}$ is a linearly dependent set of vectors, there are scalars $\alpha_1, \alpha_2, \ldots, \alpha_n$, *not all zero*, such that $\alpha_1\boldsymbol{x}_1 + \alpha_2\boldsymbol{x}_2 + \cdots + \alpha_n\boldsymbol{x}_n = \boldsymbol{0}$.

Thus, in $\boldsymbol{R}^2$, the vectors

$$\begin{bmatrix} 1 \\ 1 \end{bmatrix}, \quad \begin{bmatrix} 1 \\ 0 \end{bmatrix}, \quad \begin{bmatrix} 0 \\ -2 \end{bmatrix}$$

are linearly dependent, for we have

$$(1)\begin{bmatrix} 1 \\ 1 \end{bmatrix} + (-1)\begin{bmatrix} 1 \\ 0 \end{bmatrix} + \tfrac{1}{2}\begin{bmatrix} 0 \\ -2 \end{bmatrix} = \begin{bmatrix} 0 \\ 0 \end{bmatrix}$$

As another example, in $\boldsymbol{R}^3$, the vectors

$$\begin{bmatrix} 1 \\ -3 \\ 7 \end{bmatrix}, \quad \begin{bmatrix} 2 \\ 0 \\ 1 \end{bmatrix}, \quad \begin{bmatrix} 6 \\ -6 \\ 16 \end{bmatrix}$$

are dependent, since

$$(-2)\begin{bmatrix} 1 \\ -3 \\ 7 \end{bmatrix} + (-2)\begin{bmatrix} 2 \\ 0 \\ 1 \end{bmatrix} + (+1)\begin{bmatrix} 6 \\ -6 \\ 16 \end{bmatrix} = \begin{bmatrix} 0 \\ 0 \\ 0 \end{bmatrix}$$

The following evinces the relationship between span and linear independence.

Theorem Let $\{\boldsymbol{x}_1, \boldsymbol{x}_2, \ldots, \boldsymbol{x}_n\}$ be a collection of vectors in a vector space V. Then $\{\boldsymbol{x}_1, \boldsymbol{x}_2, \ldots, \boldsymbol{x}_n\}$ is a linearly dependent set of vectors if and only if one of the vectors is a linear combination of the remaining vectors in the set.

PROOF Suppose $\{\boldsymbol{x}_1, \boldsymbol{x}_2, \ldots, \boldsymbol{x}_n\}$ is linearly dependent. Then $\alpha_1\boldsymbol{x}_1 + \cdots + \alpha_n\boldsymbol{x}_n = \boldsymbol{0}$, where $\alpha_1, \alpha_2, \ldots, \alpha_n$ are not all zero. Suppose $\alpha_i \neq 0$. Then

$$\alpha_i\boldsymbol{x}_i = (-\alpha_1)\boldsymbol{x}_1 + \cdots + (-\alpha_{i-1}\boldsymbol{x}_{i-1}) + (-\alpha_{i+1}\boldsymbol{x}_{i+1}) + \cdots + (-\alpha_n\boldsymbol{x}_n)$$

$$\boldsymbol{x}_i = (-\alpha_i^{-1}\alpha_1)\boldsymbol{x}_1 + \cdots + (-\alpha_i^{-1}\alpha_{i-1}\boldsymbol{x}_{i-1}) + (-\alpha_i^{-1}\alpha_{i+1}\boldsymbol{x}_{i+1})$$

$$+ \cdots + (-\alpha_i^{-1}\alpha_n)\boldsymbol{x}_n$$

Hence, $\boldsymbol{x}_i$ is a linear combination of $\boldsymbol{x}_1, \ldots, \boldsymbol{x}_{i-1}, \boldsymbol{x}_{i+1}, \ldots, \boldsymbol{x}_n$.

On the other hand, suppose one of the vectors, say $\boldsymbol{x}_i$, is a linear combination of $\{\boldsymbol{x}_1, \ldots, \boldsymbol{x}_{i-1}, \boldsymbol{x}_{i+1}, \ldots, \boldsymbol{x}_n\}$. Then

$$\boldsymbol{x}_i = \beta_1\boldsymbol{x}_1 + \cdots + \beta_{i-1}\boldsymbol{x}_{i-1} + \beta_{i+1}\boldsymbol{x}_{i+1} + \cdots + \beta_n\boldsymbol{x}_n$$

or

$$\beta_1\boldsymbol{x}_1 + \cdots + \beta_{i-1}\boldsymbol{x}_{i-1} + (-1)\boldsymbol{x}_i + \beta_{i+1}\boldsymbol{x}_{i+1} + \cdots + \beta_n\boldsymbol{x}_n = \boldsymbol{0}$$

Since the coefficient of $\boldsymbol{x}_i$ is $-1 \neq 0$, the vectors $\{\boldsymbol{x}_1, \boldsymbol{x}_2, \ldots, \boldsymbol{x}_n\}$ are linearly dependent. ■

For example, in P_2, the polynomials 1, $1 + x$, $(1 + x)^2$, and x^2 are linearly dependent, since $(1 + x)^2 = 1 \cdot x^2 + 2 \cdot (1 + x) + (-1) \cdot 1$.

Let us now consider the geometric meaning of linear independence. First, suppose that $\boldsymbol{a}$ and $\boldsymbol{b}$ are two linearly dependent vectors in $\boldsymbol{R}^3$. By the last theorem, one is a scalar multiple of the other. Thus, $\boldsymbol{a}$ and $\boldsymbol{b}$ both lie on some line through the origin. On the other hand, if $\boldsymbol{a}$ and $\boldsymbol{b}$ both lie on some line through the origin, one is a scalar multiple of the other. By the last theorem, $\boldsymbol{a}$ and $\boldsymbol{b}$ are linearly dependent. Thus, $\boldsymbol{a}$ and $\boldsymbol{b}$ are linearly independent if and only if they are noncollinear. (See Figure 4-7.)

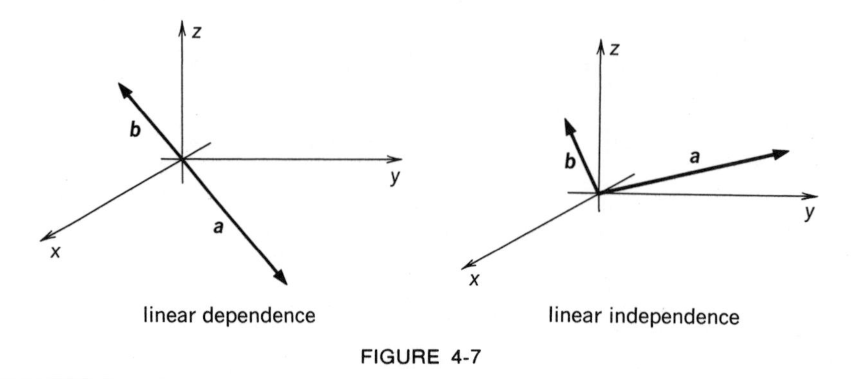

FIGURE 4-7

Thus, for example, the pairs of vectors $\{\boldsymbol{i}-\boldsymbol{j}+\boldsymbol{k}, \boldsymbol{i}+\boldsymbol{j}\}$ and $\{\boldsymbol{i}-\boldsymbol{j}, \boldsymbol{j}+\boldsymbol{k}\}$ are linearly independent. (See Figure 4-8.) On the other hand, the pairs of vectors $\{\boldsymbol{i}+\boldsymbol{j}, -\boldsymbol{i}-\boldsymbol{j}\}$ and $\{\boldsymbol{i}+2\boldsymbol{j}+\boldsymbol{k}, 2\boldsymbol{i}+4\boldsymbol{j}+2\boldsymbol{k}\}$ are linearly dependent. (See Figure 4-9.)

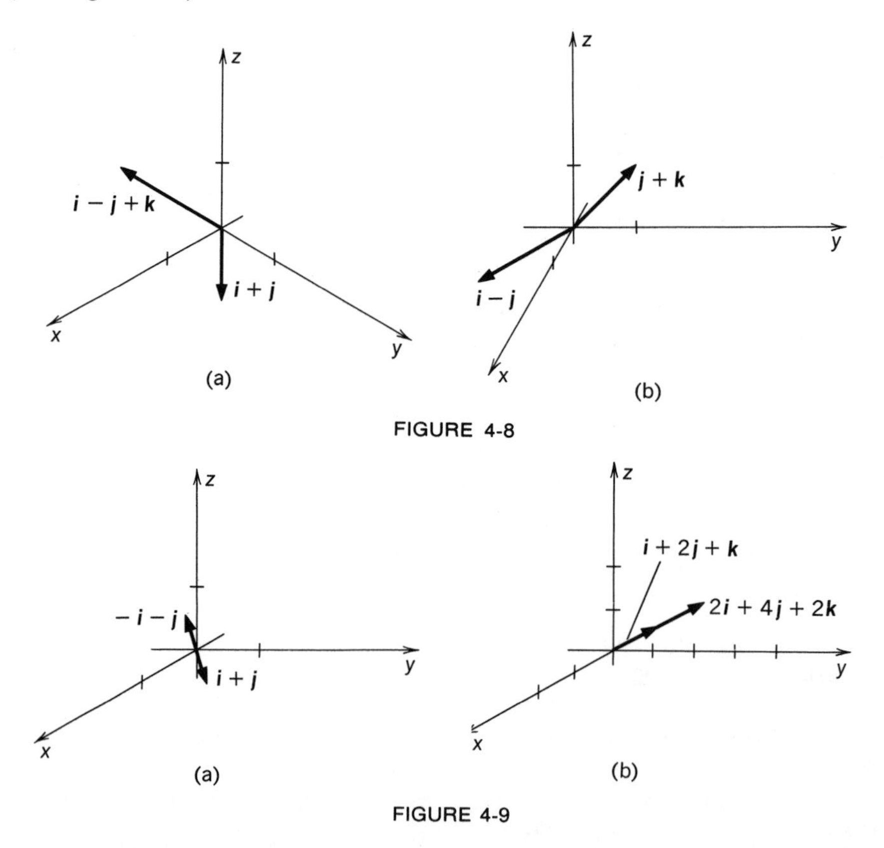

FIGURE 4-9

There is also a geometric criterion for visualizing whether a set of three vectors in $\boldsymbol{R}^3$ is linearly dependent: A set of three vectors in $\boldsymbol{R}^3$ is linearly dependent if and only if they all lie in some plane passing through the origin. We see this as follows.

Suppose we are given three linearly dependent vectors $\boldsymbol{x}_1$, $\boldsymbol{x}_2$, and $\boldsymbol{x}_3$ in $\boldsymbol{R}^3$. (See Figure 4-10.) We must show they all lie in some plane. By the theorem of this section, one of the vectors, say $\boldsymbol{x}_1$, is a linear combination of $\boldsymbol{x}_2$ and $\boldsymbol{x}_3$, i.e., $\boldsymbol{x}_1 = \alpha_2\boldsymbol{x}_2 + \alpha_3\boldsymbol{x}_3$. There are two cases. Either $\boldsymbol{x}_2$ and $\boldsymbol{x}_3$ are collinear or they are not collinear. If $\boldsymbol{x}_2$ and $\boldsymbol{x}_3$ are not collinear, we have seen in an example of the previous section that they span a plane through the origin. Since $\boldsymbol{x}_1$ is a linear combination of $\boldsymbol{x}_2$ and $\boldsymbol{x}_3$, it necessarily lies on the plane spanned by $\boldsymbol{x}_2$ and $\boldsymbol{x}_3$. Thus, $\boldsymbol{x}_1$, $\boldsymbol{x}_2$, and $\boldsymbol{x}_3$ all lie in some plane through the origin. If $\boldsymbol{x}_2$ and $\boldsymbol{x}_3$ are collinear, they span a line through the origin. $\boldsymbol{x}_1$, being a linear combination of $\boldsymbol{x}_2$ and $\boldsymbol{x}_3$, necessarily lies on this line. Since $\boldsymbol{x}_1$, $\boldsymbol{x}_2$, and $\boldsymbol{x}_3$ all lie on some line, they necessarily lie in some plane, indeed, any plane containing the line in which they lie, and are necessarily coplanar.

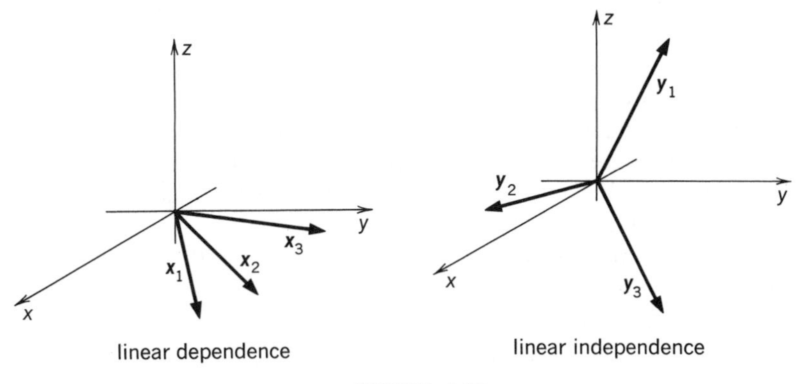

FIGURE 4-10

Next, we see that if $\boldsymbol{x}_1$, $\boldsymbol{x}_2$, and $\boldsymbol{x}_3$ all lie in some plane, they are linearly dependent. If two of the vectors, say $\boldsymbol{x}_2$ and $\boldsymbol{x}_3$, are not collinear, by the example of the previous section, they span the plane in which they lie. But by hypothesis $\boldsymbol{x}_1$ belongs to this plane, and is, therefore, a linear combination of $\boldsymbol{x}_2$ and $\boldsymbol{x}_3$. Thus, $\boldsymbol{x}_1$, $\boldsymbol{x}_2$, and $\boldsymbol{x}_3$ are linearly dependent. If, on the other hand, each pair of vectors in the set $\{\boldsymbol{x}_1, \boldsymbol{x}_2, \boldsymbol{x}_3\}$ is collinear, each vector must lie on some line passing through the origin. Thus, each is a scalar multiple of a single vector, and so $\boldsymbol{x}_1$, $\boldsymbol{x}_2$, and $\boldsymbol{x}_3$ are linearly dependent.

Using this criterion we see that the sets of vectors $\{\boldsymbol{i}, \boldsymbol{i}+\boldsymbol{j}, \boldsymbol{i}+\boldsymbol{j}+\boldsymbol{k}\}$, $\{\boldsymbol{i}+\boldsymbol{j}, -\boldsymbol{i}+\boldsymbol{j}, -\boldsymbol{j}+\boldsymbol{k}\}$, and $\{\boldsymbol{i}+\boldsymbol{j}-\boldsymbol{k}, \boldsymbol{i}-\boldsymbol{j}-\boldsymbol{k}, \boldsymbol{k}\}$ are independent (see Figure 4-11) while the sets $\{\boldsymbol{i}+\boldsymbol{j}, \boldsymbol{i}-\boldsymbol{j}, \boldsymbol{i}\}$ and $\{\boldsymbol{i}+\boldsymbol{j}+\boldsymbol{k}, -\boldsymbol{i}+\boldsymbol{j}+\boldsymbol{k}, -\boldsymbol{j}-\boldsymbol{k}\}$ are dependent (see Figure 4-12).

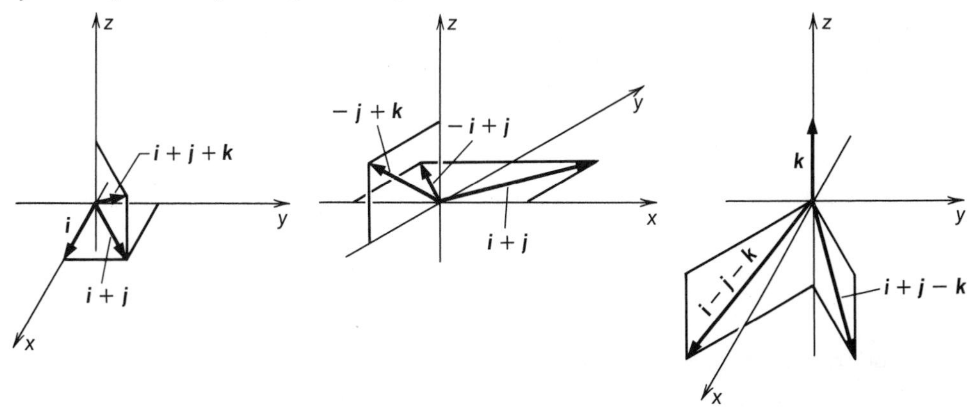

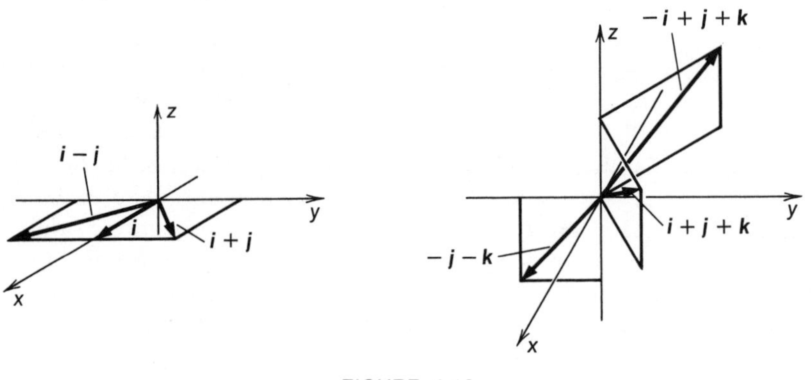

FIGURE 4-12

To recapitulate, three vectors are linearly dependent if they are coplanar. Two vectors are linearly dependent if they are collinear. We point this out primarily to facilitate geometric intuition. In attempting to solve specific problems involving linear independence, it is probably easier to solve linear equations than it is to draw pictures.

Our next example is an important illustration of the concepts of span and linear independence. In $\boldsymbol{R}^n$ we consider $\{\boldsymbol{e}_1, \boldsymbol{e}_2, \ldots, \boldsymbol{e}_n\}$, where $\boldsymbol{e}_i$ is the vector that has zeros in all components except the ith and a 1 in the ith component:

$$\mathbf{e}_i = \begin{bmatrix} 0 \\ \vdots \\ 0 \\ 1 \\ 0 \\ \vdots \\ 0 \end{bmatrix} \left.\begin{matrix} \\ \\ \\ \\ \end{matrix}\right\} i$$

First, we observe that these vectors are linearly independent. For suppose $\alpha_1\boldsymbol{e}_1 + \alpha_2\boldsymbol{e}_2 + \cdots + \alpha_n\boldsymbol{e}_n = \mathbf{0}$. Then

$$\alpha_1\begin{bmatrix} 1 \\ 0 \\ \vdots \\ 0 \end{bmatrix} + \alpha_2\begin{bmatrix} 0 \\ 1 \\ \vdots \\ 0 \end{bmatrix} + \cdots + \alpha_n\begin{bmatrix} 0 \\ 0 \\ \vdots \\ 1 \end{bmatrix} = \begin{bmatrix} 0 \\ 0 \\ \vdots \\ 0 \end{bmatrix}$$

or

$$\begin{bmatrix} \alpha_1 \\ \alpha_2 \\ \vdots \\ \alpha_n \end{bmatrix} = \begin{bmatrix} 0 \\ 0 \\ \vdots \\ 0 \end{bmatrix}$$

So $\alpha_1 = \alpha_2 = \cdots = \alpha_n = 0$.

It is also of interest to note that the vectors $e_1, e_2, \ldots, e_n$ span R^n. For if $x \in R^n$,

$$x = \begin{bmatrix} \alpha_1 \\ \alpha_2 \\ \vdots \\ \alpha_n \end{bmatrix} = \alpha_1 \begin{bmatrix} 1 \\ 0 \\ \vdots \\ 0 \end{bmatrix} + \alpha_2 \begin{bmatrix} 0 \\ 1 \\ \vdots \\ 0 \end{bmatrix} + \cdots + \alpha_n \begin{bmatrix} 0 \\ 0 \\ \vdots \\ 1 \end{bmatrix}$$

$$= \alpha_1 e_1 + \alpha_2 e_2 + \cdots + \alpha_n e_n$$

Since every vector x is a suitable linear combination of $e_1, e_2, \ldots, e_n$, we see that $e_1, e_2, \ldots, e_n$ span R^n.

EXERCISES

1. Which of the following sets of vectors in R^2 are linearly independent? Illustrate geometrically.

   (a) $\begin{bmatrix} -1 \\ 2 \end{bmatrix}, \begin{bmatrix} 0 \\ 1 \end{bmatrix}$ (b) $\begin{bmatrix} 3 \\ 1 \end{bmatrix}, \begin{bmatrix} 1 \\ 3 \end{bmatrix}$ (c) $\begin{bmatrix} 3 \\ 3 \end{bmatrix}, \begin{bmatrix} -5 \\ -5 \end{bmatrix}$ (d) $\begin{bmatrix} 1 \\ 1 \end{bmatrix}, \begin{bmatrix} 1 \\ -1 \end{bmatrix}$

   (e) $\begin{bmatrix} 1 \\ 2 \end{bmatrix}, \begin{bmatrix} 3 \\ 1 \end{bmatrix}, \begin{bmatrix} 2 \\ -2 \end{bmatrix}$ (f) $\begin{bmatrix} 3 \\ 0 \end{bmatrix}, \begin{bmatrix} 6 \\ 1 \end{bmatrix}, \begin{bmatrix} 4 \\ 2 \end{bmatrix}$ (g) $\begin{bmatrix} 3 \\ 1 \end{bmatrix}, \begin{bmatrix} 3 \\ 2 \end{bmatrix}, \begin{bmatrix} -1 \\ 1 \end{bmatrix}$

2. In the space of 2×2 matrices, M_{22}, which of the sets are independent?

   (a) $\begin{bmatrix} 1 & 0 \\ 0 & 1 \end{bmatrix} \begin{bmatrix} 1 & 1 \\ 1 & 1 \end{bmatrix} \begin{bmatrix} 0 & 1 \\ 1 & 0 \end{bmatrix}$ (b) $\begin{bmatrix} 1 & 0 \\ 0 & 1 \end{bmatrix} \begin{bmatrix} 0 & 1 \\ 0 & 0 \end{bmatrix} \begin{bmatrix} 0 & 0 \\ 1 & 0 \end{bmatrix} \begin{bmatrix} 0 & 0 \\ 0 & 1 \end{bmatrix}$

3. In the space, P_2, which of the sets are independent?

   (a) $1 + t, t, t^2$ (b) $1 + t^2, t^2, 3 + t^2$ (c) $1 + t, t + t^2, 1 + t^2$

4. Let V be a vector space and x, y, and z be linearly independent vectors of V. Show that $x + y$, $x + z$, and $y + z$ are linearly independent.

5. Which of the following sets of vectors in R^3 are linearly independent?

   (a) $\begin{bmatrix} 1 \\ 1 \\ 1 \end{bmatrix}, \begin{bmatrix} 0 \\ 1 \\ 1 \end{bmatrix}, \begin{bmatrix} 0 \\ 0 \\ 1 \end{bmatrix}$ (b) $\begin{bmatrix} 1 \\ 2 \\ 1 \end{bmatrix}, \begin{bmatrix} 0 \\ 2 \\ 3 \end{bmatrix}, \begin{bmatrix} 1 \\ 5 \\ -1 \end{bmatrix}$ (c) $\begin{bmatrix} 1 \\ -1 \\ 3 \end{bmatrix}, \begin{bmatrix} 0 \\ 1 \\ 2 \end{bmatrix}, \begin{bmatrix} -1 \\ 2 \\ -1 \end{bmatrix}$

6. Find four vectors in R^3 such that each subset of two vectors is linearly independent, while any subset of three vectors is linearly dependent.

7. Let $A_1, A_2, \ldots, A_l$ be $m \times n$ matrices and B be an $n \times p$ matrix. If A_1B, $A_2B, \ldots, A_lB$ are linearly independent in the space of $m \times p$ matrices, show that $A_1, A_2, \ldots, A_l$ are linearly independent in the space of $m \times n$ matrices.

8. If x and y are linearly independent vectors, show that the vectors

$$\alpha_1 x + \alpha_2 y \quad \text{and} \quad \beta_1 x + \beta_2 y$$

   are independent if and only if

$$\begin{vmatrix} \alpha_1 & \alpha_2 \\ \beta_1 & \beta_2 \end{vmatrix} \neq 0$$

9. Show that no linearly independent set of vectors can contain the zero vector.

10. Show that any subset of a linearly independent set of vectors is linearly independent.

11. Show that the vectors

$$\boldsymbol{e}_1, \boldsymbol{e}_1 + \boldsymbol{e}_2, \boldsymbol{e}_1 + \boldsymbol{e}_2 + \boldsymbol{e}_3, \ldots, \boldsymbol{e}_1 + \boldsymbol{e}_2 + \boldsymbol{e}_3 + \cdots + \boldsymbol{e}_n$$

are linearly independent in $\boldsymbol{R}^n$.

12. Let S be a collection of vectors in a vector space with the property that any subset of two elements is linearly dependent. Show that all vectors in S are scalar multiples of a single vector.

13. Let $\{\boldsymbol{x}_1, \boldsymbol{x}_2, \ldots, \boldsymbol{x}_n\}$ be a set of linearly independent vectors in a vector space V. Let

$$\boldsymbol{y} = \alpha_1\boldsymbol{x}_1 + \alpha_2\boldsymbol{x}_2 + \cdots + \alpha_n\boldsymbol{x}_n$$

What condition on the scalars α_i will guarantee that for each i, the vectors $\boldsymbol{x}_1, \boldsymbol{x}_2, \ldots, \boldsymbol{x}_{i-1}, \boldsymbol{y}, \boldsymbol{x}_{i+1}, \ldots, \boldsymbol{x}_n$ are linearly independent?

14. Let $X_1, X_2, \ldots, X_k$ be $m \times n$ matrices that are linearly independent in the space of $m \times n$ matrices. If A is an invertible $m \times m$ matrix and B is an invertible $n \times n$ matrix, show that the matrices $AX_1B, AX_2B, \ldots, AX_kB$ are linearly independent in the space of $m \times n$ matrices.

15. Let V be a vector space and $\boldsymbol{x}_1, \boldsymbol{x}_2, \ldots, \boldsymbol{x}_n$ be vectors in V. If $\boldsymbol{x}_1 \neq \boldsymbol{0}$, $\boldsymbol{x}_2 \notin \text{sp}(\{\boldsymbol{x}_1\})$, $\boldsymbol{x}_3 \notin \text{sp}(\{\boldsymbol{x}_1, \boldsymbol{x}_2\}), \ldots, \boldsymbol{x}_n \notin \text{sp}(\{\boldsymbol{x}_1, \ldots, \boldsymbol{x}_{n-1}\})$, show that the vectors $\boldsymbol{x}_1, \boldsymbol{x}_2, \ldots, \boldsymbol{x}_n$ are linearly independent.

16. Let f_1 and f_2 be two polynomials and suppose there are points x_1 and x_2 such that

$$f_1(x_1) = 1, \qquad f_2(x_1) = 0$$
$$f_1(x_2) = 0, \qquad f_2(x_2) = 1$$

Show that f_1 and f_2 are linearly independent in the space of all polynomials.

17. If $A \neq 0$ is a symmetric matrix and $B \neq 0$ is a skew-symmetric matrix in the space of $n \times n$ matrices, show that A and B are linearly independent.

18. If f and g are polynomials and

$$\begin{vmatrix} f(0) & g(0) \\ f'(0) & g'(0) \end{vmatrix} \neq 0$$

show that f and g are linearly independent in the space of polynomials.

6 BASIS

In the last example of the previous section, we noted that the set of vectors $\{\boldsymbol{e}_1, \boldsymbol{e}_2, \ldots, \boldsymbol{e}_n\}$ is linearly independent and spans $\boldsymbol{R}^n$. Such sets are very important in the theory of vector spaces, and hence we have the following:

Definition Let V be a vector space and $\{\boldsymbol{x}_1, \boldsymbol{x}_2, \ldots, \boldsymbol{x}_n\}$ be a collection of vectors in V. The set $\{\boldsymbol{x}_1, \boldsymbol{x}_2, \ldots, \boldsymbol{x}_n\}$ is said to be a **basis** for V if

(i) $\{\boldsymbol{x}_1, \boldsymbol{x}_2, \ldots, \boldsymbol{x}_n\}$ is a linearly independent set of vectors, and
(ii) $\{\boldsymbol{x}_1, \boldsymbol{x}_2, \ldots, \boldsymbol{x}_n\}$ spans V

According to this definition, the vectors $\boldsymbol{e}_1, \boldsymbol{e}_2, \ldots, \boldsymbol{e}_n$ form a basis for $\boldsymbol{R}^n$. This particular basis is of such frequent occurrence that it is called the **standard basis** for $\boldsymbol{R}^n$. As another example, consider the set of vectors $\{\boldsymbol{x}_1, \boldsymbol{x}_2, \boldsymbol{x}_3\}$, where

$$\boldsymbol{x}_1 = \begin{bmatrix} 1 \\ -1 \\ 1 \end{bmatrix}, \quad \boldsymbol{x}_2 = \begin{bmatrix} 0 \\ 1 \\ 1 \end{bmatrix}, \quad \boldsymbol{x}_3 = \begin{bmatrix} 2 \\ 3 \\ 0 \end{bmatrix}$$

To show that this set of vectors spans $\boldsymbol{R}^3$, we must show that given any vector

$$\boldsymbol{x} = \begin{bmatrix} x \\ y \\ z \end{bmatrix} \text{ in } \boldsymbol{R}^3$$

there exist scalars α_1, α_2, and α_3, such that $\boldsymbol{x} = \alpha_1\boldsymbol{x}_1 + \alpha_2\boldsymbol{x}_2 + \alpha_3\boldsymbol{x}_3$. Referring back to components, this becomes

$$\begin{bmatrix} x \\ y \\ z \end{bmatrix} = \begin{bmatrix} \alpha_1 \quad\quad + 2\alpha_3 \\ -\alpha_1 + \alpha_2 + 3\alpha_3 \\ \alpha_1 + \alpha_2 \end{bmatrix}$$

In other words, we must solve the following system of linear equations:

$$\begin{aligned} x &= \alpha_1 \quad\quad + 2\alpha_3 \\ y &= -\alpha_1 + \alpha_2 + 3\alpha_3 \\ z &= \alpha_1 + \alpha_2 \end{aligned}$$

for α_1, α_2, and α_3.

Since the determinant of the system

$$\begin{vmatrix} 1 & 0 & 2 \\ -1 & 1 & 3 \\ 1 & 1 & 0 \end{vmatrix} = -7$$

we see that the system of equations is solvable, and from this that $\{\boldsymbol{x}_1, \boldsymbol{x}_2, \boldsymbol{x}_3\}$ spans $\boldsymbol{R}^3$.

To prove independence, we suppose there are scalars, α_1, α_2, α_3, such that $\alpha_1\boldsymbol{x}_1 + \alpha_2\boldsymbol{x}_2 + \alpha_3\boldsymbol{x}_3 = \boldsymbol{0}$. Converting this to a system of linear equations, we obtain

$$\begin{aligned} \alpha_1 \quad\quad + 2\alpha_3 &= 0 \\ -\alpha_1 + \alpha_2 + 3\alpha_3 &= 0 \\ \alpha_1 + \alpha_2 \quad\quad &= 0 \end{aligned}$$

Since the determinant of the coefficient matrix is nonzero, the only solution to the system is $\alpha_1 = 0$, $\alpha_2 = 0$, $\alpha_3 = 0$. Thus, the vectors $\boldsymbol{x}_1, \boldsymbol{x}_2, \boldsymbol{x}_3$ are linearly independent, and taken together with the fact that they span $\boldsymbol{R}^3$, we see that $\{\boldsymbol{x}_1, \boldsymbol{x}_2, \boldsymbol{x}_3\}$ is a basis for $\boldsymbol{R}^3$.

An interesting property of bases is indicated in the next result.

Theorem A set $\{\boldsymbol{x}_1, \boldsymbol{x}_2, \ldots, \boldsymbol{x}_n\}$ in a vector space V is a basis for V if and only if for each $\boldsymbol{x}$ in V, there are unique scalars, $\alpha_1, \alpha_2, \ldots, \alpha_n$ such that $\boldsymbol{x} = \alpha_1\boldsymbol{x}_1 + \alpha_2\boldsymbol{x}_2 + \cdots + \alpha_n\boldsymbol{x}_n$.

PROOF Let us suppose $\{\boldsymbol{x}_1, \boldsymbol{x}_2, \ldots, \boldsymbol{x}_n\}$ is a basis. Then $\{\boldsymbol{x}_1, \boldsymbol{x}_2, \ldots, \boldsymbol{x}_n\}$ spans V. Thus, there exist scalars $\alpha_1, \alpha_2, \ldots, \alpha_n$, such that

$$\boldsymbol{x} = \alpha_1\boldsymbol{x}_1 + \alpha_2\boldsymbol{x}_2 + \cdots + \alpha_n\boldsymbol{x}_n$$

To show that these scalars are unique, suppose we have, as well,

$$\boldsymbol{x} = \beta_1\boldsymbol{x}_1 + \beta_2\boldsymbol{x}_2 + \cdots + \beta_n\boldsymbol{x}_n$$

Then

$$\alpha_1\boldsymbol{x}_1 + \cdots + \alpha_n\boldsymbol{x}_n = \beta_1\boldsymbol{x}_1 + \cdots + \beta_n\boldsymbol{x}_n$$

or

$$(\alpha_1 - \beta_1)\boldsymbol{x}_1 + \cdots + (\alpha_n - \beta_n)\boldsymbol{x}_n = \boldsymbol{0}$$

Since $\{\boldsymbol{x}_1, \boldsymbol{x}_2, \ldots, \boldsymbol{x}_n\}$ is an independent set, we have

$$\alpha_1 - \beta_1 = 0, \ldots, \alpha_n - \beta_n = 0$$

or

$$\alpha_1 = \beta_1, \ldots, \alpha_n = \beta_n$$

thus demonstrating the uniqueness of the scalars $\alpha_1, \ldots, \alpha_n$.

On the other hand, suppose $\{\boldsymbol{x}_1, \boldsymbol{x}_2, \ldots, \boldsymbol{x}_n\}$ is a set such that every $\boldsymbol{x}$ in V can be expressed uniquely in the form $\boldsymbol{x} = \alpha_1\boldsymbol{x}_1 + \cdots + \alpha_n\boldsymbol{x}_n$. It is then clear that $\{\boldsymbol{x}_1, \boldsymbol{x}_2, \ldots, \boldsymbol{x}_n\}$ spans V. To see that $\{\boldsymbol{x}_1, \ldots, \boldsymbol{x}_n\}$ is independent, suppose $\alpha_1\boldsymbol{x}_1 + \cdots + \alpha_n\boldsymbol{x}_n = \boldsymbol{0}$. We know that $0 \cdot \boldsymbol{x}_1 + \cdots + 0 \cdot \boldsymbol{x}_n = \boldsymbol{0}$. Thus, the zero vector is expressed as two linear combinations of $\boldsymbol{x}_1, \ldots, \boldsymbol{x}_n$, and since the coefficients in these linear combinations are, by hypothesis, uniquely determined, we must have $\alpha_1 = 0$, $\alpha_2 = 0, \ldots, \alpha_n = 0$. Thus, we have demonstrated linear independence. ■

For example, in P_n every element may be expressed as a unique linear combination of $\{1, x, x^2, \ldots, x^n\}$. Thus, we see that $\{1, x, \ldots, x^n\}$ forms a basis for P_n.

Let L be a subspace of $\boldsymbol{R}^3$ consisting of those vectors that lie on some fixed plane passing through the origin.

In §4.4 on span, we saw that any two noncollinear vectors in L span L. In §4.5 on linear independence, we saw that any two noncollinear vectors in $\boldsymbol{R}^3$ are linearly independent. Thus, if $\boldsymbol{a}$ and $\boldsymbol{b}$ are noncollinear vectors in L, we see that $\boldsymbol{a}$ and $\boldsymbol{b}$ span L and are linearly independent. Thus, in order to obtain a basis for L, we need only to choose two noncollinear vectors in L.

For example, in $\boldsymbol{R}^3$, consider the plane consisting of the points (x, y, z), where $y = z$. (See Figure 4-13.)

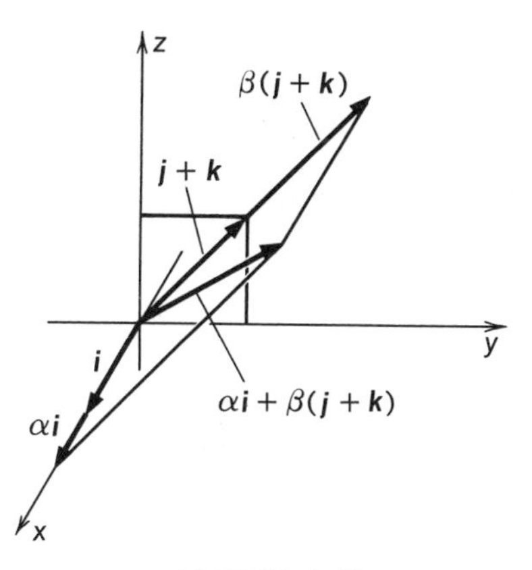

FIGURE 4-13

A pair of noncollinear vectors on this plane is $\{\boldsymbol{i}, \boldsymbol{j} + \boldsymbol{k}\}$. Thus, any vector whatsoever lying on the plane $y = z$ may be expressed uniquely as $\alpha\boldsymbol{i} + \beta(\boldsymbol{j} + \boldsymbol{k})$.

EXERCISES

1. Which of the following subsets are bases for $\boldsymbol{R}^2$?

(a) $\begin{bmatrix}1\\1\end{bmatrix}, \begin{bmatrix}3\\2\end{bmatrix}$ (b) $\begin{bmatrix}6\\3\end{bmatrix}, \begin{bmatrix}14\\7\end{bmatrix}$ (c) $\begin{bmatrix}2\\1\end{bmatrix}, \begin{bmatrix}1\\2\end{bmatrix}$ (d) $\begin{bmatrix}2\\7\end{bmatrix}, \begin{bmatrix}-1\\4\end{bmatrix}$

(e) $\begin{bmatrix}1\\1\end{bmatrix}, \begin{bmatrix}-1\\0\end{bmatrix}, \begin{bmatrix}8\\2\end{bmatrix}$ (f) $\begin{bmatrix}4\\3\end{bmatrix}, \begin{bmatrix}2\\2\end{bmatrix}$ (g) $\begin{bmatrix}3\\1\end{bmatrix}, \begin{bmatrix}4\\7\end{bmatrix}, \begin{bmatrix}0\\3\end{bmatrix}$

2. Which of the following subsets are bases for $\boldsymbol{R}^3$?

(a) $\begin{bmatrix}1\\3\\2\end{bmatrix}, \begin{bmatrix}0\\5\\2\end{bmatrix}, \begin{bmatrix}1\\0\\2\end{bmatrix}$ (b) $\begin{bmatrix}1\\0\\-1\end{bmatrix}, \begin{bmatrix}0\\1\\-1\end{bmatrix}, \begin{bmatrix}-1\\1\\0\end{bmatrix}$ (c) $\begin{bmatrix}1\\2\\0\end{bmatrix}, \begin{bmatrix}0\\2\\1\end{bmatrix}, \begin{bmatrix}2\\1\\0\end{bmatrix}$

(d) $\begin{bmatrix}1\\2\\1\end{bmatrix}, \begin{bmatrix}3\\0\\4\end{bmatrix}$ (e) $\begin{bmatrix}1\\0\\1\end{bmatrix}, \begin{bmatrix}2\\1\\2\end{bmatrix}, \begin{bmatrix}3\\-1\\4\end{bmatrix}$ (f) $\begin{bmatrix}1\\0\\1\end{bmatrix}, \begin{bmatrix}0\\1\\1\end{bmatrix}, \begin{bmatrix}1\\1\\0\end{bmatrix}$

3. Show that the given subsets are bases for the indicated subspaces of the space of 2×2 matrices.
   (a) $\begin{bmatrix} 1 & 0 \\ 0 & 1 \end{bmatrix}, \begin{bmatrix} 0 & 0 \\ 0 & 1 \end{bmatrix}, \begin{bmatrix} 0 & 1 \\ 1 & 0 \end{bmatrix}$, 2×2 symmetric matrices.
   (b) $\begin{bmatrix} 0 & 1 \\ -1 & 0 \end{bmatrix}$, 2×2 skew-symmetric matrices.
   (c) $\begin{bmatrix} 1 & 0 \\ 0 & 1 \end{bmatrix}, \begin{bmatrix} 1 & 0 \\ 0 & -1 \end{bmatrix}, \begin{bmatrix} 1 & 1 \\ 1 & 1 \end{bmatrix}$, 2×2 symmetric matrices.
   (d) $\begin{bmatrix} 1 & 0 \\ 0 & 0 \end{bmatrix}, \begin{bmatrix} 0 & 1 \\ 0 & 0 \end{bmatrix}, \begin{bmatrix} 0 & 0 \\ 0 & 1 \end{bmatrix}$, 2×2 upper triangular matrices.
   (e) $\begin{bmatrix} 1 & 0 \\ 0 & 0 \end{bmatrix}, \begin{bmatrix} 0 & 1 \\ 0 & 0 \end{bmatrix}, \begin{bmatrix} 0 & 0 \\ 1 & 0 \end{bmatrix}, \begin{bmatrix} 0 & 0 \\ 0 & 1 \end{bmatrix}$, all 2×2 matrices.
   (f) $\begin{bmatrix} 0 & 1 \\ 0 & 0 \end{bmatrix}, \begin{bmatrix} 0 & 0 \\ 0 & 1 \end{bmatrix}$, 2×2 matrices, A, such that $A\begin{bmatrix} 1 \\ 0 \end{bmatrix} = \begin{bmatrix} 0 \\ 0 \end{bmatrix}$.
   (g) $\begin{bmatrix} 1 & 0 \\ 0 & 1 \end{bmatrix}, \begin{bmatrix} 0 & 1 \\ 1 & 0 \end{bmatrix}$, 2×2 matrices that commute with $\begin{bmatrix} 0 & 1 \\ 1 & 0 \end{bmatrix}$.
4. Show that the indicated subsets are bases for P_3, the polynomials in a variable t.
   (a) $1, t, t^2, 1 + t + t^2 + t^3$
   (b) $1, 1 + t, (1 + t)^2, (1 + t)^3$
   (c) $1, 1 + t, 1 + t^2, 1 + t^3$
   (d) $1, 1 + t, 1 + t + t^2, 1 + t + t^2 + t^3$
   (e) $1 + t, 1 - t, t^2 + t^3, t^2 - t^3$
5. Consider the collection of all polynomials in two variables of degree less than or equal to 2, i.e., those of the form

$$a_0 + b_1 x + b_2 y + c_1 x^2 + c_2 xy + c_3 y^2$$

   With the usual definitions of addition and scalar multiplication, show that these polynomials form a vector space and determine a basis for this space.
6. Find a basis for each of the following spaces.
   (a) 3×3 symmetric matrices.
   (b) 3×3 skew-symmetric matrices.
   (c) 2×3 matrices.
   (d) 3×2 matrices.
   (e) Even polynomials of degree less than or equal to n.
   (f) Subspace of the space of polynomials of degree less than or equal to 3 that vanish at $x = 1$.
   (g) Subspace of $\boldsymbol{R}^n$ consisting of those vectors whose first two components vanish.
   (h) 4×4 diagonal matrices.
7. In the space of $n \times m$ matrices, let E_{ij} denote that matrix with 0 in all entries except in position (i, j), and 1 in position (i, j).
   (a) Show that the matrices E_{ij} form a basis for the space of $n \times m$ matrices.
   (b) By choosing a suitable subcollection of the E_{ij}'s, find a basis for the upper triangular matrices and diagonal matrices.
8. Find a basis for the space of 2×2 matrices that consists of
   (a) Only matrices such that $A^2 = A$ (a matrix is said to be **idempotent** if $A^2 = A$);
   (b) Only invertible matrices.

9. Show that it is not possible to find a basis for the space of $n \times n$ matrices such that each pair of elements in the basis commute.

10. The matrices

$$\sigma_x = \begin{bmatrix} 0 & 1 \\ 1 & 0 \end{bmatrix}, \quad \sigma_y = \begin{bmatrix} 0 & -i \\ i & 0 \end{bmatrix}, \quad \sigma_z = \begin{bmatrix} 1 & 0 \\ 0 & -1 \end{bmatrix}$$

are called the Pauli-spin matrices.

(a) Show that

$$\sigma_x\sigma_y = -\sigma_y\sigma_x$$
$$\sigma_x\sigma_z = -\sigma_z\sigma_x$$
$$\sigma_y\sigma_z = -\sigma_z\sigma_y$$

(b) Show that along with the identity matrix these matrices form a basis for the space of 2×2 matrices with complex entries.

11. Is it possible to find a basis for P_n, such that every element of the basis is divisible by the polynomial $f(x) = x$?

12. Let $A_1, A_2, \ldots, A_m$ be a collection of $n \times n$ matrices. If $\boldsymbol{x} \neq 0$ is an n-vector such that $A_1\boldsymbol{x} = A_2\boldsymbol{x} = \cdots = A_m\boldsymbol{x} = 0$, show that the matrices $A_1, A_2, \ldots, A_m$ do not form a basis for the space of $n \times n$ matrices.

13. If $\boldsymbol{x}_1$, $\boldsymbol{x}_2$, and $\boldsymbol{x}_3$ form a basis for some real vector space V, show that $(1 + t)\boldsymbol{x}_1 + (1 + t^2)\boldsymbol{x}_2 + (t^4 + t^2)\boldsymbol{x}_3$ is nonzero for all real t.

7 DIMENSION

For purposes of further study, we single out a collection of vector spaces that are in some sense "small" and particularly suited for applications. Thus, we say a vector space V is **finite dimensional** if it can be spanned by a finite number of vectors. Previous examples show that both the vector spaces $\boldsymbol{R}^n$ and P_n are finite dimensional.

Our first objective is to show that any space spanned by finitely many vectors has a finite basis. Preliminary to this we prove:

Lemma Let V be a vector space. Suppose $\boldsymbol{x}_1, \boldsymbol{x}_2, \ldots, \boldsymbol{x}_n$ span V, and that the vectors $\boldsymbol{x}_1, \boldsymbol{x}_2, \ldots, \boldsymbol{x}_n$ are linearly dependent. Then by deleting a suitable vector of the set $\{\boldsymbol{x}_1, \boldsymbol{x}_2, \ldots, \boldsymbol{x}_n\}$, say $\boldsymbol{x}_i$, we may obtain a set $\{\boldsymbol{x}_1, \boldsymbol{x}_2, \ldots, \boldsymbol{x}_{i-1}, \boldsymbol{x}_{i+1}, \ldots, \boldsymbol{x}_n\}$ that still spans V.

PROOF Since the vectors $\boldsymbol{x}_1, \boldsymbol{x}_2, \ldots, \boldsymbol{x}_n$ are linearly dependent, by the theorem in §4.5, one of the vectors, say $\boldsymbol{x}_i$, is a linear combination of the remaining vectors in the set:

$$\boldsymbol{x}_i = \beta_1\boldsymbol{x}_1 + \cdots + \beta_{i-1}\boldsymbol{x}_{i-1} + \beta_{i+1}\boldsymbol{x}_{i+1} + \cdots + \beta_n\boldsymbol{x}_n$$

for some scalars $\beta_1, \ldots, \beta_{i-1}, \beta_{i+1}, \ldots, \beta_n$.

Now, let $\boldsymbol{x}$ be any vector in V. Since $\{\boldsymbol{x}_1, \boldsymbol{x}_2, \ldots, \boldsymbol{x}_n\}$ spans V, there are scalars, $\alpha_1, \ldots, \alpha_n$, such that

$$x = \alpha_1 x_1 + \cdots + \alpha_{i-1} x_{i-1} + \alpha_i x_i + \alpha_{i+1} x_{i+1} + \cdots + \alpha_n x_n$$

By substituting the above expression for x_i into the linear combination that equals x, we see that x may be expressed as a linear combination of $x_1, \ldots, x_{i-1}, x_{i+1}, \ldots, x_n$.

Thus, every vector in V is a linear combination of $x_1, \ldots, x_{i-1}, x_{i+1}, \ldots, x_n$, showing that this set spans V. ■

Using this, we may demonstrate the following theorem.

Theorem 1 Let V be a vector space spanned by finitely many vectors. Then V has a finite basis.

PROOF Suppose $\{x_1, x_2, \ldots, x_n\}$ spans V. If $\{x_1, x_2, \ldots, x_n\}$ is a linearly independent set, it is a basis, and we are finished. So suppose it is not linearly independent. By the preceding lemma, we may delete a vector in the set and obtain a smaller set that still spans V. If this new set is linearly independent, we are finished, since we then have a finite basis. If it is not linearly independent, we may obtain a yet smaller set that spans V. By repeating this process it must eventually come to pass that we obtain a linearly independent set that spans V, demonstrating our theorem. ■

Actually our proof yielded the stronger statement: If $\{x_1, \ldots, x_n\}$ spans a vector space V, some subset of $\{x_1, \ldots, x_n\}$ is a basis for V.

Let H be a subspace of $\boldsymbol{R}^3$ consisting of those vectors lying on some fixed plane that passes through the origin. Let us determine the number of vectors in some basis for H. In §4.5 we saw that any set of three coplanar vectors in $\boldsymbol{R}^3$ is necessarily linearly dependent. Thus, a basis for H, since it must consist of linearly independent vectors, must have at most two vectors. Since the vectors in any basis for H spans a plane, no basis for H can have only one vector. Thus, any basis for H must have exactly two vectors.

This remarkable and useful fact, namely, that any two bases have the same number of vectors, is true in any finite-dimensional vector space. This follows immediately from the next result.

Theorem 2 Let V be a finite-dimensional vector space and $\{x_1, x_2, \ldots, x_n\}$ be a basis for V with n elements. If $\{y_1, y_2, \ldots, y_m\}$ is a set of m linearly independent vectors in V, then $m \leqslant n$.

We may restate this: If V is a finite-dimensional space having a basis of n elements and $\{y_1, \ldots, y_m\}$ is a set of vectors in V with $m > n$, then the vectors $y_1, y_2, \ldots, y_m$ are linearly dependent. We prove the second statement.

PROOF Let $\{x_1, x_2, \ldots, x_n\}$ be a basis for V. In order to show that $\{y_1, y_2, \ldots, y_m\}$ is a linearly dependent set, we must exhibit scalars

$\beta_1, \beta_2, \ldots, \beta_m$ not all zero, such that

$$\sum_{j=1}^{m} \beta_j y_j = \beta_1 y_1 + \beta_2 y_2 + \cdots + \beta_m y_m = \mathbf{0}$$

Since $\{x_1, x_2, \ldots, x_n\}$ is a basis for V, there are scalars α_{ij}, such that

$$y_j = \sum_{i=1}^{n} \alpha_{ij} x_i$$

Thus, we wish to find $\beta_1, \beta_2, \ldots, \beta_m$, such that

$$\sum_{j=1}^{m} \beta_j \left(\sum_{i=1}^{n} \alpha_{ij} x_i \right) = \mathbf{0}$$

or

$$\sum_{i=1}^{n} \left(\sum_{j=1}^{m} \alpha_{ij} \beta_j \right) x_i = \mathbf{0}$$

Thus, if we can find a nontrivial solution to the system of homogeneous equations

$$\begin{aligned}
\alpha_{11}\beta_1 + \cdots + \alpha_{1m}\beta_m &= 0 \\
\alpha_{21}\beta_1 + \cdots + \alpha_{2m}\beta_m &= 0 \\
&\vdots \\
\alpha_{n1}\beta_1 + \cdots + \alpha_{nm}\beta_m &= 0
\end{aligned}$$

our theorem will be demonstrated. Since $m > n$, the number of unknowns in the system is greater than the number of equations, hence, we know from §1.4 that a nontrivial solution exists. ■

Immediately we obtain the next result.

Theorem 3 In a finite-dimensional space, any two bases have the same number of elements.

PROOF Let $\{x_1, x_2, \ldots, x_m\}$ and $\{y_1, y_2, \ldots, y_n\}$ be bases for V, a finite-dimensional space. Since $\{y_1, y_2, \ldots, y_n\}$ is a linearly independent set and $\{x_1, x_2, \ldots, x_m\}$ is a basis, by the preceding theorem $m \geqslant n$. Since $\{x_1, x_2, \ldots, x_m\}$ is a linearly independent set and $\{y_1, y_2, \ldots, y_n\}$ is a basis, we also have $n \geqslant m$. Thus, $m = n$. ■

In $\boldsymbol{R}^n$ we exhibited the standard basis $\{e_1, e_2, \ldots, e_n\}$ with n elements.

By Theorem 3, we see that any other basis whatsoever must also have n elements.

If V is a finite-dimensional space, we define the **dimension** of V to be the number of elements in some basis for V. By Theorem 3, this number is independent of the basis we choose. The dimension of V is often denoted by dim V. According to this definition dim $\boldsymbol{R}^n = n$.

We have seen in $\boldsymbol{R}^3$ that the vectors on any line passing through the origin all consist of scalar multiples of a single vector. Thus, if H denotes a subspace of $\boldsymbol{R}^3$ consisting solely of vectors on some line passing through the origin, we see that dim $H = 1$. Thus, our definition of dimension makes precise the statement that a line is a "one-dimensional object." (See Figure 4-14.)

If, in a similar manner, we let K denote the vectors that lie on some plane passing through the origin, then we have seen that K has a basis of two vectors, and so dim $K = 2$. Thus, we see why it makes perfectly good sense to say that a plane is a "two-dimensional object." (See Figure 4-15.)

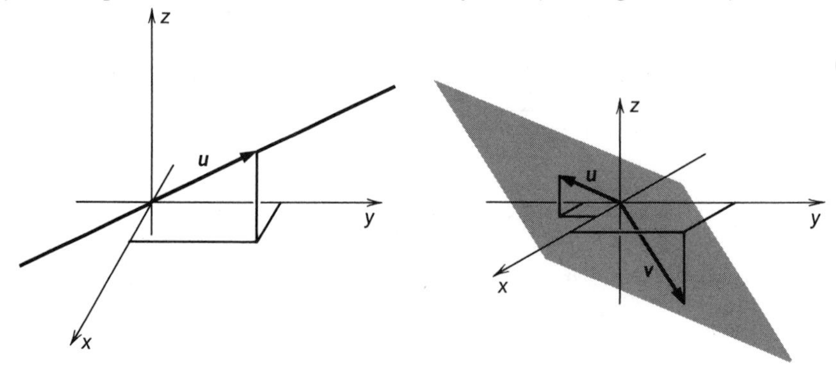

FIGURE 4-14

FIGURE 4-15

Example 1 Let P_n be the space of polynomials of degree less than or equal to n. We have seen that $1, x, x^2, \ldots, x^n$ is a basis for P_n. Since the set $\{1, x, \ldots, x^i, \ldots, x^n\}$ has $n + 1$ elements, we see that dim $P_n = n + 1$.

Example 2 Consider the 2×2 matrices with real coefficients, M_{22}. Observe that the matrices

$$\begin{bmatrix} 1 & 0 \\ 0 & 0 \end{bmatrix}, \quad \begin{bmatrix} 0 & 1 \\ 0 & 0 \end{bmatrix}, \quad \begin{bmatrix} 0 & 0 \\ 1 & 0 \end{bmatrix}, \quad \begin{bmatrix} 0 & 0 \\ 0 & 1 \end{bmatrix}$$

form a basis.

To see this, note that

$$\begin{bmatrix} a & b \\ c & d \end{bmatrix} = \begin{bmatrix} a & 0 \\ 0 & 0 \end{bmatrix} + \begin{bmatrix} 0 & b \\ 0 & 0 \end{bmatrix} + \begin{bmatrix} 0 & 0 \\ c & 0 \end{bmatrix} + \begin{bmatrix} 0 & 0 \\ 0 & d \end{bmatrix}$$

$$= a\begin{bmatrix} 1 & 0 \\ 0 & 0 \end{bmatrix} + b\begin{bmatrix} 0 & 1 \\ 0 & 0 \end{bmatrix} + c\begin{bmatrix} 0 & 0 \\ 1 & 0 \end{bmatrix} + d\begin{bmatrix} 0 & 0 \\ 0 & 1 \end{bmatrix}$$

Thus, the given set spans the space of 2×2 matrices. If

$$a\begin{bmatrix} 1 & 0 \\ 0 & 0 \end{bmatrix} + b\begin{bmatrix} 0 & 1 \\ 0 & 0 \end{bmatrix} + c\begin{bmatrix} 0 & 0 \\ 1 & 0 \end{bmatrix} + d\begin{bmatrix} 0 & 0 \\ 0 & 1 \end{bmatrix} = \begin{bmatrix} 0 & 0 \\ 0 & 0 \end{bmatrix}$$

then

$$\begin{bmatrix} a & b \\ c & d \end{bmatrix} = \begin{bmatrix} 0 & 0 \\ 0 & 0 \end{bmatrix}$$

So $a = b = c = d = 0$.

Thus, the given set of matrices is linearly independent. Since the space of 2×2 matrices with real coefficients admits a basis with four elements, it is of dimension 4.

Example 3 Consider the system of homogeneous equations

$$\begin{aligned} x_1 + 2x_2 - x_3 + x_4 &= 0 \\ x_1 - x_2 + x_3 + 2x_4 &= 0 \end{aligned}$$

In matrix notation,

$$\begin{bmatrix} 1 & 2 & -1 & 1 \\ 1 & -1 & 1 & 2 \end{bmatrix} \begin{bmatrix} x_1 \\ x_2 \\ x_3 \\ x_4 \end{bmatrix} = \begin{bmatrix} 0 \\ 0 \end{bmatrix}$$

We have seen in example 2, §4.3, that the set of solutions forms a subspace of $\boldsymbol{R}^4$. Let us find a basis for this subspace and calculate its dimension.

We apply Gaussian elimination.

$$\begin{aligned} x_1 + 2x_2 - x_3 + x_4 &= 0 \\ x_1 - x_2 + x_3 + 2x_4 &= 0 \end{aligned}$$

Use x_1 and the first equation.

$$\downarrow$$

$$\begin{aligned} x_1 + 2x_2 - x_3 + x_4 &= 0 \\ -3x_2 + 2x_3 + x_4 &= 0 \end{aligned}$$

Use x_4 and the second equation.

$$\downarrow$$

$$\begin{aligned} x_1 + 5x_2 - 3x_3 \qquad &= 0 \\ -3x_2 + 2x_3 + x_4 &= 0 \end{aligned}$$

The procedure comes to a halt, since there are no unused equations.

If we choose $x_2 = c$, $x_3 = d$, where c and d are arbitrary numbers, and then choose

$$\begin{aligned} x_1 &= -5c + 3d \\ x_4 &= 3c - 2d \end{aligned}$$

we see that any solution to the system is of the form

$$\begin{bmatrix} -5c+3d \\ c \\ d \\ 3c-2d \end{bmatrix} = c\begin{bmatrix} -5 \\ 1 \\ 0 \\ 3 \end{bmatrix} + d\begin{bmatrix} 3 \\ 0 \\ 1 \\ -2 \end{bmatrix}$$

Thus, we see that the vectors

$$\begin{bmatrix} -5 \\ 1 \\ 0 \\ 3 \end{bmatrix} \quad \text{and} \quad \begin{bmatrix} 3 \\ 0 \\ 1 \\ -2 \end{bmatrix}$$

span the space of solutions. Since the two vectors are clearly linearly independent, it follows that taken together they provide a basis for the space of solutions. Since the space of solutions is a two-dimensional vector space, we see precisely what it means to say that the solutions form a "two-parameter family."

EXERCISES

1. Calculate the dimension of the following subspaces of the space of 2×2 matrices.
   (a) The diagonal matrices.
   (b) The symmetric matrices.
   (c) The skew-symmetric matrices.
   (d) The matrices of the form $\begin{bmatrix} a & b \\ -b & a \end{bmatrix}$.
   (e) The matrices of the form $\begin{bmatrix} 0 & a \\ 0 & b \end{bmatrix}$.
   (f) The upper triangular matrices.
   (g) The matrices of the form $\begin{bmatrix} a & b \\ -b & c \end{bmatrix}$.

2. Calculate the dimension of the following subspaces of P_3.

   (a) $\{f \mid f(0) = 0\}$ (b) $\{f \mid f(1) = 0\}$
   (c) $\{f \mid f(0) = f'(0) = 0\}$ (d) $\{f \mid f'(0) + f(0) = 0\}$

3. Let $i_1 < i_2 < \cdots < i_n$ be integers. Calculate the dimension of the subspace of $\boldsymbol{R}^m$ that consists of all vectors such that i_1th, i_2th, . . . , i_nth components vanish. $(m \geqslant n.)$

4. Find a basis and the dimension of the space of solutions to each system of linear homogeneous equations.

   (a) $x + y + z = 0$
   $x - y + z = 0$

   (b) $x + y + z = 0$
   $x + 2y + 3z = 0$

   (c) $x_1 + x_2 + x_3 - x_4 = 0$
   $x_1 - x_2 + x_3 - x_4 = 0$

   (d) $x_1 + x_2 + 2x_3 - 4x_4 = 0$
   $x_1 + x_2 - x_3 - x_4 = 0$
   $2x_1 + 2x_2 - x_3 - 3x_4 = 0$

   (e) $x_1 + x_2 - 3x_3 + x_4 + x_5 = 0$
   $x_1 - x_2 + x_3 + x_4 + x_5 = 0$
   $x_1 + 2x_2 + 3x_3 + 2x_4 + x_5 = 0$

5. Show that the complex numbers with the usual addition and multiplication by scalars form a two-dimensional vector space over the reals.

6. What is the dimension of the space of $n \times n$ diagonal matrices?

7. If f is a polynomial of degree n, show that $f, f', f'', \ldots, f^{(n)}$ form a basis for P_n.

8. Show that the dimension of the space of $n \times m$ matrices is nm.

9. Let H_e be the subspace of P_n consisting of the even polynomials, and H_0 be the subspace of odd polynomials. What is the dimension of H_0 and H_e?

10. What is the dimension of the space of upper triangular $n \times n$ matrices?

11. Show that the dimension of the space of $n \times n$ symmetric matrices is $\frac{1}{2}n(n+1)$, while the dimension of the space of skew-symmetric matrices is $\frac{1}{2}n(n-1)$.

12. Let V be an n-dimensional vector space. If V is spanned by $\{\boldsymbol{x}_1, \boldsymbol{x}_2, \ldots, \boldsymbol{x}_n\}$, a set of n vectors, show that $\{\boldsymbol{x}_1, \boldsymbol{x}_2, \ldots, \boldsymbol{x}_n\}$ are linearly independent.

13. Show that the space of polynomials in two variables of degree less than or equal to n, i.e., all polynomials of the form

$$\sum_{\substack{0 \leq i+j \leq n \\ 0 \leq i \\ 0 \leq j}} a_{ij} x^i y^j$$

form a vector space under ordinary addition and scalar multiplication, and show that the dimension of this space is $\frac{1}{2}(n+1)(n+2)$.

14. If $\boldsymbol{x}$, $\boldsymbol{y}$, and $\boldsymbol{z}$ form a basis for a three-dimensional vector space, show also that $\boldsymbol{x}+\boldsymbol{y}, \boldsymbol{y}+\boldsymbol{z}, \boldsymbol{x}+\boldsymbol{z}$ form a basis as well.

15. If $f_0, f_1, \ldots, f_{n+1}$ are polynomials of degree less than or equal to n, show that there are scalars, $\alpha_0, \alpha_1, \ldots, \alpha_{n+1}$, not all zero, such that $\alpha_0 f_0 + \alpha_1 f_1 + \cdots + \alpha_{n+1} f_{n+1} = 0$.

16. Show that any $n \times n$ matrix with real coefficients satisfies a polynomial equation $f(A) = 0$, where $f(x)$ is a nonzero polynomial with real coefficients. [Hint: Show that the matrices $I_n, A, A^2, \ldots, A^{n^2}$, must be linearly dependent.]

17. Let A and B be fixed real numbers. Let V be the collection of all real infinite sequences $(c_0, c_1, c_2, c_3, \ldots)$ that satisfy the recursion relation $c_{k+2} = Ac_k + Bc_{k+1}$.
    (a) With the obvious (i.e., termwise) addition and scalar multiplication, show that V is a real vector space.
    (b) Show that c_0 and c_1 determine all the remaining terms of the sequence.
    (c) Show that every element of V is a linear combination of the sequence $(0, 1, \ldots)$ and $(1, 0, \ldots)$.
    (d) Show that $\dim V = 2$.
    (e) Let t be a root of the polynomial $x^2 - Bx - A = 0$. Show that the sequence $(1, t, t^2, t^3, \ldots)$ lies in V (if t is real).
    (f) If $x^2 - Bx - A$ has two distinct real roots, s and t, show that every element of V is a linear combination of $(1, s, s^2, s^3, \ldots)$ and $(1, t, t^2, t^3, \ldots)$.
    (g) The Fibonacci sequence is defined in the following way: The 0th term is 0, the 1st is 1, and, thereafter, every term is the sum of the two previous terms. It looks like: $(0, 1, 1, 2, 3, 5, 8, \ldots)$. Show that

$$c_n = \frac{1}{\sqrt{5}}\left(\frac{1+\sqrt{5}}{2}\right)^n - \frac{1}{\sqrt{5}}\left(\frac{1-\sqrt{5}}{2}\right)^n$$

18. Given that the sequence (0, 1, 3, 13, 51, . . .) is determined by the rule of the last exercise. Find A and B and give a general formula for the nth term of the sequence.

8 ADDITIONAL PROPERTIES OF FINITE-DIMENSIONAL SPACES

In this section, we note some of the consequences of theorems in previous sections.

Theorem 1 Let V be a vector space and suppose $\boldsymbol{x}_1, \boldsymbol{x}_2, \ldots, \boldsymbol{x}_m$ span V. Then $\dim V \leqslant m$.

PROOF By Theorem 1, §4.7, some subset of $\boldsymbol{x}_1, \boldsymbol{x}_2, \ldots, \boldsymbol{x}_m$ is a basis for V. But $\dim V$ is the number of elements in any subset that is a basis. Consequently, $\dim V \leqslant m$. ■

Preliminary to our next results, we have

Lemma Let V be a vector space and H be a subspace of V. Let $\{\boldsymbol{x}_1, \boldsymbol{x}_2, \ldots, \boldsymbol{x}_n\}$ be a basis for H and suppose $\boldsymbol{y}$ does not belong to H. Then the set $\{\boldsymbol{y}, \boldsymbol{x}_1, \boldsymbol{x}_2, \ldots, \boldsymbol{x}_n\}$ is linearly independent.

PROOF We show that if $\{\boldsymbol{y}, \boldsymbol{x}_1, \boldsymbol{x}_2, \ldots, \boldsymbol{x}_m\}$ is linearly dependent, then $\boldsymbol{y}$ belongs to H, proving our assertion.

So suppose $\{\boldsymbol{y}, \boldsymbol{x}_1, \boldsymbol{x}_2, \ldots, \boldsymbol{x}_n\}$ is linearly dependent. Then there are scalars $\beta, \alpha_1, \alpha_2, \ldots, \alpha_n$, not all zero, such that

$$\beta\boldsymbol{y} + \alpha_1\boldsymbol{x}_1 + \alpha_2\boldsymbol{x}_2 + \cdots + \alpha_n\boldsymbol{x}_n = \boldsymbol{0}$$

Now, β is not zero. For if $\beta = 0$, $\alpha_1\boldsymbol{x}_1 + \alpha_2\boldsymbol{x}_2 + \cdots + \alpha_n\boldsymbol{x}_n = \boldsymbol{0}$. Since not all of the scalars $\alpha_1, \alpha_2, \ldots, \alpha_n$ are zero, it follows that $\boldsymbol{x}_1, \ldots, \boldsymbol{x}_n$ are linearly dependent. But by hypothesis, $\boldsymbol{x}_1, \boldsymbol{x}_2, \ldots, \boldsymbol{x}_n$ are linearly independent. Consequently, $\beta \neq 0$.

Since $\beta \neq 0$, $\boldsymbol{y} = -(\beta^{-1}\alpha_1)\boldsymbol{x}_1 + \cdots + (-\beta^{-1}\alpha_n)\boldsymbol{x}_n$. Thus, $\boldsymbol{y}$ is a linear combination of vectors in H and so belongs to H. ■

Let H be a subspace of $\boldsymbol{R}^3$ consisting of those vectors lying on some fixed plane through the origin. We saw in §4.4 that the linear combinations of two noncollinear vectors in H fill out H. Restated in algebraic terminology, two linearly independent vectors in H span H and therefore form a basis for H. The generalization of this result to arbitrary finite-dimensional spaces is formulated in the next theorem.

Theorem 2 Let V be a vector space of dimension n. If $\{\boldsymbol{x}_1, \boldsymbol{x}_2, \ldots, \boldsymbol{x}_n\}$ is a set of n linearly independent vectors in V, then $\{\boldsymbol{x}_1, \boldsymbol{x}_2, \ldots, \boldsymbol{x}_n\}$ is a basis for V.

PROOF Let $L = \text{sp}(\boldsymbol{x}_1, \boldsymbol{x}_2, \ldots, \boldsymbol{x}_n)$. We desire to show $L = V$.

If $L \neq V$, there is some vector, say $\boldsymbol{y}$, in V, such that $\boldsymbol{y} \notin L$. By the lemma, the vectors in the set $\{\boldsymbol{y}, \boldsymbol{x}_1, \boldsymbol{x}_2, \ldots, \boldsymbol{x}_n\}$ are linearly independent. But the set $\{\boldsymbol{y}, \boldsymbol{x}_1, \ldots, \boldsymbol{x}_n\}$ has $n + 1$ elements, and by Theorem 2 of §4.7, we know that in any n-dimensional vector space, we can find at most n linearly independent vectors. Thus, it was false to assume that $L \neq V$, and so $L = V$. ■

We can interpret this theorem in $\boldsymbol{R}^3$. If $\boldsymbol{x}_1$, $\boldsymbol{x}_2$, and $\boldsymbol{x}_3$ are any three linearly independent vectors in $\boldsymbol{R}^3$, then $\boldsymbol{x}_1, \boldsymbol{x}_2, \boldsymbol{x}_3$ form a basis for $\boldsymbol{R}^3$. In geometric terminology, any three noncoplanar (i.e., linearly independent) vectors in $\boldsymbol{R}^3$ span $\boldsymbol{R}^3$. Figure 4-16 illustrates this principle. The reader may find it a worthwhile exercise to demonstrate the same result from geometric considerations by using a procedure analogous to that of constructing Cartesian coordinates in three-dimensional space. (See §2.2.)

We have another intuitively reasonable result.

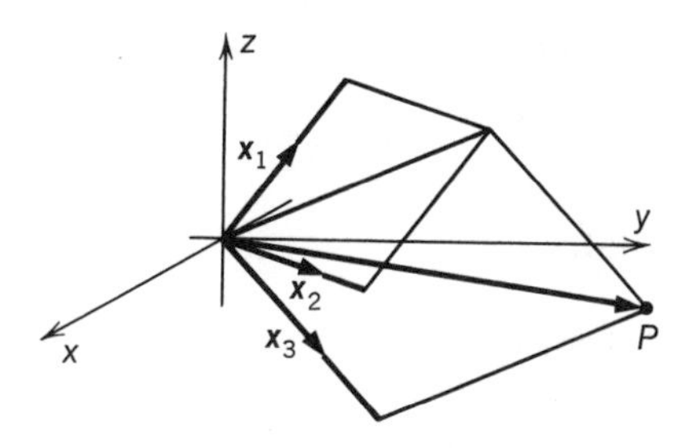

FIGURE 4-16

Theorem 3 Let V be a finite-dimensional vector space. If H is a subspace of V, H is finite dimensional and $\dim H \leqslant \dim V$.

PROOF If $H = \{\boldsymbol{0}\}$, the zero subspace, then $\dim H = 0 \leqslant \dim V$. If H is not the zero subspace, it has some nonzero element, say $\boldsymbol{x}_1$. Consider $\text{sp}(\boldsymbol{x}_1)$. If $\text{sp}(\boldsymbol{x}_1) = H$, H is finite dimensional. If $\text{sp}(\boldsymbol{x}_1) \neq H$, then there is some vector $\boldsymbol{x}_2$ in H but not in $\text{sp}(\boldsymbol{x}_1)$. By the lemma of this section $\{\boldsymbol{x}_1, \boldsymbol{x}_2\}$ is a linearly independent set. If $H = \text{sp}(\boldsymbol{x}_1, \boldsymbol{x}_2)$, H is finite dimensional. If $H \neq \text{sp}(\boldsymbol{x}_1, \boldsymbol{x}_2)$, there is some vector $\boldsymbol{x}_3$ in H but not in $\text{sp}(\boldsymbol{x}_1, \boldsymbol{x}_2)$. By the same lemma $\{\boldsymbol{x}_1, \boldsymbol{x}_2, \boldsymbol{x}_3\}$ is a linearly independent set.

By continuing this process, we obtain sets of linearly dependent vectors $\{\boldsymbol{x}_1, \boldsymbol{x}_2, \ldots, \boldsymbol{x}_i\}$. As long as $\text{sp}(\boldsymbol{x}_1, \boldsymbol{x}_2, \ldots, \boldsymbol{x}_i)$ is properly contained in H, we can find a vector $\boldsymbol{x}_{i+1}$ in H but not in $\text{sp}(\boldsymbol{x}_1, \boldsymbol{x}_2, \ldots, \boldsymbol{x}_i)$. By the lemma we then obtain a larger set of independent vectors $\{\boldsymbol{x}_1, \boldsymbol{x}_2, \ldots, \boldsymbol{x}_i, \boldsymbol{x}_{i+1}\}$. However, we know that V is a finite-dimensional space and that any linearly independent set of vectors in V has at most $\dim V$ elements.

Thus, eventually we must have $\text{sp}(\boldsymbol{x}_1, \boldsymbol{x}_2, \ldots, \boldsymbol{x}_i) = H$, and so H is finite dimensional. Choosing a basis for H, we know, by definition of dimension, that it contains $\dim H$ elements. By Theorem 2, §4.7, $\dim H \leqslant \dim V$. ■

Note that the bulk of the previous theorem consisted in demonstrating that H was finite dimensional. As soon as we obtained finite dimensionality, the fact that $\dim H \leqslant \dim V$ was immediately apparent.

As an application of these theorems, let us determine all subspaces of $\boldsymbol{R}^3$. We know, by Theorem 3, that any such subspace H is finite dimensional and satisfies $0 \leqslant \dim H \leqslant 3$.

If $\dim H = 0$, $H = \{\mathbf{0}\}$, the zero subspace.

If $\dim H = 3$, we choose a basis for H with three elements. By Theorem 2, this basis, having three elements, is also a basis for $\boldsymbol{R}^3$. Thus, $H = \boldsymbol{R}^3$.

From these two facts, we see that any proper subspace H of $\boldsymbol{R}^3$ must satisfy $1 \leqslant \dim H \leqslant 2$.

If $\dim H = 1$, H consists of all scalar multiples of a single vector, or in other words, H consists of all vectors lying on some line through the origin.

If $\dim H = 2$, H consists of all linear combinations of two noncollinear vectors, or, as we have seen earlier, H consists of all vectors lying on some plane that passes through the origin.

Thus, we see, as is intuitively plausible, that all proper subspaces of $\boldsymbol{R}^3$ are formed by lines or planes passing through the origin.

EXERCISES

1. If H is a subspace of the space of 2×2 matrices, show that $\dim H = 0, 1, 2, 3$, or 4.

2. Show that the polynomials $1, x - \alpha, (x - \alpha)^2, \ldots, (x - \alpha)^n$ form a basis for P_n, if α is any real number.

3. If f is a polynomial of degree n and g is a polynomial of degree less than or equal to n, show there are constants $\alpha_0, \alpha_1, \ldots, \alpha_n$, such that $g = \alpha_0 f + \alpha_1 f' + \alpha_2 f'' + \cdots + \alpha_n f^{(n)}$.

4. If $\boldsymbol{x}$ is a nonzero vector in a finite-dimensional vector space, show that $\boldsymbol{x}$ belongs to some basis.

5. If a vector space V is spanned by $\boldsymbol{x}_1, \boldsymbol{x}_2, \ldots, \boldsymbol{x}_k$, show that a set of $k + 1$ vectors in V is linearly dependent.

6. Find a set of $n + 1$ vectors in $\boldsymbol{R}^n$ such that any subset of n elements is linearly independent.

7. Let $\{\boldsymbol{x}_1, \boldsymbol{x}_2, \ldots, \boldsymbol{x}_n\}$ be a set of independent vectors in a vector space V. If $\{\boldsymbol{x}_1, \boldsymbol{x}_2, \ldots, \boldsymbol{x}_n\}$ is contained in no larger set of independent vectors in V, show that $\{\boldsymbol{x}_1, \boldsymbol{x}_2, \ldots, \boldsymbol{x}_n\}$ is a basis for V.

8. If V and W are subspaces of a vector space, $V \subset W$, and $\dim V = \dim W$, show that $V = W$.

9. Show that any independent set $\{\boldsymbol{x}_1, \ldots, \boldsymbol{x}_m\}$ of vectors in a finite-dimensional space V can be extended to a basis for V, i.e., there are vectors $\{\boldsymbol{y}_1, \ldots, \boldsymbol{y}_n\}$, such that $\{\boldsymbol{x}_1, \ldots, \boldsymbol{x}_m, \boldsymbol{y}_1, \ldots, \boldsymbol{y}_n\}$ is a basis for V.

10. Let S be a set of vectors all of whose subsets of $k + 1$ vectors are linearly

dependent but some subset of k vectors is linearly independent. Show that each vector in S is a linear combination of those k vectors.

11. If V and W are subspaces of a vector space, with $V \subset W$ and $\dim V \leqslant l \leqslant \dim W$, show that there is a subspace U, such that $V \subset U \subset W$ and $\dim U = l$.

12. If $f_0, f_1, \ldots, f_n$ are polynomials in P_n, such that $\deg f_i = i$, show that $f_0, f_1, \ldots, f_n$ form a basis for P_n.

13. If $\{\boldsymbol{x}_1, \boldsymbol{x}_2, \ldots, \boldsymbol{x}_n\}$ spans V, but no smaller subset spans V, show that $\{\boldsymbol{x}_1, \boldsymbol{x}_2, \ldots, \boldsymbol{x}_n\}$ is a basis for V.

14. Let $\alpha_0, \alpha_1, \ldots, \alpha_n$ be $n+1$ distinct real numbers. Consider the polynomials $p_i(x) = (x - \alpha_0) \ldots (x - \alpha_{i-1})(x - \alpha_{i+1}) \ldots (x - \alpha_n)$. Show that $p_0, p_1, \ldots, p_n$ form a basis for P_n.

15. If H is a subspace of a finite-dimensional vector space V, show that there is a subspace K, such that

$$H \cap K = 0 \quad \text{and} \quad H + K = V$$

16. Let V be a finite-dimensional space.
    (a) Let $H_0 \leqslant H_1 \leqslant H_2 \leqslant H_3 \leqslant \cdots$ be an increasing collection of subspaces of V. Show that we must eventually have $H_n = H_{n+1} = H_{n+2} = \cdots$.
    (b) Let $H_0 \geqslant H_1 \geqslant H_2 \geqslant H_3 \geqslant \cdots$ be a decreasing family of subspaces of V. Show that we must eventually have $H_n = H_{n+1} = H_{n+2} = \cdots$.

17. Let V be a vector space and $\boldsymbol{x}_1, \ldots, \boldsymbol{x}_n$ be a collection of vectors in V. Suppose $y_1, y_2, \ldots, y_n$ belongs to $\mathrm{sp}(\boldsymbol{x}_1, \boldsymbol{x}_2, \ldots, \boldsymbol{x}_n)$. Let μ be the maximum number of linearly independent vectors in $\{\boldsymbol{x}_1, \boldsymbol{x}_2, \ldots, \boldsymbol{x}_n\}$ and ν be the maximum number of linearly independent vectors in $\{y_1, \ldots, y_n\}$. Show that $\nu \leqslant \mu$.

18. If $\boldsymbol{x}_1, \boldsymbol{x}_2, \ldots, \boldsymbol{x}_n$ are linearly independent vectors in a vector space V and if $\boldsymbol{x} \notin \mathrm{sp}(\boldsymbol{x}_1, \boldsymbol{x}_2, \ldots, \boldsymbol{x}_n)$, show that $\{\boldsymbol{x}_1 + \boldsymbol{x}, \boldsymbol{x}_2 + \boldsymbol{x}, \ldots, \boldsymbol{x}_n + \boldsymbol{x}\}$ are also linearly independent in V.

19. Let S be the set of vectors in $\boldsymbol{R}^n$ that have exactly two nonzero components and these nonzero components are both 1. Show that S is a linearly independent set of vectors if and only if $n \leqslant 3$.

20. If a is a complex number that is not real, show that a and a^2 form a basis for the complexes, regarded as a vector space over the reals.

9 CHANGE OF COORDINATES

There are many problems in mathematics and the sciences for which the standard coordinate basis $\{\boldsymbol{e}_1, \boldsymbol{e}_2, \ldots, \boldsymbol{e}_n\}$ is not the most convenient for dealing with a particular problem. For example, in studying the conic sections in plane analytic geometry, it is often convenient to bring the equations representing the conics into a standard form by means of a rotation of axes. It is to facilitate such change of coordinates that this section is presented.

Thus, let V be an n-dimensional vector space and $\{\boldsymbol{x}_1, \boldsymbol{x}_2, \ldots, \boldsymbol{x}_n\}$ be a basis for V. As we have seen, given any $\boldsymbol{x}$ in V, there are unique scalars $\alpha_1, \alpha_2, \ldots, \alpha_n$, such that $\boldsymbol{x} = \alpha_1\boldsymbol{x}_1 + \alpha_2\boldsymbol{x}_2 + \cdots + \alpha_n\boldsymbol{x}_n$. This fact enables us to "coordinatize" the space relative to the given basis. This procedure is analogous to the by now familiar method of "coordinatizing" $\boldsymbol{R}^n$ by means of the standard basis $\{\boldsymbol{e}_1, \boldsymbol{e}_2, \ldots, \boldsymbol{e}_n\}$. For if $\boldsymbol{x}$ is a vector in $\boldsymbol{R}^n$,

$$\begin{bmatrix} \alpha_1 \\ \alpha_2 \\ \vdots \\ \alpha_n \end{bmatrix} = \alpha_1 \begin{bmatrix} 1 \\ 0 \\ 0 \\ \vdots \\ 0 \end{bmatrix} + \alpha_2 \begin{bmatrix} 0 \\ 1 \\ 0 \\ \vdots \\ 0 \end{bmatrix} + \cdots + \alpha_n \begin{bmatrix} 0 \\ 0 \\ 0 \\ \vdots \\ 1 \end{bmatrix}$$
$$= \alpha_1 \boldsymbol{e}_1 + \alpha_2 \boldsymbol{e}_2 + \cdots + \alpha_n \boldsymbol{e}_n$$

In this case, the ith coordinate of the vector $\boldsymbol{x}$ is just the coefficient of $\boldsymbol{e}_i$ in the linear expression for $\boldsymbol{x}$ relative to the basis $\{\boldsymbol{e}_1, \boldsymbol{e}_2, \ldots, \boldsymbol{e}_n\}$.

Thus, if $\mathcal{B} = \{\boldsymbol{x}_1, \boldsymbol{x}_2, \ldots, \boldsymbol{x}_n\}$ is a basis for V, and if $\boldsymbol{x}$ is a vector in V and $\boldsymbol{x} = \alpha_1 \boldsymbol{x}_1 + \alpha_2 \boldsymbol{x}_2 + \cdots + \alpha_n \boldsymbol{x}_n$, we say that α_i is the ith coordinate of $\boldsymbol{x}$ relative to the basis $\mathcal{B}$. We may thus associate with the vector $\boldsymbol{x}$ its coordinate n-tuple relative to the basis $\mathcal{B}$:

$$\begin{bmatrix} \alpha_1 \\ \alpha_2 \\ \vdots \\ \alpha_n \end{bmatrix} \underset{\mathcal{B}}{\leftrightarrow} \boldsymbol{x}$$

The $\mathcal{B}$ below the arrow indicates that $\alpha_1, \alpha_2, \ldots, \alpha_n$ are coordinates of $\boldsymbol{x}$ relative to $\mathcal{B}$. Since the expression $\boldsymbol{x} = \alpha_1 \boldsymbol{x}_1 + \alpha_2 \boldsymbol{x}_2 + \cdots + \alpha_n \boldsymbol{x}_n$ for $\boldsymbol{x}$ relative to the basis is unique, the coordinates of $\boldsymbol{x}$ relative to $\mathcal{B}$ are uniquely determined.

As an example, let us calculate the coordinates of a vector $\boldsymbol{x} = \begin{bmatrix} \alpha_1 \\ \alpha_2 \end{bmatrix}$ relative to a basis $\mathcal{B} = \{\boldsymbol{x}_1, \boldsymbol{x}_2\}$, where $\boldsymbol{x}_1 = \begin{bmatrix} 1 \\ 1 \end{bmatrix}$ and $\boldsymbol{x}_2 = \begin{bmatrix} 1 \\ -1 \end{bmatrix}$. Observe that $\boldsymbol{e}_1 = \frac{1}{2}(\boldsymbol{x}_1 + \boldsymbol{x}_2)$ and $\boldsymbol{e}_2 = \frac{1}{2}(\boldsymbol{x}_1 - \boldsymbol{x}_2)$. From this, it follows that

$$\boldsymbol{x} = \begin{bmatrix} \alpha_1 \\ \alpha_2 \end{bmatrix} = \alpha_1 \boldsymbol{e}_1 + \alpha_2 \boldsymbol{e}_2$$
$$= \tfrac{1}{2}\alpha_1(\boldsymbol{x}_1 + \boldsymbol{x}_2) + \tfrac{1}{2}\alpha_2(\boldsymbol{x}_1 - \boldsymbol{x}_2) = \tfrac{1}{2}(\alpha_1 + \alpha_2)\boldsymbol{x}_1 + \tfrac{1}{2}(\alpha_1 - \alpha_2)\boldsymbol{x}_2.$$

Thus, $\boldsymbol{x} \underset{\mathcal{B}}{\leftrightarrow} \begin{bmatrix} \frac{1}{2}(\alpha_1 + \alpha_2) \\ \frac{1}{2}(\alpha_1 - \alpha_2) \end{bmatrix}$. A geometric means of visualizing this process is depicted in Figure 4-17.

There is a useful rule for computing the coordinates of a vector with respect to one basis, say $\mathcal{B} = \{\boldsymbol{x}_1, \ldots, \boldsymbol{x}_n\}$, from its coordinates with respect to another basis $\mathcal{B}' = \{\boldsymbol{x}_1', \ldots, \boldsymbol{x}_n'\}$. Relative to the bases $\mathcal{B}$ and $\mathcal{B}'$, we

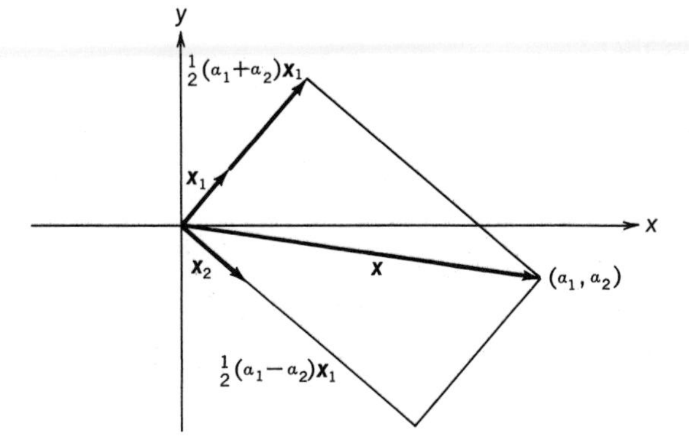

FIGURE 4-17

have:

$$\boldsymbol{x} \underset{\mathfrak{B}}{\leftrightarrow} \begin{bmatrix} \alpha_1 \\ \alpha_2 \\ \vdots \\ \alpha_n \end{bmatrix} \quad \text{and} \quad \boldsymbol{x} \underset{\mathfrak{B}'}{\leftrightarrow} \begin{bmatrix} \alpha_1' \\ \alpha_2' \\ \vdots \\ \alpha_n' \end{bmatrix}$$

Next, find the coordinates of $\boldsymbol{x}_1', \ldots, \boldsymbol{x}_n'$ with respect to $\mathfrak{B}$. Suppose that $\boldsymbol{x}_j' = \sum_{i=1}^{n} a_{ij}\boldsymbol{x}_i = a_{1j}\boldsymbol{x}_1 + a_{2j}\boldsymbol{x}_2 + \cdots + a_{nj}\boldsymbol{x}_n$. Now let A be the $n \times n$ matrix, $A = [a_{ij}]_{(nn)}$. (In words, the jth column of A is the coordinate n-tuple of $\boldsymbol{x}_j'$ with respect to $\mathfrak{B}$.) Then,

$$\begin{bmatrix} \alpha_1 \\ \alpha_2 \\ \vdots \\ \alpha_n \end{bmatrix} = A \begin{bmatrix} \alpha_1' \\ \alpha_2' \\ \vdots \\ \alpha_n' \end{bmatrix}$$

To prove this, observe

$$\begin{aligned} \boldsymbol{x} &= \sum_{j=1}^{n} \alpha_j' \boldsymbol{x}_j' \\ &= \sum_{j=1}^{n} \alpha_j' \left(\sum_{i=1}^{n} a_{ij}\boldsymbol{x}_i \right) \\ &= \sum_{j=1}^{n} \sum_{i=1}^{n} a_{ij}\alpha_j' \boldsymbol{x}_i \\ &= \sum_{i=1}^{n} \left(\sum_{j=1}^{n} a_{ij}\alpha_j' \right) \boldsymbol{x}_i \\ &= \sum_{i=1}^{n} \alpha_i \boldsymbol{x}_i \end{aligned}$$

Thus, it follows that

$$\sum_{j=1}^{n} a_{ij}\alpha_j' = \alpha_i, \quad \text{for } i = 1, 2, \ldots, n$$

If we interpret this result using matrix multiplication, the desired assertion follows.

For example in $\boldsymbol{R}^3$, let $\boldsymbol{x}_1 = \begin{bmatrix} 1 \\ 0 \\ 0 \end{bmatrix}$, $\boldsymbol{x}_2 = \begin{bmatrix} 0 \\ 1 \\ 0 \end{bmatrix}$, $\boldsymbol{x}_3 = \begin{bmatrix} 0 \\ 0 \\ 1 \end{bmatrix}$, and $\boldsymbol{x}_1' = \begin{bmatrix} 1 \\ 2 \\ 3 \end{bmatrix}$, $\boldsymbol{x}_2' = \begin{bmatrix} -1 \\ 0 \\ 3 \end{bmatrix}$, $\boldsymbol{x}_3' = \begin{bmatrix} 2 \\ 1 \\ 2 \end{bmatrix}$. In this case the coordinates $\alpha_1, \alpha_2, \alpha_3, \alpha_1', \alpha_2', \alpha_3'$ are related by the rule $\begin{bmatrix} \alpha_1 \\ \alpha_2 \\ \alpha_3 \end{bmatrix} = \begin{bmatrix} 1 & -1 & 2 \\ 2 & 0 & 1 \\ 3 & 3 & 2 \end{bmatrix} \begin{bmatrix} \alpha_1' \\ \alpha_2' \\ \alpha_3' \end{bmatrix}$.

A matrix obtained by a change of coordinates is, interestingly enough, always invertible.

To see this, suppose $\mathcal{B} = \{\boldsymbol{x}_1, \boldsymbol{x}_2, \ldots, \boldsymbol{x}_n\}$ is a basis and $\mathcal{B}' = \{\boldsymbol{x}_1', \boldsymbol{x}_2', \ldots, \boldsymbol{x}_n'\}$ is a second basis. Let

$$\begin{bmatrix} \alpha_1' \\ \alpha_2' \\ \vdots \\ \alpha_n' \end{bmatrix} \underset{\mathcal{B}'}{\leftrightarrow} \boldsymbol{x} \underset{\mathcal{B}}{\leftrightarrow} \begin{bmatrix} \alpha_1 \\ \alpha_2 \\ \vdots \\ \alpha_n \end{bmatrix}$$

be coordinate n-tuples of the vector $\boldsymbol{x}$ relative to the bases $\mathcal{B}$ and $\mathcal{B}'$.

Let $\boldsymbol{x}_j' = \sum_{i=1}^{n} a_{ij}\boldsymbol{x}_i$ be the expression of $\boldsymbol{x}_j'$ relative to the basis $\mathcal{B}$, $A = [a_{ij}]_{(nn)}$ and $\boldsymbol{x}_j = \sum_{i=1}^{n} b_{ij}\boldsymbol{x}_i'$ be the expression of $\boldsymbol{x}_j$ relative to the basis $\mathcal{B}'$, $B = [b_{ij}]_{(nn)}$.

We saw above that

$$\begin{bmatrix} \alpha_1 \\ \alpha_2 \\ \vdots \\ \alpha_n \end{bmatrix} = A \begin{bmatrix} \alpha_1' \\ \alpha_2' \\ \vdots \\ \alpha_n' \end{bmatrix} \quad \text{and} \quad \begin{bmatrix} \alpha_1' \\ \alpha_2' \\ \vdots \\ \alpha_n' \end{bmatrix} = B \begin{bmatrix} \alpha_1 \\ \alpha_2 \\ \vdots \\ \alpha_n \end{bmatrix}$$

Therefore, for all n-tuples

$$\begin{bmatrix} \alpha_1 \\ \alpha_2 \\ \vdots \\ \alpha_n \end{bmatrix} = AB \begin{bmatrix} \alpha_1 \\ \alpha_2 \\ \vdots \\ \alpha_n \end{bmatrix} \quad \text{and} \quad \begin{bmatrix} \alpha_1' \\ \alpha_2' \\ \vdots \\ \alpha_n' \end{bmatrix} = BA \begin{bmatrix} \alpha_1' \\ \alpha_2' \\ \vdots \\ \alpha_n' \end{bmatrix}$$

By the lemma of §3.7, it follows that

$$AB = I_n \quad \text{and} \quad BA = I_n$$

Thus, the matrix A is invertible and its inverse is the matrix B.

NOTE Again we mention that the ith column of A is just the coordinate n-tuple of x_i' relative to the basis $\mathfrak{B}$. The ith column of B is the coordinate n-tuple of x_i relative to the basis $\mathfrak{B}'$.

Example 1 Let P_4 be, as usual, the space of polynomials in the variable x, of degree less than or equal to 4.

Consider the two bases for P_4: $1, x, x^2, x^3, x^4$ and $1, 1 + x, (1 + x)^2, (1 + x)^3, (1 + x)^4$. Suppose we have f in P_4. Then

$$\begin{aligned} f &= \alpha_0 + \alpha_1 x + \alpha_2 x^2 + \alpha_3 x^3 + \alpha_4 x^4 \\ &= \beta_0 + \beta_1(1 + x) + \beta_2(1 + x)^2 + \beta_3(1 + x)^3 + \beta_4(1 + x)^4 \end{aligned}$$

If we form the matrix A, having as its columns the coordinates of $1, 1 + x, (1 + x)^2, (1 + x)^3, (1 + x)^4$ relative to the basis $1, x, x^2, x^3, x^4$, we obtain

$$A = \begin{bmatrix} 1 & 1 & 1 & 1 & 1 \\ 0 & 1 & 2 & 3 & 4 \\ 0 & 0 & 1 & 3 & 6 \\ 0 & 0 & 0 & 1 & 4 \\ 0 & 0 & 0 & 0 & 1 \end{bmatrix}$$

Thus,

$$\begin{bmatrix} \alpha_0 \\ \alpha_1 \\ \alpha_2 \\ \alpha_3 \\ \alpha_4 \end{bmatrix} = \begin{bmatrix} 1 & 1 & 1 & 1 & 1 \\ 0 & 1 & 2 & 3 & 4 \\ 0 & 0 & 1 & 3 & 6 \\ 0 & 0 & 0 & 1 & 4 \\ 0 & 0 & 0 & 0 & 1 \end{bmatrix} \begin{bmatrix} \beta_0 \\ \beta_1 \\ \beta_2 \\ \beta_3 \\ \beta_4 \end{bmatrix}$$

If we form the matrix B, having as its columns the coordinates of $1, x, x^2, x^3, x^4$, relative to the basis, $1, 1 + x, (1 + x)^2, (1 + x)^3, (1 + x)^4$, by observing that $x^k = ((1 + x) - 1)^k$, we see

$$B = \begin{bmatrix} 1 & -1 & 1 & -1 & 1 \\ 0 & 1 & -2 & 3 & -4 \\ 0 & 0 & 1 & -3 & 6 \\ 0 & 0 & 0 & 1 & -4 \\ 0 & 0 & 0 & 0 & 1 \end{bmatrix}$$

These matrices afford a convenient means of passing from the expression of a polynomial relative to one basis to its expression relative to another basis. As an auxiliary result we see

$$\begin{bmatrix} 1 & 1 & 1 & 1 & 1 \\ 0 & 1 & 2 & 3 & 4 \\ 0 & 0 & 1 & 3 & 6 \\ 0 & 0 & 0 & 1 & 4 \\ 0 & 0 & 0 & 0 & 1 \end{bmatrix}^{-1} = \begin{bmatrix} 1 & -1 & 1 & -1 & 1 \\ 0 & 1 & -2 & 3 & -4 \\ 0 & 0 & 1 & -3 & 6 \\ 0 & 0 & 0 & 1 & -4 \\ 0 & 0 & 0 & 0 & 1 \end{bmatrix}$$

Example 2 Using the methods of this section, we derive the standard formulas for rotation of axes encountered in plane analytic geometry.

Let

$$\boldsymbol{i} = \begin{bmatrix} 1 \\ 0 \end{bmatrix}, \quad \boldsymbol{j} = \begin{bmatrix} 0 \\ 1 \end{bmatrix}$$

be the standard basis for $\boldsymbol{R}^2$.

Let

$$\boldsymbol{i}_\theta = (\cos\theta)\boldsymbol{i} + (\sin\theta)\boldsymbol{j} \qquad \boldsymbol{j}_\theta = -(\sin\theta)\boldsymbol{i} + (\cos\theta)\boldsymbol{j}$$

be the vectors obtained by rotating $\boldsymbol{i}$ and $\boldsymbol{j}$ by θ degrees, respectively.

We wish to find a relationship between the coordinates of a poin determined by the $\{\boldsymbol{i}, \boldsymbol{j}\}$ basis and its coordinates as determined by the $\{\boldsymbol{i}_\theta, \boldsymbol{j}_\theta\}$ basis. Note that the matrix $[\boldsymbol{i}_\theta, \boldsymbol{j}_\theta]$ is

$$\begin{bmatrix} \cos\theta & -\sin\theta \\ \sin\theta & \cos\theta \end{bmatrix}$$

Let $\boldsymbol{v} = x\boldsymbol{i} + y\boldsymbol{j}$ be the linear expression for the vector $\boldsymbol{v}$ relative to the $\{\boldsymbol{i}, \boldsymbol{j}\}$ basis, and $\boldsymbol{v} = x_\theta \boldsymbol{i}_\theta + y_\theta \boldsymbol{j}_\theta$ be the expression relative to the $\{\boldsymbol{i}_\theta, \boldsymbol{j}_\theta\}$ basis. Then by an earlier result, we have

$$\begin{bmatrix} x \\ y \end{bmatrix} = \begin{bmatrix} \cos\theta & -\sin\theta \\ \sin\theta & \cos\theta \end{bmatrix} \begin{bmatrix} x_\theta \\ y_\theta \end{bmatrix}$$

Since the $\{\boldsymbol{i}, \boldsymbol{j}\}$ basis may be obtained by rotating the $\{\boldsymbol{i}_\theta, \boldsymbol{j}_\theta\}$ basis by $-\theta$ degrees, it follows that

$$\begin{bmatrix} x_\theta \\ y_\theta \end{bmatrix} = \begin{bmatrix} \cos\theta & -\sin\theta \\ \sin\theta & \cos\theta \end{bmatrix} \begin{bmatrix} x \\ y \end{bmatrix}$$

$$= \begin{bmatrix} \cos\theta & \sin\theta \\ -\sin\theta & \cos\theta \end{bmatrix} \begin{bmatrix} x \\ y \end{bmatrix}$$

Writing out the two matrix equations explicitly, we obtain

$$x = (\cos\theta)x_\theta - (\sin\theta)y_\theta$$

$$y = (\sin\theta)x_\theta + (\cos\theta)y_\theta$$

$$x_\theta = (\cos\theta)x + (\sin\theta)y$$

$$y_\theta = (-\sin\theta)x + (\cos\theta)y$$

which are, of course, the standard formulas for rotation of axes in analytic geometry.

Our previous results imply (as the reader may verify by other means) that

$$\begin{bmatrix} \cos\theta & -\sin\theta \\ \sin\theta & \cos\theta \end{bmatrix}^{-1} = \begin{bmatrix} \cos\theta & \sin\theta \\ -\sin\theta & \cos\theta \end{bmatrix}$$

As another application of the ideas of this section, we have

Theorem 1 Let $\boldsymbol{x}_1, \boldsymbol{x}_2, \ldots, \boldsymbol{x}_n$ be vectors in $\boldsymbol{R}^n$ (or $\boldsymbol{C}^n$). Let $A = [\boldsymbol{x}_1, \boldsymbol{x}_2, \ldots, \boldsymbol{x}_n]$ be the matrix whose jth column is the vector $\boldsymbol{x}_j$. Then the following are equivalent:

(i) The vectors $\boldsymbol{x}_1, \boldsymbol{x}_2, \ldots, \boldsymbol{x}_n$ are linearly independent.
(ii) The matrix A is invertible.
(iii) $\det A \neq 0$.

PROOF We already know from chapter 3 that (ii) and (iii) are equivalent. Thus, it will suffice to show that (i) and (ii) are equivalent.

First, suppose the vectors $\boldsymbol{x}_1, \boldsymbol{x}_2, \ldots, \boldsymbol{x}_n$ are linearly independent. By Theorem 2, §4.8, we know that $\boldsymbol{x}_1, \boldsymbol{x}_2, \ldots, \boldsymbol{x}_n$ form a basis for $\boldsymbol{R}^n$. Thus, the matrix A represents a change of coordinates from the standard basis $\{\boldsymbol{e}_1, \boldsymbol{e}_2, \ldots, \boldsymbol{e}_n\}$ to the basis $\{\boldsymbol{x}_1, \boldsymbol{x}_2, \ldots, \boldsymbol{x}_n\}$, and we have seen above that any such matrix is invertible. Thus (i) implies (ii).

Second, suppose the vectors $\boldsymbol{x}_1, \boldsymbol{x}_2, \ldots, \boldsymbol{x}_n$ are linearly dependent. Then one of the vectors, say $\boldsymbol{x}_1$, is a linear combination of the remaining vectors

$$\boldsymbol{x}_1 = \alpha_2\boldsymbol{x}_2 + \alpha_3\boldsymbol{x}_3 + \cdots + \alpha_n\boldsymbol{x}_n$$

Then

$$\begin{aligned} \det A &= \det[\boldsymbol{x}_1, \boldsymbol{x}_2, \ldots, \boldsymbol{x}_n] \\ &= \det[\boldsymbol{x}_1 - \alpha_2\boldsymbol{x}_2 - \alpha_3\boldsymbol{x}_3 \cdots - \alpha_n\boldsymbol{x}_n, \boldsymbol{x}_2, \ldots, \boldsymbol{x}_n] \quad [\text{by (D5)}] \\ &= \det[\boldsymbol{0}, \boldsymbol{x}_2, \ldots, \boldsymbol{x}_n] \\ &= 0 \quad [\text{by (D7)}] \end{aligned}$$

Thus, the matrix A is not invertible. From this we see that (ii) implies (i). ■

This theorem provides a convenient means for testing whether a given set of vectors forms a basis.

Example 3 Do the vectors

$$\begin{bmatrix} 1 \\ 2 \\ 4 \\ 8 \end{bmatrix}, \quad \begin{bmatrix} 1 \\ 0 \\ 1 \\ 7 \end{bmatrix}, \quad \begin{bmatrix} 6 \\ 1 \\ 0 \\ 1 \end{bmatrix}, \quad \begin{bmatrix} -1 \\ -1 \\ 0 \\ 0 \end{bmatrix}$$

form a basis for $\boldsymbol{R}^4$? By the previous theorem, it is only necessary to calculate the determinant

$$\begin{vmatrix} 1 & 1 & 6 & -1 \\ 2 & 0 & 1 & -1 \\ 4 & 1 & 0 & 0 \\ 8 & 7 & 1 & 0 \end{vmatrix} = \begin{vmatrix} 1 & 0 & 0 & 0 \\ 2 & -2 & -11 & 1 \\ 4 & -3 & -24 & 4 \\ 8 & -1 & -47 & 8 \end{vmatrix} = \begin{vmatrix} 1 & 0 & 0 & 0 \\ 0 & 0 & 0 & 1 \\ -4 & 5 & 20 & 4 \\ -8 & 15 & 41 & 8 \end{vmatrix}$$

$$= \begin{vmatrix} 5 & 20 \\ 15 & 41 \end{vmatrix} = 205 - 300 = -95 \neq 0$$

Thus, the vectors do indeed form a basis.

Example 4 Are the vectors

$$\begin{bmatrix} 1 \\ 3 \\ -1 \end{bmatrix}, \quad \begin{bmatrix} 2 \\ 1 \\ 3 \end{bmatrix}, \quad \begin{bmatrix} 3 \\ 4 \\ 2 \end{bmatrix}$$

linearly independent? We consider the determinant

$$\begin{vmatrix} 1 & 2 & 3 \\ 3 & 1 & 4 \\ -1 & 3 & 2 \end{vmatrix} = \begin{vmatrix} 1 & 0 & 0 \\ 3 & -5 & -5 \\ -1 & 5 & 5 \end{vmatrix} = \begin{vmatrix} -5 & -5 \\ 5 & 5 \end{vmatrix} = 0$$

Thus, the vectors are linearly dependent.

Actually, the method of proof of Theorem 1 yields the stronger result:

Let V be an n-dimensional vector space and let $\boldsymbol{x}_1, \boldsymbol{x}_2, \ldots, \boldsymbol{x}_n$ be a basis for V. Let $\boldsymbol{x}'_1, \boldsymbol{x}'_2, \ldots, \boldsymbol{x}'_n$ be n vectors in V, and suppose $\boldsymbol{x}'_j = \sum_{i=1}^{n} \alpha_{ij}\boldsymbol{x}_j$ is the expression for $\boldsymbol{x}'_j$ relative to the basis $\boldsymbol{x}_1, \boldsymbol{x}_2, \ldots, \boldsymbol{x}_n$.

Then the following are equivalent statements.

(1) $\boldsymbol{x}'_1, \boldsymbol{x}'_2, \boldsymbol{x}'_3, \ldots, \boldsymbol{x}'_n$ are linearly independent.

(2)

$$A = \begin{bmatrix} \alpha_{11} & \alpha_{12} & \cdots & \alpha_{1n} \\ \alpha_{21} & \alpha_{22} & \cdots & \alpha_{2n} \\ & & \vdots & \\ \alpha_{n1} & \alpha_{n2} & \cdots & \alpha_{nn} \end{bmatrix} \text{ is invertible.}$$

(3) $\det A \neq 0$.

Theorem 2 Let

$$\begin{aligned}
a_{11}x_1 + a_{12}x_2 + \cdots + a_{1n}x_n &= 0 \\
a_{21}x_1 + a_{22}x_2 + \cdots + a_{2n}x_n &= 0 \\
&\vdots \\
a_{n1}x_1 + a_{n2}x_2 + \cdots + a_{nn}x_n &= 0
\end{aligned}$$

be a system of n homogeneous linear equations in n variables. The system admits a nontrivial solution, i.e., a solution in which not all the x_i's are 0, if and only if

$$\begin{vmatrix} a_{11} & a_{12} & \cdots & a_{1n} \\ a_{21} & a_{22} & \cdots & a_{2n} \\ & & \vdots & \\ a_{n1} & a_{n2} & \cdots & a_{nn} \end{vmatrix} = 0$$

PROOF We let $A = [a_{ij}]$ be the matrix of coefficients of the system. Then, $A = [A_1, A_2, \ldots, A_n]$, where A_j is the jth column of A. If $\det A = 0$, by Theorem 1, we know that the columns of the matrix A are linearly dependent. Thus, we have $\alpha_1 A_1 + \alpha_2 A_2 + \cdots + \alpha_n A_n = 0$, for scalars $\alpha_1, \alpha_2, \ldots, \alpha_n$ not all zero. But

$$\begin{bmatrix} a_{11} & a_{12} & \cdots & a_{1n} \\ a_{21} & a_{22} & \cdots & a_{2n} \\ & & \vdots & \\ a_{n1} & a_{n2} & \cdots & a_{nn} \end{bmatrix} \begin{bmatrix} \alpha_1 \\ \alpha_2 \\ \vdots \\ \alpha_n \end{bmatrix} = \begin{bmatrix} \alpha_1 a_{11} + \alpha_2 a_{12} + \cdots + \alpha_n a_{1n} \\ \alpha_1 a_{21} + \alpha_2 a_{22} + \cdots + \alpha_n a_{2n} \\ \vdots \\ \alpha_1 a_{n1} + \alpha_2 a_{n2} + \cdots + \alpha_n a_{nn} \end{bmatrix}$$

$$= \left[\alpha_1 A_1 + \alpha_2 A_2 + \cdots + \alpha_n A_n \right] = 0$$

Thus, $x_1 = \alpha_1$, $x_2 = \alpha_2, \ldots, x_n = \alpha_n$ is a nontrivial solution to the system.

If, on the other hand, $\det A \neq 0$, then A^{-1} exists. So if $A\boldsymbol{x} = 0$, then $A^{-1}(A\boldsymbol{x}) = \boldsymbol{x} = \mathbf{0}$. Thus, no nontrivial solution exists. ■

Example 5 In the system

$$\begin{aligned}
x + 2y - z &= 0 \\
x + 3y + 2z &= 0 \\
x + 5y + 8z &= 0
\end{aligned}$$

the determinant of the coefficient matrix is

$$\begin{vmatrix} 1 & 2 & -1 \\ 1 & 3 & 2 \\ 1 & 5 & 8 \end{vmatrix} = \begin{vmatrix} 1 & 0 & 0 \\ 1 & 1 & 3 \\ 1 & 3 & 9 \end{vmatrix} = 0$$

Thus, the system admits nontrivial solutions, for example, $x = -7$, $y = 3$, $z = -1$.

EXERCISES

1. Determine the coordinates of the vector $\begin{bmatrix} x \\ y \end{bmatrix}$ in $\boldsymbol{R}^2$, relative to each of the following bases for $\boldsymbol{R}^2$.

$$\text{(a)}\ \begin{bmatrix} 1 \\ 0 \end{bmatrix}, \begin{bmatrix} 1 \\ 1 \end{bmatrix} \qquad \text{(b)}\ \begin{bmatrix} 1 \\ 1 \end{bmatrix}, \begin{bmatrix} 1 \\ -1 \end{bmatrix}$$
$$\text{(c)}\ \begin{bmatrix} 0 \\ 1 \end{bmatrix}, \begin{bmatrix} -1 \\ -1 \end{bmatrix} \qquad \text{(d)}\ \begin{bmatrix} 2 \\ 1 \end{bmatrix}, \begin{bmatrix} 5 \\ 3 \end{bmatrix}$$

2. Determine the coordinates of the vector $\begin{bmatrix} x \\ y \\ z \end{bmatrix}$ relative to each of the following bases for $\boldsymbol{R}^3$.

$$\text{(a)}\ \begin{bmatrix} 1 \\ 0 \\ 0 \end{bmatrix}, \begin{bmatrix} 1 \\ 2 \\ 1 \end{bmatrix}, \begin{bmatrix} 1 \\ 5 \\ 3 \end{bmatrix} \qquad \text{(b)}\ \begin{bmatrix} 1 \\ 1 \\ 0 \end{bmatrix}, \begin{bmatrix} 0 \\ 1 \\ 1 \end{bmatrix}, \begin{bmatrix} 1 \\ 0 \\ 1 \end{bmatrix}$$
$$\text{(c)}\ \begin{bmatrix} 1 \\ 3 \\ 0 \end{bmatrix}, \begin{bmatrix} 2 \\ 1 \\ 3 \end{bmatrix}, \begin{bmatrix} -1 \\ 1 \\ 2 \end{bmatrix} \qquad \text{(d)}\ \begin{bmatrix} 1 \\ 1 \\ -1 \end{bmatrix}, \begin{bmatrix} 1 \\ -1 \\ 1 \end{bmatrix}, \begin{bmatrix} -1 \\ 1 \\ 1 \end{bmatrix}$$

3. Determine which of the following sets of vectors are independent in the indicated vector spaces.

(a) $\begin{bmatrix} -1 & 1 \\ 1 & 1 \end{bmatrix}, \begin{bmatrix} 1 & -1 \\ 1 & 1 \end{bmatrix}, \begin{bmatrix} 1 & 1 \\ -1 & 1 \end{bmatrix}, \begin{bmatrix} 1 & 1 \\ 1 & -1 \end{bmatrix}$
in the space of 2×2 matrices.

(b) $\begin{bmatrix} 1 & -1 \\ 0 & 1 \end{bmatrix}, \begin{bmatrix} -1 & 1 \\ 0 & 1 \end{bmatrix}, \begin{bmatrix} 1 & 1 \\ 0 & -1 \end{bmatrix}$
in the space of 2×2 upper triangular matrices.

(c) $1 + x + x^2 + x^3$, $1 - x + x^2 + x^3$, $1 + x - x^2 + x^3$, $1 + x + x^2 - x^3$, in P_3.

4. For what values of λ do the following sets of vectors form a basis for $\boldsymbol{R}^3$?

$$\text{(a)}\ \begin{bmatrix} 1 \\ \lambda \\ 0 \end{bmatrix}, \begin{bmatrix} \lambda \\ 1 \\ \lambda \end{bmatrix}, \begin{bmatrix} 0 \\ \lambda \\ 1 \end{bmatrix} \qquad \text{(b)}\ \begin{bmatrix} 1 \\ 0 \\ \lambda \end{bmatrix}, \begin{bmatrix} 0 \\ 1 \\ 0 \end{bmatrix}, \begin{bmatrix} \lambda \\ 0 \\ 1 \end{bmatrix}$$
$$\text{(c)}\ \begin{bmatrix} 1 \\ 0 \\ 0 \end{bmatrix}, \begin{bmatrix} \lambda \\ 1 \\ 0 \end{bmatrix}, \begin{bmatrix} \lambda^2 \\ \lambda \\ 1 \end{bmatrix} \qquad \text{(d)}\ \begin{bmatrix} 0 \\ \lambda \\ \lambda \end{bmatrix}, \begin{bmatrix} \lambda \\ 0 \\ \lambda \end{bmatrix}, \begin{bmatrix} \lambda \\ \lambda \\ 0 \end{bmatrix}$$

5. For what real values of λ do the following systems of equations admit nontrivial solutions?

$$\text{(a)}\ \begin{aligned} \lambda x + y + z &= 0 \\ x + \lambda y + z &= 0 \\ x + y + \lambda z &= 0 \end{aligned} \qquad \text{(b)}\ \begin{aligned} x + y + z &= 0 \\ \lambda x + y + z &= 0 \\ \lambda^2 x + y + z &= 0 \end{aligned}$$

6. Let (x_1, y_1), (x_2, y_2), (x_3, y_3) be the coordinates of three points in the plane. Show that the points lie on a line if and only if $\begin{vmatrix} x_1 & y_1 & 1 \\ x_2 & y_2 & 1 \\ x_3 & y_3 & 1 \end{vmatrix} = 0.$

7. Let (x_1, y_1), (x_2, y_2), (x_3, y_3), (x_4, y_4) be the coordinates of four points in the plane.

Show that the points lie on a line or circle if and only if

$$\begin{vmatrix} x_1^2+y_1^2 & x_1 & y_1 & 1 \\ x_2^2+y_2^2 & x_2 & y_2 & 1 \\ x_3^2+y_3^2 & x_3 & y_3 & 1 \\ x_4^2+y_4^2 & x_4 & y_4 & 1 \end{vmatrix} = 0$$

8. In P_3 determine the matrices associated with the change from the basis $1, x, x^2, x^3$ to the basis $1, 1+x, (1+x)^2, (1+x)^3$.

9. If $\boldsymbol{x}_1, \boldsymbol{x}_2, \ldots, \boldsymbol{x}_n$ form a basis for the vector space V, show that the following sets of vectors also form a basis for V.
   (a) $\boldsymbol{x}_1, \boldsymbol{x}_1+\boldsymbol{x}_2, \boldsymbol{x}_1+\boldsymbol{x}_2+\boldsymbol{x}_3, \ldots, \boldsymbol{x}_1+\boldsymbol{x}_2+\boldsymbol{x}_3+\cdots+\boldsymbol{x}_n$
   (b) $\boldsymbol{x}_1+\boldsymbol{x}_2+\cdots+\boldsymbol{x}_n, \boldsymbol{x}_1-\boldsymbol{x}_2+\boldsymbol{x}_3+\cdots+\boldsymbol{x}_n,$
   $\boldsymbol{x}_1+\boldsymbol{x}_2-\boldsymbol{x}_3+\cdots+\boldsymbol{x}_n, \ldots, \boldsymbol{x}_1+\boldsymbol{x}_2+\boldsymbol{x}_3+\cdots+(-\boldsymbol{x}_n)$
   (c) $\boldsymbol{x}_1, \boldsymbol{x}_1+\boldsymbol{x}_2, \boldsymbol{x}_1+\boldsymbol{x}_3, \ldots, \boldsymbol{x}_1+\boldsymbol{x}_n$

10. Show that any set of three distinct vectors whose endpoints lie on the parametrized parabolic curve

$$x(t)=1, \qquad y(t)=t, \qquad z(t)=t^2$$

form a basis for $\boldsymbol{R}^3$.

11. Let

$$\begin{bmatrix} a_1 \\ a_2 \\ a_3 \\ \vdots \\ a_n \end{bmatrix}, \quad \begin{bmatrix} b_1 \\ b_2 \\ b_3 \\ \vdots \\ b_n \end{bmatrix}, \quad \begin{bmatrix} c_1 \\ c_2 \\ c_3 \\ \vdots \\ c_n \end{bmatrix}$$

be three vectors in $\boldsymbol{R}^n$. If

$$\begin{vmatrix} a_1 & b_1 & c_1 \\ a_2 & b_2 & c_2 \\ a_3 & b_3 & c_3 \end{vmatrix} \neq 0$$

show that the three vectors are linearly independent.

12. If p_0, p_1, p_2 are independent polynomials in P_2 and x_0, x_1, and x_2 are distinct real numbers, show that the vectors

$$\begin{bmatrix} p_0(x_0) \\ p_1(x_0) \\ p_2(x_0) \end{bmatrix}, \quad \begin{bmatrix} p_0(x_1) \\ p_1(x_1) \\ p_2(x_1) \end{bmatrix}, \quad \begin{bmatrix} p_0(x_2) \\ p_1(x_2) \\ p_2(x_2) \end{bmatrix}$$

are linearly independent.

13. If $\boldsymbol{x}_1, \boldsymbol{x}_2, \ldots, \boldsymbol{x}_n$ is a basis for a vector space V, show that the vectors $\boldsymbol{x}_1 - \boldsymbol{x}, \boldsymbol{x}_2 - \boldsymbol{x}, \ldots, \boldsymbol{x}_n - \boldsymbol{x}$ form a basis for V if and only if $\boldsymbol{x}$ cannot be expressed in the form

$$\boldsymbol{x} = \alpha_1\boldsymbol{x}_1 + \alpha_2\boldsymbol{x}_2 + \cdots + \alpha_n\boldsymbol{x}_n$$

with

$$\alpha_1 + \alpha_2 + \cdots + \alpha_n = 1$$

14. A function f on $\boldsymbol{R}^3$ is defined by the following rule: If $\boldsymbol{v} = \begin{bmatrix} x \\ y \\ z \end{bmatrix}$, then $f(\boldsymbol{v}) = x^2 + 2xy - 4xz + y^2 + z^2$. Let $\mathcal{B}$ be the basis $\begin{bmatrix} 0 \\ 1 \\ 1 \end{bmatrix}, \begin{bmatrix} 1 \\ 0 \\ 1 \end{bmatrix}, \begin{bmatrix} 1 \\ 1 \\ 0 \end{bmatrix}$. Let $\boldsymbol{v} \underset{\mathcal{B}}{\leftrightarrow} \begin{bmatrix} x' \\ y' \\ z' \end{bmatrix}$. Show that $f(\boldsymbol{v}) = 2(x')^2 - (y')^2 + 4(z')^2$. (This shows how a formula may be simplified by judicious choice of coordinates.)

15. Show that $\begin{bmatrix} 1 \\ 2 \\ 3 \end{bmatrix}, \begin{bmatrix} 2 \\ 1 \\ 3 \end{bmatrix}, \begin{bmatrix} x \\ y \\ z \end{bmatrix}$ is linearly independent if and only if $3x + 3y - 3z \neq 0$.

linear transformations

1 LINEAR TRANSFORMATIONS DEFINED

In the last chapter, we studied vector spaces. In this chapter, we study linear transformations. In essence, a linear transformation is a function from one vector space to another that preserves algebraic structure. In the last chapter, our model and most common vector space was the space of column vectors of a fixed size. In this chapter, matrices provide our models for linear transformations. We shall also see that, in a very precise sense, all vector spaces may be realized as spaces of column vectors and all linear transformations as matrices.

Definition Let V and W be vector spaces and T be a function from V into W with the properties

(i) $T(\boldsymbol{x}+\boldsymbol{y}) = T(\boldsymbol{x}) + T(\boldsymbol{y})$, for all vectors $\boldsymbol{x}$ and $\boldsymbol{y}$ in V.
(ii) $T(\alpha\boldsymbol{x}) = \alpha T(\boldsymbol{x})$, for all vectors $\boldsymbol{x}$ in V and scalars α.

Then, T is said to be a **linear transformation** from V to W.

Often we write $T : V \to W$ to indicate that T is a linear transformation from the vector space V into the vector space W.

Example 1 Let A be an $m \times n$ matrix with real entries. Consider the function T_A from $\boldsymbol{R}^n$ into $\boldsymbol{R}^m$ defined by

$$T_A(\boldsymbol{x}) = A\boldsymbol{x}$$

where $\boldsymbol{x}$ is an n-vector.

If $\boldsymbol{x}$ is an n-vector, the product $A\boldsymbol{x}$ is, of course, defined and is an m-vector. Hence, T_A defines a function from $\boldsymbol{R}^n$ into $\boldsymbol{R}^m$.

To verify linearity, observe that if $\boldsymbol{x}$ and $\boldsymbol{y}$ belong to $\boldsymbol{R}^n$ and α is a scalar, we have

$$\begin{aligned} T_A(\boldsymbol{x}+\boldsymbol{y}) &= A(\boldsymbol{x}+\boldsymbol{y}) && \text{(by definition of } T_A\text{)} \\ &= A\boldsymbol{x} + A\boldsymbol{y} && \text{(by the distributive law of matrix multiplication)} \\ &= T_A(\boldsymbol{x}) + T_A(\boldsymbol{y}) && \text{(by definition of } T_A\text{)} \end{aligned}$$

and

$$\begin{aligned} T_A(\alpha\boldsymbol{x}) &= A(\alpha\boldsymbol{x}) \\ &= \alpha A\boldsymbol{x} \\ &= \alpha T_A(\boldsymbol{x}) \end{aligned}$$

For example, the function T, from $\boldsymbol{R}^3$ to $\boldsymbol{R}^2$, defined by

$$T\left(\begin{bmatrix} x \\ y \\ z \end{bmatrix}\right) = \begin{bmatrix} 1 & -1 & 1 \\ 0 & 1 & 1 \end{bmatrix}\begin{bmatrix} x \\ y \\ z \end{bmatrix} = \begin{bmatrix} x-y+z \\ y+z \end{bmatrix}$$

is linear.

Example 2 If A is an $m \times n$ matrix with complex entries, the function T_A, defined by $T_A(\boldsymbol{x}) = A\boldsymbol{x}$, for $\boldsymbol{x}$ belonging to $\boldsymbol{C}^n$, defines a linear transformation from $\boldsymbol{C}^n$ to $\boldsymbol{C}^m$.

Example 3 Let T_θ be a function from $\boldsymbol{R}^2$ into $\boldsymbol{R}^2$ defined as follows: If $\boldsymbol{v}$ is a vector in the plane, $T_\theta(\boldsymbol{v})$ is obtained by rotating $\boldsymbol{v}$ θ degrees in the counterclockwise direction. (See Figure 5-1.)

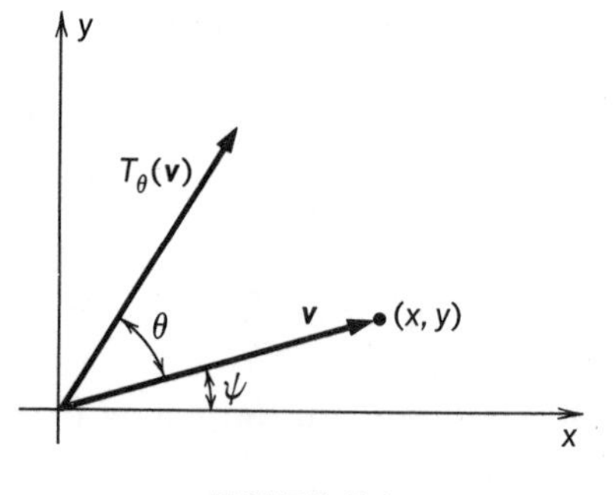

FIGURE 5-1

Let r be the length of the vector $\boldsymbol{v}$ and ψ be the angle the vector $\boldsymbol{v}$ makes with the x axis. If the vector $\boldsymbol{v}$ ends at the point (x, y) in the plane, we have $\boldsymbol{v} = x\boldsymbol{i} + y\boldsymbol{j}$.

By trigonometry, it follows that $x = r\cos\psi$ and $y = r\sin\psi$. Thus, $\boldsymbol{v} = (r\cos\psi)\boldsymbol{i} + (r\sin\psi)\boldsymbol{j}$.

Since the vector $T_\theta(\boldsymbol{v})$ is obtained by rotating $\boldsymbol{v}$ by θ degrees, $T_\theta(\boldsymbol{v})$ has the same length as $\boldsymbol{v}$, namely r, and makes an angle of $\theta + \psi$ degrees with the

x axis. Thus,

$$\begin{aligned} T_\theta(\boldsymbol{v}) &= (r\cos(\theta+\psi))\boldsymbol{i} + (r\sin(\theta+\psi))\boldsymbol{j} \\ &= (r\cos\theta\cos\psi - r\sin\theta\sin\psi)\boldsymbol{i} \\ &\quad + (r\cos\theta\sin\psi + r\sin\theta\cos\psi)\boldsymbol{j} \\ &= (x\cos\theta - y\sin\theta)\boldsymbol{i} + (y\cos\theta + x\sin\theta)\boldsymbol{j} \end{aligned}$$

In terms of matrices and column vectors,

$$T_\theta\left(\begin{bmatrix} x \\ y \end{bmatrix}\right) = \begin{bmatrix} \cos\theta & -\sin\theta \\ \sin\theta & \cos\theta \end{bmatrix}\begin{bmatrix} x \\ y \end{bmatrix}$$

Now, it follows from example 1 that T_θ is a linear transformation from $\boldsymbol{R}^2$ into itself.

In this example, we used geometry to derive an algebraic expression for T_θ. Then, by manipulating the algebraic expression, we showed that T_θ was linear. However, the linearity could also have been demonstrated from purely geometrical considerations.

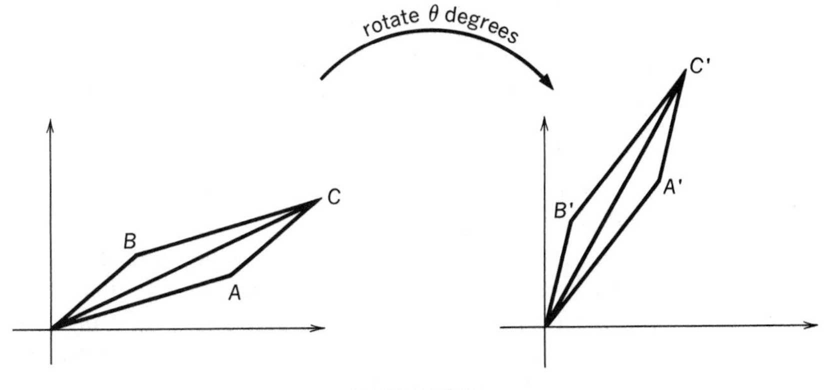

FIGURE 5-2

For example, we show that $T_\theta(\boldsymbol{u}+\boldsymbol{v}) = T_\theta(\boldsymbol{u}) + T_\theta(\boldsymbol{v})$. In Figure 5-2, let $\boldsymbol{u}$ and $\boldsymbol{v}$ be the vectors ending at A and B, respectively. Then, $\boldsymbol{u}+\boldsymbol{v}$ ends at C. Also, $T_\theta(\boldsymbol{u})$, $T_\theta(\boldsymbol{v})$, and $T_\theta(\boldsymbol{u}+\boldsymbol{v})$ end at A', B', and C', respectively. The important thing to note is that since T_θ is a rotation and $OABC$ is a parallelogram, $OA'B'C'$ is still a parallelogram. Thus, by vector addition it follows that $T_\theta(\boldsymbol{u}+\boldsymbol{v}) = T_\theta(\boldsymbol{u}) + T_\theta(\boldsymbol{v})$. Similarly, one can show that $T_\theta(\alpha\boldsymbol{u}) = \alpha T_\theta(\boldsymbol{u})$.

Example 4 A certain product is manufactured by three companies, L, M, and N, that completely control the market. Each year each company retains a certain percentage of its customers and a certain percentage switch to its competitors. We indicate the amount of retention and loss with the matrix

$$A = \begin{array}{c} \\ \\ \\ \end{array} \begin{array}{c} \begin{array}{ccc} L & M & N \end{array} \\ \begin{bmatrix} \frac{1}{2} & \frac{1}{3} & \frac{1}{6} \\ \frac{1}{4} & \frac{1}{2} & \frac{1}{6} \\ \frac{1}{4} & \frac{1}{6} & \frac{2}{3} \end{bmatrix} \end{array} \begin{array}{c} \\ L \\ M \\ N \end{array}$$

Reading down the first column, we see that L retains $\frac{1}{2}$ its customers, loses $\frac{1}{4}$ to M, and $\frac{1}{4}$ to N, with similar interpretations for other columns. We define a vector $\mathbf{v} = \begin{bmatrix} x \\ y \\ z \end{bmatrix}$, where x is the fraction of the market controlled by L at some particular time, y that of M, and z that of N. We say that $\mathbf{v}$ represents the state of the market. Let $T(\mathbf{v})$ be the state of the market one year later. Then $T(\mathbf{v}) = A\mathbf{v}$. For example, we check M's share of the market after one year. M acquires $\frac{1}{4}$ of L's customers, getting $\frac{1}{4}x$ share of the market from this source. It retains $\frac{1}{2}$ its own customers for $\frac{1}{2}y$ share of the market. From N, M picks up $\frac{1}{6}z$ part of the market. Altogether then, after one year M's share of the market is $\frac{1}{4}x + \frac{1}{2}y + \frac{1}{6}z$. This is, of course, just the second component of the vector $A\mathbf{v}$.

In this case, it is perfectly natural to regard T as being defined on all of $\mathbf{R}^3$ by the formula $T(\mathbf{v}) = A\mathbf{v}$. Example 1 above shows that T is a linear transformation from $\mathbf{R}^3$ into itself. In this example, T transforms the state of the market one year to its state the next year.

The matrix, A, is a particular example of a stochastic matrix. Any matrix whose column sums are all 1 and whose entries are nonnegative is called a stochastic matrix. Earlier in the test, we encountered several examples of such matrices. In example 2 of §2.4, the matrices A_1, A_2, A_1A_2, and A_2A_1 are all stochastic matrices. Such matrices are of great importance in the study of Markov processes, of which both this example and the example of §2.4 are instances. In the study of these processes involving stochastic matrices, the theory of linear transformations is very useful.

Examples of this type are often encountered in mathematical models in the behavioral and social sciences.

In all of the above examples, matrices were used to define the linear transformations. This need not be the case, as the following show.

Example 5 Let M_{mn} denote the space of $m \times n$ matrices with real entries and M_{nm} the space of $n \times m$ matrices with real entries. Consider the function T from M_{mn} to M_{nm} defined by $T(A) = A^T$, where A is an $m \times n$ matrix.

Since the transpose of an $m \times n$ matrix is an $n \times m$ matrix, T is a well-defined function from M_{mn} to M_{nm}. If A and B are $m \times n$ matrices and α

is a scalar,

$$\begin{aligned} T(A+B) &= (A+B)^T && \text{(by definition of } T) \\ &= A^T + B^T && (\S 2.7) \\ &= T(A) + T(B) && \text{(by definition of } T) \end{aligned}$$

and

$$\begin{aligned} T(\alpha A) &= (\alpha A)^T && \text{(by definition of } T) \\ &= \alpha A^T && (\S 2.7) \\ &= \alpha T(A) && \text{(by definition of } T) \end{aligned}$$

It follows that T is a linear transformation.

Example 6 Let D be the function from P_n into P_n defined by $D(f) = f'$, where f' denotes the derivative of f. Since the derivative of a polynomial of degree less than or equal to n is a polynomial of degree less than or equal to n, D is a well-defined function from P_n into P_n.

To demonstrate linearity, observe that

$$\begin{aligned} D(f+g) &= (f+g)' && \text{(by definition of } D) \\ &= f' + g' && \text{(by Calculus)} \\ &= D(f) + D(g) && \text{(by definition of } D) \end{aligned}$$

and

$$\begin{aligned} D(\alpha f) &= (\alpha f)' && \text{(by definition of } D) \\ &= \alpha f' && \text{(by Calculus)} \\ &= \alpha D(f) && \text{(by definition of } D) \end{aligned}$$

If $\boldsymbol{x}$ is a vector in a vector space V and T is a linear transformation from V into a vector space W, and if $T(\boldsymbol{x}) = \boldsymbol{y}$, we say that the vector $\boldsymbol{y}$ is the **image** under T of the vector $\boldsymbol{x}$. We may also say that $\boldsymbol{x}$ goes into $\boldsymbol{y}$ under T, or that $\boldsymbol{x}$ is sent into $\boldsymbol{y}$ by T. Thus, for example, the image of $1 + x + 3x^2$ under the differentiation operator D of example 4 is $1 + 6x$.

In many cases a linear transformation of a vector space into itself is called a **linear operator**.

If T is a linear operator on $\boldsymbol{R}^2$, the image of an arbitrary vector of $\boldsymbol{R}^2$ can be determined if the images under T of the basis vectors $\boldsymbol{i}$ and $\boldsymbol{j}$ are known. Suppose that $T(\boldsymbol{i}) = \boldsymbol{v}$ and $T(\boldsymbol{j}) = \boldsymbol{w}$. Then, $T(x\boldsymbol{i} + y\boldsymbol{j}) = T(x\boldsymbol{i}) + T(y\boldsymbol{j}) = xT(\boldsymbol{i}) + yT(\boldsymbol{j}) = x\boldsymbol{v} + y\boldsymbol{w}$. Thus, the image of $x\boldsymbol{i} + y\boldsymbol{j}$ can be computed using $\boldsymbol{v}$ and $\boldsymbol{w}$. This is illustrated in Figure 5-3.

In fact, similar considerations show that the image of any two linearly independent vectors under T determines the images of all vectors of $\boldsymbol{R}^2$. In some applications, this may be useful.

Example 7 An energy company produces coal and natural gas. A certain fraction of the coal which the company mines is consumed within the company in the production of more coal and natural gas. Likewise, some of the gas tapped is consumed internally.

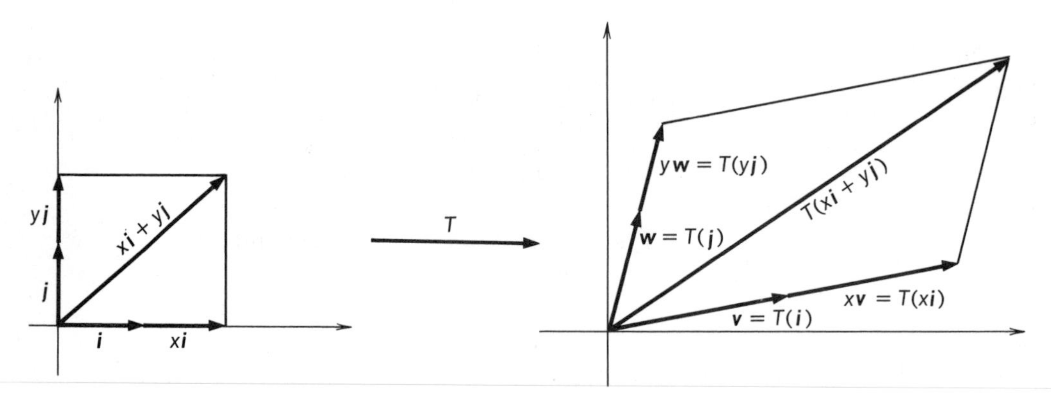

FIGURE 5-3

We represent the amount of coal mined and gas tapped during a week with the vector $\begin{bmatrix} u \\ v \end{bmatrix}$. Here u is the number of units of coal mined during one week and v the number of units of gas tapped. We call this vector the gross output.

The amount of coal and gas available for sale during a week is some function of the gross output for that week. Accordingly, we define $T\left(\begin{bmatrix} u \\ v \end{bmatrix}\right) = \begin{bmatrix} s \\ t \end{bmatrix}$, where s and t are the number of units of coal and gas that the company produces for sale when the gross output is given by the vector $\begin{bmatrix} u \\ v \end{bmatrix}$. We call $\begin{bmatrix} s \\ t \end{bmatrix}$ the net output. In this case, T is a function that transforms gross to net output.

Our intuition might lead us to think that the function T is linear. For example, it would seem plausible that doubling the gross output would double the net output. In any case, in many mathematical models of economic systems this assumption would be made. Accordingly, we suppose that T is linear.

Suppose that for one week the gross and net outputs are, respectively, $\begin{bmatrix} 1600 \\ 1200 \end{bmatrix}$ and $\begin{bmatrix} 900 \\ 700 \end{bmatrix}$. In another week, they are $\begin{bmatrix} 2400 \\ 2000 \end{bmatrix}$ and $\begin{bmatrix} 1300 \\ 1200 \end{bmatrix}$. If the gross output is $\begin{bmatrix} 2000 \\ 1600 \end{bmatrix}$, what is the net?

Observing that $\begin{bmatrix} 2000 \\ 1600 \end{bmatrix} = \frac{1}{2}\begin{bmatrix} 2400 \\ 2000 \end{bmatrix} + \frac{1}{2}\begin{bmatrix} 1600 \\ 1200 \end{bmatrix}$, and using the linearity of T, we have $T\left(\begin{bmatrix} 2000 \\ 1600 \end{bmatrix}\right) = \frac{1}{2}\begin{bmatrix} 1300 \\ 1200 \end{bmatrix} + \frac{1}{2}\begin{bmatrix} 900 \\ 700 \end{bmatrix} = \begin{bmatrix} 1100 \\ 950 \end{bmatrix}$.

This is a simple example of a Leontief input–output model used in studying economic systems.

As an illustration of the geometric nature of linear transformations, we show that a linear operator on $\boldsymbol{R}^3$ carries lines into points or lines.

To see this, let $\boldsymbol{r}(t) = \boldsymbol{a} + t\boldsymbol{b}$ be the parametric equation of a line, where $\boldsymbol{a}$ and $\boldsymbol{b}$ are vectors in $\boldsymbol{R}^3$ and t is a real parameter. If T is a linear operator on $\boldsymbol{R}^3$, the image of $\boldsymbol{r}(t)$ is $\boldsymbol{s}(t) = T(\boldsymbol{r}(t))$. Since T is linear, we have $\boldsymbol{s}(t) = T(\boldsymbol{r}(t)) = T(\boldsymbol{a} + t\boldsymbol{b}) = T(\boldsymbol{a}) + T(t\boldsymbol{b}) = T(\boldsymbol{a}) + tT(\boldsymbol{b})$.

Now, observe that if $T(\boldsymbol{b}) \neq \boldsymbol{0}$, $\boldsymbol{s}(t)$ is the equation of the line through the point $T(\boldsymbol{a})$ in the direction of the vector $T(\boldsymbol{b})$. If $T(\boldsymbol{b}) = \boldsymbol{0}$, $\boldsymbol{s}(t) = T(\boldsymbol{a})$ is constant. Thus, $\boldsymbol{r}(t)$ is carried into a point by T.

The following shows that both situations may occur. Consider the linear operator on $\boldsymbol{R}^2$ defined by $T\left(\begin{bmatrix} x \\ y \end{bmatrix}\right) = \begin{bmatrix} 1 & 0 \\ 0 & 0 \end{bmatrix}\begin{bmatrix} x \\ y \end{bmatrix} = \begin{bmatrix} x \\ 0 \end{bmatrix}$.

The parametric form of the line $x + y = 1$ is

$$\boldsymbol{r}(t) = t\boldsymbol{i} + (1 - t)\boldsymbol{j}$$

Its image is

$$\boldsymbol{s}(t) = T(t\boldsymbol{i} + (1 - t)\boldsymbol{j}) = t\boldsymbol{i}$$

which is the parametric form of the line $y = 0$. (See Figure 5-4.)

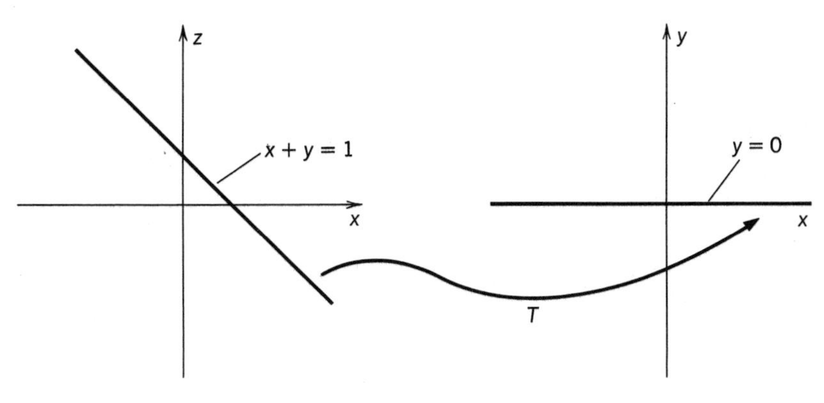

FIGURE 5-4

On the other hand, consider the line $x = 1$. Its parametric form is

$$\boldsymbol{r}(t) = \boldsymbol{i} + t\boldsymbol{j}$$

which is carried into

$$\boldsymbol{s}(t) = T(\boldsymbol{r}(t)) = T(\boldsymbol{i} + t\boldsymbol{j}) = \boldsymbol{i}$$

which is the point (1, 0). (See Figure 5-5.)

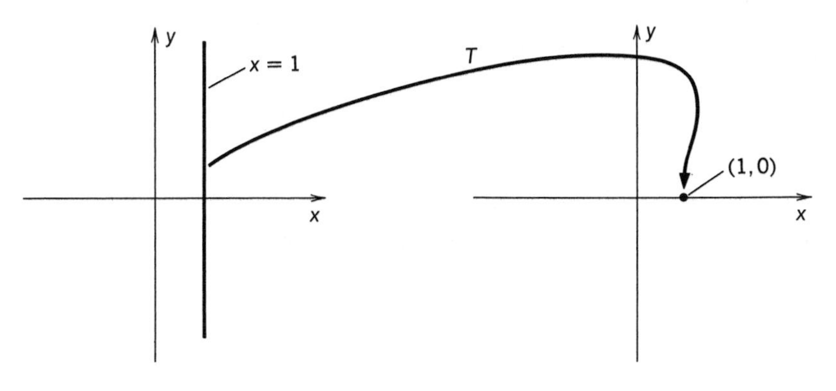

FIGURE 5-5

The reader may verify that the linear transformation T carries each line parallel to the y axis into a single point, while all other lines are carried into the line $y = 0$.

EXERCISES

1. Determine which of the following functions from $\boldsymbol{R}^2$ to $\boldsymbol{R}^2$ are linear.

(a) $T\left(\begin{bmatrix} x \\ y \end{bmatrix}\right) = \begin{bmatrix} y \\ x \end{bmatrix}$ (b) $T\left(\begin{bmatrix} x \\ y \end{bmatrix}\right) = \begin{bmatrix} x+1 \\ y \end{bmatrix}$

(c) $T\left(\begin{bmatrix} x \\ y \end{bmatrix}\right) = \begin{bmatrix} x^2 \\ y \end{bmatrix}$ (d) $T\left(\begin{bmatrix} x \\ y \end{bmatrix}\right) = \begin{bmatrix} x+y \\ x-y \end{bmatrix}$

(e) $T\left(\begin{bmatrix} x \\ y \end{bmatrix}\right) = \begin{bmatrix} 2x-3y \\ y \end{bmatrix}$ (f) $T\left(\begin{bmatrix} x \\ y \end{bmatrix}\right) = \begin{bmatrix} x \\ (x^2+y^2)^{\frac{1}{2}} \end{bmatrix}$

2. Let R be the function from $\boldsymbol{R}^2$ into $\boldsymbol{R}^2$ that reflects points about the x axis. Show that R is linear and that $R\left(\begin{bmatrix} x \\ y \end{bmatrix}\right) = \begin{bmatrix} x \\ -y \end{bmatrix}$.

3. Show that the following functions from $\boldsymbol{R}^3$ to $\boldsymbol{R}^3$ are linear.

(a) $T\left(\begin{bmatrix} x \\ y \\ z \end{bmatrix}\right) = \begin{bmatrix} x+3y+2z \\ x+y \\ 2x+y+z \end{bmatrix}$

(b) $T\left(\begin{bmatrix} x \\ y \\ z \end{bmatrix}\right) = \begin{bmatrix} x-y+z \\ x+z \\ x+y \end{bmatrix}$

4. Let R_θ be the function that assigns to each vector $\boldsymbol{v}$ the vector $R_\theta(\boldsymbol{v})$, obtained by rotating the vector $\boldsymbol{v}$ through an angle of θ degrees about the z axis. The endpoint of $\boldsymbol{v}$ is to remain in the same plane perpendicular to the z axis throughout the rotation.
   (a) Show that R_θ is linear and that

$$R_\theta\left(\begin{bmatrix} x \\ y \\ z \end{bmatrix}\right) = \begin{bmatrix} \cos\theta & -\sin\theta & 0 \\ \sin\theta & \cos\theta & 0 \\ 0 & 0 & 1 \end{bmatrix}\begin{bmatrix} x \\ y \\ z \end{bmatrix}$$

   (b) If in the above instructions we replace the z axis by the y axis throughout, show that

$$R_\theta\left(\begin{bmatrix} x \\ y \\ z \end{bmatrix}\right) = \begin{bmatrix} \cos\theta & 0 & -\sin\theta \\ 0 & 1 & 0 \\ \sin\theta & 0 & \cos\theta \end{bmatrix}\begin{bmatrix} x \\ y \\ z \end{bmatrix}$$

5. In example 7 of the text, let v be the gross output vector. Find a matrix A such that $T(v) = Av$.

6. In example 7 of the text, let v be the gross output vector. Show that $v - T(v)$ gives the internal consumption of coal and gas as a function of $\boldsymbol{v}$. Show that the mapping $S(\boldsymbol{v}) = \boldsymbol{v} - T(\boldsymbol{v})$ is linear.

7. Two beakers containing water are put on a table. The amount of water in each beaker is described by a vector $v = \begin{bmatrix} x \\ y \end{bmatrix}$, where x is the amount of water in the first beaker and y the amount in the second. A two-step operation is performed on the beakers.
   (1) $\frac{2}{3}$ of the water in the first beaker is placed in the second.
   (2) $\frac{1}{2}$ of the water in the second beaker is placed in the first.
   Let $T(v)$ be the vector describing the amount of water in each beaker after both steps are performed.
   (a) Find a matrix A such that $T(v) = Av$.
   (b) Show that T is linear.

8. Let T be a linear operator on $\boldsymbol{R}^2$ such that $T(\boldsymbol{i} + \boldsymbol{j}) = \boldsymbol{j}$ and $T(2\boldsymbol{i} - \boldsymbol{j}) = \boldsymbol{i} + \boldsymbol{j}$. Find $T(\boldsymbol{i})$ and $T(\boldsymbol{j})$.

9. Two men agree to the following income-leveling scheme. The first man gives half his money to the second, and the second gives a third of his money to the first. Let $v = \begin{bmatrix} x \\ y \end{bmatrix}$ be the vector whose first component, x, is the amount of money the first man has before the exchange, and whose second component, y, is the amount of money the second man has before the exchange. Let $T(v)$ be the vector giving the amount each has after the exchange. Find a matrix A so that $T(v) = Av$.

10. If A is an $m \times n$ matrix, consider the function T from M_{np} into M_{mp},

$$T(B) = AB, \quad \text{with } B \text{ in } M_{np}$$

    Show that T is linear.

11. Give an example of a function T from $\boldsymbol{R}^3$ to $\boldsymbol{R}^3$ that satisfies $T(\alpha v) = \alpha T(v)$ for all v in $\boldsymbol{R}^3$ and all scalars α, which is not linear.

12. If B is an invertible $n \times n$ matrix, show that the function from M_{nn} into itself defined by $T(A) = BAB^{-1}$ is a linear transformation.

13. If B is a fixed $n \times n$ matrix, show that the following functions from M_{nn} into M_{nn} are linear.
    (a) $T(A) = AB - BA$ (b) $T(A) = AB + BA$
    (c) $T(A) = AB - B^2A$

14. If V is a vector space and α is a fixed scalar, show that the function $T : V \to V$, defined by $T(\boldsymbol{x}) = \alpha \boldsymbol{x}$ is linear. Interpret this function geometrically if $V = \boldsymbol{R}^2$.

15. Let V be a vector space and $\boldsymbol{x}_1, \boldsymbol{x}_2, \ldots, \boldsymbol{x}_n$ be a basis for V. Then if $\boldsymbol{x}$ is a vector in V, there are unique scalars $\alpha_1, \alpha_2, \alpha_3, \ldots, \alpha_n$, such that $\boldsymbol{x} = \alpha_1\boldsymbol{x}_1 + \alpha_2\boldsymbol{x}_2 + \cdots + \alpha_n\boldsymbol{x}_n$. Consider the function $f(\boldsymbol{x}) = \alpha_1$, with α_1 the first coordinate of $\boldsymbol{x}$ relative to the basis $\boldsymbol{x}_1, \boldsymbol{x}_2, \ldots, \boldsymbol{x}_n$. Show that f is linear.

16. Let $f : V \to \boldsymbol{R}$ and $g : V \to \boldsymbol{R}$ be linear transformations from a vector space V to $\boldsymbol{R}$. Show that

$$T(\boldsymbol{x}) = \begin{bmatrix} f(\boldsymbol{x}) \\ g(\boldsymbol{x}) \end{bmatrix}$$

    is a linear function from V into $\boldsymbol{R}^2$. Generalize to the case of n functions.

17. Let P_n be the space of polynomials of degree less than or equal to n in a variable x with real coefficients. Show that the following functions from P_n into P_n are linear transformations.

(a) $(T(f))(x) = f(x + \alpha)$, where α is a fixed real number. In other words, T carries $f(x)$ into $f(x + \alpha)$.
(b) $T(f) = a_0 f + (b_0 + b_1 x)f' + (c_0 + c_1 x + c_2 x^2)f''$, where f', f'' are first and second derivatives of f, respectively, and $a_0, b_0, b_1, c_0, c_1, c_2$ are real numbers.
(c) $(T(f))(x) = \int_0^x t f''(t)\, dt$.
(d) $(T(f))(x) = f(\alpha x)$, α a real number.
(e) $T(f) = f(\alpha)$, α a real number.

18. Is the function $f(A) = \det A$ from M_{nn} into the reals, a linear transformation?

19. Let $T : M_{33} \to M_{22}$ be the function

$$T\left(\begin{bmatrix} a_{11} & a_{12} & a_{13} \\ a_{21} & a_{22} & a_{23} \\ a_{31} & a_{32} & a_{33} \end{bmatrix}\right) = \begin{bmatrix} a_{11} & a_{12} \\ a_{21} & a_{22} \end{bmatrix}$$

Show that T is linear.

20. Let P be the space of all polynomials with real coefficients in a variable x. Show that the functions S and T, from P to P, defined by

$$S(f) = xf \quad \text{and} \quad (T(f))(x) = \int_0^x f(t)\, dt$$

are linear transformations from P into itself.

21. Let T be a function from V into V, a real vector space, such that
(1) $T(\mathbf{x} + \mathbf{y}) = T(\mathbf{x}) + T(\mathbf{y})$
(2) $T(\alpha \mathbf{x}) = \alpha T(\mathbf{x})$, if $\alpha \geqslant 0$
Show that T is a linear transformation.

2 ADDITIONAL PROPERTIES OF LINEAR TRANSFORMATIONS

In this section, we call attention to some additional important properties of linear transformations.

Proposition Let V and W be vector spaces, $T : V \to W$ a linear transformation. Then,

(i) $T(\mathbf{0}) = \mathbf{0}$
(ii) $T(\alpha \mathbf{x} + \beta \mathbf{y}) = \alpha T(\mathbf{x}) + \beta T(\mathbf{y})$
(iii) $T(\Sigma_{i=1}^n \alpha_i \mathbf{x}_i) = \Sigma_{i=1}^n \alpha_i T(\mathbf{x}_i)$

PROOF

(i) $T(\mathbf{0}) = T(0 \cdot \mathbf{x}) = 0 \cdot T(\mathbf{x}) = \mathbf{0}$
(ii) $T(\alpha \mathbf{x} + \beta \mathbf{y}) = T(\alpha \mathbf{x}) + T(\beta \mathbf{y})$
$= \alpha T(\mathbf{x}) + \beta T(\mathbf{y})$
(iii) The proof of (iii) is by induction on n. If $n = 1$, (iii) clearly holds by condition (ii) in the definition of a linear transformation. If $n > 1$, we have

$$T\left(\sum_{i=1}^{n} \alpha_i x_i\right) = T\left(\sum_{i=1}^{n-1} \alpha_i x_i + \alpha_n x_n\right) = T\left(\sum_{i=1}^{n-1} \alpha_i x_i\right) + T(\alpha_n x_n)$$

By the induction hypothesis $T\left(\sum_{i=1}^{n-1} \alpha_i x_i\right) = \sum_{i=1}^{n-1} \alpha_i T(x_i)$.

Thus,

$$T\left(\sum_{i=1}^{n} \alpha_i x_i\right) = \sum_{i=1}^{n-1} \alpha_i T(x_i) + \alpha_n T(x_n)$$

$$= \sum_{i=1}^{n} \alpha_i T(x_i) \blacksquare$$

There are two important linear transformations that can be defined for any vector space V. The first is the function N defined on V by $N(x) = \mathbf{0}$, for all x in V. Since $N(x + y) = \mathbf{0}$ and $N(x) = N(y) = \mathbf{0}$, we see that $N(x + y) = N(x) + N(y)$. Also, $N(\alpha x) = \mathbf{0}$ and $N(x) = \mathbf{0}$, so $N(\alpha x) = \alpha N(x)$. Thus, N is linear. Appropriately enough, N is called the **zero transformation.**

Another linear transformation defined in any vector space V is the function I_V, where $I_V(x) = x$, for all x in V. I_V is linear, since $I_V(x + y) = x + y$, by definition of I_V, and $I_V(x) = x$ and $I_V(y) = y$, thus $I_V(x + y) = I_V(x) + I_V(y)$. Also, $I_V(\alpha x) = \alpha x = \alpha I_V(x)$. I_V is called the **identity transformation** on the vector space V. When the vector space V is clear from context, I_V is often simply denoted by I.

Let T_1 and T_2 be two linear transformations from a vector space V into a vector space W. Under what circumstances are T_1 and T_2 equal? We say that T_1 and T_2 are equal if they are equal as functions. In other words, T_1 and T_2 are equal if and only if $T_1(x) = T_2(x)$ for all vectors x in V. If T_1 and T_2 are equal, we write $T_1 = T_2$.

A general method of constructing linear transformations is indicated in the following theorem.

Theorem 1 Let V be a vector space with $x_1, x_2, \ldots, x_n$ a basis for V. Let W be another vector space and $y_1, y_2, \ldots, y_n$ be n arbitrary vectors in W. Then there is one and only one linear transformation T from V into W such that $T(x_i) = y_i$, $i = 1, 2, \ldots, n$.

PROOF First, we show that such a linear transformation exists.

Since $x_1, x_2, \ldots, x_n$ is a basis, if x is a vector in V there are unique scalars $\alpha_1, \alpha_2, \ldots, \alpha_n$ such that $x = \alpha_1 x_1 + \alpha_2 x_2 + \cdots + \alpha_n x_n$.

We define $T(x)$ by $T(x) = \alpha_1 y_1 + \alpha_2 y_2 + \cdots + \alpha_n y_n$. Since the scalars $\alpha_1, \alpha_2, \ldots, \alpha_n$ are uniquely determined by x, the function T is well-defined.

We prove T is linear. Suppose x and y are vectors in V, then for suitable scalars $\alpha_1, \alpha_2, \ldots, \alpha_n, \beta_1, \beta_2, \ldots, \beta_n$, we have

$$x = \alpha_1 x_1 + \alpha_2 x_2 + \cdots + \alpha_n x_n$$

and

$$y = \beta_1 x_1 + \beta_2 x_2 + \cdots + \beta_n x_n$$

So $x + y = (\alpha_1 + \beta_1)x_1 + (\alpha_2 + \beta_2)x_2 + \cdots + (\alpha_n + \beta_n)x_n$.

By definition of T, we have

$$T(x) = \alpha_1 y_1 + \alpha_2 y_2 + \cdots + \alpha_n y_n$$

$$T(y) = \beta_1 y_1 + \beta_2 y_2 + \cdots + \beta_n y_n$$

and

$$T(x + y) = (\alpha_1 + \beta_1) y_1 + (\alpha_2 + \beta_2) y_2 + \cdots + (\alpha_n + \beta_n) y_n$$

Thus, $T(x + y) = T(x) + T(y)$.

If x belongs to V and α is a scalar, $x = \alpha_1 x_1 + \alpha_2 x_2 + \cdots + \alpha_n x_n$ for suitable scalars $\alpha_1, \alpha_2, \ldots, \alpha_n$. Then

$$\alpha x = (\alpha\alpha_1)x_1 + (\alpha\alpha_2)x_2 + \cdots + (\alpha\alpha_n)x_n$$

Thus, by definition of T,

$$T(x) = \alpha_1 y_1 + \alpha_2 y_2 + \cdots + \alpha_n y_n$$

$$T(\alpha x) = (\alpha\alpha_1) y_1 + (\alpha\alpha_2) y_2 + \cdots + (\alpha\alpha_n) y_n$$

Thus, $T(\alpha x) = \alpha T(x)$.

Hence we can conclude that the function T is linear. By definition of T, $T(x_i) = y_i$, and so the first half of the theorem is proved.

The second half of the theorem asserts that there is only one linear transformation such that $T(x_1) = y_1, T(x_2) = y_2, \ldots, T(x_n) = y_n$.

To prove this suppose S and T are two linear operators such that $T(x_i) = y_i$ and $S(x_i) = y_i$ for $i = 1, 2, \ldots, n$. We desire to show that $T = S$, i.e., that $T(x) = S(x)$ for all x in V.

If x belongs to V, there are scalars $\alpha_1, \alpha_2, \ldots, \alpha_n$ such that $x = \alpha_1 x_1 + \alpha_2 x_2 + \cdots + \alpha_n x_n$. Then

$$T(x) = \alpha_1 T(x_1) + \alpha_2 T(x_2) + \cdots + \alpha_n T(x_n) = \alpha_1 y_1 + \alpha_2 y_2 + \cdots + \alpha_n y_n$$

and

$$S(x) = \alpha_1 S(x_1) + \alpha_2 S(x_2) + \cdots + \alpha_n S(x_n) = \alpha_1 y_1 + \alpha_2 y_2 + \cdots + \alpha_n y_n$$

Thus, $T(x) = S(x)$. Since x was an arbitrary vector, it follows that $T = S$. ■

Notice that the theorem has two parts. The first part enables us to construct linear transformations by suitably transforming the basis vectors, while the images of the remaining vectors are determined by linearity. For example, the zero transformation might have been constructed by specifying $\boldsymbol{y}_1 = \mathbf{0}, \boldsymbol{y}_2 = \mathbf{0}, \ldots, \boldsymbol{y}_n = \mathbf{0}$. The identity transformation might be constructed by specifying $V = W$ and $\boldsymbol{y}_1 = \boldsymbol{x}_1, \boldsymbol{y}_2 = \boldsymbol{x}_2, \ldots, \boldsymbol{y}_n = \boldsymbol{x}_n$. On P_3, we may construct a linear transformation by insisting that, under T,

$$1 \to 0, \qquad x \to 1, \qquad x^2 \to 2x, \qquad x^3 \to 3x^2$$

The resulting linear operator is just the differentiation operator of example 6, §5.1.

The second half of the theorem has an alternative formulation. If T and S are linear transformations from a vector space V into a vector space W and T and S agree for the vectors in some basis for V, then T and S agree on all of V. This statement generalizes to arbitrary finite-dimensional spaces the fact we demonstrated in the last section. There we saw that a linear transformation from $\boldsymbol{R}^2$ into $\boldsymbol{R}^2$ is completely determined by the images of the basis vectors $\boldsymbol{i}$ and $\boldsymbol{j}$.

We are now in a position to generalize a result from the previous section and show that any linear transformation from $\boldsymbol{R}^n$ into $\boldsymbol{R}^m$ may be induced by multiplication by a suitable matrix.

Theorem 2 Let $T : \boldsymbol{R}^n \to \boldsymbol{R}^m$ be a linear transformation. Then there is an $m \times n$ matrix A, with real entries, such that $T(\boldsymbol{x}) = A\boldsymbol{x}$ for all $\boldsymbol{x}$ in $\boldsymbol{R}^n$.

PROOF Let $\boldsymbol{e}_1, \ldots, \boldsymbol{e}_n$ be the standard basis for $\boldsymbol{R}^n$, and $\boldsymbol{e}'_1, \ldots, \boldsymbol{e}'_m$ be the standard basis for $\boldsymbol{R}^m$.

Since $\boldsymbol{e}'_1, \ldots, \boldsymbol{e}'_m$ is a basis for $\boldsymbol{R}^m$, there are scalars a_{ij} such that $T(\boldsymbol{e}_j) = \sum_{i=1}^{m} a_{ij}\boldsymbol{e}'_i$. If $\boldsymbol{x} = \sum_{j=1}^{n} x_j\boldsymbol{e}_j$ is a vector in $\boldsymbol{R}^n$, we have $T(\boldsymbol{x}) = T\left(\sum_{j=1}^{n} x_j\boldsymbol{e}_j\right)$

$$= \sum_{j=1}^{n} x_j T(\boldsymbol{e}_j) = \sum_{j=1}^{n} x_j\left(\sum_{i=1}^{m} a_{ij}\boldsymbol{e}'_i\right) = \sum_{j=1}^{n}\sum_{i=1}^{m} a_{ij}x_j\boldsymbol{e}'_i = \sum_{i=1}^{m}\left(\sum_{j=1}^{n} a_{ij}x_j\right)\boldsymbol{e}'_i.$$

Thus, the ith component of the vector $T(\boldsymbol{x})$ is just $\sum_{j=1}^{n} a_{ij}x_j$.

If we let $A = [a_{ij}]_{(mn)}$, then the ith component of the product $A\boldsymbol{x}$ is precisely $\sum_{j=1}^{n} a_{ij}x_i$.

Thus, for $i = 1, 2, \ldots, m$, the vectors $A\boldsymbol{x}$ and $T(\boldsymbol{x})$ have the same ith components, and so $T(\boldsymbol{x}) = A\boldsymbol{x}$. Hence, the linear transformation T is precisely that induced by multiplication by the matrix A. ■

If A is an $m \times n$ matrix, we often denote the linear transformation from $\boldsymbol{R}^n$ to $\boldsymbol{R}^m$ induced by A simply by T_A. It is important to note that the matrix A has as its jth column the vector $T(\boldsymbol{e}_j)$. Thus, if T is a linear operator from $\boldsymbol{R}^n$ to $\boldsymbol{R}^m$, in order to calculate its associated matrix we need only calculate the image of the basis vectors $\boldsymbol{e}_1, \boldsymbol{e}_2, \ldots, \boldsymbol{e}_n$ and form the matrix $A = [T(\boldsymbol{e}_1), T(\boldsymbol{e}_2), \ldots, T(\boldsymbol{e}_n)]$. It is also clear that the matrix A is uniquely determined by the image under T of the basis vectors $\boldsymbol{e}_1, \boldsymbol{e}_2, \ldots, \boldsymbol{e}_n$.

As an example, recall the linear transformation T_θ which rotates the vectors in the plane by θ degrees. (See Figure 5-6.)

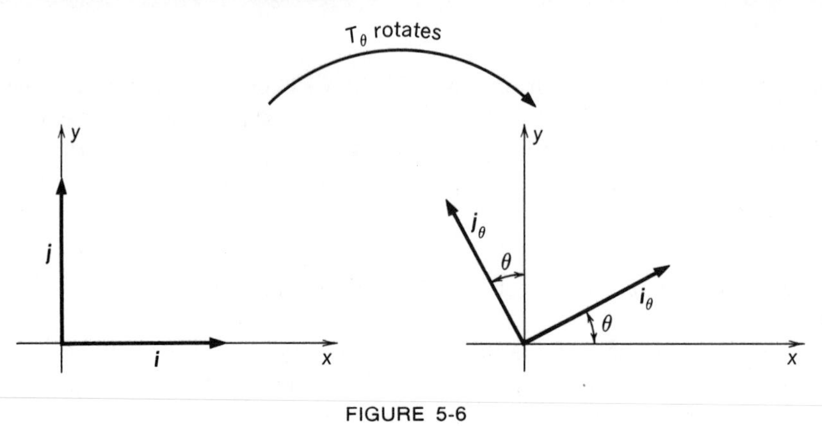

FIGURE 5-6

Since

$$i_\theta = T_\theta(i) = (\cos\theta)i + (\sin\theta)j$$

$$j_\theta = T_\theta(j) = -(\sin\theta)i + (\cos\theta)j$$

we see that the matrix associated with T_θ is

$$A_\theta = \begin{bmatrix} \cos\theta & -\sin\theta \\ \sin\theta & \cos\theta \end{bmatrix}$$

This formula is obtained by writing the vector i_θ in the first and j_θ in the second columns of the matrix A_θ.

Example 1 Suppose that in example 7 of §5.1, the gross output vectors $\begin{bmatrix} 500 \\ 1000 \end{bmatrix}$ and $\begin{bmatrix} 800 \\ 1200 \end{bmatrix}$ had given the respective net output vectors $\begin{bmatrix} 200 \\ 400 \end{bmatrix}$ and $\begin{bmatrix} 400 \\ 400 \end{bmatrix}$. Find a matrix A so that $T(v) = Av$.

By Theorem 2, we know there is such an A. Let $A = \begin{bmatrix} a & b \\ c & d \end{bmatrix}$. Since $T\left(\begin{bmatrix} 500 \\ 1000 \end{bmatrix}\right) = \begin{bmatrix} 200 \\ 400 \end{bmatrix}$, we obtain the equations $500a + 1000b = 200$ and $500c + 1000d = 400$. Since $T\left(\begin{bmatrix} 800 \\ 1200 \end{bmatrix}\right) = \begin{bmatrix} 400 \\ 400 \end{bmatrix}$, we find $800a + 1200b = 400$ and $800c + 1200d = 400$. These give the systems (after canceling 100):

$$5a + 10b = 2 \qquad 5c + 10d = 4$$

$$8a + 12b = 4 \qquad 8c + 12d = 4$$

Solving, we find that

$$a = \tfrac{4}{5},\ b = -\tfrac{1}{5},\ c = -\tfrac{2}{5},\ d = \tfrac{3}{5}, \quad \text{or} \quad A = \begin{bmatrix} \frac{4}{5} & -\frac{1}{5} \\ -\frac{2}{5} & \frac{3}{5} \end{bmatrix}$$

Example 2 As another example, we consider the following projection transformation. Let l be a line in the plane through the origin that makes an angle of θ degrees with the x axis. If $\boldsymbol{v}$ is a vector in the plane ending at the point R, let $P_\theta(\boldsymbol{v})$ be the vector along the line l ending at the point S, where S is obtained by dropping a perpendicular from the point R to the line l. (See Figure 5-7.)

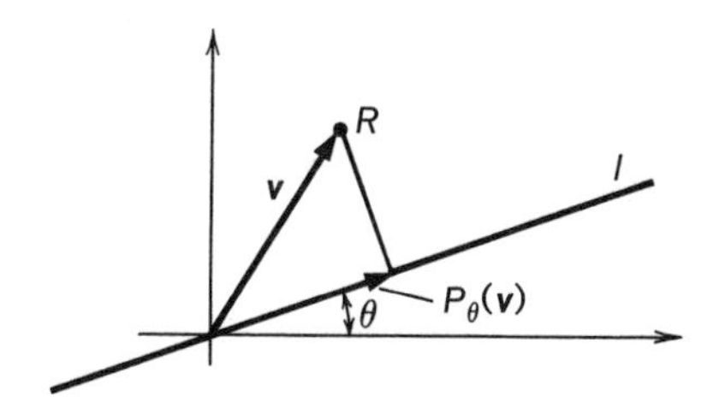

FIGURE 5-7

The linearity of the geometrically defined function P_θ may be demonstrated in a manner completely analogous to the proof, in §2.2 that $\boldsymbol{v}(x, y) + \boldsymbol{v}(x', y') = \boldsymbol{v}(x + x', y + y')$, that is, that the geometric and algebraic definitions of vector addition are equivalent. Indeed, if $\theta = 0$, i.e., the line l is the x axis, the proofs are completely identical, for in this case $P_\theta(\boldsymbol{v}(x, y)) = x\boldsymbol{i}$.

Having obtained the linearity of the function P_θ from geometric considerations, we see that there is some matrix, say B_θ, such that $P_\theta(\boldsymbol{v}) = B_\theta \boldsymbol{v}$.

We complete our determination of the function P_θ by finding the matrix B_θ. In order to obtain the matrix B_θ it is only necessary to determine $P_\theta(\boldsymbol{i})$ and $P_\theta(\boldsymbol{j})$. (See Figure 5-8.)

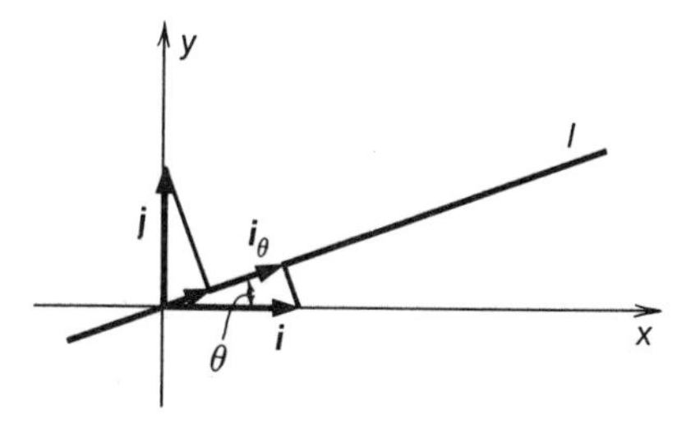

FIGURE 5-8

If $\boldsymbol{i}_\theta$ is the vector of unit length in the direction of the line l, it is clear that $\boldsymbol{i}_\theta = (\cos\theta)\boldsymbol{i} + (\sin\theta)\boldsymbol{j}$.

From trigonometry the length of the vector $P_\theta(\boldsymbol{i})$ is $\cos\theta$ and the length of $P_\theta(\boldsymbol{j})$ is $\sin\theta$. Thus,

$$P_\theta(\boldsymbol{i}) = (\cos\theta)\boldsymbol{i}_\theta = (\cos^2\theta)\boldsymbol{i} + (\cos\theta\sin\theta)\boldsymbol{j}$$

$$P_\theta(\boldsymbol{j}) = (\sin\theta)\boldsymbol{i}_\theta = (\sin\theta\cos\theta)\boldsymbol{i} + (\sin^2\theta)\boldsymbol{j}$$

It follows that

$$P_\theta\left(\begin{bmatrix} x \\ y \end{bmatrix}\right) = \begin{bmatrix} \cos^2\theta & \cos\theta\sin\theta \\ \cos\theta\sin\theta & \sin^2\theta \end{bmatrix}\begin{bmatrix} x \\ y \end{bmatrix}$$

Example 3 A tent manufacturer produces three types of tents. The first type requires 4 square yards of canvas and 10 feet of rope in the production of one unit. The second requires 6 square yards of canvas and 16 feet of rope; the third, 10 square yards of canvas and 30 feet of rope. Let $\boldsymbol{v} = \begin{bmatrix} x_1 \\ x_2 \\ x_3 \end{bmatrix}$ be the vector whose components are the number of units of each type of tent produced. Let $T(\boldsymbol{v}) = \begin{bmatrix} a \\ b \end{bmatrix}$, where a is the number of square yards of canvas required to produce x_1 tents of type 1, x_2 tents of type 2, and x_3 tents of type 3, and b is the number of feet of rope in this production. Then, $T(\boldsymbol{v}) = \begin{bmatrix} 4 & 6 & 10 \\ 10 & 16 & 30 \end{bmatrix}\boldsymbol{v}$. Note that the ith column of this matrix gives the amount of canvas and rope necessary to produce one tent of the ith type.

In this example, the transformation T transforms the vector of production $\boldsymbol{v}$ into the vector of requirements $T(\boldsymbol{v})$.

EXERCISES

1. Find all linear operators on $\boldsymbol{R}^2$ that carry vectors on the line $x = 0$ into vectors on the line $x = 0$ and vectors on the line $y = 0$ into vectors on the line $y = 0$.

2. T is a linear operator on $\boldsymbol{R}^2$. Find a matrix A such that $T(\boldsymbol{v}) = A\boldsymbol{v}$ if
   (a) $T(\boldsymbol{i}) = \boldsymbol{i} + \boldsymbol{j}$, $T(\boldsymbol{j}) = \boldsymbol{i} - \boldsymbol{j}$ (b) $T(\boldsymbol{i} + \boldsymbol{j}) = \boldsymbol{i}$, $T(\boldsymbol{i} - \boldsymbol{j}) = \boldsymbol{j}$.

3. Suppose T is a linear operator on $\boldsymbol{R}^2$, and $\boldsymbol{u}$ and $\boldsymbol{v}$ are linearly independent vectors in $\boldsymbol{R}^2$. If $T(\boldsymbol{u}) = \boldsymbol{u}$ and $T(\boldsymbol{v}) = \boldsymbol{v}$, show that T is the identity.

4. If T is a linear transformation between two vector spaces V and W, and $\boldsymbol{x}_1, \boldsymbol{x}_2, \ldots, \boldsymbol{x}_n$ are vectors in V such that $T(\boldsymbol{x}_1), T(\boldsymbol{x}_2), \ldots, T(\boldsymbol{x}_n)$ are linearly independent, show that the $\boldsymbol{x}_i$'s are linearly independent in V.

5. Suppose $T : V \to W$ and $S : V \to W$ are linear transformations. If $\boldsymbol{x}_1, \boldsymbol{x}_2, \ldots, \boldsymbol{x}_n$ are vectors in V such that $\operatorname{sp}(\boldsymbol{x}_1, \boldsymbol{x}_2, \ldots, \boldsymbol{x}_n) = V$, and if
$$T(\boldsymbol{x}_1) = S(\boldsymbol{x}_1), T(\boldsymbol{x}_2) = S(\boldsymbol{x}_2), \ldots, T(\boldsymbol{x}_n) = S(\boldsymbol{x}_n)$$
   show that $T = S$.

6. Give an example of a linear transformation T from $\boldsymbol{R}^2$ into $\boldsymbol{R}^2$ such that $\boldsymbol{x}_1$ and $\boldsymbol{x}_2$ are linearly independent but $T(\boldsymbol{x}_1)$ and $T(\boldsymbol{x}_2)$ are linearly dependent.

7. Find all linear transformations from $\boldsymbol{R}^2$ into $\boldsymbol{R}^2$ that
   (a) Carry the line $x = 0$ into the line $x = 0$.
   (b) Carry the line $y = 0$ into the line $y = 0$.
   (c) Carry the line $x = y$ into the line $x = y$.

8. In a certain animal species, there are three age groups. We represent the number of animals in each age group with a vector $v = \begin{bmatrix} x \\ y \\ z \end{bmatrix}$. Let $T(v)$ represent the population in each age group after one year has elapsed. As in example 3, of §2.5, we suppose that T is linear.

   In four successive years, it is observed that the population vectors are

   $$\begin{bmatrix} 1600 \\ 800 \\ 400 \end{bmatrix}, \begin{bmatrix} 1200 \\ 1200 \\ 400 \end{bmatrix}, \begin{bmatrix} 1400 \\ 900 \\ 600 \end{bmatrix}, \begin{bmatrix} 1325 \\ 1050 \\ 450 \end{bmatrix}. \text{ Show that } T(v) = \begin{bmatrix} \frac{1}{4} & \frac{3}{4} & \frac{1}{2} \\ \frac{3}{4} & 0 & 0 \\ 0 & \frac{1}{2} & 0 \end{bmatrix} \begin{bmatrix} x \\ y \\ z \end{bmatrix},$$

   and interpret in the spirit of example 3 of §2.5. [Hint: With a little thought, one need not solve any linear equations.]

9. Two beakers containing water are put on a table. Let $v = \begin{bmatrix} x \\ y \end{bmatrix}$ represent the amount of water in each beaker. A two-step operation is performed on the two beakers. (1) An ath part of beaker 1 is put into beaker 2. (2) a bth part of beaker 2 is put into beaker 1. Let $T(v)$ represent the amount in each beaker after the operation is performed.

   (a) Show that $T(v) = \begin{bmatrix} 1 - a + ba & b \\ a - ba & 1 - b \end{bmatrix} \begin{bmatrix} x \\ y \end{bmatrix}$.

   (b) This operation is performed twice. The results are $\begin{bmatrix} 180 \\ 30 \end{bmatrix} \rightarrow \begin{bmatrix} 160 \\ 50 \end{bmatrix}$ and $\begin{bmatrix} 90 \\ 90 \end{bmatrix} \rightarrow \begin{bmatrix} 130 \\ 50 \end{bmatrix}$. What are a and b?

10. Let V be a two-dimensional vector space and let $\boldsymbol{x}_1, \boldsymbol{x}_2, \boldsymbol{x}_3$ be vectors in V, any two of which are linearly independent. Let $T : V \rightarrow V$ be a linear transformation such that

    $$T(\boldsymbol{x}_1) = \alpha_1 \boldsymbol{x}_1, \qquad T(\boldsymbol{x}_2) = \alpha_2 \boldsymbol{x}_2, \qquad T(\boldsymbol{x}_3) = \alpha_3 \boldsymbol{x}_3$$

    for scalars α_1, α_2, and α_3. Show that there is a scalar α such that $T(\boldsymbol{x}) = \alpha \boldsymbol{x}$ for all $\boldsymbol{x}$ in V. Interpret geometrically if $V = \boldsymbol{R}^2$.

11. Let

    $$\begin{bmatrix} x_1 \\ x_2 \end{bmatrix} \text{ and } \begin{bmatrix} x_1' \\ x_2' \end{bmatrix}$$

    be linearly independent vectors in $\boldsymbol{R}^2$. Let T be a linear transformation from $\boldsymbol{R}^2$ into $\boldsymbol{R}^2$ such that

    $$T\left(\begin{bmatrix} x_1 \\ x_2 \end{bmatrix}\right) = \begin{bmatrix} y_1 \\ y_2 \end{bmatrix}, \qquad T\left(\begin{bmatrix} x_1' \\ x_2' \end{bmatrix}\right) = \begin{bmatrix} y_1' \\ y_2' \end{bmatrix}$$

    Show that $T(\boldsymbol{x}) = A\boldsymbol{x}$, where A is the 2×2 matrix

    $$A = \begin{bmatrix} y_1 & y_1' \\ y_2 & y_2' \end{bmatrix} \begin{bmatrix} x_1 & x_1' \\ x_2 & x_2' \end{bmatrix}^{-1}$$

12. Let T be a linear transformation of $\boldsymbol{R}^3$ into itself. Show that T carries any plane through the origin into a plane through the origin, a line through the origin, or the origin itself. Give an example of each of the three cases.

13. The population of a certain metropolitan area is represented by a vector $v = \begin{bmatrix} x \\ y \end{bmatrix}$, where x is the population of the city and y that of the suburbs. Over ten-year periods, it is found that a certain fraction, a, of the city dwellers move to the suburbs, and a certain fraction, b, of the suburbanites move back to the city. The population otherwise remains constant. In three successive ten-year periods the population vectors are $\begin{bmatrix} 1{,}000{,}000 \\ 400{,}000 \end{bmatrix}$, $\begin{bmatrix} 920{,}000 \\ 480{,}000 \end{bmatrix}$, $\begin{bmatrix} 852{,}000 \\ 548{,}000 \end{bmatrix}$. Find a function T such that $T(v)$ is the population after ten years. What are a and b?

14. If $f : \boldsymbol{R}^n \rightarrow \boldsymbol{R}$ is a linear transformation, show that there are scalars $a_1, a_2, \ldots, a_n$ such that

$$f\left(\begin{bmatrix} x_1 \\ x_2 \\ \vdots \\ x_n \end{bmatrix}\right) = a_1x_1 + a_2x_2 + \cdots + a_nx_n$$

15. If $e_1, e_2, \ldots, e_n$ is the standard basis for $\boldsymbol{R}^n$, obtain the matrix representation of the linear operator T, where
(a) $T(e_1) = e_2,\ T(e_2) = e_3, \ldots, T(e_{n-1}) = e_n,\ T(e_n) = \mathbf{0}$
(b) $T(e_1) = e_2,\ T(e_2) = e_3, \ldots, T(e_{n-1}) = e_n,\ T(e_n) = e_1$
(c) $T(e_1) = e_1,\ T(e_2) = e_2 + e_1, \ldots, T(e_{n-1}) = e_{n-1} + e_{n-2},\ T(e_n) = e_n + e_{n-1}$

16. Find the matrix representation of the linear transformation from $\boldsymbol{R}^n$ into $\boldsymbol{R}^n$ that carries the vector $\boldsymbol{x}$ into the vector $\alpha\boldsymbol{x}$ where α is a fixed scalar.

3 RANGE SPACE

Let T be a linear transformation from a vector space V into a vector space W. We wish to study those vectors in W, each of which is the image of some vector in V. Let us call this collection of vectors the **range space** of the linear transformation T. We denote this set of vectors by R_T. Thus, $R_T = \{\boldsymbol{y} \mid \boldsymbol{y} \in W \text{ and for some } \boldsymbol{x} \text{ in } V, \boldsymbol{y} = T(\boldsymbol{x})\}$. In order to justify this terminology we prove the following:

Theorem 1 Let $T : V \rightarrow W$ be a linear transformation of a vector space V into a vector space W. Then the set of vectors $R_T = \{\boldsymbol{y} \mid \boldsymbol{y} \in W \text{ and } \boldsymbol{y} = T(\boldsymbol{x}), \text{ for some } \boldsymbol{x} \text{ in } V\}$ is a subspace of W.

PROOF Suppose $\boldsymbol{y}_1$ and $\boldsymbol{y}_2$ are vectors belonging to R_T. By definition of R_T, there are vectors $\boldsymbol{x}_1$ and $\boldsymbol{x}_2$ in V such that $T(\boldsymbol{x}_1) = \boldsymbol{y}_1$ and $T(\boldsymbol{x}_2) = \boldsymbol{y}_2$. By the linearity of T, $T(\boldsymbol{x}_1 + \boldsymbol{x}_2) = T(\boldsymbol{x}_1) + T(\boldsymbol{x}_2) = \boldsymbol{y}_1 + \boldsymbol{y}_2$, and so since $\boldsymbol{y}_1 + \boldsymbol{y}_2$ is the image under T of some vector in V, namely, $\boldsymbol{x}_1 + \boldsymbol{x}_2$, $\boldsymbol{y}_1 + \boldsymbol{y}_2$ belongs to R_T.

Next, suppose $\boldsymbol{y}$ belongs to R_T and α is a scalar. Since $\boldsymbol{y}$ is in R_T, there is some vector $\boldsymbol{x}$ in V such that $T(\boldsymbol{x}) = \boldsymbol{y}$. By the linearity of T, $T(\alpha\boldsymbol{x}) = \alpha T(\boldsymbol{x}) = \alpha\boldsymbol{y}$. Since $\alpha\boldsymbol{y}$ is the T image of some vector in V, namely, $\alpha\boldsymbol{x}$, we see that $\alpha\boldsymbol{y}$ belongs to R_T. Since the subset R_T is closed under the algebraic operations of addition and scalar multiplication, it follows that R_T is a subspace of W. ■

For example, the range space of N, the zero transformation, is precisely the zero subspace. If $I_V : V \to V$ is the identity operator on the vector space V, the range space is V, the whole space.

Using the fact that R_T is a subspace of the vector space W, and Theorem 3, §4.8, we can immediately conclude that $\dim R_T \leqslant \dim W$.

The quantity $\dim R_T$ is of such importance that it is given a special name. It is called the **rank** of T and is denoted by $r(T)$.

Example 1 In example 2 of §5.2 we defined a projection transformation P_θ from $\boldsymbol{R}^2$ into itself. $P_\theta(\boldsymbol{v})$ is the vector obtained by projecting $\boldsymbol{v}$ perpendicularly on the line making the angle θ with the x axis. (See Figure 5-9.) From the definition of P_θ, it immediately follows that any vector in the range space of P_θ is a scalar multiple of the vector $\boldsymbol{i}_\theta = (\cos\theta)\boldsymbol{i} + (\sin\theta)\boldsymbol{j}$. Thus, R_{P_θ} is the one-dimensional subspace of $\boldsymbol{R}^2$ spanned by the vector $\boldsymbol{i}_\theta$. Moreover, $r(P_\theta) = 1$.

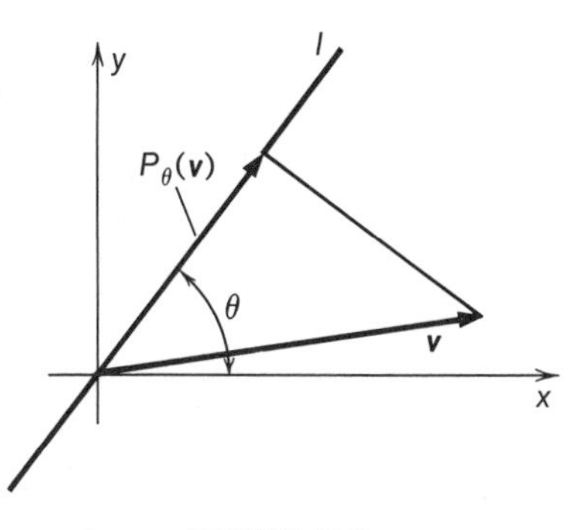

FIGURE 5-9

Example 2 Let

$$\begin{aligned} a_{11}x_1 + a_{12}x_2 + \cdots + a_{1n}x_n &= y_1 \\ a_{21}x_1 + a_{22}x_2 + \cdots + a_{2n}x_n &= y_2 \\ &\vdots \\ a_{m1}x_1 + a_{m2}x_2 + \cdots + a_{mn}x_n &= y_m \end{aligned}$$

be a system of m linear equations in n variables. Using matrix notation, with $A = [a_{ij}]_{(mn)}$, $\boldsymbol{x} = [x_i]_{(n1)}$, $\boldsymbol{y} = [y_j]_{(m1)}$, the system of equations becomes $A\boldsymbol{x} = \boldsymbol{y}$.

Let $T_A : \boldsymbol{R}^n \to \boldsymbol{R}^m$ be the linear transformation defined by $T_A(\boldsymbol{x}) = A\boldsymbol{x}$. The range space of T_A consists of all those vectors $\boldsymbol{y}$ in $\boldsymbol{R}^m$ for which there is some vector $\boldsymbol{x}$ in $\boldsymbol{R}^n$ with $A\boldsymbol{x} = \boldsymbol{y}$. In other words, the range space is exactly that collection of vectors for which the linear system $A\boldsymbol{x} = \boldsymbol{y}$ is solvable.

The procedure of Gaussian elimination discussed in chapter 1 provides a computational means of determining the range space of the linear transformation associated with a matrix. If

$$A = \begin{bmatrix} 1 & -1 & 0 \\ 3 & 1 & 7 \\ 4 & 0 & 7 \end{bmatrix}$$

we can consider the linear transformation from $\boldsymbol{R}^3$ to $\boldsymbol{R}^3$, defined by $T_A(\boldsymbol{x}) = A\boldsymbol{x}$. In order to obtain the range space of T_A, we must find those column vectors with components u, v, w for which the system of equations

$$\begin{aligned} x - y \quad\quad &= u \\ 3x + y + 7z &= v \\ 4x \quad\; + 7z &= w \end{aligned}$$

is solvable.

We follow the elimination procedure

$$\begin{aligned} x - y \quad\quad &= u \\ 3x + y + 7z &= v \\ 4x \quad\; + 7z &= w \end{aligned}$$

Use x and the first equation.

$$\downarrow$$

$$\begin{aligned} x - y \quad\quad &= u \\ 4y + 7z &= v - 3u \\ 4y + 7z &= w - 4u \end{aligned}$$

Use z and the second equation.

$$\downarrow$$

$$\begin{aligned} x - y \quad\quad &= u \\ 4y + 7z &= v - 3u \\ 0 &= w - v - u \end{aligned}$$

The procedure comes to a halt, since in the only unused equation, the third, all variables have coefficient 0. Thus, in order that the system be satisfied, it is necessary to have $w - v - u = 0$. On the other hand, if we let $y = 0$, and find x and z using the first two equations, we see that the system indeed admits a solution. Thus, the range space of the linear transformation T_A consists exactly of those vectors for which $w = v + u$.

In order to determine $r(T_A)$, we must find the dimension of the range space. To begin with, it is clear that the vectors $\boldsymbol{a} = \begin{bmatrix} 1 \\ 0 \\ 1 \end{bmatrix}$ and $\boldsymbol{b} = \begin{bmatrix} 0 \\ 1 \\ 1 \end{bmatrix}$ are both members of R_{T_A}. Since neither $\boldsymbol{a}$ nor $\boldsymbol{b}$ is a multiple of the other, $\boldsymbol{a}$ and $\boldsymbol{b}$ are linearly independent. Let $\boldsymbol{c}$ be a vector in R_{T_A} having components u, v, and w. Then, $w = u + v$. Thus,

$$\boldsymbol{c} = \begin{bmatrix} u \\ v \\ w \end{bmatrix} = \begin{bmatrix} u \\ v \\ u + v \end{bmatrix} = u \begin{bmatrix} 1 \\ 0 \\ 1 \end{bmatrix} + v \begin{bmatrix} 0 \\ 1 \\ 1 \end{bmatrix} = u\boldsymbol{a} + v\boldsymbol{b}$$

Thus, $\boldsymbol{a}$ and $\boldsymbol{b}$ span R_{T_A}. Thus, $\dim R_{T_A} = 2$, and $r(T_A) = 2$.

In subsequent sections we shall develop other procedures for determining the rank and range space of a linear transformation associated with a matrix. As further explication of example 2 above we state and prove Theorem 2.

Theorem 2 Let A be an $m \times n$ matrix and T_A be the linear transformation from $\boldsymbol{R}^n$ into $\boldsymbol{R}^m$ induced by A. Then, the columns of the matrix A span the range space of T_A.

PROOF Let $\boldsymbol{e}_1, \boldsymbol{e}_2, \ldots, \boldsymbol{e}_n$ be the standard basis for $\boldsymbol{R}^n$. If $\boldsymbol{x}$ is in $\boldsymbol{R}^n$, there are scalars $\alpha_1, \alpha_2, \ldots, \alpha_n$ such that $\boldsymbol{x} = \alpha_1\boldsymbol{e}_1 + \alpha_2\boldsymbol{e}_2 + \cdots + \alpha_n\boldsymbol{e}_n$. Thus,

$$\begin{aligned} T_A(\boldsymbol{x}) &= T_A(\alpha_1\boldsymbol{e}_1 + \alpha_2\boldsymbol{e}_2 + \cdots + \alpha_n\boldsymbol{e}_n) \\ &= \alpha_1 T_A(\boldsymbol{e}_1) + \alpha_2 T_A(\boldsymbol{e}_2) + \cdots + \alpha_n T_A(\boldsymbol{e}_n) \\ &= \alpha_1(A\boldsymbol{e}_1) + \alpha_2(A\boldsymbol{e}_2) + \cdots + \alpha_n(A\boldsymbol{e}_n) \end{aligned}$$

But $A\boldsymbol{e}_i$ is just the ith column of the matrix A. Since every vector in R_{T_A} can be expressed as a linear combination of $A\boldsymbol{e}_1, A\boldsymbol{e}_2, \ldots, A\boldsymbol{e}_n$, we have the desired result. ■

For example, if A is the 2×4 matrix

$$\begin{bmatrix} -1 & 0 & -1 & 3 \\ 1 & 2 & 0 & 4 \end{bmatrix}$$

our theorem tells us that the vectors

$$\begin{bmatrix} -1 \\ 1 \end{bmatrix}, \quad \begin{bmatrix} 0 \\ 2 \end{bmatrix}, \quad \begin{bmatrix} -1 \\ 0 \end{bmatrix}, \quad \begin{bmatrix} 3 \\ 4 \end{bmatrix}$$

span the range of T_A. Since the vectors

$$\begin{bmatrix} -1 \\ 1 \end{bmatrix} \quad \text{and} \quad \begin{bmatrix} 0 \\ 2 \end{bmatrix}$$

form a basis for $\boldsymbol{R}^2$, we see that $R_{T_A} = \boldsymbol{R}^2$, and $r(T_A) = 2$.

Example 3 Let A be an invertible $n \times n$ matrix. If T_A is the linear operator on $\boldsymbol{R}^n$ induced by A, $R_{T_A} = \boldsymbol{R}^n$.

For under these circumstances, the equation $T_A(\boldsymbol{x}) = \boldsymbol{y}$, i.e., $A\boldsymbol{x} = \boldsymbol{y}$ is always solvable. Indeed, its solution is $\boldsymbol{x} = A^{-1}\boldsymbol{y}$.

Example 4 Let S be the function from M_{nn} to M_{nn}, defined by $S(A) = A + A^T$.

First, S is linear, since if A and B are $n \times n$ matrices and α is a scalar then $S(A + B) = A + B + (A + B)^T = A + B + A^T + B^T = A + A^T + B + B^T = S(A) + S(B)$ and $S(\alpha A) = \alpha A + (\alpha A)^T = \alpha(A + A^T) = \alpha S(A)$.

We claim that the range of S is precisely the space of symmetric matrices. If B belongs to R_S, then $B = A + A^T$, for some matrix A. It follows that $B^T = (A + A^T)^T = A^T + A = B$. Thus, any matrix in R_S is symmetric.

Next, we show that any symmetric matrix belongs to R_S. If B is symmetric, $S(\frac{1}{2}B) = \frac{1}{2}B + (\frac{1}{2}B)^T = (\frac{1}{2} + \frac{1}{2})B = B$. Thus, B belongs to R_S.

It follows that R_S is precisely the space of symmetric matrices. Since the space of symmetric matrices is of dimension $\frac{1}{2}n(n+1)$ (§4.7, exercise 11), we see that $r(S) = \frac{1}{2}n(n+1)$.

Given a linear transformation $T : V \to W$, if $R_T = W$, T is said to be **onto**. For example, I_V, the identity operator on some vector space V, is onto. In example 3 above, the linear transformation from $\boldsymbol{R}^n$ into $\boldsymbol{R}^n$ induced by an invertible matrix was seen to be onto. If A is an $m \times n$ matrix, the linear transformation T_A from $\boldsymbol{R}^n$ to $\boldsymbol{R}^m$ induced by the matrix A is onto if and only if the systems of equations $A\boldsymbol{x} = \boldsymbol{y}$ is always solvable.

EXERCISES

1. Find a basis for the range space and the rank of the linear transformations induced by the following matrices.

   (a) $\begin{bmatrix} -1 & 1 & 3 \\ 0 & 2 & 0 \end{bmatrix}$ (b) $\begin{bmatrix} 1 & 0 \\ 3 & 1 \\ 0 & 2 \end{bmatrix}$ (c) $\begin{bmatrix} 0 & 1 & 2 \\ 1 & 2 & 5 \\ 0 & -1 & -2 \end{bmatrix}$

   (d) $\begin{bmatrix} 1 & 2 & 3 \\ 0 & 0 & 0 \\ -3 & 1 & -2 \\ 1 & 3 & 4 \end{bmatrix}$ (e) $\begin{bmatrix} 1 & -1 & 0 & 3 \\ 0 & 3 & 2 & -1 \\ 2 & 7 & 1 & -2 \end{bmatrix}$

2. Let $D : P_n \to P_n$ be the differentiation operator on P_n. Show that $R_D = P_{n-1}$, and $r(D) = n$.
3. If $T : V \to W$ is a linear transformation and if $R_T = 0$, show that T is the zero operator.
4. Let A be an $n \times n$ matrix. If $T : M_{nn} \to M_{nn}$ is the linear transformation defined by $T(B) = AB$, show that T is onto if and only if A is invertible.
5. Let T be a linear transformation between two vector spaces V and W. If $\boldsymbol{x}_1, \boldsymbol{x}_2, \ldots, \boldsymbol{x}_n$ are vectors in V such that $\text{sp}(\boldsymbol{x}_1, \boldsymbol{x}_2, \ldots, \boldsymbol{x}_n) = V$, show that $T(\boldsymbol{x}_1), T(\boldsymbol{x}_2), \ldots, T(\boldsymbol{x}_n)$ span R_T.
6. Let $T : M_{nn} \to M_{nn}$ be the linear operator defined on the space of $n \times n$ matrices by $T(A) = A^T$. Show that T is onto.
7. Calculate a basis for the range space and the rank of the following linear operators on P_n.
   (a) $T(f) = xf'$, f' is the derivative of f.
   (b) $(T(f))(x) = \int_0^x tf''(t)\,dt$, f'' is the second derivative of f.
   (c) $(T(f))(x) = f(x+1)$.
8. Let V be a vector space, and let $\boldsymbol{x}_1, \boldsymbol{x}_2, \ldots, \boldsymbol{x}_n$ be a basis for V. Let T be the function from V into $\boldsymbol{R}^n$ that carries the vector $\boldsymbol{x}$ in V into its coordinate n-tuple in $\boldsymbol{R}^n$ relative to the basis $\boldsymbol{x}_1, \boldsymbol{x}_2, \ldots, \boldsymbol{x}_n$. Show that T is a linear transformation from V onto $\boldsymbol{R}^n$.
9. Let A be a 2×2 diagonal matrix that is not a scalar multiple of the identity. Let $C : M_{22} \to M_{22}$ be the linear operator defined by $C(B) = AB - BA$. Show that the matrices

   $$\begin{bmatrix} 0 & 1 \\ 0 & 0 \end{bmatrix} \quad \text{and} \quad \begin{bmatrix} 0 & 0 \\ 1 & 0 \end{bmatrix}$$

   form a basis for R_C.

10. Let $x \neq 0$ be a fixed vector in R^n. Show that the function from the space of $n \times n$ matrices into R^n defined by $P(A) = Ax$ for each $n \times n$ matrix A is a linear transformation onto R^n.

11. Let $S : M_{nn} \to M_{nn}$ be the linear operator defined by $S(A) = A - A^T$. Show that R_S consists precisely of the skew-symmetric matrices, i.e., those matrices such that $B^T = -B$.

12. Let D be a diagonal $n \times n$ matrix. Let $T(x) = Dx$ be the linear transformation from R^n to R^n induced by D. Show that the rank of D is precisely the number of nonzero entries on the diagonal of D.

13. Let A and B be $n \times n$ matrices with B invertible. Let $T(x) = Ax$, $S(x) = (AB)x$ be the linear transformations of R^n induced by A and AB, respectively. Show that T and S have the same range space.

14. Let $T : R^n \to R^n$ be a linear operation of rank 1. Show that there are scalars $a_1, a_2, \ldots, a_n, b_1, b_2, \ldots, b_n$ such that $T(x) = Ax$, where

$$A = \begin{bmatrix} b_1a_1 & b_2a_1 & \cdots & b_na_1 \\ b_1a_2 & b_2a_2 & \cdots & b_na_2 \\ & & \vdots & \\ b_1a_n & b_2a_n & \cdots & b_na_n \end{bmatrix}$$

Show that conversely any matrix of this sort induces a linear transformation from R^n into itself of rank 1.

15. Let $T : R^n \to R^n$ be a linear transformation of rank r. Suppose $x_1, x_2, \ldots, x_r$ is a basis for R_T. If $T(e_j) = a_{j1}x_1 + a_{j2}x_2 + \cdots + a_{jr}x_r$, show that

$$T(x) = [x_1, x_2, \ldots, x_r]\begin{bmatrix} a_{11} & \cdots & a_{n1} \\ & \vdots & \\ a_{1r} & \cdots & a_{nr} \end{bmatrix} x$$

where $[x_1, x_2, \ldots, x_r]$ is the $n \times r$ matrix whose columns are $x_1, x_2, \ldots, x_n$ successively.

This shows that a matrix of rank r can be expressed as a product of an $n \times r$ matrix and an $r \times n$ matrix. Why does this imply exercise 14 above?

16. Let A and B be $n \times n$ matrices T_A, T_B, T_{A+B} the linear transformations of R^n induced by the matrices A, B, and $A + B$, respectively. Show that

$$r(T_A) + r(T_B) \geqslant r(T_{A+B})$$

17. Find a linear transformation that maps the space of 3×3 matrices onto the space of 2×2 matrices.

18. Let A be a fixed $n \times n$ matrix. Let T be the function from P_n into M_{nn}, which sends the polynomial $f(x)$ into the matrix $f(A)$, i.e., if $f(x) = \alpha_0 + \alpha_1 x + \cdots + \alpha_n x^n$, then $T(f) = \alpha_0 I_n + \alpha_1 A + \cdots + \alpha_n A^n$. Show that T is a linear transformation from P_n into M_{nn}. If $n > 1$, why is T not onto?

4 NULLSPACE

Let T be a linear transformation from a vector space V into a vector space W. There is a subspace of V associated with the linear transformation T that is, in some sense, complementary to the range space discussed in the previous section. We denote by N_T the subset of V that consists of those vectors $\boldsymbol{x}$ such that $T(\boldsymbol{x})=\boldsymbol{0}$. In other words, the elements of N_T are just those vectors sent into $\boldsymbol{0}$ by the linear transformation T. N_T is called the **nullspace** of T. This terminology is reasonable, in view of the following:

Theorem 1 Let $T: V \rightarrow W$ be a linear transformation from a vector space V into a vector space W. Then $N_T=\{\boldsymbol{x} \mid \boldsymbol{x} \in V \text{ and } T(\boldsymbol{x})=\boldsymbol{0}\}$ is a subspace of V.

PROOF Suppose $\boldsymbol{x}$ and $\boldsymbol{y}$ belong to N_T. Then $T(\boldsymbol{x}+\boldsymbol{y})=T(\boldsymbol{x})+T(\boldsymbol{y})=\boldsymbol{0}+\boldsymbol{0}=\boldsymbol{0}$. Thus, $\boldsymbol{x}+\boldsymbol{y}$ belongs to N_T. If $\boldsymbol{x}$ is in N_T and α is a scalar, $T(\alpha \boldsymbol{x})=\alpha T(\boldsymbol{x})=\alpha \cdot \boldsymbol{0}=\boldsymbol{0}$, and so $\alpha \boldsymbol{x}$ belongs to N_T. Thus, N_T is a subspace of V. ■

If N is the zero operator from V to W, since N sends all vectors in V into $\boldsymbol{0}$, we see that $N_T=V$. If I_V is the identity operator on V and $I_V(\boldsymbol{x})=\boldsymbol{0}$, it follows that $\boldsymbol{x}=\boldsymbol{0}$ since $I(\boldsymbol{x})=\boldsymbol{x}$. Thus $N_{I_V}=0$, the zero subspace.

The quantity $\dim N_T$ is, like $\dim R_T$, of interest. It is called the **nullity** of T and is denoted by $n(T)$.

Example 1 Let P_θ be the projection transformation of $\boldsymbol{R}^2$ defined in §5.2. Recall that $P_\theta(\boldsymbol{v})$ is obtained by dropping a perpendicular from the endpoint of $\boldsymbol{v}$ to the line l making an angle of θ degrees with the $\boldsymbol{x}$ axis. $P_\theta(\boldsymbol{v})$ is then the vector ending at the point where the perpendicular from the endpoint of $\boldsymbol{v}$ intersects l. (See Figure 5-10.) It is clear from this geometric definition that a vector $\boldsymbol{v}$ belongs to the nullspace of P_θ when and only when it is perpendicular to the vector $\boldsymbol{i}_\theta=(\cos\theta)\boldsymbol{i}+(\sin\theta)\boldsymbol{j}$, or in other words if and only if $\boldsymbol{v}$ is a scalar multiple of the vector $\boldsymbol{j}_\theta=-(\sin\theta)\boldsymbol{i}+(\cos\theta)\boldsymbol{j}$. Thus, the nullspace of P_θ is spanned by the vector $\boldsymbol{j}_\theta$, and so $n(P_\theta)=1$.

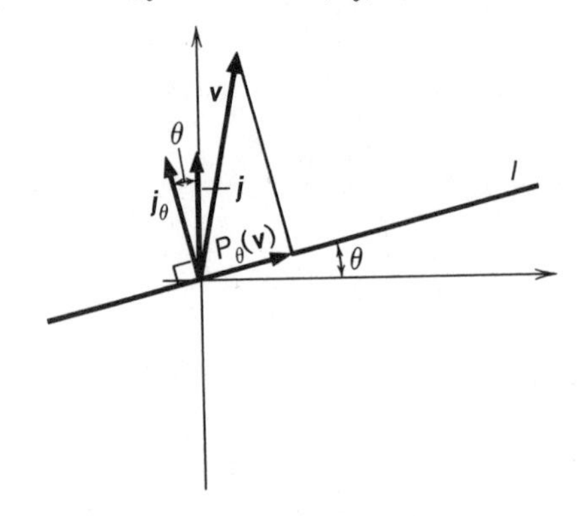

FIGURE 5-10

Example 2 Let A be an $m \times n$ matrix. Let T_A be the linear transformation from $\boldsymbol{R}^n$ to $\boldsymbol{R}^m$ induced by the matrix A. A vector $\boldsymbol{x}$ belongs to the nullspace of T_A if and only if $A\boldsymbol{x} = \boldsymbol{0}$, hence we see that the nullspace of T_A is just the set of solutions to the system of homogeneous linear equations written in matrix notation as $A\boldsymbol{x} = \boldsymbol{0}$.

Using the procedure of Gaussian elimination, it is possible to obtain a basis for the nullspace of a linear operator induced by a matrix. For example, if T is the linear transformation from $\boldsymbol{R}^5$ into $\boldsymbol{R}^3$ defined by

$$T\begin{bmatrix} x_1 \\ x_2 \\ x_3 \\ x_4 \\ x_5 \end{bmatrix} = \begin{bmatrix} 1 & -1 & 3 & 2 & 0 \\ 4 & 1 & -2 & 0 & 1 \\ 1 & 3 & 1 & 0 & 2 \end{bmatrix} \begin{bmatrix} x_1 \\ x_2 \\ x_3 \\ x_4 \\ x_5 \end{bmatrix}$$

in order to find its nullspace we must solve the following system of equations:

$$\begin{aligned} x_1 - x_2 + 3x_3 + 2x_4 \qquad &= 0 \\ 4x_1 + x_2 - 2x_3 \qquad + x_5 &= 0 \\ x_1 + 3x_2 + x_3 \qquad + 2x_5 &= 0 \end{aligned}$$

Use x_5 and equation two.

$$\downarrow$$

$$\begin{aligned} x_1 - x_2 + 3x_3 + 2x_4 \qquad &= 0 \\ 4x_1 + x_2 - 2x_3 \qquad + x_5 &= 0 \\ -7x_1 + x_2 + 5x_3 \qquad &= 0 \end{aligned}$$

Use x_2 and equation three.

$$\downarrow$$

$$\begin{aligned} -6x_1 \qquad + 8x_3 + 2x_4 \qquad &= 0 \\ 11x_1 \qquad - 7x_3 \qquad + x_5 &= 0 \\ -7x_1 + x_2 + 5x_3 \qquad &= 0 \end{aligned}$$

We pretend to use x_4 and equation one, thereby producing a system in which all equations have been used.

Letting $x_1 = c$ and $x_3 = d$, we see that the most general solution is of the form

$$\begin{bmatrix} c \\ 7c - 5d \\ d \\ 3c - 4d \\ -11c + 7d \end{bmatrix} = c\begin{bmatrix} 1 \\ 7 \\ 0 \\ 3 \\ -11 \end{bmatrix} + d\begin{bmatrix} 0 \\ -5 \\ 1 \\ -4 \\ 7 \end{bmatrix}$$

Since the vectors on the right above are clearly linearly independent and span N_T, we see that they form a basis for N_T. Since N_T admits a basis of two vectors, $n(T) = 2$.

Example 3 Let $D : P_n \to P_n$ be the differentiation operator on the space of polynomials of degree less than or equal to n, $D(f) = f'$. If a polynomial f

belongs to N_D, then $f' = 0$. By calculus or by using the standard basis 1, $x, \ldots, x^n$ for P_n, we see that f must be a constant, i.e., f is a scalar multiple of the polynomial 1. Thus, 1 forms a basis for N_D, and so $n(D) = 1$.

Example 4 Let S be the linear operator on M_{nn} of example 4 of §5.3. Thus, $S(A) = A + A^T$. Then, N_S consists precisely of those matrices A such that $S(A) = A + A^T = 0$. Thus, A belongs to N_S if and only if $A^T = -A$, i.e., A is skew-symmetric.

A linear transformation $T: V \to W$ is said to be **one-one** if $T(\boldsymbol{x}_1) = T(\boldsymbol{x}_2)$ implies $\boldsymbol{x}_1 = \boldsymbol{x}_2$. Stated otherwise, each vector of W is the image of at most one vector in V.

For example, if A is an invertible $n \times n$ matrix, the linear operator defined on $\boldsymbol{R}^n$ by $T_A(\boldsymbol{x}) = A\boldsymbol{x}$ is one-one. Indeed, if $A\boldsymbol{x}_1 = A\boldsymbol{x}_2$, multiplying both sides of the forgoing equation by A^{-1}, we have $A^{-1}(A\boldsymbol{x}_1) = A^{-1}(A\boldsymbol{x}_2)$, or $\boldsymbol{x}_1 = \boldsymbol{x}_2$.

On the other hand, as an example of a linear transformation that is not one-one, consider the differentiation operator D on P_n. We have $D(1) = D(0) = 0$. But, certainly $1 \neq 0$.

It is possible to characterize one-one transformations in terms of null space.

Theorem 2 Let $T: V \to W$ be a linear transformation between two vector spaces V and W. Then, T is one-one if and only if $N_T = 0$, i.e., $n(T) = 0$.

PROOF First, we suppose T is one-one. Let $\boldsymbol{x}$ be a vector in N_T. Then, $T(\boldsymbol{x}) = \boldsymbol{0}$. Since $T(\boldsymbol{0}) = \boldsymbol{0}$, it follows that the vectors $\boldsymbol{0}$ and $\boldsymbol{x}$ have the same image in W. Since T is one-one, these vectors must be equal, i.e., $\boldsymbol{x} = \boldsymbol{0}$. Hence, we have shown that any vector in N_T is zero, and it follows that N_T is the zero subspace.

If, on the other hand, N_T is the zero subspace, suppose $T(\boldsymbol{x}_1) = T(\boldsymbol{x}_2)$. Then $T(\boldsymbol{x}_1 - \boldsymbol{x}_2) = T(\boldsymbol{x}_1) - T(\boldsymbol{x}_2) = \boldsymbol{0}$. But, since $N_T = 0$, $\boldsymbol{x}_1 - \boldsymbol{x}_2 = \boldsymbol{0}$, or $\boldsymbol{x}_1 = \boldsymbol{x}_2$. Since $T(\boldsymbol{x}_1) = T(\boldsymbol{x}_2)$ implies $\boldsymbol{x}_1 = \boldsymbol{x}_2$, T is one-one. ■

Suppose that the linear transformation T induced by the $m \times n$ matrix A is one-one. What does this say about the system of linear equations $A\boldsymbol{x} = \boldsymbol{y}$? It does not mean that a solution to the system exists, and in fact no solution may exist; however, when the system is solvable, the solutions are unique. It is worthwhile to contrast the property of being one-one with the property of being onto. When T is onto, the system $A\boldsymbol{x} = \boldsymbol{y}$ is always solvable, but the solution need not be unique.

There is a fundamental theorem connecting the rank and nullity of a linear transformation.

Theorem 3 Let $T: V \to W$ be a linear transformation between two finite-dimensional vector spaces. Then, $\dim R_T + \dim N_T = \dim V$, or, $r(T) + n(T) = \dim V$.

PROOF Let $x_1, x_2, \ldots, x_m$ be a basis for N_T and $y_1, y_2, \ldots, y_n$ be a basis for R_T. Then $\dim N_T = m$ and $\dim R_T = n$. Since $y_1, y_2, \ldots, y_n$ belong to R_T, there are vectors $z_1, z_2, \ldots, z_n$ in V such that

$$T(z_1) = y_1, \qquad T(z_2) = y_2, \qquad \cdots, \qquad T(z_n) = y_n$$

We wish to show that $x_1, x_2, \ldots, x_m, z_1, z_2, \ldots, z_n$ form a basis for V. First, the vectors $x_1, x_2, \ldots, x_m, z_1, z_2, \ldots, z_n$ are linearly independent. For suppose we have

$$\alpha_1 x_1 + \alpha_2 x_2 + \cdots + \alpha_m x_m + \beta_1 z_1 + \beta_2 z_2 + \cdots + \beta_n z_n = \mathbf{0}$$

Then applying T,

$$\alpha_1 T(\mathbf{x}_1) + \alpha_2 T(x_2) + \cdots + \alpha_m T(x_m) + \beta_1 T(z_1) + \beta_2 T(z_2) + \cdots + \beta_n T(z_n) = \mathbf{0}$$

Since $x_1, x_2, \ldots, x_m$ belong to N_T, $T(x_1) = T(x_2) = \cdots = \mathbf{0}$. So,

$$\beta_1 T(z_1) + \beta_2 T(z_2) + \cdots + \beta_n T(z_n) = \mathbf{0}$$

But $T(z_1) = y_1, T(z_2) = y_2, \ldots, T(z_n) = y_n$ is, by hypothesis, a basis for R_T. Thus, $\beta_1 = \beta_2 = \cdots = \beta_n = 0$, and so $\alpha_1 x_1 + \alpha_2 x_2 + \cdots + \alpha_m x_m = \mathbf{0}$. Since $x_1, x_2, \ldots, x_m$ is a basis for N_T, we have $\alpha_1 = \alpha_2 = \cdots = \alpha_m = \mathbf{0}$. It, therefore, follows that the vectors $x_1, x_2, \ldots, x_m, z_1, z_2, \ldots, z_n$ are linearly independent.

Second, the vectors $x_1, x_2, \ldots, x_m, z_1, z_2, \ldots, z_n$ span V. For suppose x is a vector in V. Then $T(x)$ is a vector in R_T, and since $y_1, y_2, \ldots, y_n$ is a basis for R_T, there are scalars $\beta_1, \beta_2, \ldots, \beta_n$ such that $T(x) = \beta_1 y_1 + \beta_2 y_2 + \cdots + \beta_n y_n$.

Consider the vector $a = x - \beta_1 z_1 - \beta_2 z_2 - \cdots - \beta_n z_n$. Note that

$$\begin{aligned} T(a) &= T(x) - \beta_1 T(z_1) - \cdots - \beta_n T(z_n) \\ &= \beta_1 y_1 + \beta_2 y_2 + \cdots + \beta_n y_n - \beta_1 y_1 - \beta_2 y_2 - \cdots - \beta_n y_n \\ &= \mathbf{0} \end{aligned}$$

Thus, $a \in N_T$, and since $x_1, x_2, \ldots, x_m$ is a basis for N_T, there are scalars $\alpha_1, \alpha_2, \ldots, \alpha_m$ such that $a = \alpha_1 x_1 + \alpha_2 x_2 + \cdots + \alpha_m x_m$. Rewriting $a = x - \beta_1 z_1 - \beta_2 z_2 - \cdots - \beta_n z_n = \alpha_1 x_1 + \alpha_2 x_2 + \cdots + \alpha_m x_m$, we see that $x = \alpha_1 x_1 + \alpha_2 x_2 + \cdots + \alpha_m x_m + \beta_1 z_1 + \beta_2 z_2 + \cdots + \beta_n z_n$.

Thus, the vectors $x_1, x_2, \ldots, x_n, z_1, z_2, \ldots, z_n$ span V, and since their linear independence was demonstrated above, we see that $x_1, x_2, \ldots, x_m, z_1, z_2, \ldots, z_n$ form a basis for V. Thus,

$$\begin{aligned} \dim V &= m + n \\ &= \dim N_T + \dim R_T \; \blacksquare \end{aligned}$$

Corollary 1 If $T : V \to W$ is a linear transformation, then $\dim R_T \leqslant \dim V$.

PROOF From the above theorem, we know that $\dim V = \dim R_T + \dim N_T \geqslant \dim R_T$. ■

Now suppose that $\dim W > \dim V$. Since $\dim V \geqslant \dim R_T$, we must have $\dim W > \dim R_T$. Thus, if $T : V \to W$ is a linear transformation with $\dim V < \dim W$, T is not onto.

Example 5 Let

$$\begin{aligned} a_{11}x_1 + a_{12}x_2 + \cdots + a_{1n}x_n &= y_1 \\ a_{21}x_1 + a_{22}x_2 + \cdots + a_{2n}x_n &= y_2 \\ &\vdots \\ a_{m1}x_1 + a_{m2}x_2 + \cdots + a_{mn}x_n &= y_m \end{aligned}$$

be a system of m equations in n variables. If $m > n$, i.e., the number of equations is greater than the number of unknowns, by Corollary 1, it is possible to choose values of $y_1, y_2, \ldots, y_m$ so that the system will not be solvable.

Corollary 2 If $T : V \to W$ is a linear transformation, and $\dim W < \dim V$, then $N_T \neq \mathbf{0}$, i.e., $n(T) > 0$.

PROOF We know that $\dim N_T + \dim R_T = \dim V$. Since R_T is a subspace of W, $\dim R_T \leqslant \dim W < \dim V$. Thus, $\dim N_T = \dim V - \dim R_T > 0$. ■

Example 6 Let

$$\begin{aligned} a_{11}x_1 + a_{12}x_2 + \cdots + a_{1n}x_n &= 0 \\ a_{21}x_1 + a_{22}x_2 + \cdots + a_{2n}x_n &= 0 \\ &\vdots \\ a_{m1}x_1 + a_{m2}x_2 + \cdots + a_{mn}x_n &= 0 \end{aligned}$$

be a system of m homogeneous linear equations in n variables, where $m < n$, i.e., the number of equations is less than the number of unknowns. As usual, let $A = [a_{ij}]_{(mn)}$ and consider the linear transformation $T_A : \boldsymbol{R}^n \to \boldsymbol{R}^m$, associated with the matrix A. Since $\dim \boldsymbol{R}^m < \dim \boldsymbol{R}^n$, by Corollary 2, N_{T_A} must contain some vector other than the zero vector. This is merely a restatement of the fact we proved in chapter 1, namely, that the above system of equations admits a nontrivial solution.

Corollary 3 Let $T : V \to W$ be a linear transformation and suppose $\dim V = \dim W$. Then, T is one-one if and only if T is onto.

PROOF By Theorem 3, $\dim R_T + \dim N_T = \dim V = \dim W$. Thus, $\dim N_T = \dim W - \dim R_T$. So, $\dim N_T = 0$ (i.e., T is one-one) if and only if $\dim W = \dim R_T$ (i.e., T is onto). ■

Example 7 Let A be an $n \times n$ matrix. Consider the system of linear equations $A\boldsymbol{x} = \boldsymbol{y}$, where $\boldsymbol{x}$ and $\boldsymbol{y}$ are n-vectors. By Corollary 3, we see that the system $A\boldsymbol{x} = \boldsymbol{y}$ is solvable regardless of the value of $\boldsymbol{y}$ if and only if the system $A\boldsymbol{x} = \boldsymbol{0}$ admits only the trivial solution $\boldsymbol{x} = \boldsymbol{0}$.

Example 8 Let P_n denote the space of all real polynomials of degree at most n. We define a linear operator T on P_n by $T(f) = (c_0 + c_1x + c_2x^2)f'' + (b_0 + b_1x)f' + a_0f$, where the a's, b's, and c's are all real numbers. It is not hard to verify that T is linear and that T carries polynomials of degree n into polynomials of degree at most n. T is, in fact, an example of a linear differential operator. Such operators are of importance in the study of differential equations. We ask the question: If g is a polynomial, can we solve this differential equation?

$$(c_0 + c_1x + c_2x^2)f'' + (b_0 + b_1x)f' + a_0f = g$$

The answer in this case is quite simple.

(1) either the equation has a unique polynomial solution f for any polynomial g, or
(2) the equation $(c_0 + c_1x + c_2x^2)f'' + (b_0 + b_1x)f' + a_0f = 0$ has a nonzero polynomial solution.

To see this, suppose that g is a polynomial of degree n. Then T is a linear operator on P_n. If $N_T \neq 0$, then the second conclusion holds. Thus, we may suppose that $N_T = 0$. It follows that T is one-one. By Corollary 3, T is onto. Thus, the equation $T(f) = g$ can be solved for a polynomial f.

EXERCISES

1. Calculate a basis for the nullspace and the nullity of the linear transformation associated with the following matrices.

(a) $\begin{bmatrix} -1 & 2 \\ 3 & -6 \end{bmatrix}$ (b) $\begin{bmatrix} 1 & 0 & 3 \\ -1 & 2 & 1 \end{bmatrix}$ (c) $\begin{bmatrix} 0 & 1 & 7 \\ -1 & 2 & 3 \\ -1 & 3 & 10 \end{bmatrix}$

(d) $\begin{bmatrix} 1 & 3 & -1 & 2 \\ 3 & -4 & 0 & 1 \\ 5 & 2 & -2 & 5 \\ 4 & -1 & -1 & 3 \end{bmatrix}$ (e) $\begin{bmatrix} 0 & 1 & 3 & 1 & 4 \\ 1 & 0 & 2 & -1 & -1 \\ 1 & 0 & 1 & -1 & -2 \end{bmatrix}$

2. Let $a_1, a_2, \ldots, a_n$ be n real numbers not all of which are zero. Show that the function

$$f\left(\begin{bmatrix} x_1 \\ x_2 \\ \vdots \\ x_n \end{bmatrix}\right) = a_1x_1 + a_2x_2 + \cdots + a_nx_n$$

is linear from $\boldsymbol{R}^n$ to $\boldsymbol{R}$. Show that its nullspace is of dimension $n-1$.

3. Let D^k be the linear transformation from P_n into P_n that carries each polynomial into its kth derivative. Show that the nullspace of D^k is of dimension k.

4. Let T be the linear operator on P_2 defined by $T(f) = f - xf' + (a - x^2)f''$, where a is some number. Show that the polynomial x spans N_T.

5. Find a linear transformation T from $\boldsymbol{R}^2$ into itself such that $R_T = N_T$.

6. Let D be an $n \times n$ diagonal matrix and $T_D(\boldsymbol{x}) = D\boldsymbol{x}$ be the linear operator on $\boldsymbol{R}^n$ induced by D. Show that the nullity of T_D is just the number of zeros on the diagonal of D.

7. Let V be a vector space and $\boldsymbol{x}_1, \boldsymbol{x}_2, \ldots, \boldsymbol{x}_n$ be a basis for V. Let $T : V \to V$ be the linear transformation such that

$$T(\boldsymbol{x}_1) = \boldsymbol{x}_2, \qquad T(\boldsymbol{x}_2) = \boldsymbol{x}_3, \qquad \ldots, \qquad T(\boldsymbol{x}_{n-1}) = \boldsymbol{x}_n, \qquad T(\boldsymbol{x}_n) = \boldsymbol{0}$$

What is $n(T)$?

8. Let $T : V \to V$ be a linear transformation with the property that $N_T = V$. Show that T is the zero operator.

9. If A is an $n \times n$ matrix and there is some nonzero n-vector $\boldsymbol{x}$ such that $A\boldsymbol{x} = \boldsymbol{0}$, show that $r(A) < n$. Prove the converse.

10. Let A be a fixed $n \times n$ matrix. Suppose the nullspace of the operator on $\boldsymbol{R}^n$ induced by A is of dimension k. Let T be the linear transformation from the space of $n \times n$ matrices into itself defined by $T(B) = AB$.
    (a) Show that B belongs to N_T if and only if each column of B belongs to N_A.
    (b) Show that $n(T) = nk$.
    (c) Show that $r(T) = nr(A)$.

11. Let A be an $m \times n$ matrix. Let $T_A : \boldsymbol{R}^n \to \boldsymbol{R}^m$ be the linear transformation $T_A(\boldsymbol{x}) = A\boldsymbol{x}$, where $\boldsymbol{x}$ belongs to $\boldsymbol{R}^n$. Prove the following statements.
    (a) T_A is onto if and only if the columns of A span $\boldsymbol{R}^m$.
    (b) T_A is one-one if and only if the columns of A are linearly independent.
    (c) If T_A is one-one and onto, $m = n$.

12. Let T be a linear transformation from a vector space V to a vector space W. Let $\boldsymbol{y}$ be a vector in W. If the equation $T(\boldsymbol{x}) = \boldsymbol{y}$ is uniquely solvable for a vector $\boldsymbol{x}$ in V, show that T is one-one.

13. Let $T : P_n \to P_n$ be the linear operator on the space of polynomials of degree less than or equal to n in a variable x defined by $T(f) = f + xf'$. Show that T is one-one and onto.

14. Give an example of linear transformation on $\boldsymbol{R}^3$ of rank 1; of rank 2.

15. Let T be a linear transformation on $\boldsymbol{R}^2$, $T \neq 0$ and T not onto. What are $r(T)$ and $n(T)$?

16. Let $T : V \rightarrow W$ be a linear transformation with the property that whenever $\boldsymbol{x}_1, \boldsymbol{x}_2, \ldots, \boldsymbol{x}_n$ are linearly independent in V, $T(\boldsymbol{x}_1), T(\boldsymbol{x}_2), \ldots, T(\boldsymbol{x}_n)$ are linearly independent in W. Show that T is one-one. Show that if T is one-one, it has this property.

17. Let $T : P_n \rightarrow P_n$ be the function that carries the polynomial $f(x)$ into the polynomial $\frac{1}{2}(f(x) + f(-x))$.
    (a) Show that T is linear.
    (b) Show that R_T consists precisely of the even polynomials in P_n.
    (c) Show that N_T consists precisely of the odd polynomials in P_n.

18. Let $T : \boldsymbol{R}^n \rightarrow \boldsymbol{R}^n$ be the linear operator

$$T\begin{bmatrix} x_1 \\ x_2 \\ x_3 \\ \vdots \\ x_n \end{bmatrix} = \begin{bmatrix} 1-n & 1 & 1 & \cdots & 1 \\ 1 & 1-n & 1 & \cdots & 1 \\ 1 & 1 & 1-n & \cdots & 1 \\ & & \vdots & & \\ 1 & 1 & 1 & \cdots & 1-n \end{bmatrix}\begin{bmatrix} x_1 \\ x_2 \\ x_3 \\ \vdots \\ x_n \end{bmatrix}$$

Show that N_T is spanned by the column n-vector all of whose entries are 1. What is $r(T)$?

19. Let A be a fixed $n \times n$ matrix. Let T be the linear transformation from M_{nn} into M_{nn} defined by $T(B) = AB - BA$. Show that T is not one-one.

20. If f is a polynomial of degree less than or equal to n, show that there is a polynomial g of degree less than or equal to n such that $g + g' = f$.

21. Let T be a linear transformation between two vector spaces V and W. Let $\boldsymbol{x}_1, \boldsymbol{x}_2, \ldots, \boldsymbol{x}_n$ and $\boldsymbol{y}_1, \boldsymbol{y}_2, \ldots, \boldsymbol{y}_m$ be vectors in V. If $\text{sp}(\boldsymbol{x}_1, \boldsymbol{x}_2, \ldots, \boldsymbol{x}_n) = N_T$ and $\text{sp}(T(\boldsymbol{y}_1), T(\boldsymbol{y}_2), \ldots, T(\boldsymbol{y}_m)) = R_T$, show that $\text{sp}(\boldsymbol{x}_1, \boldsymbol{x}_2, \ldots, \boldsymbol{x}_n, \boldsymbol{y}_1, \boldsymbol{y}_2, \ldots, \boldsymbol{y}_m) = V$.

22. Let $T : V \rightarrow V$ be a linear operator on a vector space V. If $R_T = N_T$, show that $\dim V$ is even.

23. If A is an $m \times n$ matrix, show that there is an $n \times n$ matrix B, $B \neq 0$, such that $AB = 0$ if and only if $r(A) < n$.

24. If A is an $m \times n$ matrix, show that there is an $m \times m$ matrix B, $B \neq 0$, such that $BA = 0$ if and only if $r(A) < m$.

25. Let $0 \leqslant k \leqslant n$. Show that there is a linear transformation of $\boldsymbol{R}^n$ having nullity k.

5 RANK AND ELEMENTARY MATRICES

If A is an $m \times n$ matrix, A induces a linear transformation $T_A : \boldsymbol{R}^n \rightarrow \boldsymbol{R}^m$ by defining $T_A(\boldsymbol{x}) = A\boldsymbol{x}$. We defined the rank of T_A as dimension of the range space of T_A. By the **rank of the matrix** A we mean simply the rank of the linear transformation T_A. We also call the range space of T_A simply the range of A.

If we denote $A_1, A_2, \ldots, A_n$ the columns of the matrix A, we saw in §5.3 that the range space of T_A is spanned by the m-vectors, $A_1, A_2, \ldots, A_n$. For this reason, the range space of T_A is sometimes called the **column space** of the matrix A. Thus, the rank of the matrix A might well be defined to be the dimension of the span of the m-vectors $A_1, A_2, \ldots, A_n$. By Theorem 1 of §4.7, we see that the rank of A is the maximum number of linearly independent vectors in the collection $A_1, A_2, \ldots, A_n$. In this section, we simplify the problem of computing the rank of a matrix by providing an algorithm for its calculation.

The following theorem is important for our development of a process for calculating rank.

Theorem 1 Let A be an $m \times n$ matrix, B be an $m \times m$ matrix, and C be an $n \times n$ matrix. Then $r(BA) \leqslant r(A)$ and $r(AC) \leqslant r(A)$.

PROOF First, $r(BA) \leqslant r(A)$.

Let $y_1, y_2, \ldots, y_{r(A)}$ be a basis for the range of A. Thus, if $\boldsymbol{x}$ belongs to $\boldsymbol{R}^n$, there are scalars $\alpha_1, \alpha_2, \ldots, \alpha_{r(A)}$ such that

$$A\boldsymbol{x} = \alpha_1 y_1 + \alpha_2 y_2 + \cdots + \alpha_{r(A)} y_{r(A)}$$

So,

$$\begin{aligned}(BA)(\boldsymbol{x}) &= B(A\boldsymbol{x}) \\ &= B(\alpha_1 y_1 + \alpha_2 y_2 + \cdots + \alpha_{r(A)} y_{r(A)}) \\ &= \alpha_1 By_1 + \alpha_2 By_2 + \cdots + \alpha_{r(A)} By_{r(A)}\end{aligned}$$

Thus, the vectors $By_1, By_2, \ldots, By_{r(A)}$ span the range space of BA. By Theorem 1, §4.8, we see that $\dim R_{BA} \leqslant \dim R_A$. Thus, $r(BA) \leqslant r(A)$.

Next, we show that $r(AC) \leqslant r(A)$. We claim that the range space of AC is contained in the range space of A. Suppose y belongs to the range of AC, $y = (AC)(\boldsymbol{x}) = A(C\boldsymbol{x})$. Thus, y is the A image of a vector, namely $C\boldsymbol{x}$, and so y belongs to the range of A. Thus, the range space of AC is a subspace of the range space of A, and so, by Theorem 3, §4.8, $\dim R_{AC} \leqslant \dim R_A$, or $r(AC) \leqslant r(A)$. ■

Immediately, from Theorem 1, we get the following:

Theorem 2 Let A be an $m \times n$ matrix, B be an invertible $m \times m$ matrix, and C be an invertible $n \times n$ matrix. Then

$$r(BA) = r(A) \quad \text{and} \quad r(AC) = r(A)$$

PROOF By Theorem 1, $r(BA) \leqslant r(A)$. Again, by Theorem 1, $r(B^{-1}(BA)) \leqslant r(BA)$, or $r(A) \leqslant r(BA)$. Thus, $r(A) = r(BA)$. The proof that $r(AC) = r(A)$ is similar. ■

For example, let

$$A = \begin{bmatrix} 1 & 0 & -1 \\ 0 & 1 & -1 \\ 0 & 0 & 1 \end{bmatrix} \quad \text{and} \quad B = \begin{bmatrix} 1 & 0 & 1 \\ 3 & 1 & 4 \\ 7 & 2 & 9 \end{bmatrix}. \quad \text{Then } BA = \begin{bmatrix} 1 & 0 & 0 \\ 3 & 1 & 0 \\ 7 & 2 & 0 \end{bmatrix}.$$

Since $\det A = 1$, A is invertible. Thus, $r(BA) = r(B)$. Since BA has exactly two linearly independent columns, it follows that $r(B) = 2$.

Theorem 2 may be simply restated: Multiplication of a matrix by an invertible matrix does not change its rank. Our next goal is to find a simple family of invertible matrices that enables us to reduce a given matrix, by a series of matrix multiplications, to one whose rank is easily computed. The family of elementary matrices that we describe below has this property.

Let E be a matrix obtained from the identity matrix by one of the three operations:

(1) Interchanging two columns of the identity matrix.
(2) Multiplying a column of the identity matrix by a nonzero scalar.
(3) Adding a scalar multiple of one column of the identity matrix to another column.

Then, E is said to be an **elementary matrix**.

The following are all examples of elementary matrices.

$$\begin{bmatrix} 0 & 0 & 1 \\ 0 & 1 & 0 \\ 1 & 0 & 0 \end{bmatrix}, \quad \begin{bmatrix} 1 & 0 & 0 \\ 0 & 2 & 0 \\ 0 & 0 & 1 \end{bmatrix}, \quad \begin{bmatrix} 1 & 0 & -7 \\ 0 & 1 & 0 \\ 0 & 0 & 1 \end{bmatrix}$$

The first matrix was obtained by interchanging the first and third columns of the matrix I_3. The second was obtained by multiplying the second column of I_3 by 2. The third was obtained by adding -7 times the first column of I_3 to the third column of I_3.

There is a very simple way to describe what happens when we take the product of an elementary matrix and another matrix. First, we describe what happens to an $m \times n$ matrix, A, when we postmultiply it by the $n \times n$ elementary matrix, E, i.e., when we form the product, AE.

In what follows we make extensive use of the fact that if A is an $m \times n$ matrix and $\boldsymbol{e}_i$ is the ith vector in the standard basis for $\boldsymbol{R}^n$, then $A\boldsymbol{e}_i$ is just the ith column of the matrix A, which we denote by A_i.

Proposition 1 Suppose that E is an elementary matrix obtained by interchanging columns i and j, $i < j$, in the identity matrix I_n. Then the product AE is the matrix obtained from A by interchanging columns i and j in the matrix A, all other columns remaining the same.

PROOF Since E is obtained from I_n by interchanging columns i and j, the ith column of E is $\boldsymbol{e}_j$, the jth column of E is $\boldsymbol{e}_i$, and the kth column of E, for $k \neq i$

or j, is e_k. Thus, $Ee_i = e_j$, $Ee_j = e_i$, and $Ee_k = e_k$ for $k \neq i$ or j. Then

$$(AE)e_i = A(Ee_i) = Ae_j = A_j$$

$$(AE)e_j = A(Ee_j) = Ae_i = A_i$$

$$(AE)e_k = A(Ee_k) = Ae_k = A_k$$

for $k \neq i$ or j.

From this, it is clear that the ith column of AE is A_j, the jth column of AE is A_i, and the kth column of AE is A_k, if $k \neq i$ or j. Stated otherwise, AE is obtained from A by interchange of columns i and j. ■

For example, if $A = \begin{bmatrix} a_1 & b_1 & c_1 \\ a_2 & b_2 & c_2 \\ a_3 & b_3 & c_3 \end{bmatrix}$, and $E = \begin{bmatrix} 1 & 0 & 0 \\ 0 & 0 & 1 \\ 0 & 1 & 0 \end{bmatrix}$ is obtained by interchanging the second and third columns of the identity matrix, then $AE = \begin{bmatrix} a_1 & b_1 & c_1 \\ a_2 & b_2 & c_2 \\ a_3 & b_3 & c_3 \end{bmatrix}\begin{bmatrix} 1 & 0 & 0 \\ 0 & 0 & 1 \\ 0 & 1 & 0 \end{bmatrix} = \begin{bmatrix} a_1 & c_1 & b_1 \\ a_2 & c_2 & b_2 \\ a_3 & c_3 & b_3 \end{bmatrix}$ is obtained by interchanging the second and third columns of the matrix A.

Proposition 2 Suppose the elementary matrix E is obtained by multiplying the ith column of the identity matrix by a nonzero scalar α. The product AE is the matrix obtained by multiplying the ith column of the matrix A by the scalar α, all other columns remaining the same.

PROOF It is clear that the ith column of E is αe_i, while the kth column of E, for $k \neq i$ is e_k. Thus, $Ee_i = \alpha e_i$ and $Ee_k = e_k$, for $k \neq i$.

$$(AE)e_k = A(Ee_k) = Ae_k = A_k, \quad \text{for } k \neq i$$

and

$$(AE)e_i = A(Ee_i) = A(\alpha e_i) = \alpha A e_i = \alpha A_i$$

From this, we see that the kth column of AE is A_k, if $k \neq i$. The ith column of AE is αA_i, that is, AE is obtained by multiplying the ith column of A by α. ■

As an example, if $A = \begin{bmatrix} a_1 & b_1 & c_1 \\ a_2 & b_2 & c_2 \\ a_3 & b_3 & c_3 \end{bmatrix}$ and $E = \begin{bmatrix} 1 & 0 & 0 \\ 0 & x & 0 \\ 0 & 0 & 1 \end{bmatrix}$, then $AE = \begin{bmatrix} a_1 & xb_1 & c_1 \\ a_2 & xb_2 & c_2 \\ a_3 & xb_3 & c_3 \end{bmatrix}$ is obtained by multiplying the second column of A by x.

Proposition 3 Suppose that the elementary matrix E is obtained from I_n by adding α times the jth column of I_n to the ith column of I_n, for $i \neq j$. Then the matrix AE is obtained from the matrix A by adding α times the jth column of A to the ith column of A, all other columns remaining the same.

PROOF It is clear that the kth column of E, for $k \neq i$, is $\boldsymbol{e}_k$ and that the ith column of E is $\boldsymbol{e}_i + \alpha\boldsymbol{e}_j$. Thus,

$$E\boldsymbol{e}_k = \boldsymbol{e}_k, \quad \text{if } k \neq i$$

$$E\boldsymbol{e}_i = \boldsymbol{e}_i + \alpha\boldsymbol{e}_j$$

Therefore,

$$(AE)\boldsymbol{e}_k = A(E\boldsymbol{e}_k) = A\boldsymbol{e}_k = A_k, \quad \text{for } k \neq i$$

$$(AE)\boldsymbol{e}_i = A(E\boldsymbol{e}_i) = A(\boldsymbol{e}_i + \alpha\boldsymbol{e}_j) = A_i + \alpha A_j$$

Hence we see that the kth column of the matrix AE, for $k \neq i$, is A_k, and the ith column of AE is $A_i + \alpha A_j$. So AE can be obtained from A by adding α times the jth column of A to the ith column of A. ■

For example, if $A = \begin{bmatrix} a_1 & b_1 & c_1 \\ a_2 & b_2 & c_2 \\ a_3 & b_3 & c_3 \end{bmatrix}$ and $E = \begin{bmatrix} 1 & 0 & \alpha \\ 0 & 1 & 0 \\ 0 & 0 & 1 \end{bmatrix}$ is obtained by adding α times the first column of the identity matrix to the third column of the identity,

$$AE = \begin{bmatrix} a_1 & b_1 & c_1 \\ a_2 & b_2 & c_2 \\ a_3 & b_3 & c_3 \end{bmatrix} \begin{bmatrix} 1 & 0 & \alpha \\ 0 & 1 & 0 \\ 0 & 0 & 1 \end{bmatrix} = \begin{bmatrix} a_1 & b_1 & c_1 + \alpha a_1 \\ a_2 & b_2 & c_2 + \alpha a_2 \\ a_3 & b_3 & c_3 + \alpha a_3 \end{bmatrix}$$

is obtained by adding α times the first column of A to the third column of A.

We may combine Propositions 1, 2, and 3 into one rule: Postmultiplication of A by an elementary matrix E performs an operation on A that is the same type as that operation used to obtain E from the identity matrix.

As a consequence of Propositions 1, 2, and 3 we have

Theorem 3 Any elementary matrix is invertible and its inverse is a matrix of the same type.

PROOF (i) Let E be an elementary matrix obtained by interchanging columns of i and j of the identity for $i < j$. Then, EE, the result of postmultiplying E by E, is a matrix obtained by interchanging the ith and jth columns of E. Thus, $EE = I_n$. So E is invertible and its inverse is E, again an elementary matrix.

(ii) Let E be an elementary matrix obtained by multiplying the ith column of the identity matrix by the nonzero scalar α. Then, E^{-1} is just the matrix obtained by multiplying the ith column of the identity matrix by α^{-1}.

(iii) Let E be an elementary matrix obtained by adding α times the jth column of the identity matrix to the ith column of the identity matrix. Then, E^{-1} is the matrix obtained by adding $-\alpha$ times the jth column of the identity matrix to the ith column of the identity matrix. ■

For example,

$$\begin{bmatrix} 1 & 0 & 0 \\ 0 & 0 & 1 \\ 0 & 1 & 0 \end{bmatrix}^{-1} = \begin{bmatrix} 1 & 0 & 0 \\ 0 & 0 & 1 \\ 0 & 1 & 0 \end{bmatrix} \qquad \begin{bmatrix} 1 & 0 & \alpha \\ 0 & 1 & 0 \\ 0 & 0 & 1 \end{bmatrix}^{-1} = \begin{bmatrix} 1 & 0 & -\alpha \\ 0 & 1 & 0 \\ 0 & 0 & 1 \end{bmatrix}$$

In a similar way it is possible to determine the effect of premultiplying an $m \times n$ matrix A by an $m \times m$ elementary matrix E, that is, forming the product EA. Note that the matrix obtained by interchanging two rows of the identity matrix, multiplying some row by a nonzero scalar, or adding a scalar multiple of one row to another is again elementary.

(i′) If E is an elementary matrix obtained from the identity matrix by interchanging rows i and j, EA is obtained from A by interchanging rows i and j.

(ii′) If E is an elementary matrix obtained from the identity matrix by multiplying the ith row of the identity matrix by the nonzero scalar α, EA is obtained from A by multiplying the ith row of A by the scalar α.

(iii′) If E is an elementary matrix obtained by adding α times the jth row of the identity matrix to the ith row of the identity matrix, for $i \neq j$, then the matrix EA is obtained by adding α times the jth row of A to the ith row of A.

As examples, we see

$$\begin{bmatrix} 1 & 0 & 0 \\ 0 & 0 & 1 \\ 0 & 1 & 0 \end{bmatrix} \begin{bmatrix} a_1 & b_1 & c_1 \\ a_2 & b_2 & c_2 \\ a_3 & b_3 & c_3 \end{bmatrix} = \begin{bmatrix} a_1 & b_1 & c_1 \\ a_3 & b_3 & c_3 \\ a_2 & b_2 & c_2 \end{bmatrix}$$

$$\begin{bmatrix} 1 & 0 & 0 \\ 0 & \alpha & 0 \\ 0 & 0 & 1 \end{bmatrix} \begin{bmatrix} a_1 & b_1 & c_1 \\ a_2 & b_2 & c_2 \\ a_3 & b_3 & c_3 \end{bmatrix} = \begin{bmatrix} a_1 & b_1 & c_1 \\ \alpha a_2 & \alpha b_2 & \alpha c_2 \\ a_3 & b_3 & c_3 \end{bmatrix}$$

$$\begin{bmatrix} 1 & 0 & \alpha \\ 0 & 1 & 0 \\ 0 & 0 & 1 \end{bmatrix} \begin{bmatrix} a_1 & b_1 & c_1 \\ a_2 & b_2 & c_2 \\ a_3 & b_3 & c_3 \end{bmatrix} = \begin{bmatrix} a_1 + \alpha a_3 & b_1 + \alpha b_3 & c_1 + \alpha c_3 \\ a_2 & b_2 & c_2 \\ a_3 & b_3 & c_3 \end{bmatrix}$$

If A is a matrix we sometimes use the word **line** to denote either a row or a column of A.

If A is a matrix, we define an **elementary operation** on A to be any of the following three transformations.

(1) Interchanging two parallel lines of A.
(2) Multiplication of a line of A by a nonzero constant.
(3) Addition of a scalar multiple of a line of A to another parallel line of A.

We have seen above that any elementary operation may be performed by premultiplying or postmultiplying A by a suitable elementary matrix. The object of introducing elementary operations is to facilitate the process of calculating the rank of a matrix. The significance of elementary operations in this procedure stems for the most part from the following theorem.

Theorem 4 Let A and A' be $m \times n$ matrices and suppose that A' is obtained from A by an elementary operation. Then $r(A) = r(A')$.

PROOF Since A' is obtained from A by means of an elementary operation, we have $A' = EA$ or $A' = AE$, where E is a suitable elementary matrix. Since any elementary matrix is invertible by Theorem 3 above, and since by Theorem 2, multiplication of A by an invertible matrix does not change its rank, we see that $r(A') = r(A)$. ■

If the matrix B can be obtained from the matrix A by a sequence of elementary operations, we say that the matrices A and B are **equivalent** and write $A \sim B$. By Theorem 4, $A \sim B$ implies $r(A) = r(B)$. Note the following properties of equivalence:

(1) $A \sim A$.
(2) $A \sim B$ implies $B \sim A$.
(3) $A \sim B$ and $B \sim C$ implies $A \sim C$.

The first property is obvious. To prove (2), note that if $A \sim B$, there are elementary matrices $E_1, E_2, \ldots, E_m, F_1, F_2, \ldots, F_n$, such that $B = E_1E_2 \cdots E_mAF_1F_2 \cdots F_n$. Therefore, $A = E_m^{-1}E_{m-1}^{-1} \cdots E_1^{-1}BF_n^{-1}F_{n-1}^{-1} \cdots F_1^{-1}$. Since the inverse of an elementary matrix is again an elementary matrix by Theorem 3, we see that A can be obtained from B by elementary operations and so $B \sim A$. Property (3) says that if B is obtained from A by elementary operations and C is obtained from B by elementary operations, then C can be obtained from A by elementary operations.

As an example of the use of elementary operations, we have,

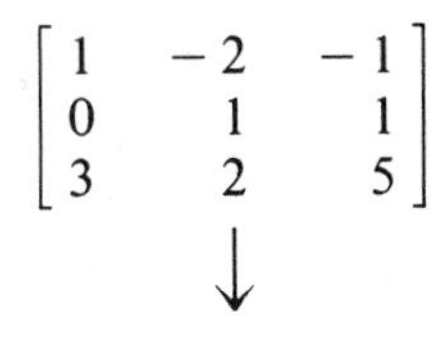

Add twice the first column to the second column.

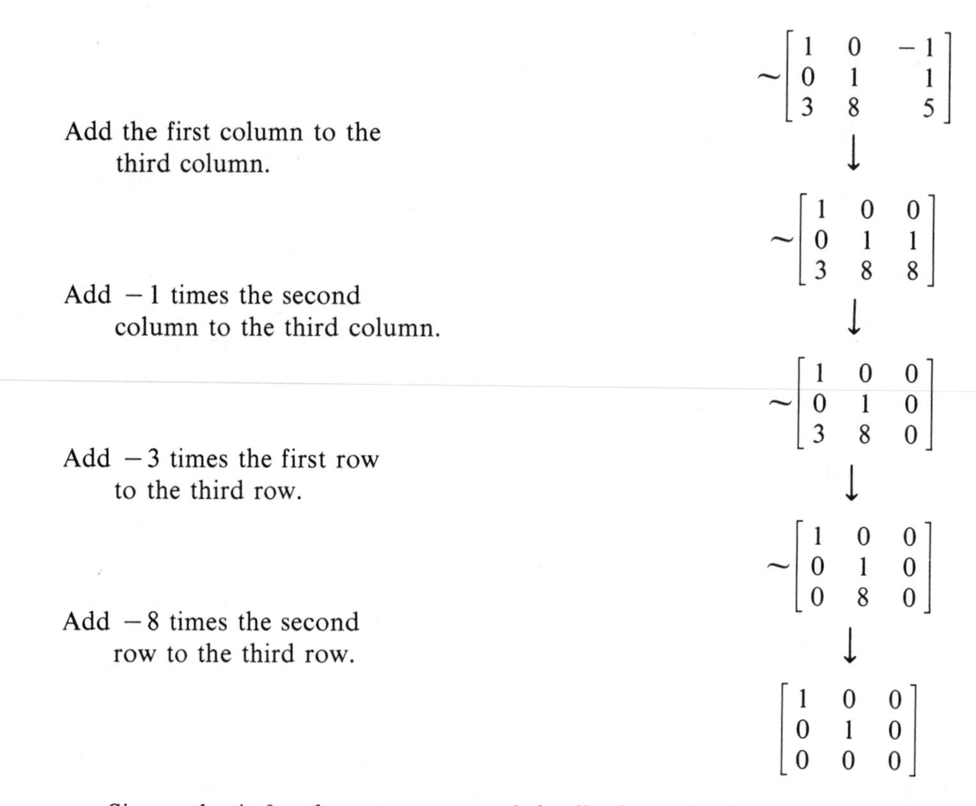

$$\sim\begin{bmatrix} 1 & 0 & -1 \\ 0 & 1 & 1 \\ 3 & 8 & 5 \end{bmatrix}$$

Add the first column to the third column.

$$\sim\begin{bmatrix} 1 & 0 & 0 \\ 0 & 1 & 1 \\ 3 & 8 & 8 \end{bmatrix}$$

Add -1 times the second column to the third column.

$$\sim\begin{bmatrix} 1 & 0 & 0 \\ 0 & 1 & 0 \\ 3 & 8 & 0 \end{bmatrix}$$

Add -3 times the first row to the third row.

$$\sim\begin{bmatrix} 1 & 0 & 0 \\ 0 & 1 & 0 \\ 0 & 8 & 0 \end{bmatrix}$$

Add -8 times the second row to the third row.

$$\begin{bmatrix} 1 & 0 & 0 \\ 0 & 1 & 0 \\ 0 & 0 & 0 \end{bmatrix}$$

Since a basis for the range space of the final matrix above is

$$\begin{bmatrix} 1 \\ 0 \\ 0 \end{bmatrix}, \quad \begin{bmatrix} 0 \\ 1 \\ 0 \end{bmatrix}$$

we see that the rank of that matrix is 2. Since equivalent matrices have the same rank, the rank of the first, and indeed of all intervening matrices, is 2.

Using a procedure analogous to Gaussian elimination and suggested by the previous example, it is possible to show that any matrix is equivalent to a matrix of one of the following types:

$$\left[\begin{array}{c} I_r \\ \hline 0 \end{array}\right], \quad [\, I_r \mid 0 \,], \quad \left[\begin{array}{c|c} I_r & 0 \\ \hline 0 & 0 \end{array}\right], \quad I_r$$

I_r is the identity matrix of order r and 0 stands for a block of 0's. This equivalence may be effected by the following procedure:

(1) Using interchanges of columns and rows, obtain a nonzero element (preferably a 1) in the first row and column.
(2) Divide the first column by this element.
(3) By adding appropriate multiples of the first column to each of the remaining columns and then appropriate multiples of the first row to

each of the remaining rows, obtain an equivalent matrix of the following form

$$\left[\begin{array}{c|cccc} 1 & 0 & 0 & \cdots & 0 \\ \hline 0 & a_{22} & a_{23} & \cdots & a_{2n} \\ & & \vdots & & \\ 0 & a_{m2} & a_{m3} & \cdots & a_{mn} \end{array}\right]$$

(4) Repeat the procedure on the submatrix

$$\begin{bmatrix} a_{22} & a_{23} & \cdots & a_{2n} \\ & & \vdots & \\ a_{m2} & a_{m3} & \cdots & a_{mn} \end{bmatrix}$$

(5) As asserted, one of the following types of matrices is obtained

$$\left[\begin{array}{c} I_r \\ \hline 0 \end{array}\right], \quad \left[\begin{array}{c|c} I_r & 0 \end{array}\right], \quad \left[\begin{array}{c|c} I_r & 0 \\ \hline 0 & 0 \end{array}\right], \quad I_r$$

Since each of the last matrices admits the vectors $\boldsymbol{e}_1, \boldsymbol{e}_2, \ldots, \boldsymbol{e}_r$ as a basis for its range space, it is clear that each of the last matrices is of rank r. Matrices of the type depicted in (5) are said to be of **normal form**. Thus, every matrix is equivalent to some matrix of normal form.

Example 1 Calculate the rank of the matrix:

$$\begin{bmatrix} 3 & 1 & 3 & 7 \\ -1 & -3 & -1 & -5 \\ 7 & 0 & 2 & 9 \\ 0 & 1 & 3 & 4 \\ 2 & 0 & 1 & 3 \end{bmatrix}$$

Interchange columns one and two.

$$\downarrow$$

$$\sim \begin{bmatrix} 1 & 3 & 3 & 7 \\ -3 & -1 & -1 & -5 \\ 0 & 7 & 2 & 9 \\ 1 & 0 & 3 & 4 \\ 0 & 2 & 1 & 3 \end{bmatrix}$$

Add suitable multiples of column one to each remaining column.

$$\downarrow$$

$$\sim \begin{bmatrix} 1 & 0 & 0 & 0 \\ -3 & 8 & 8 & 16 \\ 0 & 7 & 2 & 9 \\ 1 & -3 & 0 & -3 \\ 0 & 2 & 1 & 3 \end{bmatrix}$$

Add suitable multiples of row one to each remaining row.

$$\downarrow$$

$$\sim \begin{bmatrix} 1 & 0 & 0 & 0 \\ 0 & 8 & 8 & 16 \\ 0 & 7 & 2 & 9 \\ 0 & -3 & 0 & -3 \\ 0 & 2 & 1 & 3 \end{bmatrix}$$

Interchange columns two and three, rows two and five.

$$\downarrow$$

$$\sim \begin{bmatrix} 1 & 0 & 0 & 0 \\ 0 & 1 & 2 & 3 \\ 0 & 2 & 7 & 9 \\ 0 & 0 & -3 & -3 \\ 0 & 8 & 8 & 16 \end{bmatrix}$$

Add suitable multiples of column two to each remaining column. Then, add suitable multiples of row two to each remaining row.

$$\downarrow$$

$$\sim \begin{bmatrix} 1 & 0 & 0 & 0 \\ 0 & 1 & 0 & 0 \\ 0 & 0 & 3 & 3 \\ 0 & 0 & -3 & -3 \\ 0 & 0 & -8 & -8 \end{bmatrix}$$

Add (-1) times column three to column four.

$$\downarrow$$

$$\sim \begin{bmatrix} 1 & 0 & 0 & 0 \\ 0 & 1 & 0 & 0 \\ 0 & 0 & 3 & 0 \\ 0 & 0 & -3 & 0 \\ 0 & 0 & -8 & 0 \end{bmatrix}$$

Multiply row three by $\frac{1}{3}$ and add suitable multiples to each remaining row.

$$\downarrow$$

$$\sim \begin{bmatrix} 1 & 0 & 0 & 0 \\ 0 & 1 & 0 & 0 \\ 0 & 0 & 1 & 0 \\ 0 & 0 & 0 & 0 \\ 0 & 0 & 0 & 0 \end{bmatrix}$$

Thus, the rank of the initial matrix is 3.

The computational procedure used to calculate the rank of a matrix may be used to solve other problems involving dimension.

Example 2 Calculate the dimension of the subspace of solutions to the system of linear homogeneous equations

$$\begin{aligned} x_1 + 2x_2 - 3x_3 + x_4 + x_5 &= 0 \\ -3x_1 + x_2 + 7x_3 - x_4 + x_5 &= 0 \\ -2x_1 + 3x_2 + 4x_3 \phantom{{}- x_4} + 2x_5 &= 0 \end{aligned}$$

In other words, we wish to calculate the dimension of the nullspace of the linear transformation $T : \boldsymbol{R}^5 \to \boldsymbol{R}^3$. We can write

$$T\begin{bmatrix} x_1 \\ x_2 \\ x_3 \\ x_4 \\ x_5 \end{bmatrix} = \begin{bmatrix} 1 & 2 & -3 & 1 & 1 \\ -3 & 1 & 7 & -1 & 1 \\ -2 & 3 & 4 & 0 & 2 \end{bmatrix}\begin{bmatrix} x_1 \\ x_2 \\ x_3 \\ x_4 \\ x_5 \end{bmatrix}$$

Since $\dim R_T + \dim N_T = 5$, in order to determine $\dim N_T$ we need only determine the rank of T, which is the rank of the matrix

$$\begin{bmatrix} 1 & 2 & -3 & 1 & 1 \\ -3 & 1 & 7 & -1 & 1 \\ -2 & 3 & 4 & 0 & 2 \end{bmatrix}$$

Add suitable multiples of the first column to each of the remaining columns.

$$\downarrow$$

$$\sim\begin{bmatrix} 1 & 0 & 0 & 0 & 0 \\ -3 & 7 & -2 & 2 & 4 \\ 2 & 7 & -2 & 2 & 4 \end{bmatrix}$$

Add suitable multiples of the first row to each remaining row.

$$\downarrow$$

$$\sim\begin{bmatrix} 1 & 0 & 0 & 0 & 0 \\ 0 & 7 & -2 & 2 & 4 \\ 0 & 7 & -2 & 2 & 4 \end{bmatrix}$$

Add (-1) times the second row to the third row.

$$\downarrow$$

$$\sim\begin{bmatrix} 1 & 0 & 0 & 0 & 0 \\ 0 & 7 & -2 & 2 & 4 \\ 0 & 0 & 0 & 0 & 0 \end{bmatrix}$$

Multiply the second column by $\frac{1}{7}$ and add suitable multiples to the remaining columns.

$$\downarrow$$

$$\sim\begin{bmatrix} 1 & 0 & 0 & 0 & 0 \\ 0 & 1 & 0 & 0 & 0 \\ 0 & 0 & 0 & 0 & 0 \end{bmatrix}$$

Thus, $r(T) = 2$. So $\dim N_T = 3$.

Example 3 Find the dimension of the subspace of $\boldsymbol{R}^4$ spanned by the vectors

$$\begin{bmatrix} 1 \\ -1 \\ 1 \\ -1 \end{bmatrix}, \quad \begin{bmatrix} 0 \\ 1 \\ 3 \\ 2 \end{bmatrix}, \quad \begin{bmatrix} 1 \\ 1 \\ 7 \\ 3 \end{bmatrix}$$

Observe that the subspace of $\boldsymbol{R}^4$ spanned by these vectors is precisely the column space of the matrix

$$A = \begin{bmatrix} 1 & 0 & 1 \\ -1 & 1 & 1 \\ 1 & 3 & 7 \\ -1 & 2 & 3 \end{bmatrix}$$

Hence, it suffices to determine $r(A)$.

$$\begin{bmatrix} 1 & 0 & 1 \\ -1 & 1 & 1 \\ 1 & 3 & 7 \\ -1 & 2 & 3 \end{bmatrix} \sim \begin{bmatrix} 1 & 0 & 0 \\ -1 & 1 & 2 \\ 1 & 3 & 6 \\ -1 & 2 & 4 \end{bmatrix} \sim \begin{bmatrix} 1 & 0 & 0 \\ 0 & 1 & 2 \\ 0 & 3 & 6 \\ 0 & 2 & 4 \end{bmatrix}$$

$$\sim \begin{bmatrix} 1 & 0 & 0 \\ 0 & 1 & 0 \\ 0 & 3 & 0 \\ 0 & 2 & 0 \end{bmatrix} \sim \begin{bmatrix} 1 & 0 & 0 \\ 0 & 1 & 0 \\ 0 & 0 & 0 \\ 0 & 0 & 0 \end{bmatrix}$$

Thus, the given collection of vectors spans a space of dimension 2.

Using the procedure of reduction to normal form we obtain Theorem 5.

Theorem 5 Let A be an $m \times n$ matrix of rank r and N be the $m \times n$ matrix of rank r in normal form. Then, there are $m \times m$ elementary matrices $E_1, E_2, \ldots, E_k$ and $n \times n$ elementary matrices $F_1, F_2, \ldots, F_l$ such that $A = E_1E_2 \ldots E_kNF_1F_2 \ldots F_l$.

PROOF We know that $A \sim N$. Thus, there are $m \times m$ elementary matrices $G_1, G_2, \ldots, G_k$ and $n \times n$ elementary matrices $H_1, H_2, \ldots, H_l$ such that

$$N = G_1G_2 \ldots G_kAH_1H_2 \ldots H_l$$

or

$$A = G_k^{-1} \ldots G_1^{-1}NH_l^{-1} \ldots H_1^{-1}$$

Since the inverse of an elementary matrix is elementary by Theorem 3, renaming $G_k^{-1}, \ldots, G_1^{-1}, H_l^{-1}, \ldots, H_1^{-1}$, we obtain the desired result. ∎

Corollary 1 Any two matrices of the same rank are equivalent.

PROOF If A and B are $m \times n$ matrices of rank r, then $A = E_1 \ldots E_k N F_1 F_2 \ldots F_l$, $B = G_1 \ldots G_p N H_1 \ldots H_q$, for elementary matrices E_i, F_i, G_i, H_i. Thus, $G_1 \ldots G_p E_k^{-1} \ldots E_1^{-1} A F_l^{-1} \ldots F_1^{-1} H_1 \ldots H_q = B$. So, A and B are equivalent. ∎

Corollary 2 Any invertible $n \times n$ matrix can be factored into the product of elementary matrices.

PROOF Since A is invertible, it is of rank n, and its normal form is just the matrix I_n. Thus, there are elementary matrices $E_1, E_2, \ldots, E_k, F_1, F_2, \ldots, F_l$ such that $A = (E_1 E_2 \ldots E_k) I_n (F_1 F_2 \ldots F_l)$. ∎

EXERCISES

1. Find the rank of the following matrices.

(a) $\begin{bmatrix} 1 & -3 & -2 \\ 2 & 1 & 3 \\ 0 & 3 & 3 \end{bmatrix}$ (b) $\begin{bmatrix} 1 & 3 & 5 & 8 \\ -2 & 1 & -3 & -2 \\ 0 & 7 & 7 & 14 \end{bmatrix}$

(c) $\begin{bmatrix} -1 & 7 & 0 \\ -3 & 0 & 1 \\ 0 & -1 & 2 \\ 2 & 3 & 0 \end{bmatrix}$ (d) $\begin{bmatrix} 1 & 7 \\ -3 & 5 \\ 0 & 6 \\ 7 & 1 \end{bmatrix}$

(e) $\begin{bmatrix} 4 & 1 & 0 \\ 0 & 0 & 1 \\ 7 & -1 & 3 \\ 3 & 2 & 7 \end{bmatrix}$ (f) $\begin{bmatrix} -1 & 0 & 2 & 3 \\ 3 & 7 & 1 & 4 \end{bmatrix}$

2. Determine the dimension of the solution space for each system of homogeneous linear equations.

(a)
$$\begin{aligned} -7x + 2y - 3z &= 0 \\ x + 3y + 5z &= 0 \\ -2x - y - 4z &= 0 \end{aligned}$$

(b)
$$\begin{aligned} 3x_1 - 5x_2 + 4x_3 - x_4 &= 0 \\ x_1 - x_2 + 3x_3 + 2x_4 &= 0 \\ x_1 + x_2 - 3x_3 + 3x_4 &= 0 \end{aligned}$$

3. Invert the following elementary matrices.

(a) $\begin{bmatrix} 1 & \alpha & 0 \\ 0 & 1 & 0 \\ 0 & 0 & 1 \end{bmatrix}$ (b) $\begin{bmatrix} 1 & 0 & 0 & 0 \\ 0 & 0 & 1 & 0 \\ 0 & 1 & 0 & 0 \\ 0 & 0 & 0 & 1 \end{bmatrix}$

(c) $\begin{bmatrix} \alpha & 0 & 0 \\ 0 & 1 & 0 \\ 0 & 0 & 1 \end{bmatrix}, \alpha \neq 0$ (d) $\begin{bmatrix} 1 & 0 & 0 \\ 0 & 1 & 0 \\ \alpha & 0 & 1 \end{bmatrix}$

4. Express as a product of elementary matrices.

(a) $\begin{bmatrix} 1 & 0 & 1 \\ 3 & 1 & 4 \\ -1 & 2 & 3 \end{bmatrix}$ (b) $\begin{bmatrix} -1 & -2 \\ 3 & 5 \end{bmatrix}$

(c) $\begin{bmatrix} 1 & 0 & 3 \\ 0 & 2 & 2 \\ -1 & 1 & 0 \end{bmatrix}$ (d) $\begin{bmatrix} 1 & 3 \\ 2 & 7 \end{bmatrix}$

5. If A is an $n \times n$ matrix, show that there are invertible matrices B and C such that
   (a) AB is lower triangular. (b) CA is upper triangular.

6. If A is an $m \times n$ matrix and B is an $n \times p$ matrix, show that $r(AB) \leqslant \min(r(A), r(B))$.

7. If A is an $m \times n$ matrix, show that there is an invertible $m \times m$ matrix B and an invertible $n \times n$ matrix C such that BAC is a matrix in normal form.

8. Prove in two different ways that premultiplication of a matrix A by an elementary matrix E performs on A an elementary operation of the type described in the text.
   (a) Use the fact that the transpose of an elementary matrix is an elementary matrix and $(EA)^T = A^T E^T$.
   (b) Use a standard basis for the space of row n-vectors.

9. Show that if A is an $m \times n$ matrix, $r(A) = r(A^T)$.

10. Let A be the $2n + 1$ matrix

$$A = \begin{bmatrix} 1 & 0 & 0 & \dots & 0 & 0 & 1 \\ 0 & 1 & 0 & \dots & 0 & 1 & 0 \\ 0 & 0 & 1 & \dots & 1 & 0 & 0 \\ & & & \vdots & & & \\ 0 & 0 & 1 & \dots & 1 & 0 & 0 \\ 0 & 1 & 0 & \dots & 0 & 1 & 0 \\ 1 & 0 & 0 & \dots & 0 & 0 & 1 \end{bmatrix}$$

    Show that $r(A) = n + 1$.

11. Let A be a matrix of order $2n$ obtained from the identity matrix by interchanging rows 1 and 2, rows 3 and 4, . . . , rows $2n - 1$ and $2n$. Show that $A^{-1} = A$.

12. Let A be an $m \times n$ matrix, H be the subspace of $\boldsymbol{R}^n$ spanned by the columns of A, and K be the subspace of $\boldsymbol{R}^m$ spanned by the rows of A. Show $\dim H = \dim K$. [Hint: Use exercise 9.]

13. If A is an $m \times n$ matrix of rank $r(A)$, and $0 \leqslant k \leqslant r(A)$, show that there is an $m \times m$ matrix B such that $r(BA) = k$, and an $n \times n$ matrix C such that $r(AC) = k$.

14. Determine the dimension of the subspace of $\boldsymbol{R}^n$ spanned by the following vectors.

    (a) $\begin{bmatrix} -1 \\ 3 \\ 2 \end{bmatrix}, \begin{bmatrix} 1 \\ 7 \\ 3 \end{bmatrix}, \begin{bmatrix} 1 \\ 0 \\ 2 \end{bmatrix}$ (b) $\begin{bmatrix} -1 \\ 0 \\ 2 \\ 1 \end{bmatrix}, \begin{bmatrix} 3 \\ 1 \\ -2 \\ -3 \end{bmatrix}, \begin{bmatrix} 2 \\ 1 \\ 0 \\ -2 \end{bmatrix}$

    (c) $\begin{bmatrix} 1 \\ -5 \\ 0 \\ 3 \end{bmatrix}, \begin{bmatrix} 1 \\ 0 \\ -3 \\ 2 \end{bmatrix}, \begin{bmatrix} 0 \\ 5 \\ 3 \\ 5 \end{bmatrix}, \begin{bmatrix} 1 \\ 5 \\ 6 \\ 1 \end{bmatrix}$

15. If A and B are $m \times n$ matrices, show that the following are equivalent statements.
    (a) $r(A) = r(B)$.
    (b) There exists an $m \times m$ matrix C and an $n \times n$ matrix D, both invertible, such that $A = CBD$.

16. If A and B are $m \times n$ matrices, show that the following statements are equivalent.
    (a) The range spaces of A and B are the same subspaces of $\boldsymbol{R}^m$.
    (b) There exists an invertible $n \times n$ matrix C such that $A = BC$.

17. If A is an $m \times n$ matrix with the property that r of its columns are linearly independent but any $r + 1$ columns are linearly dependent, show that $r(A) = r$.

18. Show that any rectangular matrix of rank r is the sum of r matrices of rank 1.

19. Show that any invertible 2×2 matrix can be expressed as a product of matrices of the form

$$\begin{bmatrix} 1 & t \\ 0 & 1 \end{bmatrix}, \quad \begin{bmatrix} 1 & 0 \\ t & 1 \end{bmatrix}, \quad \begin{bmatrix} t & 0 \\ 0 & 1 \end{bmatrix}, \quad \begin{bmatrix} 1 & 0 \\ 0 & t \end{bmatrix}, \quad \begin{bmatrix} 0 & 1 \\ 1 & 0 \end{bmatrix}$$

20. Consider the system of linear equations

$$\begin{aligned} a_{11}x_1 + a_{12}x_2 + \cdots + a_{1n}x_n &= y_1 \\ a_{21}x_1 + a_{22}x_2 + \cdots + a_{2n}x_n &= y_2 \\ &\vdots \\ a_{m1}x_1 + a_{m2}x_2 + \cdots + a_{mn}x_n &= y_m \end{aligned}$$

Show that the system admits a solution if and only if the rank of the matrix

$$\begin{bmatrix} a_{11} & a_{12} & \cdots & a_{1n} \\ a_{21} & a_{22} & \cdots & a_{2n} \\ & & \vdots & \\ a_{m1} & a_{m2} & \cdots & a_{mn} \end{bmatrix}$$

is the same as that of the matrix

$$\begin{bmatrix} a_{11} & a_{12} & \cdots & a_{1n} & y_1 \\ a_{21} & a_{22} & \cdots & a_{2n} & y_2 \\ & & \vdots & & \\ a_{m1} & a_{m2} & \cdots & a_{mn} & y_m \end{bmatrix}$$

21. Let $A_i x + B_i y + C_i = 0$, $i = 1, 2, \ldots, k$ be a set of k lines. Show that the lines intersect at a point or are parallel if and only if the rank of the matrix

$$\begin{bmatrix} A_1 & B_1 & C_1 \\ A_2 & B_2 & C_2 \\ & \vdots & \\ A_k & B_k & C_k \end{bmatrix}$$

is less than or equal to 2.

22. Determine the rank of the matrix $\begin{bmatrix} 1 & x & x \\ x & 1 & x \\ x & x & 1 \end{bmatrix}$ as a function of x.

6 ISOMORPHISM

Let $T : V \to W$ be a linear transformation from a vector space V to a vector space W. If T is both one-one and onto, then T is said to be an isomorphism. In other words, T is an isomorphism if $R_T = W$ and $N_T = 0$.

Example 1 Let A be an $n \times n$ matrix, and T_A the linear operator on R^n defined by $T_A(\boldsymbol{x}) = A\boldsymbol{x}$.

Suppose that A is invertible. Then, by example 3 of §5.3, T_A is onto. In §5.4 we saw that T_A is one-one. Thus, T_A is an isomorphism.

On the other hand, if A is not invertible, $\det A = 0$, and, by Theorem 2 of §4.9, $A\boldsymbol{x} = 0$ has a nonzero solution. Thus, T_A is not one-one.

Thus, T_A is an isomorphism if and only if A is invertible.

Example 2 Consider the linear transformation T from $\boldsymbol{R}^3$ into P_2 defined by $T(\boldsymbol{e}_1) = 1$, $T(\boldsymbol{e}_2) = x$, $T(\boldsymbol{e}_3) = x^2$. By the Theorem of §5.2 such a linear transformation exists. By definition of T, $T(a\boldsymbol{e}_1 + b\boldsymbol{e}_2 + c\boldsymbol{e}_3) = a + bx + cx^2$. It follows easily that T is one-one and onto. Thus, T is an isomorphism from $\boldsymbol{R}^3$ into P_2.

Example 3 In chapter 2, we defined vectors in $\boldsymbol{R}^3$ in two different ways. In the algebraic definition, vectors were regarded as columns of numbers. In the geometric definition, vectors were realized as directed line segments beginning at the origin. With our algebraic definition, we gave algebraic rules for vector addition and scalar multiplication. With the geometric definition of vectors, the definition of vector addition was by means of the parallelogram law, and scalar multiplication was thought of as signed magnification of length.

Let us now consider the transformation T from the space of column vectors into the space of directed line segments beginning at the origin, namely, $T\left(\begin{bmatrix} x \\ y \\ z \end{bmatrix}\right) = \boldsymbol{v}(x, y, z)$. In other words, T assigns to the column vector with components x, y, z the directed line segment ending at (x, y, z). Because of the correspondence between triples of real numbers and points in space, the transformation T is one-one and onto. In §2.2 we saw that $\boldsymbol{v}(x, y, z) + \boldsymbol{v}(x', y', z') = \boldsymbol{v}(x + x', y + y', z + z')$ and $\alpha\boldsymbol{v}(x, y, z) = \boldsymbol{v}(\alpha x, \alpha y, \alpha z)$.

If we reinterpret this using the transformation T defined above, we have

$$T\left(\begin{bmatrix} x \\ y \\ z \end{bmatrix}\right) + T\left(\begin{bmatrix} x' \\ y' \\ z' \end{bmatrix}\right) = T\left(\begin{bmatrix} x \\ y \\ z \end{bmatrix} + \begin{bmatrix} x' \\ y' \\ z' \end{bmatrix}\right), \text{ and } T\left(\alpha\begin{bmatrix} x \\ y \\ z \end{bmatrix}\right) = \alpha T\left(\begin{bmatrix} x \\ y \\ z \end{bmatrix}\right).$$

In other words, T defines a linear transformation from vectors defined algebraically to vectors defined geometrically. Because T is one-one and onto, T is an isomorphism.

As a consequence of this correspondence between the algebraically and geometrically defined vectors and vector operations, we agreed to regard algebraic and geometric vectors as equivalent. This correspondence helps to illuminate the essential meaning of the concept of isomorphism.

The word "isomorphism" is derived from the Greek "iso = same" and "morphos = form." It is appropriate to ask why, in this case, the two spaces have the "same form." The two spaces have the same form in the sense that T induces a one-one correspondence that renders the two definitions of addition and scalar multiplication equivalent. Using this equivalence, we may prove results in whichever space is the more convenient and use the isomorphism T

to transfer the results to the other space, thereby exhibiting a correspondence between algebraic and geometric statements.

For example, by virtue of the isomorphism T, the geometric statement "If $\boldsymbol{v}_1$ and $\boldsymbol{v}_2$ are two noncollinear directed line segments in the plane beginning at the origin, then every directed line segment in the plane that begins at the origin is the diagonal of some parallelogram with adjacent sides lying on the lines determined by extending $\boldsymbol{v}_1$ and $\boldsymbol{v}_2$, respectively," is equivalent to the algebraic statement "If $\boldsymbol{v}_1$ and $\boldsymbol{v}_2$ are linearly independent vectors in $\boldsymbol{R}^2$, their linear combinations span $\boldsymbol{R}^2$." (See Figure 5-11.)

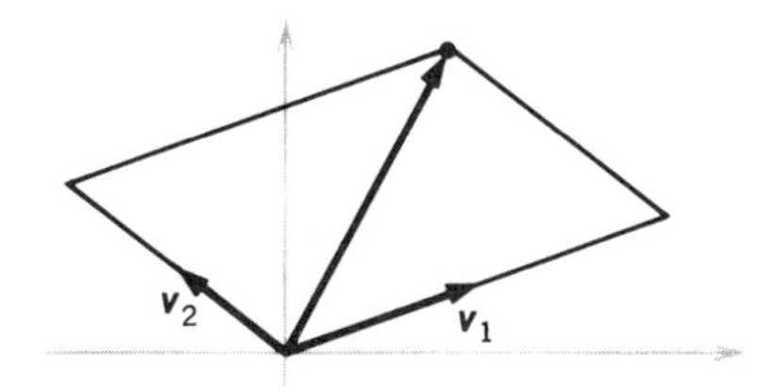

FIGURE 5-11

In the case of an isomorphism $T : V \to W$ between two vector spaces, T renders the algebraic operations in V equivalent to those of W. The next theorem helps make this statement more precise.

Theorem 1 Let $T : V \to W$ be an isomorphism of two vector spaces. Then,

(i) If $\boldsymbol{x}_1, \boldsymbol{x}_2, \ldots, \boldsymbol{x}_n$ are linearly independent in V, $T(\boldsymbol{x}_1), T(\boldsymbol{x}_2), \ldots, T(\boldsymbol{x}_n)$ are linearly independent in W.
(ii) If $\boldsymbol{x}_1, \boldsymbol{x}_2, \ldots, \boldsymbol{x}_n$ span V, $T(\boldsymbol{x}_1), T(\boldsymbol{x}_2), \ldots, T(\boldsymbol{x}_n)$ span W.
(iii) If $\boldsymbol{x}_1, \boldsymbol{x}_2, \ldots, \boldsymbol{x}_n$ form a basis for V, $T(\boldsymbol{x}_1), T(\boldsymbol{x}_2), \ldots, T(\boldsymbol{x}_n)$ form a basis for W.
(iv) $\dim V = \dim W$.

PROOF (i) Suppose $\boldsymbol{x}_1, \boldsymbol{x}_2, \ldots, \boldsymbol{x}_n$ are independent in V. If $\alpha_1 T(\boldsymbol{x}_1) + \alpha_2 T(\boldsymbol{x}_2) + \cdots + \alpha_n T(\boldsymbol{x}_n) = \boldsymbol{0}$ in W, then by linearity of T, $T(\alpha_1\boldsymbol{x}_1 + \alpha_2\boldsymbol{x}_2 + \cdots + \alpha_n\boldsymbol{x}_n) = \boldsymbol{0}$. Because T is one-one, we conclude that $\alpha_1\boldsymbol{x}_1 + \alpha_2\boldsymbol{x}_2 + \cdots + \alpha_n\boldsymbol{x}_n = \boldsymbol{0}$. Since $\boldsymbol{x}_1, \boldsymbol{x}_2, \ldots, \boldsymbol{x}_n$ are independent in V, $\alpha_1 = \alpha_2 = \cdots = \alpha_n = 0$. Thus, $\alpha_1 T(\boldsymbol{x}_1) + \alpha_2 T(\boldsymbol{x}_2) + \cdots + \alpha_n T(\boldsymbol{x}_n) = \boldsymbol{0}$ implies that $\alpha_1 = \alpha_2 = \cdots = \alpha_n = 0$, so $T(\boldsymbol{x}_1), T(\boldsymbol{x}_2), \ldots, T(\boldsymbol{x}_n)$ are linearly independent.

(ii) Suppose $\boldsymbol{x}_1, \boldsymbol{x}_2, \ldots, \boldsymbol{x}_n$ span V and $\boldsymbol{y}$ is a vector in W. Since T is onto, there is a vector $\boldsymbol{x}$ in V, such that $T(\boldsymbol{x}) = \boldsymbol{y}$. Since $\boldsymbol{x}_1, \boldsymbol{x}_2, \ldots, \boldsymbol{x}_n$ span V, there are scalars $\alpha_1, \alpha_2, \ldots, \alpha_n$ such that $\boldsymbol{x} = \alpha_1\boldsymbol{x}_1 + \alpha_2\boldsymbol{x}_2 + \cdots + \alpha_n\boldsymbol{x}_n$. Thus, $\boldsymbol{y} = T(\boldsymbol{x}) = T(\alpha_1\boldsymbol{x}_1 + \cdots + \alpha_n\boldsymbol{x}_n) = \alpha_1 T(\boldsymbol{x}_1) + \cdots + \alpha_n T(\boldsymbol{x}_n)$. Thus, if $\boldsymbol{y}$ belongs to W, $\boldsymbol{y}$ is a linear combination of $T(\boldsymbol{x}_1), T(\boldsymbol{x}_2), \ldots, T(\boldsymbol{x}_n)$. Therefore, $T(\boldsymbol{x}_1), T(\boldsymbol{x}_2), \ldots, T(\boldsymbol{x}_n)$ span W.

(iii) Suppose $\boldsymbol{x}_1, \boldsymbol{x}_2, \ldots, \boldsymbol{x}_n$ is a basis for V. By (a), $T(\boldsymbol{x}_1), T(\boldsymbol{x}_2), \ldots, T(\boldsymbol{x}_n)$ are linearly independent in W. By (b), $T(\boldsymbol{x}_1)$,

$T(\boldsymbol{x}_2), \ldots, T(\boldsymbol{x}_n)$ span W. Thus, $T(\boldsymbol{x}_1), T(\boldsymbol{x}_2), \ldots, \mathrm{I}(\boldsymbol{x}_n)$ forms a basis for W.

(iv) If $\dim V = n$, then there is some basis, say $\boldsymbol{x}_1, \boldsymbol{x}_2, \ldots, \boldsymbol{x}_n$ for V with n elements. Since $T(\boldsymbol{x}_1), T(\boldsymbol{x}_2), \ldots, T(\boldsymbol{x}_n)$ is a basis for W with n elements, we see that $\dim V = \dim W$. ■

In checking whether a linear transformation is an isomorphism, it may be helpful to keep the following statement in mind.

Suppose that $T : V \to W$ is a linear transformation between two finite-dimensional vector spaces with $\dim V = \dim W$. Then the following are equivalent:

(1) T is an isomorphism.
(2) T is one-one.
(3) T is onto.

We prove this statement. Clearly, if T is an isomorphism, T is one-one and onto. So suppose that T is one-one. Then, $N_T = 0$. Since $\dim N_T + \dim R_T = \dim V$, it follows that $\dim R_T = \dim V$. Since $\dim V = \dim W$, $\dim R_T = \dim W$. Since R_T is a subspace of W, $R_T = W$, and T is onto.

Similarly, if T is onto, $\dim R_T = \dim W = \dim V$. Since $\dim N_T + \dim R_T = \dim V$, it follows that $\dim N_T = 0$. Thus, $N_T = 0$, and T is one-one.

Thus, in checking whether a linear operator is an isomorphism, it suffices to check that it is either one-one or onto.

Example 4 Consider the linear transformation from P_n to itself defined by $H(f) = f + f'$, where f' denotes the derivative of f.

In this case, it is not hard to see that H is linear. We wish to show that H is an isomorphism. By what we have just shown, it suffices to prove that H is one-one.

So suppose that $H(f) = 0$. Then, $f + f' = 0$. It follows that $f = -f'$. If f is a nonzero polynomial of degree k, then f' is a polynomial of degree $k - 1$. Thus, if $f \neq 0$, f cannot possibly equal f'. Thus, $f = 0$, and H is one-one. It follows that H is an isomorphism.

By using Theorem 1 above, it follows that if $f_0, f_1, \ldots, f_n$ is a basis for P_n, then another basis is $f_0 + f_0', f_1 + f_1', \ldots, f_n + f_n'$. For example, $1, x, \ldots, x^n$ is a basis for P_n. It follows that $1, x + 1, x^2 + 2x, \ldots, x^n + nx^{n-1}$ is also a basis for P_n.

By Theorem 1, if there is an isomorphism between two vector spaces, then the vector spaces have the same dimension. What is more interesting is that if two vector spaces have the same dimension, then there is an isomorphism between them.

Theorem 2 Let V and W be two n-dimensional real vector spaces. Then, there is an isomorphism between V and W.

PROOF Take $\boldsymbol{x}_1, \ldots, \boldsymbol{x}_n$ as a basis for V and $\boldsymbol{y}_1, \ldots, \boldsymbol{y}_n$ as a basis for W. By the theorem of §5.2, there is a linear transformation T from V to W such that $T(\boldsymbol{x}_i) = \boldsymbol{y}_i$ for $i = 1, \ldots, n$.

We claim that T is an isomorphism. Since V and W have the same dimension, it suffices by the above to show that T is one-one.

Let $\boldsymbol{x}$ be a vector in V with $T(\boldsymbol{x}) = \boldsymbol{0}$. Since $\boldsymbol{x}_1, \ldots, \boldsymbol{x}_n$ is a basis for V, there are scalars $\alpha_1, \ldots, \alpha_n$ such that $\boldsymbol{x} = \alpha_1\boldsymbol{x}_1 + \cdots + \alpha_n\boldsymbol{x}_n$. Then, $T(\boldsymbol{x}) = T(\alpha_1\boldsymbol{x}_1 + \cdots + \alpha_n\boldsymbol{x}_n) = \alpha_1 T(\boldsymbol{x}_1) + \cdots + \alpha_n T(\boldsymbol{x}_n) = \boldsymbol{0}$. Since $T(\boldsymbol{x}_i) = \boldsymbol{y}_i$, it follows that $\alpha_1\boldsymbol{y}_1 + \cdots + \alpha_n\boldsymbol{y}_n = \boldsymbol{0}$. Since $\boldsymbol{y}_1, \ldots, \boldsymbol{y}_n$ is a basis for W, it follows that $\alpha_1 = \alpha_2 = \cdots = \alpha_n = 0$. Thus, $\boldsymbol{x} = \boldsymbol{0}$. It follows that T is one-one. ■

Often, when there is an isomorphism between two vector spaces, we say that the spaces are isomorphic.

Corollary If V is an n-dimensional real vector space, V is isomorphic to $\boldsymbol{R}^n$.

This corollary means that there is an isomorphism from V to $\boldsymbol{R}^n$. It is in this sense that $\boldsymbol{R}^n$ is the model for real finite-dimensional spaces. Using a proof similar to that of Theorem 2, we can conclude that any two complex n-dimensional vector spaces are isomorphic, and any complex n-dimensional vector space is isomorphic to $\boldsymbol{C}^n$.

We have already seen that the space P_n is a vector space of dimension $n+1$. Therefore, P_n and $\boldsymbol{R}^{n+1}$ are isomorphic. In a similar manner, we see that the space of 2×2 real matrices, which is of dimension 4, is isomorphic to $\boldsymbol{R}^4$.

EXERCISES

1. Which of the following matrices induce isomorphisms of $\boldsymbol{R}^n$?

   (a) $\begin{bmatrix} -1 & 0 & 3 \\ 1 & 2 & 0 \\ 0 & 2 & 3 \end{bmatrix}$ (b) $\begin{bmatrix} 7 & 5 \\ 4 & 3 \end{bmatrix}$

   (c) $\begin{bmatrix} 1 & 1 & 3 \\ 3 & 2 & 5 \\ 7 & 5 & 12 \end{bmatrix}$ (d) $\begin{bmatrix} 1 & 5 \\ 3 & 0 \end{bmatrix}$

2. Let V be a vector space with basis $\boldsymbol{x}_1, \boldsymbol{x}_2, \ldots, \boldsymbol{x}_n$. Let T be the function from V to $\boldsymbol{R}^n$ that assigns to each vector $\boldsymbol{x}$ its coordinate n-tuple relative to the basis $\boldsymbol{x}_1, \boldsymbol{x}_2, \ldots, \boldsymbol{x}_n$. Show that T is an isomorphism of V and $\boldsymbol{R}^n$.

3. Let P_e be the vector space of even polynomials in a variable x of degree less than or equal to $2n$ and let P_o be the space of odd polynomials of degree less than or equal to $2n+1$. Show that the function $T : P_e \to P_o$, defined by $T(f) = xf$, is an isomorphism of P_e onto P_o.

4. Show that the function from the space of $n \times n$ matrices into itself that carries A into A^T is an isomorphism.

5. Let T be the mapping from P_n into P_n defined by

   $$T(f) = f + \alpha_1 f' + \alpha_2 f'' + \cdots + \alpha_n f^{(n)}, \qquad \alpha_1, \alpha_2, \ldots, \alpha_n, \text{ real numbers}$$

   Show that T is an isomorphism.

6. Let A be an invertible $n \times n$ matrix. Show that the following linear transformations from the space of $n \times n$ matrices into itself are isomorphisms.

   (a) $T(B) = AB$ (b) $T(B) = BA$
   (c) $T(B) = ABA$ (d) $T(B) = ABA^{-1}$

7. Let $T: V \to W$ be a linear transformation between two finite-dimensional vector spaces. Show that the following statements are equivalent.
   (a) T is an isomorphism.
   (b) If $\boldsymbol{x}_1, \boldsymbol{x}_2, \ldots, \boldsymbol{x}_n$ are linearly independent in V, then $T(\boldsymbol{x}_1), T(\boldsymbol{x}_2), \ldots, T(\boldsymbol{x}_n)$ are linearly independent in W, and if $\boldsymbol{y}_1, \boldsymbol{y}_2, \ldots, \boldsymbol{y}_n$ span V, then $T(\boldsymbol{y}_1), T(\boldsymbol{y}_2), \ldots, T(\boldsymbol{y}_n)$ span W.
   (c) If $\boldsymbol{x}_1, \boldsymbol{x}_2, \ldots, \boldsymbol{x}_n$ is a basis for V, then $T(\boldsymbol{x}_1), T(\boldsymbol{x}_2), \ldots, T(\boldsymbol{x}_n)$ is a basis for W.

8. Show that the linear transformation of P_n that carries $f(x)$ into $f(x + \alpha)$, with α real, is an isomorphism of P_n.

9. Let T be an isomorphism of $\boldsymbol{R}^3$. Show that T carries planes through the origin into planes through the origin and lines through the origin into lines through the origin.

10. Construct an isomorphism between the space of 3×4 matrices and the space of 2×6 matrices. When, in general, is it possible to find an isomorphism between the space of $k \times l$ matrices and the space of $m \times n$ matrices?

11. Show that the space of upper triangular matrices is isomorphic to the space of lower triangular matrices.

12. Show that the function from the complex numbers (regarded here as a real vector space) to the space of 2×2 matrices of the form

$$\begin{bmatrix} a & b \\ -b & a \end{bmatrix} \quad \text{defined by} \quad T(a + bi) = \begin{bmatrix} a & b \\ -b & a \end{bmatrix}$$

   is an isomorphism. Show that this isomorphism preserves products, i.e., $T(a + bi)T(c + di) = T((a + bi)(c + di))$.

13. Let V be a vector space and T be a fixed vector in V. Define new operations on V by

$$\boldsymbol{x} \oplus \boldsymbol{y} = \boldsymbol{x} + \boldsymbol{y} + \boldsymbol{t} \quad \text{and} \quad \alpha * \boldsymbol{x} = \alpha \boldsymbol{x} + (1 - \alpha)\boldsymbol{t}$$

   Show that V with the operations $\oplus$ and $*$ is a vector space. Show that the function $T(\boldsymbol{x}) = \boldsymbol{x} + \boldsymbol{t}$ which carries V, regarded as a vector space under $+$ and $\cdot$, into V, regarded as a vector space under $\oplus$ and $*$ induces an isomorphism.

14. For what values of λ is the function from P_n into P_n defined by $T(f) = \lambda f - xf'$ an isomorphism of P_n?

15. Let $\boldsymbol{x}_1, \boldsymbol{x}_2, \boldsymbol{x}_3$ be one set of linearly independent vectors in $\boldsymbol{R}^3$ and let $\boldsymbol{y}_1, \boldsymbol{y}_2, \boldsymbol{y}_3$ be another set of linearly independent vectors in $\boldsymbol{R}^3$. Suppose T is a linear operator on $\boldsymbol{R}^3$ such that $T(\boldsymbol{x}_i) = \boldsymbol{y}_i$, $i = 1, 2, 3$.
    (a) Show that T is an isomorphism.
    (b) Show that $T(\boldsymbol{x}) = A\boldsymbol{x}$, where A is the 3×3 matrix,

$$[\boldsymbol{y}_1, \boldsymbol{y}_2, \boldsymbol{y}_3][\boldsymbol{x}_1, \boldsymbol{x}_2, \boldsymbol{x}_3]^{-1}$$

    where $[\boldsymbol{y}_1, \boldsymbol{y}_2, \boldsymbol{y}_3]$ is the 3×3 matrix with columns $\boldsymbol{y}_1, \boldsymbol{y}_2, \boldsymbol{y}_3$ and $[\boldsymbol{x}_1, \boldsymbol{x}_2, \boldsymbol{x}_3]$ is the 3×3 matrix with columns $\boldsymbol{x}_1, \boldsymbol{x}_2, \boldsymbol{x}_3$.

16. Which of the following pairs of spaces are isomorphic?
    (a) 3×3 skew-symmetric matrices with real entries and $\boldsymbol{R}^3$.
    (b) $n \times n$ diagonal matrices with real entries and P_n.
    (c) $(n-1) \times (n-1)$ symmetric matrices with real entries and $n \times n$ skew-symmetric matrices with real entries.

7 ALGEBRA OF LINEAR TRANSFORMATIONS

In chapter 2, we discussed addition and multiplication of matrices. In view of the close connection between linear transformations and matrices, it is reasonable to ask whether similar operations can be performed on linear transformations. In this section, we define such operations and see that they are completely analogous to the corresponding operations on matrices.

If V and W are two vector spaces, let $L(V, W)$ denote the family of all linear transformations from V into W. There are certain natural algebraic operations defined in $L(V, W)$ that we study in this section. Suppose T_1 and T_2 are two members of $L(V, W)$, that is, T_1 and T_2 are linear transformations from the vector space V into the vector space W. We define the **sum** of T_1 and T_2, denoted by $T_1 + T_2$, to be $(T_1 + T_2)(\boldsymbol{x}) = T_1(\boldsymbol{x}) + T_2(\boldsymbol{x})$.

In order for this definition to be of interest, the sum should again be a linear transformation. To see that this is indeed the case, let $\boldsymbol{x}$ and $\boldsymbol{y}$ belong to V, then

$$\begin{aligned}
(T_1 + T_2)(\boldsymbol{x} + \boldsymbol{y}) &= T_1(\boldsymbol{x} + \boldsymbol{y}) + T_2(\boldsymbol{x} + \boldsymbol{y}) && \text{(by definition of } T_1 + T_2\text{)} \\
&= T_1(\boldsymbol{x}) + T_1(\boldsymbol{y}) + T_2(\boldsymbol{x}) + T_2(\boldsymbol{y}) && \text{(by linearity of } T_1, T_2\text{)} \\
&= T_1(\boldsymbol{x}) + T_2(\boldsymbol{x}) + T_1(\boldsymbol{y}) + T_2(\boldsymbol{y}) && \text{[by (V1)]} \\
&= (T_1 + T_2)(\boldsymbol{x}) + (T_1 + T_2)(\boldsymbol{y}) && \text{(by definition of } T_1 + T_2\text{)}
\end{aligned}$$

Similarly, if $\boldsymbol{x}$ belongs to V and α is a scalar, we have

$$\begin{aligned}
(T_1 + T_2)(\alpha\boldsymbol{x}) &= T_1(\alpha\boldsymbol{x}) + T_2(\alpha\boldsymbol{x}) && \text{(by definition of } T_1 + T_2\text{)} \\
&= \alpha T_1(\boldsymbol{x}) + \alpha T_2(\boldsymbol{x}) && \text{(by linearity of } T_1, T_2\text{)} \\
&= \alpha(T_1(\boldsymbol{x}) + T_2(\boldsymbol{x})) && \text{[by (V6)]} \\
&= \alpha(T_1 + T_2)(\boldsymbol{x}) && \text{(by definition of } T_1 + T_2\text{)}
\end{aligned}$$

There is also a natural operation of scalar multiplication on $L(V, W)$. If T belongs to $L(V, W)$ we define the **scalar multiple** of T by the scalar α, denoted by αT, to be $(\alpha T)(\boldsymbol{x}) = \alpha(T(\boldsymbol{x}))$.

αT is again a linear transformation, since if $\boldsymbol{x}$ and $\boldsymbol{y}$ belong to V,

$$\begin{aligned}
(\alpha T)(\boldsymbol{x} + \boldsymbol{y}) &= \alpha(T(\boldsymbol{x} + \boldsymbol{y})) && \text{(by definition of } \alpha T\text{)} \\
&= \alpha(T(\boldsymbol{x}) + T(\boldsymbol{y})) && \text{(by linearity of } T\text{)} \\
&= \alpha T(\boldsymbol{x}) + \alpha T(\boldsymbol{y}) && \text{[by (V6)]} \\
&= (\alpha T)(\boldsymbol{x}) + (\alpha T)(\boldsymbol{y}) && \text{(by definition of } \alpha T\text{)}
\end{aligned}$$

Moreover, if $\boldsymbol{x}$ is a vector in V and β is a scalar, we have

$$\begin{aligned}
(\alpha T)(\beta \boldsymbol{x}) &= \alpha(T(\beta \boldsymbol{x})) && \text{(by definition of } \alpha T\text{)} \\
&= \alpha(\beta T(\boldsymbol{x})) && \text{(by linearity of } T\text{)} \\
&= (\alpha\beta)(T(\boldsymbol{x})) && \text{[by (V7)]} \\
&= \beta(\alpha T(\boldsymbol{x})) && \text{[by (V7)]} \\
&= \beta(\alpha T)(\boldsymbol{x}) && \text{(by definition of } \alpha T\text{)}
\end{aligned}$$

Using linear combinations of known linear transformations, we can construct many new examples of linear transformations. Using the above laws, we can avoid the bother of verifying the transformations we obtain are linear. Many examples we constructed earlier were obtained this way. In the last section, for example, we considered the linear operator on P_n defined by $H(f) = f + f'$. Earlier, we defined $I(f) = f$ and $D(f) = f'$. Thus, $H = I + D$. It follows that H is linear. Likewise, $K(f) = 3f - f'$ and $L(f) = f + 7f'$ are linear operators on P_n.

In view of the notation we have chosen, perhaps the following theorem is not surprising.

Theorem 1 $L(V, W)$, with the addition and scalar multiplication defined above, is a vector space.

PROOF Let T_1, T_2, T_3 be members of $L(V, W)$, α, β be scalars, and $\boldsymbol{x}$ be an arbitrary vector in V.

(V1)

$$\begin{aligned}
(T_1 + T_2)(\boldsymbol{x}) &= T_1(\boldsymbol{x}) + T_2(\boldsymbol{x}) && \text{(by definition of } T_1 + T_2\text{)} \\
&= T_2(\boldsymbol{x}) + T_1(\boldsymbol{x}) && \text{[using (V1) in } W\text{]} \\
&= (T_2 + T_1)(\boldsymbol{x}) && \text{(by definition of } T_2 + T_1\text{)}
\end{aligned}$$

(V2)

$$\begin{aligned}
((T_1 + T_2) + T_3)(\boldsymbol{x}) &= (T_1 + T_2)(\boldsymbol{x}) + T_3(\boldsymbol{x}) && \text{[by definition of } (T_1 + T_2) + T_3\text{]} \\
&= (T_1(\boldsymbol{x}) + T_2(\boldsymbol{x})) + T_3(\boldsymbol{x}) && \text{(by definition of } T_1 + T_2\text{)} \\
&= T_1(\boldsymbol{x}) + (T_2(\boldsymbol{x}) + T_3(\boldsymbol{x})) && \text{[using (V2) in } W\text{]} \\
&= T_1(\boldsymbol{x}) + (T_2 + T_3)(\boldsymbol{x}) && \text{(by definition of } T_2 + T_3\text{)} \\
&= (T_1 + (T_2 + T_3))(\boldsymbol{x}) && \text{[by definition of } T_1 + (T_2 + T_3)\text{]}
\end{aligned}$$

(V3)

Let 0 denote the zero transformation from V into W, i.e., the operator sending all vectors of V into $\mathbf{0}$. Then

$$\begin{aligned}
(0 + T_1)(\boldsymbol{x}) &= 0(\boldsymbol{x}) + T_1(\boldsymbol{x}) && \text{(by definition of } 0 + T_1\text{)} \\
&= \mathbf{0} + T_1(\boldsymbol{x}) && \text{(by definition of 0)} \\
&= T_1(\boldsymbol{x}) && \text{[using (V3) in } W\text{]}
\end{aligned}$$

(V4)

For T in $L(V, W)$, define the negative of T by $(-T)(x) = -(T(x))$. Then

$$\begin{aligned}
(T+(-T))(x) &= T(x)+(-T)(x) && \text{[by definition of } T+(-T)\text{]}\\
&= T(x)-T(x) && \text{[by definition of } (-T)(x)\text{]}\\
&= \mathbf{0} && \text{[using (V4) in } W\text{]}\\
&= 0(x) && \text{(by definition of the zero operator)} \blacksquare
\end{aligned}$$

Verifying the remaining laws (V5)–(V8) is very similar, and we omit it.

Example 1 Let A and B be $m \times n$ matrices having real entries, and T_A and T_B the linear transformations from $\boldsymbol{R}^n$ into $\boldsymbol{R}^m$ defined by $T_A(x) = Ax$ and $T_B(x) = Bx$, for x in $\boldsymbol{R}^n$. Then it follows that $(T_A + T_B)(x) = T_A(x) + T_B(x) = Ax + Bx = (A + B)x = T_{A+B}(x)$. Since x is an arbitrary vector in $\boldsymbol{R}^n$, it follows that $T_A + T_B = T_{A+B}$, i.e., addition of transformations corresponds to addition of matrices. Likewise, it is not difficult to demonstrate that $\alpha T_A = T_{\alpha A}$, which means scalar multiplication of linear transformations corresponds to scalar multiplication of matrices.

Let L denote the function from M_{mn}, the space of $m \times n$ matrices with real entries, into $L(\boldsymbol{R}^n, \boldsymbol{R}^m)$, defined by $L(A) = T_A$. In other words, L carries the matrix A into the linear transformation induced by multiplication by A. Since

$$\begin{aligned}
L(A+B) &= T_{A+B}\\
&= T_A + T_B\\
&= L(A) + L(B)
\end{aligned}$$

and

$$\begin{aligned}
L(\alpha A) &= T_{\alpha A}\\
&= \alpha T_A\\
&= \alpha L(A)
\end{aligned}$$

we see that L is linear.

We saw in Theorem 2, §5.2 that any linear transformation from $\boldsymbol{R}^n$ into $\boldsymbol{R}^m$ is induced by multiplication by a suitable $m \times n$ matrix. That is, given a linear transformation $T : \boldsymbol{R}^n \to \boldsymbol{R}^m$, there is some matrix A, such that $T = T_A$. In other words, the function L from M_{mn} into $L(\boldsymbol{R}^n, \boldsymbol{R}^m)$ is onto.

The function L is also one-one. For if $L(A) = 0$, then $T_A(x) = Ax = \mathbf{0}$, for all vectors x in $\boldsymbol{R}^n$. By letting $x = e_1, e_2, \ldots, e_n$, we see that all columns of A are zero, and so $A = 0$.

Thus, the function L defined above induces an isomorphism from the space of $m \times n$ matrices into the space of linear transformations from $\boldsymbol{R}^n$ to $\boldsymbol{R}^m$.

Just as addition and scalar multiplication of linear transformations are analogous to the operations with matrices, there is also a method of combining linear transformations that corresponds to matrix multiplication.

Let $T : U \to V$ and $S : V \to W$ be linear transformations, with U, V, and W vector spaces. We define the **composition** of S and T, denoted by $S \circ T$, or more often simply ST, by

$$(S \circ T)(\boldsymbol{x}) = S(T(\boldsymbol{x})), \quad \text{for } \boldsymbol{x} \text{ in } U$$

It seems reasonable to expect that the composition of two linear transformations would again be linear. This is indeed the case, since if $\boldsymbol{x}_1$ and $\boldsymbol{x}_2$ are vectors in U, we have

$$\begin{aligned}
(S \circ T)(\boldsymbol{x}_1 + \boldsymbol{x}_2) &= S(T(\boldsymbol{x}_1 + \boldsymbol{x}_2)) && \text{(by definition of } S \circ T\text{)} \\
&= S(T(\boldsymbol{x}_1) + T(\boldsymbol{x}_2)) && \text{(by linearity of } T\text{)} \\
&= S(T(\boldsymbol{x}_1)) + S(T(\boldsymbol{x}_2)) && \text{(by linearity of } S\text{)} \\
&= (S \circ T)(\boldsymbol{x}_1) + (S \circ T)(\boldsymbol{x}_2) && \text{(by definition of } S \circ T\text{)}
\end{aligned}$$

Similarly, we have, with $\boldsymbol{x}$ in U and α a scalar,

$$\begin{aligned}
(S \circ T)(\alpha\boldsymbol{x}) &= S(T(\alpha\boldsymbol{x})) && \text{(by definition of } S \circ T\text{)} \\
&= S(\alpha T(\boldsymbol{x})) && \text{(by linearity of } T\text{)} \\
&= \alpha S(T(\boldsymbol{x})) && \text{(by linearity of } S\text{)} \\
&= \alpha(S \circ T)(\boldsymbol{x}) && \text{(by definition of } S \circ T\text{)}
\end{aligned}$$

Since we have just verified conditions (i) and (ii) in the definition of a linear transformation, it follows that $S \circ T$ is a linear transformation from U to W.

Example 2 Our purpose here is to show precisely how composition of linear transformations corresponds to multiplication of matrices.

Suppose $T_A : \boldsymbol{R}^p \to \boldsymbol{R}^n$ and $T_B : \boldsymbol{R}^n \to \boldsymbol{R}^m$ are linear transformations induced by the matrices A, an $n \times p$ matrix, and B, an $m \times n$ matrix, respectively. That is $T_A(\boldsymbol{x}) = A\boldsymbol{x}$, for $\boldsymbol{x}$ in $\boldsymbol{R}^p$, and $T_B(\boldsymbol{y}) = B\boldsymbol{y}$, for $\boldsymbol{y}$ in $\boldsymbol{R}^n$.

Then for all $\boldsymbol{x}$ in $\boldsymbol{R}^p$

$$\begin{aligned}
(T_B \circ T_A)(\boldsymbol{x}) &= T_B(T_A(\boldsymbol{x})) && \text{(by definition of } T_B \circ T_A\text{)} \\
&= T_B(A\boldsymbol{x}) && \text{(by definition of } T_A\text{)} \\
&= B(A\boldsymbol{x}) && \text{(by definition of } T_B\text{)} \\
&= (BA)(\boldsymbol{x}) && \text{(by associativity of matrix multiplication)}
\end{aligned}$$

Since $\boldsymbol{x}$ was arbitrary $T_B \circ T_A = T_{BA}$, where T_{BA} is the linear transformation induced by the $m \times p$ matrix BA.

This natural correspondence between composition of linear functions and multiplication of matrices often enables us to give a geometric interpretation of a matrix product. For example, consider the linear operator T_θ of $\boldsymbol{R}^2$

into itself, discussed in §5.1. $T_\theta(v)$ was defined to be the vector obtained by rotating a given vector v by θ degrees. If θ_1 and θ_2 are two numbers, what is the composition $T_{\theta_1} \circ T_{\theta_2}$?

If v is a vector in the plane, $T_{\theta_2}(v)$ is obtained by rotating v by θ_2 degrees. $T_{\theta_1}(T_{\theta_2}(v))$ is then found by rotating $T_{\theta_2}(v)$ by θ_1 degrees. All together then, $T_{\theta_1}(T_{\theta_2}(v))$ is that vector resulting from rotating the vector v by $\theta_1 + \theta_2$ degrees. In other words, $T_{\theta_1+\theta_2} = T_{\theta_1} \circ T_{\theta_2}$.

Since

$$T_\theta\left(\begin{bmatrix} x \\ y \end{bmatrix}\right) = \begin{bmatrix} \cos\theta & -\sin\theta \\ \sin\theta & \cos\theta \end{bmatrix}\begin{bmatrix} x \\ y \end{bmatrix}$$

we see that

$$\begin{bmatrix} \cos\theta_1 & -\sin\theta_1 \\ \sin\theta_1 & \cos\theta_1 \end{bmatrix}\begin{bmatrix} \cos\theta_2 & -\sin\theta_2 \\ \sin\theta_2 & \cos\theta_2 \end{bmatrix} = \begin{bmatrix} \cos(\theta_1+\theta_2) & -\sin(\theta_1+\theta_2) \\ \sin(\theta_1+\theta_2) & \cos(\theta_1+\theta_2) \end{bmatrix}$$

This result may, of course, be obtained by an explicit multiplication of the two matrices in question.

Example 3 In example 3 of §5.2 we considered the canvas and rope requirements of a tent manufacturer as a function of the number of units of each type of tent the manufacturer produces. In that example, we let $v = \begin{bmatrix} x_1 \\ x_2 \\ x_3 \end{bmatrix}$ represent the number of units of each type of tent. Then $T(v) = \begin{bmatrix} 4 & 6 & 10 \\ 10 & 16 & 30 \end{bmatrix} v$ is the vector giving the number of units of canvas and rope used in the production of x_1 units of tent 1, x_2 units of tent 2, and x_3 units of tent 3.

Suppose also that the manufacturer makes his own canvas and rope from cotton, hemp, and flax. One unit of canvas requires 5 units of cotton, 2 units of hemp, and 3 units of flax. One unit of rope requires 1 unit of cotton, 2 units of hemp, and 0 units of flax.

Let $\boldsymbol{u} = \begin{bmatrix} a \\ b \end{bmatrix}$ be the vector whose components are the number of units of canvas and rope to be produced. Then, $S(\boldsymbol{u}) = \begin{bmatrix} 5 & 1 \\ 2 & 2 \\ 3 & 0 \end{bmatrix}\begin{bmatrix} a \\ b \end{bmatrix}$ is the vector whose components are, successively, the number of units of cotton, hemp, and flax required in the production of a units of canvas and b units of rope.

We wish to find the vector of cotton, hemp, and flax requirements as a function of the vector v of tent production.

In this case, T transforms the vector v of tent production into the vector $\boldsymbol{u} = T(v)$ of canvas and rope requirements. S transforms the vector $\boldsymbol{u}$ of canvas and rope requirements into the vector $S(\boldsymbol{u})$ of cotton, hemp, and flax requirements. Thus, $S(T(v))$ gives the cotton, hemp, and flax requirements as a function of v. Then, we have

$$S(T(v)) = \begin{bmatrix} 5 & 1 \\ 2 & 2 \\ 3 & 0 \end{bmatrix} \begin{bmatrix} 4 & 6 & 10 \\ 10 & 16 & 30 \end{bmatrix} v = \begin{bmatrix} 30 & 46 & 80 \\ 28 & 44 & 80 \\ 12 & 18 & 30 \end{bmatrix} v.$$

We can read from this matrix the requirements for the production of one tent of each type by reading down the appropriate column. Thus, a tent of type 2 requires 46 units of cotton, 44 units of hemp, and 18 units of flax.

The identity transformation plays a role in composition of linear transformations analogous to that played by the identity matrix in matrix multiplication.

Theorem 2 Let $T : V \to W$ be a linear transformation between two vector spaces V and W. Let $I_V : V \to V$ and $I_W : W \to W$ be the identity transformations on V and W, respectively. Then

$$T \circ I_V = T \quad \text{and} \quad I_W \circ T = T$$

PROOF Let $\boldsymbol{x}$ be a vector in V. Then

$$\begin{aligned} (T \circ I_V)(\boldsymbol{x}) &= T(I_V(\boldsymbol{x})) && \text{(by definition of } T \circ I_V) \\ &= T(\boldsymbol{x}) && \text{(by definition of } I_V) \end{aligned}$$

Since this holds for all $\boldsymbol{x}$, we have $T \circ I_V = T$. The proof of $I_W \circ T = T$ is similar. ■

Next we verify that composition of linear transformations satisfies the associative law.

Theorem 3 Let $T_1 : V_1 \to V_2$, $T_2 : V_2 \to V_3$, and $T_3 : V_3 \to V_4$ be linear transformations, where V_1, V_2, V_3, V_4 are vector spaces. Then

$$T_3 \circ (T_2 \circ T_1) = (T_3 \circ T_2) \circ T_1$$

PROOF Let x be any vector in V_1. Then

$$\begin{aligned} (T_3 \circ (T_2 \circ T_1))(\boldsymbol{x}) &= T_3((T_2 \circ T_1)(\boldsymbol{x})) && \text{[by definition of } T_3 \circ (T_2 \circ T_1)] \\ &= T_3(T_2(T_1(\boldsymbol{x}))) && \text{(by definition of } T_2 \circ T_1) \\ &= (T_3 \circ T_2)(T_1(\boldsymbol{x})) && \text{(by definition of } T_3 \circ T_2) \\ &= ((T_3 \circ T_2) \circ T_1)(\boldsymbol{x}) && \text{[by definition of } (T_3 \circ T_2) \circ T_1] \end{aligned}$$

Since for all $\boldsymbol{x}$ in V_1, $(T_3 \circ (T_2 \circ T_1))(\boldsymbol{x}) = ((T_3 \circ T_2) \circ T_1)(\boldsymbol{x})$, we have proved that $T_3 \circ (T_2 \circ T_1) = (T_3 \circ T_2) \circ T_1$. ■

Theorem 3 implies that the multiplication of matrices is associative. For if A_1, A_2, and A_3 are matrices of appropriate order, we have, by Theorem 3,

$$T_{A_3} \circ (T_{A_2} \circ T_{A_1}) = (T_{A_3} \circ T_{A_2}) \circ T_{A_1}$$

Using $T_{AB} = T_A \circ T_B$ repeatedly,

$$T_{A_3} \circ T_{A_2A_1} = T_{A_3A_2} \circ T_{A_1}$$

$$T_{A_3(A_2A_1)} = T_{(A_3A_2)A_1}$$

Since $T_A = T_B$ implies $A = B$, we have $A_3(A_2A_1) = (A_3A_2)A_1$, a result proved in chapter 2 by an entirely different method.

In view of the close connection between matrix multiplication and composition of linear transformations, it is not surprising that the following distributive laws are also valid.

Theorem 4 Let $T_2 : V_2 \to V_3$, $T_2' : V_2 \to V_3$, $T_1 : V_1 \to V_2$, and $T_3 : V_3 \to V_4$ be linear transformations on vector spaces V_1, V_2, V_3, V_4. Then,

$$(T_2 + T_2') \circ T_1 = T_2 \circ T_1 + T_2' \circ T_1$$

and

$$T_3 \circ (T_2 + T_2') = T_3 \circ T_2 + T_3 \circ T_2'$$

PROOF Let $\boldsymbol{x}$ be any vector in V_1. Then

$$\begin{aligned}
&((T_2 + T_2') \circ T_1)(\boldsymbol{x}) \\
&= (T_2 + T_2')(T_1(\boldsymbol{x})) && [\text{by definition of } (T_2 + T_2') \circ T_1] \\
&= T_2(T_1(\boldsymbol{x})) + T_2'(T_1(\boldsymbol{x})) && (\text{by definition of } T_2 + T_2') \\
&= (T_2 \circ T_1)(\boldsymbol{x}) + (T_2' \circ T_1)(\boldsymbol{x}) && (\text{by definition of } T_2 \circ T_1 \text{ and } T_2' \circ T_1) \\
&= (T_2 \circ T_1 + T_2' \circ T_1)(\boldsymbol{x}) && (\text{by definition of } T_2 \circ T_1 + T_2' \circ T_1)
\end{aligned}$$

Since the above equality holds for all $\boldsymbol{x}$ in V_1, we have shown that

$$(T_2 + T_2') \circ T_1 = T_2 \circ T_1 + T_2' \circ T_1$$

The other distributive law may be proved analogously. ■

If $T : V \to V$ is a linear operator on V, a vector space, we define powers of V in the usual way:

$$T^2(\boldsymbol{x}) = T(T(\boldsymbol{x})),\ T^3(\boldsymbol{x}) = T(T^2(\boldsymbol{x})),\ \ldots,\ T^n(\boldsymbol{x}) = T(T^{n-1}(\boldsymbol{x}))$$

Because of the associative law, T^n is just T composed with itself n times, regardless of the order in which these operations take place.

Example 4 Let D be the differentiation operator on P_n, $D(f) = f'$. Then $D^2(f) = D(D(f)) = D(f') = f''$. Thus, D^2 carries f into its second derivative. In general, $D^k(f) = f^{(k)}$, i.e., D^k carries f into its kth derivative. Since the $(n+1)$st derivative of a polynomial in P_n is 0, we see that $D^{n+1} = 0$. Operators having the property that some power of the operator is 0 are said to be **nilpotent.**

Example 5 Let P_θ be the projection transformation on $\boldsymbol{R}^2$ (see §5.2). Recall that $P_\theta(\boldsymbol{v})$ is the vector that lies on a line l making an angle θ with the x axis and $P_\theta(\boldsymbol{v})$ obtained by dropping a perpendicular from the endpoint of $\boldsymbol{v}$ to l. (See Figure 5-12.) If $\boldsymbol{u}$ is any vector lying on l, we see that $P_\theta(\boldsymbol{u}) = \boldsymbol{u}$. Thus, since $P_\theta(\boldsymbol{v})$ lies on l, $P_\theta(P_\theta(\boldsymbol{v})) = P_\theta(\boldsymbol{v})$, or $P_\theta^2(\boldsymbol{v}) = P_\theta(\boldsymbol{v})$. Since this is true for all vectors $\boldsymbol{v}$ in $\boldsymbol{R}^2$, we obtain $P_\theta^2 = P_\theta$.

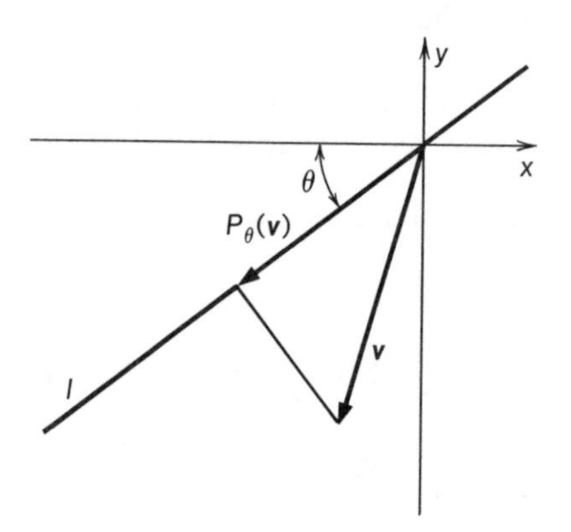

FIGURE 5-12

Operators of this sort, that is, those operators that equal their square, are said to be **idempotent**.

In example 4 of § 5.1, we considered a situation in which three companies controlled the market for a certain product. In this example, there was a linear transformation that predicted the shifts of the market from year to year. In the next example, we see what happens if such a process is allowed to continue for a long period of time. For simplicity, we suppose that there are two companies instead of three.

Example 6 Two companies, A and B, completely control the market for a certain product. Each year A retains $\frac{7}{10}$ of its customers while $\frac{3}{10}$ switch to B. B holds $\frac{6}{10}$ of its customers while $\frac{4}{10}$ switch to A. Describe the long term prospects of each company.

Let $\boldsymbol{v} = \begin{bmatrix} x \\ y \end{bmatrix}$ represent the state of the market at a given time. Here x is the fraction of the market held by A and y that held by B. For example, $x = \frac{2}{5}$ and $y = \frac{3}{5}$, if A holds $\frac{2}{5}$ of the market. We let $T(\boldsymbol{v})$ be the state of the market one year later. Then $T(v) = \begin{bmatrix} 0.7 & 0.4 \\ 0.3 & 0.6 \end{bmatrix}\begin{bmatrix} x \\ y \end{bmatrix}$, as in example 4 of §5.1. T is the linear operator that transforms the state of the market to its state one year later. If $T(v)$ is the state one year later, $T^2(v)$ is the state two years later. In general, the state after k years is $T^k(v)$. Thus, to solve the problem we must compute powers of the matrix $A = \begin{bmatrix} 0.7 & 0.4 \\ 0.3 & 0.6 \end{bmatrix}$. A few powers are:

$$A^2 = \begin{bmatrix} 0.61 & 0.52 \\ 0.39 & 0.48 \end{bmatrix}, A^3 = \begin{bmatrix} 0.583 & 0.556 \\ 0.417 & 0.444 \end{bmatrix}, A^4 = \begin{bmatrix} 0.5749 & 0.5668 \\ 0.4251 & 0.4332 \end{bmatrix},$$

$$A^5 = \begin{bmatrix} 0.57247 & 0.57004 \\ 0.42753 & 0.42996 \end{bmatrix}$$

A cursory glance would seem to indicate that this sequence of matrices is approaching a limit, and this is indeed the case. Here the limit is in fact $\begin{bmatrix} \frac{4}{7} & \frac{4}{7} \\ \frac{3}{7} & \frac{3}{7} \end{bmatrix}$. (Note: $\frac{4}{7} = 0.5714285\ldots$.) What this means is that in the limit the state of the market will be given by $\begin{bmatrix} \frac{4}{7} & \frac{4}{7} \\ \frac{3}{7} & \frac{3}{7} \end{bmatrix}\begin{bmatrix} x \\ y \end{bmatrix} = \begin{bmatrix} \frac{4}{7}(x+y) \\ \frac{3}{7}(x+y) \end{bmatrix} = \begin{bmatrix} \frac{4}{7} \\ \frac{3}{7} \end{bmatrix}$.

Even after five years this distribution is correct to within 0.001.

Two things are worthy to note about this example. First, the market approaches the distribution $\begin{bmatrix} \frac{4}{7} \\ \frac{3}{7} \end{bmatrix}$ quite rapidly no matter what its initial state is. Moreover, if the initial state is $\begin{bmatrix} \frac{4}{7} \\ \frac{3}{7} \end{bmatrix}$, then since $T\left(\begin{bmatrix} \frac{4}{7} \\ \frac{3}{7} \end{bmatrix}\right) = \begin{bmatrix} \frac{4}{7} \\ \frac{3}{7} \end{bmatrix}$, the state of the market does not change.

As in example 4 of §5.1, A is an instance of a stochastic matrix. Most stochastic matrices have the property that the sequence of powers of the matrix converges. (Note: For the reader who is skeptical about the limit of the sequence $A, A^2, A^3, \ldots,$ a simple proof by induction will show that

$$A^n = \begin{bmatrix} \frac{4}{7} & \frac{4}{7} \\ \frac{3}{7} & \frac{3}{7} \end{bmatrix} + \left(\tfrac{1}{7}\right)\left(\tfrac{3}{10}\right)^n \begin{bmatrix} 3 & -4 \\ -3 & 4 \end{bmatrix}).$$

In view of the important role played by the inverse of a matrix in certain questions involving matrices, it is appropriate to ask if there is a corresponding notion for linear transformations.

Definition Let $T: V \to W$ be a linear transformation between two vector spaces, V and W. If there is a linear transformation $S: W \to V$ such that $T \circ S = I_W$ and $S \circ T = I_V$, then T is **invertible** and S is said to be the **inverse** of T.

If a linear transformation T has an inverse, it is unique. To see this, suppose that S_1 and S_2 are two linear transformations from W to V, which satisfy the conditions of the inverse in the above definition. Then

$$\begin{aligned} S_1 &= S_1 \circ I_W && \text{(by Theorem 2)} \\ &= S_1 \circ (T \circ S_2) && \text{(since } T \circ S_2 = I_W \text{, by hypothesis)} \\ &= (S_1 \circ T) \circ S_2 && \text{(by Theorem 3)} \\ &= I_V \circ S_2 && \text{(since } S_1 \circ T = I_V \text{, by hypothesis)} \\ &= S_2 && \text{(by Theorem 2)} \end{aligned}$$

Thus, if the inverse exists, it is unique. The inverse of a transformation T is generally denoted by T^{-1}.

Example 7 Let A be an invertible $n \times n$ matrix and let T_A be the linear operator on $\boldsymbol{R}^n$ defined by $T_A(\boldsymbol{x}) = A\boldsymbol{x}$. Then, there is also the operator $T_{A^{-1}}$ on $\boldsymbol{R}^n$ defined by $T_{A^{-1}}(\boldsymbol{x}) = A^{-1}\boldsymbol{x}$. We compute $T_A \circ T_{A^{-1}}$. $(T_A \circ T_{A^{-1}})(\boldsymbol{x}) = T_A(T_{A^{-1}}(\boldsymbol{x})) = T_A(A^{-1}\boldsymbol{x}) = A(A^{-1}\boldsymbol{x}) = \boldsymbol{x}$. Thus, we see that $T_A \circ T_{A^{-1}} = I$. Similarly, it can be shown that $T_{A^{-1}} \circ T_A = I$. It follows that T_A is invertible and $(T_A)^{-1} = T_{A^{-1}}$.

Example 8 Let $S : P_n \rightarrow P_n$ be the linear operator that carries the polynomial $f(x)$ into the polynomial $f(x + 1)$, or $(S(f))(x) = f(x + 1)$. Thus, x is transformed into $x + 1$, x^2 into $(x + 1)^2$, and so on.

Now, if we let $T : P_n \rightarrow P_n$ be the linear operator that carries $f(x)$ into $f(x - 1)$, or $(T(f))(x) = f(x - 1)$, then S is invertible and $S^{-1} = T$.

For we have

$$\begin{aligned} ((S \circ T)(f))(x) &= (S(T(f)))(x) && \text{(by definition of } S \circ T) \\ &= (T(f))(x + 1) && \text{(by definition of } S) \\ &= f((x + 1) - 1) && \text{(by definition of } T) \\ &= f(x) \end{aligned}$$

Thus, for all f, $(S \circ T)f = f$, so $S \circ T = I$. In a like manner, it can be shown that $T \circ S = I$.

The concept of the inverse of a linear operator is closely related to the concept of isomorphism. This relationship is made explicit in the following theorem.

Theorem 5 Let $T : V \rightarrow W$ be a linear transformation between two vector spaces, V and W. Then, T is invertible if and only if T is an isomorphism.

PROOF Suppose T is invertible and $T^{-1} : W \rightarrow V$ is its inverse. If $T(\boldsymbol{x}) = \boldsymbol{0}$, then $T^{-1}(T(\boldsymbol{x})) = T^{-1}(\boldsymbol{0}) = \boldsymbol{0}$. But $T^{-1}(T(\boldsymbol{x})) = \boldsymbol{x}$. Thus, $\boldsymbol{x} = \boldsymbol{0}$. Since $T(\boldsymbol{x}) = \boldsymbol{0}$ implies $\boldsymbol{x} = \boldsymbol{0}$, we see that $N_T = \boldsymbol{0}$. Next, we show T is onto. If $\boldsymbol{y}$ is a vector in W, let $\boldsymbol{x} = T^{-1}(\boldsymbol{y})$. Then $T(\boldsymbol{x}) = T(T^{-1}(\boldsymbol{y})) = \boldsymbol{y}$. Thus, $\boldsymbol{y}$ is the image under T of the vector $\boldsymbol{x}$. From this, it follows that T is onto. Since T is one-one and onto, T is an isomorphism.

Next suppose T is an isomorphism. We must construct an inverse of T.

Since T is onto, given $\boldsymbol{y}$ in W there is some vector $\boldsymbol{x}$ in V such that $T(\boldsymbol{x}) = \boldsymbol{y}$. We define $S(\boldsymbol{y}) = \boldsymbol{x}$. Then S is well-defined, since if $T(\boldsymbol{x}_1) = y$ and $T(\boldsymbol{x}_2) = \boldsymbol{y}$, the assumption that T is one-one implies $\boldsymbol{x}_1 = \boldsymbol{x}_2$. Thus, S is a function from W to V.

Next we must show that S is linear. Let $\boldsymbol{y}_1$ and $\boldsymbol{y}_2$ be vectors in W. Then there are vectors $\boldsymbol{x}_1$ and $\boldsymbol{x}_2$ in V such that $T(\boldsymbol{x}_1) = \boldsymbol{y}_1$ and $T(\boldsymbol{x}_2) = \boldsymbol{y}_2$. Then $T(\boldsymbol{x}_1 + \boldsymbol{x}_2) = \boldsymbol{y}_1 + \boldsymbol{y}_2$ by linearity of T. Hence, $S(\boldsymbol{y}_1) = \boldsymbol{x}_1$, $S(\boldsymbol{y}_2) = \boldsymbol{x}_2$, and $S(\boldsymbol{y}_1 + \boldsymbol{y}_2) = \boldsymbol{x}_1 + \boldsymbol{x}_2$, by definition of S. Therefore, $S(\boldsymbol{y}_1 + \boldsymbol{y}_2) = S(\boldsymbol{y}_1) + S(\boldsymbol{y}_2)$.

Next, suppose $\boldsymbol{y}$ belongs to W and α is a scalar. Since T is onto, $T(\boldsymbol{x}) = \boldsymbol{y}$, for some $\boldsymbol{x}$ in V. Then $T(\alpha\boldsymbol{x}) = \alpha T(\boldsymbol{x}) = \alpha\boldsymbol{y}$ by linearity of T. By definition of S, $S(\boldsymbol{y}) = \boldsymbol{x}$ and $S(\alpha\boldsymbol{y}) = \alpha\boldsymbol{x}$. Thus, $S(\alpha\boldsymbol{y}) = \alpha S(\boldsymbol{y})$. So it follows that S is linear.

Now we claim $S \circ T = I_V$ and $T \circ S = I_W$. Pick any vector $\boldsymbol{x}$ in V and let $T(\boldsymbol{x}) = \boldsymbol{y}$. By definition of S, $S(\boldsymbol{y}) = \boldsymbol{x}$. Thus,

$$\begin{aligned}(S \circ T)(\boldsymbol{x}) &= S(T(\boldsymbol{x})) && \text{(by definition of } S \circ T)\\ &= S(\boldsymbol{y}) && [\text{since } \boldsymbol{y} = T(\boldsymbol{x})]\\ &= \boldsymbol{x} && \text{(by definition of } S)\end{aligned}$$

Therefore, we have shown that $S \circ T = I_V$.

Next, pick any vector $\boldsymbol{y}$ in W. Since T is onto there is some $\boldsymbol{x}$ in V such that $T(\boldsymbol{x}) = \boldsymbol{y}$. By definition of S, $S(\boldsymbol{y}) = \boldsymbol{x}$.

Then

$$\begin{aligned}(T \circ S)(\boldsymbol{y}) &= T(S(\boldsymbol{y})) && \text{(by definition of } T \circ S)\\ &= T(\boldsymbol{x}) && \text{(by definition of } S)\\ &= \boldsymbol{y} && [\text{since } T(\boldsymbol{x}) = \boldsymbol{y}]\end{aligned}$$

It follows that $T \circ S = I_W$.

Thus, we have demonstrated that any isomorphism has an inverse. ■

In some types of problems the procedure just described can be used to find the inverse of a given transformation. Indeed, if $T : V \to W$ is a linear isomorphism between V and W, and $\boldsymbol{x}_1, \boldsymbol{x}_2, \ldots, \boldsymbol{x}_n$ is a basis for V, it follows by Theorem 1, §5.6 that $T(\boldsymbol{x}_1) = \boldsymbol{y}_1$, $T(\boldsymbol{x}_2) = \boldsymbol{y}_2, \ldots, T(\boldsymbol{x}_n) = \boldsymbol{y}_n$ is a basis for W. Thus, T^{-1} is the linear function defined by $T^{-1}(\boldsymbol{y}_1) = \boldsymbol{x}_1$, $T^{-1}(\boldsymbol{y}_2) = \boldsymbol{x}_2, \ldots, T^{-1}(\boldsymbol{y}_n) = \boldsymbol{x}_n$. An example may help to clarify the method.

Example 9 Consider the linear transformation $T(f) = f + f'$ from P_4 to P_4 defined in example 4 of §5.6. We saw that T is an isomorphism of P_4 with itself. We calculate the images of the basis vectors $1, x, x^2, x^3, x^4$: $T(1) = 1$, $T(x) = 1 + x$, $T(x^2) = 2x + x^2$, $T(x^3) = 3x^2 + x^3$, $T(x^4) = 4x^3 + x^4$. It follows that

$$T^{-1}(1) = 1,\ T^{-1}(1 + x) = x,\ T^{-1}(2x + x^2) = x^2,$$

$$T^{-1}(3x^2 + x^3) = x^3, T^{-1}(4x^3 + x^4) = x^4$$

Then using the linearity of T^{-1},

$$T^{-1}(1) = 1,\ T^{-1}(x) = T^{-1}((x+1) - 1) = x - 1,$$

$$T^{-1}(x^2) = T^{-1}\big((x^2 + 2x) - 2x\big) = x^2 - 2(x-1) = x^2 - 2x + 2,$$

$$T^{-1}(x^3) = T^{-1}\big((x^3 + 3x^2) - 3x^2\big)$$

$$= x^3 - 3(x^2 - 2x + 2) = x^3 - 3x^2 + 6x - 6,$$

$$T^{-1}(x^4) = T^{-1}\big((x^4 + 4x^3) - 4x^3\big) = x^4 - 4x^3 + 12x^2 - 24x + 24$$

Since we know the T^{-1} image of each basis element, using linearity we can determine the T^{-1} image of any element.

EXERCISES

1. Let S, T, and U be linear operators on $\boldsymbol{R}^2$ defined by

$$S\left(\begin{bmatrix} x \\ y \end{bmatrix}\right) = \begin{bmatrix} 1 & 0 \\ 1 & 1 \end{bmatrix}\begin{bmatrix} x \\ y \end{bmatrix},\ T\left(\begin{bmatrix} x \\ y \end{bmatrix}\right) = \begin{bmatrix} 0 & 1 \\ 1 & 0 \end{bmatrix}\begin{bmatrix} x \\ y \end{bmatrix},\ U\left(\begin{bmatrix} x \\ y \end{bmatrix}\right) = \begin{bmatrix} 1 & 0 \\ 0 & -1 \end{bmatrix}\begin{bmatrix} x \\ y \end{bmatrix}$$

   Calculate the matrices associated with the following linear transformations.
   (a) $S \circ T$ (b) $T \circ S$ (c) S^2 (d) S^3 (e) S^{-1} (f) $S \circ (T \circ U)$
   (g) $(S \circ T) \circ U$ (h) $S \circ T - T \circ S$ (i) $T \circ U - U \circ T$ (j) $U^2 - I_2$

2. Let S, T, and U be linear operators on P_3 defined by $S(f) = f' + f$, $T(f) = f + xf'$, $U(f) = f''$. Calculate the following linear operators.
   (a) $S \circ U$ (b) $U \circ S$ (c) $U \circ T$ (d) $T \circ U$ (e) $T \circ U - U \circ T$
   (f) U^2 (g) S^2 (h) S^3 (i) $T \circ S$ (j) $S \circ T$

3. Which of the following matrices induce invertible linear operators of R^n? Determine the inverse if it exists.

   (a) $\begin{bmatrix} 3 & 2 \\ 4 & 3 \end{bmatrix}$ (b) $\begin{bmatrix} -3 & 4 \\ 6 & -8 \end{bmatrix}$ (c) $\begin{bmatrix} 1 & 0 & 3 \\ 2 & 1 & 0 \\ 7 & 4 & 2 \end{bmatrix}$ (d) $\begin{bmatrix} -1 & 0 & 1 \\ 6 & 2 & -4 \\ 3 & 4 & 1 \end{bmatrix}$

4. Decide which of the following operators on P_3 are invertible.

   (a) $D(f) = f'$ (b) $S(f)(x) = f(x) + f(-x)$ (c) $T(f) = f + 2f'$

5. Let a, b, and c be real numbers and let T be the linear operator of P_n defined by $T(f) = af + bxf' + cx^2 f''$.
   (a) Show that $T(x^n) = (a + bn + cn(n-1))x^n$.
   (b) If a, b, and c are positive, show that T is invertible.
   (c) Choose nonzero values of a, b, and c, so that T is not invertible.

6. A linear transformation from P_2 to $\boldsymbol{R}^3$ is defined $T(f) = \begin{bmatrix} f(1) \\ f(0) \\ f(-1) \end{bmatrix}$. Show that T is invertible.

7. A furniture manufacturer produces two types of bookcases. The first type requires 10 square feet of wood, 40 nails, and 2 hours of labor. The second type requires 30 square feet of wood, 60 nails, and 3 hours of labor.

   Let $\boldsymbol{v} = \begin{bmatrix} x_1 \\ x_2 \end{bmatrix}$ be the vector whose ith component is the number of bookcases of each type to be made. Let $T(\boldsymbol{v}) = \boldsymbol{u} = \begin{bmatrix} a \\ b \\ c \end{bmatrix}$ represent the number of units of wood, nails, and labor required to produce x_1 cases of type 1 and x_2 cases of type 2.

   (a) Find a matrix A such that $T(\boldsymbol{v}) = A\boldsymbol{v}$.

   The cost of one unit of wood, nails, and labor is respectively \$.10, \$.02, and \$3.50. Let $S(\boldsymbol{u})$ be the cost of a units of wood, b units of nails, and c units of labor.

   (b) Find a matrix B such that $S(\boldsymbol{u}) = B\boldsymbol{u}$.

   (c) Find the function $S \circ T(\boldsymbol{v})$ and interpret it.

   A bookcase of type 1 sells for \$10.00 and one of type 2 for \$16.00. Let $R(\boldsymbol{v})$ be the total sale price of x_1 units of type 1 and x_2 units of type 2.

   (d) Find a matrix C such that $R(\boldsymbol{v}) = C\boldsymbol{v}$.

   (e) Compute and interpret the function $(R - S \circ T)(\boldsymbol{v})$.

8. Prove by induction that

$$\begin{bmatrix} 1-a & b \\ a & 1-b \end{bmatrix}^n = \frac{1}{a+b}\begin{bmatrix} b & b \\ a & a \end{bmatrix} + \frac{(1-a-b)^n}{a+b}\begin{bmatrix} a & -b \\ -a & b \end{bmatrix}$$

9. Let a and b be real numbers with $0 < a < 1$ and $0 < b < 1$. Let

$$A = \begin{bmatrix} 1-a & b \\ a & 1-b \end{bmatrix}$$

   Show that the sequence of matrices, $A, A^2, A^3, \ldots,$ approaches $\dfrac{1}{a+b}\begin{bmatrix} b & b \\ a & a \end{bmatrix}$.

10. Suppose that in example 6 of this section, A retains $1 - a$ of its customers and a switch to B. B retains $1 - b$ of its customers and b switch to A. Suppose that $0 < a < 1$ and $0 < b < 1$. Determine the long term prospects of each company. (Hint: Use exercises 8 and 9.)

11. Let A be an $n \times n$ matrix. Let $T_1(A) = \frac{1}{2}(A + A^T)$, $T_2(A) = \frac{1}{2}(A - A^T)$.

    (a) Show that $T_1^2 = T_1$, i.e., T_1 is idempotent.

    (b) Show that $T_2^2 = T_2$, i.e., T_2 is idempotent.

    (c) Show that $T_1T_2 = T_2T_1 = 0$.

12. Let $T: V \to V$ be a linear operator on a vector space V such that $T^2 = I_V$. Show that

    (a) The operator $T_1 = \frac{1}{2}(1 + T)$ satisfies $T_1^2 = T_1$, i.e., T_1 is idempotent.

    (b) The operator $T_2 = \frac{1}{2}(1 - T)$ satisfies $T_2^2 = T_2$, i.e., T_2 is idempotent.

    (c) $T_1T_2 = T_2T_1 = 0$, $T_1 + T_2 = I_V$.

    (d) $R_{T_1} = N_{T_2}$, $N_{T_2} = R_{T_1}$.

    (e) If $V = P_n$, the space of polynomials of degree less than or equal to n in a variable x, and $T(f(x)) = f(-x)$, show that $T^2 = I_{P_n}$. What are the operators T_1 and T_2 of (a) and (b)? What polynomials belong to R_{T_1} and R_{T_2}?

    (f) If $V = P_n$, as in part (e), and $(T(f))(x) = f(1 - x)$, show that $T^2 = I_{P_n}$.

13. If V is a vector space and $x_1, x_2, \ldots, x_n$ is a basis for V show that the operator $T(x_1) = x_2$, $T(x_2) = x_3, \ldots, T(x_{n-1}) = x_n$, $T(x_n) = \alpha x_1$, satisfies the relation $T^n = \alpha I_V$. If $\alpha \neq 0$, what is T^{-1}?

14. Let V_1, V_2, and V_3 be vector spaces, and let $S : V_1 \to V_2$ and $T : V_2 \to V_3$ be linear transformations.
    (a) If S is one-one and T is one-one, show that $T \circ S$ is one-one.
    (b) If S is onto and T is onto, show that $T \circ S$ is onto.
    (c) If S and T are isomorphisms, show that $T \circ S$ is an isomorphism and $(T \circ S)^{-1} = S^{-1} \circ T^{-1}$.

15. Let $T : V \to V$ and $S : V \to V$ be linear operators on a vector space V of finite dimension.
    (a) If $T \circ S$ is one-one, show that T is an isomorphism.
    (b) If $T \circ S$ is onto, show that T is an isomorphism.
    (c) If T^k is an isomorphism for some positive integer k, show that T is an isomorphism.
    (d) If T^k is invertible for some positive integer k, and $(T^k)^{-1} = S$, show that $T^{-1} = T^{k-1} \circ S$.

16. If T and S are linear operators on a vector space V and $(T + S)^2 = T^2 + 2S \circ T + S^2$, show that S and T commute, i.e., $S \circ T = T \circ S$.

17. If T is a linear transformation from a finite-dimensional vector space V onto a vector space W, show that
    (a) dim W is finite, and dim $W \leqslant$ dim V.
    (b) There is a linear transformation $S : W \to V$, such that $T \circ S = I_W$.

18. If T is a one-one linear transformation from a vector space V into a finite-dimensional vector space W, show that
    (a) V is finite-dimensional.
    (b) There is a linear transformation $S : W \to V$, such that $S \circ T = I_V$.

19. If T is a linear operator on a vector space V and if $T^2 - T + I_V = 0$, show that T is invertible.

20. Let $T : \boldsymbol{R}^n \to \boldsymbol{R}^n$ be a linear operator on $\boldsymbol{R}^n$. If $m < n$, and there are linear transformations $S : \boldsymbol{R}^n \to \boldsymbol{R}^m$ and $U : \boldsymbol{R}^m \to \boldsymbol{R}^n$, such that $T = U \circ S$, show that $r(T) \leqslant m$ and T is not invertible.

21. If T is a linear operator on a finite-dimensional vector space V, show that $r(T) = r(T^2)$ if and only if $N_T \cap R_T = 0$, i.e., the nullspace and range of T has only the zero vector in common.

22. Let D be the differentiation operator on P_n and let T_α be the operator defined by $T_\alpha(f)(x) = f(\alpha x)$. Show that $D \circ T_\alpha = \alpha T_\alpha \circ D$.

23. Let T be a linear operator on a vector space V.
    (a) If for some positive integer k, $R(T^k) = R(T^{k+1})$, show that $R(T^{k+1}) = R(T^{k+2})$.
    (b) If for some positive integer k, $N(T^k) = N(T^{k+1})$, show that $N(T^{k+1}) = N(T^{k+2})$.

24. Let V be a vector space and $x_1, x_2, \ldots, x_n$ be a basis for V. Suppose $T : V \to V$ is the linear operator $T(x_1) = \lambda_1 x_1, \ldots, T(x_n) = \lambda_n x_n$, for scalars $\lambda_1, \lambda_2, \ldots, \lambda_n$ where $\lambda_i \neq \lambda_j$ if $i \neq j$. Suppose $T \circ S = S \circ T$, where S is a linear operator on V.

Show that there are scalars $\alpha_1, \alpha_2, \ldots, \alpha_n$ such that $S(x_1) = \alpha_1 x_1$, $S(x_2) = \alpha_2 x_2, \ldots, S(x_n) = \alpha_n x_n$.

25. Let $T : V \to V$ be a linear operator on a finite-dimensional space V. Show that T commutes with all other linear operators on V if and only if $T = \alpha I_V$, i.e., T is a scalar multiple of the identity on V.

26. Let V and W be vector spaces. Let x be a fixed nonzero vector in V. Show that the function from $L(V, W)$ into W, defined by $L(T) = T(x)$, is a linear transformation from $L(V, W)$ onto W.

27. If $T : R^n \to R^n$ is a linear operator on R^n of rank r, show that there are linear transformations $S : R^n \to R^r$, where S is onto and $U : R^r \to R^n$, where U is one-one such that $T = U \circ S$. Interpret this result in terms of matrices.

28. Suppose T is a linear operator on R^3 with the property that $T \neq 0$ but $T^2 = 0$. Show that $r(T) = 1$.

29. If $T : R^n \to R^n$ is a linear operator such that $T^{n-1} \neq 0$ but $T^n = 0$, show that $r(T) = n - 1$. Give an example of such a T.

30. If $T : V \to V$ and $S : V \to V$ are linear operators on a vector space V, show that $T \circ S = 0$ if and only if $R_S \subset N_T$.

31. If $T : V \to V$ is a linear operator on a finite-dimensional vector space V and $T^2 = 0$, show that $r(T) \leqslant \frac{1}{2} \dim V$.

8 MATRIX REPRESENTATION OF A LINEAR TRANSFORMATION

Let $T : V \to W$ be a linear transformation between two finite-dimensional vector spaces V and W. If $V = R^n$ and $W = R^m$, in §5.2 we saw that it is possible, using the standard bases for R^m and R^n, to find an $m \times n$ matrix A such that $T(x) = Ax$. In this section we perform a similar construction using arbitrary bases for V and W.

Let $\mathcal{B} = \{x_1, x_2, \ldots, x_n\}$ be a basis for V and $\mathcal{C} = \{y_1, y_2, \ldots, y_m\}$ be a basis for W. Since $\mathcal{C}$ is a basis for W, there are scalars a_{ij}, such that

$$T(x_j) = \sum_{i=1}^{m} a_{ij} y_i$$

or

$$T(x_j) \underset{\mathcal{C}}{\leftrightarrow} \begin{bmatrix} a_{1j} \\ a_{2j} \\ \vdots \\ a_{mj} \end{bmatrix}$$

Define the $m \times n$ matrix A_T by $A_T = [a_{ij}]_{(mn)}$. Note that the jth column of A_T is just the m-tuple of coordinates of $T(x_j)$ relative to the basis $\mathcal{C} = \{y_1, y_2, \ldots, y_m\}$ for W.

If x belongs to V, we may associate with x its n-tuple of coordinates relative to the basis $\mathcal{B}$:

$$x \underset{\mathcal{B}}{\leftrightarrow} \begin{bmatrix} \alpha_1 \\ \alpha_2 \\ \vdots \\ \alpha_n \end{bmatrix}$$

which means, of course, $x = \alpha_1 x_1 + \alpha_2 x_2 + \cdots + \alpha_n x_n$.

Defining the m-vector by

$$\begin{bmatrix} \beta_1 \\ \beta_2 \\ \vdots \\ \beta_m \end{bmatrix} = \begin{bmatrix} a_{11} & a_{12} & \cdots & a_{1n} \\ a_{21} & a_{22} & \cdots & a_{2n} \\ & & \vdots & \\ a_{m1} & a_{m2} & \cdots & a_{mn} \end{bmatrix} \begin{bmatrix} \alpha_1 \\ \alpha_2 \\ \vdots \\ \alpha_n \end{bmatrix}$$

we claim

$$T(x) \underset{\mathcal{C}}{\leftrightarrow} \begin{bmatrix} \beta_1 \\ \beta_2 \\ \vdots \\ \beta_m \end{bmatrix}$$

or equivalently

$$T(x) = \beta_1 y_1 + \beta_2 y_2 + \cdots + \beta_m y_m$$

To prove this assertion, note that

$$x = \sum_{j=1}^{n} \alpha_j x_j$$

$$T(x) = \sum_{j=1}^{n} \alpha_j T(x_j)$$

$$= \sum_{j=1}^{n} \alpha_j \left(\sum_{i=1}^{m} a_{ij} y_i \right) = \sum_{i=1}^{m} \left(\sum_{j=1}^{n} a_{ij} \alpha_j \right) y_i$$

Thus, the ith coordinate of $T(x)$ relative to the basic $\mathcal{C} = \{y_1, y_2, \ldots, y_m\}$ is $\beta_i = \sum_{j=1}^{n} a_{ij} \alpha_j$.

In other words, as soon as a basis $\mathcal{B}$ for V and a basis $\mathcal{C}$ for W are specified, there is a matrix A_T with the property that the product of A_T with the coordinate vector of x relative to $\mathcal{B}$ is just the coordinate vector of $T(x)$ relative to $\mathcal{C}$.

If $V = W$, that is if $T : V \to V$ is a linear operator on V, it is customary to use the same basis in both the initial and final spaces (that is, assume $\mathcal{B} = \mathcal{C}$). If $I_V : V \to V$ is the identity operator on V, and if $\mathcal{B} = \{\boldsymbol{x}_1, \boldsymbol{x}_2, \ldots, \boldsymbol{x}_n\}$ is a basis for V, the fact that $I_V \boldsymbol{x}_j = \boldsymbol{x}_j$ implies that the matrix associated with I_V is just I_n, the identity matrix of order n. Likewise, the matrix associated with the zero operator is the zero matrix.

Example 1 Let M_{22} denote the space of real 2×2 matrices. Consider the linear transformation T from M_{22} into $\boldsymbol{R}^2$ defined by $T(A) = A\begin{bmatrix} 3 \\ 2 \end{bmatrix}$. Now T is linear since if A and B belong to M_{22} and α is a scalar, $T(A+B) = (A+B)\begin{bmatrix} 3 \\ 2 \end{bmatrix} = A\begin{bmatrix} 3 \\ 2 \end{bmatrix} + B\begin{bmatrix} 3 \\ 2 \end{bmatrix} = T(A) + T(B)$ and $T(\alpha A) = (\alpha A)\begin{bmatrix} 3 \\ 2 \end{bmatrix} = \alpha A\begin{bmatrix} 3 \\ 2 \end{bmatrix} = \alpha T(A)$.

Using the basis $\mathcal{B}$,

$$E_1 = \begin{bmatrix} 1 & 0 \\ 0 & 0 \end{bmatrix}, \quad E_2 = \begin{bmatrix} 0 & 1 \\ 0 & 0 \end{bmatrix}, \quad E_3 = \begin{bmatrix} 0 & 0 \\ 1 & 0 \end{bmatrix}, \quad E_4 = \begin{bmatrix} 0 & 0 \\ 0 & 1 \end{bmatrix}$$

For M_{22}, and using the basis $\mathcal{C}$,

$$y_1 = \begin{bmatrix} 1 \\ 0 \end{bmatrix}, \quad y_2 = \begin{bmatrix} 0 \\ 1 \end{bmatrix}$$

for $\boldsymbol{R}^2$, we calculate the matrix representation of T.

$$T(E_1) = \begin{bmatrix} 1 & 0 \\ 0 & 0 \end{bmatrix}\begin{bmatrix} 3 \\ 2 \end{bmatrix} = \begin{bmatrix} 3 \\ 0 \end{bmatrix}, \quad T(E_2) = \begin{bmatrix} 0 & 1 \\ 0 & 0 \end{bmatrix}\begin{bmatrix} 3 \\ 2 \end{bmatrix} = \begin{bmatrix} 2 \\ 0 \end{bmatrix}$$

$$T(E_3) = \begin{bmatrix} 0 & 0 \\ 1 & 0 \end{bmatrix}\begin{bmatrix} 3 \\ 2 \end{bmatrix} = \begin{bmatrix} 0 \\ 3 \end{bmatrix}, \quad T(E_4) = \begin{bmatrix} 0 & 0 \\ 0 & 1 \end{bmatrix}\begin{bmatrix} 3 \\ 2 \end{bmatrix} = \begin{bmatrix} 0 \\ 2 \end{bmatrix}$$

Thus, it follows that $A_T = \begin{bmatrix} 3 & 2 & 0 & 0 \\ 0 & 0 & 3 & 2 \end{bmatrix}$. Note that the matrix A_T is obtained by placing the coordinates of $T(E_1)$, $T(E_2)$, $T(E_3)$, $T(E_4)$ relative to the basis y_1, y_2 in the successive columns.

If $M = \begin{bmatrix} a & b \\ c & d \end{bmatrix}$, then $M \underset{\mathcal{B}}{\leftrightarrow} \begin{bmatrix} a \\ b \\ c \\ d \end{bmatrix}$, and so

$$T(M) \underset{\mathcal{C}}{\leftrightarrow} \begin{bmatrix} 3 & 2 & 0 & 0 \\ 0 & 0 & 3 & 2 \end{bmatrix}\begin{bmatrix} a \\ b \\ c \\ d \end{bmatrix} = \begin{bmatrix} 3a + 2b \\ 3c + 2d \end{bmatrix}$$

Example 2 Let P_3 be defined in the usual manner. Let $D : P_3 \to P_3$ be the differentiation operator, $D(f) = f'$. Choose the basis $1, x, x^2, x^3$ for P_3. Then,

$D(1) = 0$, $D(x) = 1$, $D(x^2) = 2x$, $D(x^3) = 3x^2$. Thus, the matrix A_D is

$$A_D = \begin{bmatrix} 0 & 1 & 0 & 0 \\ 0 & 0 & 2 & 0 \\ 0 & 0 & 0 & 3 \\ 0 & 0 & 0 & 0 \end{bmatrix}$$

that was found by placing the coordinates of $D(1)$, $D(x)$, $D(x^2)$, $D(x^3)$ relative to the basis $1, x, x^2, x^3$ in the successive columns. If $f = \alpha_0 + \alpha_1 x + \alpha_2 x^2 + \alpha_3 x^3$, then $f \underset{\mathcal{B}}{\leftrightarrow} \begin{bmatrix} \alpha_0 \\ \alpha_1 \\ \alpha_2 \\ \alpha_3 \end{bmatrix}$ and $D(f) \underset{\mathcal{B}}{\leftrightarrow} \begin{bmatrix} 0 & 1 & 0 & 0 \\ 0 & 0 & 2 & 0 \\ 0 & 0 & 0 & 3 \\ 0 & 0 & 0 & 0 \end{bmatrix} \begin{bmatrix} \alpha_0 \\ \alpha_1 \\ \alpha_2 \\ \alpha_3 \end{bmatrix} = \begin{bmatrix} \alpha_1 \\ 2\alpha_2 \\ 3\alpha_3 \\ 0 \end{bmatrix}$, i.e.,

$D(f) = \alpha_1 + 2\alpha_2 x + 3\alpha_3 x^2$.

The matrix representation of a linear transformation effected by choosing bases can often be used to reduce problems regarding linear transformations to problems involving matrices and linear equations, problems for which we have developed effective computational procedures.

For example, let T be a linear transformation from a vector space V to a vector space W. We wish to find the null space and rank of T. Let $\mathcal{B} = \{\boldsymbol{x}_1, \boldsymbol{x}_2, \ldots, \boldsymbol{x}_n\}$ be a basis for V, and $\mathcal{C} = \{\boldsymbol{y}_1, \boldsymbol{y}_2, \ldots, \boldsymbol{y}_m\}$ be a basis for W. Let A_T be the $m \times n$ matrix associated with T relative to the above bases. If $\boldsymbol{x}$ belongs to V, $\boldsymbol{x} \underset{\mathcal{B}}{\leftrightarrow} \begin{bmatrix} \alpha_1 \\ \alpha_2 \\ \vdots \\ \alpha_n \end{bmatrix}$, then $T(\boldsymbol{x}) \leftrightarrow A_T \begin{bmatrix} \alpha_1 \\ \alpha_2 \\ \vdots \\ \alpha_n \end{bmatrix}$. Thus, $T(\boldsymbol{x}) = \boldsymbol{0}$ if and only if $A_T \begin{bmatrix} \alpha_1 \\ \alpha_2 \\ \vdots \\ \alpha_n \end{bmatrix} = \boldsymbol{0}$. Stated otherwise, a vector $\boldsymbol{x}$ in V belongs to N_T if and only if its coordinate n-tuple belongs to the nullspace of A_T. This provides a computational means of determining the nullspace of a linear transformation. In fact, the whole problem is reduced to one of solving linear homogeneous equations. It also implies that $n(T) = n(A_T)$. Since $r(T) = \dim V - n(T)$ and $r(A_T) = \dim V - n(A_T)$, it follows that $r(T) = r(A_T)$. Thus, the rank and nullity of a linear transformation are just the rank and nullity of the associated matrix.

The rank of the matrix in example 1 above is 2. Thus, $r(T) = 2$. Since $n(T) = 4 - r(T)$, it follows that the nullspace of T is of dimension 2.

In example 2 above, the rank of the matrix A_D is 3 and $n(D) = 4 - 3 = 1$.

By associating with each linear transformation in $L(V, W)$ its matrix relative to the bases $\mathcal{B}$ and $\mathcal{C}$, we determine a function $G(T) = A_T$ from $L(V, W)$ to the space M_{mn} of $m \times n$ matrices (with real entries if V and W are real vector spaces, complex entries if V and W are complex vector spaces). In view of the examples of §5.7, the following result is not surprising.

Theorem 1 The function that assigns to each T in $L(V, W)$ its matrix relative to bases $\mathcal{B}$ for V and $\mathcal{C}$ for W is an isomorphism from $L(V, W)$ to M_{mn}, the space of $m \times n$ matrices.

PROOF We denote the function by G. First, we show that G is linear. To accomplish this let S, T be linear transformations from V to W. Then if

$$S(\boldsymbol{x}_j) = \sum_{i=1}^{m} a_{ij}\boldsymbol{y}_i \qquad T(\boldsymbol{x}_j) = \sum_{i=1}^{m} b_{ij}\boldsymbol{y}_i$$

$G(S) = A_S = [a_{ij}]_{(mn)}$ and $G(T) = A_T = [b_{ij}]_{(mn)}$. Since

$$(S + T)(\boldsymbol{x}_j) = S(\boldsymbol{x}_j) + T(\boldsymbol{x}_j)$$

$$= \sum_{i=1}^{m} (a_{ij} + b_{ij})\boldsymbol{y}_i$$

$A_{S+T} = [a_{ij} + b_{ij}]_{(mn)} = A_S + A_T$.

Thus, $G(S + T) = G(S) + G(T)$. A similar proof yields $G(\alpha S) = \alpha G(S)$.

Next, we show that G is one-one. Assume T is a linear operator such that $G(T) = A_T = 0$. Then, $T(\boldsymbol{x}_j) = \sum_{i=1}^{m} 0 \cdot \boldsymbol{y}_i = \boldsymbol{0}$. Since T sends all basis vectors into $\boldsymbol{0}$, $T = 0$. Thus, $G(T) = 0$ implies $T = 0$, and so G is one-one.

Finally, we show that G is onto. If A is an $m \times n$ matrix, $A = [a_{ij}]_{(mn)}$, we must exhibit a linear transformation T from V to W such that

$$T(\boldsymbol{x}_j) = \sum_{i=1}^{n} a_{ij}\boldsymbol{y}_i$$

for then by definition of A_T we have $A_T = A$. By Theorem 2, §5.2, such a linear transformation exists. Thus, G is onto.

Since G is linear, one-one, and onto, G is an isomorphism. ■

In particular, if $T : V \to V$ is a linear operator on a vector space V and $\boldsymbol{x}_1, \boldsymbol{x}_2, \ldots, \boldsymbol{x}_n$ is a basis for V, the function that associates with the operator T the $n \times n$ matrix

$$A_T = \left[a_{ij}\right]_{(nn)}, \quad \text{where } T\boldsymbol{x}_j = \sum_{i=1}^{n} a_{ij}\boldsymbol{x}_i$$

induces an isomorphism of $L(V, V)$ and the space of $n \times n$ matrices. By considering the case $V = \boldsymbol{R}^n$ it seems reasonable to expect that the correspondence $T \to A_T$ preserves products. The next theorem shows that our expectations are well-founded.

Theorem 2 Let $S : V \to V$ and $T : V \to V$ be linear operators on the vector space V and $\mathfrak{B} = \{\boldsymbol{x}_1, \boldsymbol{x}_2, \ldots, \boldsymbol{x}_n\}$ be a basis for V. Suppose A_S, A_T, and $A_{S \circ T}$ are the $n \times n$ matrices associated with the operators S, T, and $S \circ T$, respectively. Then, $A_S A_T = A_{S \circ T}$.

PROOF $A_S = [a_{ij}]_{(nn)}$ and $A_T = [b_{ij}]_{(nn)}$ where $S(\boldsymbol{x}_j) = \sum_{i=1}^{n} a_{ij}\boldsymbol{x}_i$ and $T(\boldsymbol{x}_k) = \sum_{j=1}^{n} b_{jk}\boldsymbol{x}_j$. These equalities follow from the definition of A_S and A_T. Thus,

$$(S \circ T)(\boldsymbol{x}_k) = S(T(\boldsymbol{x}_k)) = S\left(\sum_{j=1}^{n} b_{jk}\boldsymbol{x}_j\right) = \sum_{j=1}^{n} b_{jk}S(\boldsymbol{x}_j) = \sum_{j=1}^{n} b_{jk}\left(\sum_{i=1}^{n} a_{ij}\boldsymbol{x}_i\right)$$

$$= \sum_{i=1}^{n}\left(\sum_{j=1}^{n} a_{ij}b_{jk}\right)\boldsymbol{x}_i.$$

Therefore, $(S \circ T)(\boldsymbol{x}_k) = \sum_{i=1}^{n} c_{ik}\boldsymbol{x}_i$, where $c_{ik} = \sum_{j=1}^{n} a_{ij}b_{jk}$. Since by definition of $A_S A_T$, $A_S A_T = [c_{ik}]_{(nn)}$, it follows that $A_{S \circ T} = A_S A_T$.

Corollary A linear operator T on V is invertible if and only if the matrix A_T is invertible. Moreover, $(A_T)^{-1} = A_{T^{-1}}$.

PROOF Suppose that T is invertible. Then, by Theorem 2, $A_T A_{T^{-1}} = A_{TT^{-1}} = A_I = I$. Likewise, $A_{T^{-1}} A_T = I$. Thus, A_T is invertible, and $(A_T)^{-1} = A_{T^{-1}}$.

Suppose that A_T is invertible. Since the correspondence $T \to A_T$ is onto, there is some operator S such that $A_S = (A_T)^{-1}$. Then, $A_{S \circ T} = A_S A_T = (A_T)^{-1} A_T = I$. Since the correspondence $T \to A_T$ is one-one, and since $A_I = I$, it follows that $S \circ T = I$. Similarly, $T \circ S = I$.

In example 8 of §5.4, we showed that certain linear differential equations always admit polynomial solutions. Now we show how to find such solutions.

Example 3 Let P_2 denote the space of all polynomials in a variable x of degree less than or equal to 2, with real coefficients. Let $\mathfrak{D}$ denote the collection of linear operators on P_2 of the form

$$T(f) = a_0 f + (b_0 + b_1 x)f' + \left(c_0 + c_1 x + c_2 x^2\right)f''$$

$$= a_0 I(f) + (b_0 + b_1 x)D(f) + \left(c_0 + c_1 x + c_2 x^2\right)D^2(f)$$

where a_0, b_0, b_1, c_0, c_1, and c_2 are real numbers.

Clearly any T of this form is a linear operator on P_2. Since the sum of two operators in $\mathfrak{D}$ is again an operator in $\mathfrak{D}$ and a scalar multiple of an operator in $\mathfrak{D}$ again belongs to $\mathfrak{D}$, $\mathfrak{D}$ is a subspace of $L(P_2, P_2)$.

We calculate the 3×3 matrix associated with

$$T = a_0 I + (b_0 + b_1 x)D + \left(c_0 + c_1 x + c_2 x^2\right)D$$

relative to the basis 1, x, x^2.

Since

$$T(1) = a_0$$

$$T(x) = b_0 + (a_0 + b_1)x$$

$$T(x^2) = 2c_0 + (2c_1 + 2b_0)x + (2c_2 + 2b_1 + a_0)x^2$$

$$L(T) = A_T = \begin{bmatrix} a_0 & b_0 & 2c_0 \\ 0 & a_0 + b_1 & 2b_0 + 2c_1 \\ 0 & 0 & a_0 + 2b_1 + 2c_2 \end{bmatrix}$$

where L denotes the linear transformation from $L(P_2, P_2)$ to M_{33}, $L(T) = A_T$. If T belongs to $\mathfrak{D}$, note that A_T is an upper triangular matrix, i.e., all entries below the diagonal are 0. Denote the subspace of M_{33} consisting of the upper triangular matices by $\mathfrak{T}$. Thus, L induces a linear transformation from $\mathfrak{D}$ to $\mathfrak{T}$. We claim $L : \mathfrak{D} \to \mathfrak{T}$ is an isomorphism. Indeed, if we define $M : \mathfrak{T} \to \mathfrak{D}$ by

$$M\left(\begin{bmatrix} \alpha_0 & \beta_0 & \gamma_0 \\ 0 & \beta_1 & \gamma_1 \\ 0 & 0 & \gamma_2 \end{bmatrix}\right) = \alpha_0 I + (\beta_0 + (\beta_1 - \alpha_0)x)D$$

$$+ \left(\tfrac{1}{2}\gamma_0 + (\tfrac{1}{2}\gamma_1 - \beta_0)x + (\tfrac{1}{2}\gamma_2 + \tfrac{1}{2}\alpha_0 - \beta_1)x^2\right)D^2$$

an easy calculation shows that

$$M \circ L = I_{\mathfrak{D}}, \quad L \circ M = I_{\mathfrak{T}}, \quad \text{or } M = L^{-1}$$

It is clear that the subspace $\mathfrak{T}$ is closed under products, i.e., if A and B belong to $\mathfrak{T}$, AB belongs to $\mathfrak{T}$. Moreover, if $\alpha_0 \beta_1 \gamma_2 \neq 0$,

$$\begin{bmatrix} \alpha_0 & \beta_0 & \gamma_0 \\ 0 & \beta_1 & \gamma_1 \\ 0 & 0 & \gamma_2 \end{bmatrix}^{-1} = \frac{1}{\alpha_0 \beta_1 \gamma_2} \begin{bmatrix} \beta_1 \gamma_2 & -\beta_0 \gamma_2 & \beta_0 \gamma_1 - \beta_1 \gamma_0 \\ 0 & \alpha_0 \gamma_2 & -\alpha_0 \gamma_1 \\ 0 & 0 & \alpha_0 \beta_1 \end{bmatrix}$$

From this it follows that if A in $\mathfrak{T}$ is invertible, A^{-1} also belongs to $\mathfrak{T}$. Using these two facts along with $A_{S \circ T} = A_S A_T$, it follows that

(1) $\mathfrak{D}$ is closed under composition, i.e., if S and T belong to $\mathfrak{D}$, $S \circ T$ belongs to $\mathfrak{D}$.
(2) If S belongs to $\mathfrak{D}$ and S is invertible, then S^{-1} belongs to $\mathfrak{D}$.

Using the above formulas, we may explicitly calculate the inverses of operators in $\mathfrak{D}$. Note that T is invertible if and only if $\det A_T = a_0(a_0 + b_1)(a_0 + 2b_1 + 2c_2) \neq 0$.

For example, consider $T(f) = f + f' + (1 - x^2)f''$.

$$A_T = \begin{bmatrix} 1 & 1 & 2 \\ 0 & 1 & 2 \\ 0 & 0 & -1 \end{bmatrix} \quad \text{and} \quad (A_T)^{-1} = \begin{bmatrix} 1 & -1 & 0 \\ 0 & 1 & 2 \\ 0 & 0 & -1 \end{bmatrix}$$

Therefore,

$$T^{-1} = M(A_{T^{-1}}) = M\left((A_T)^{-1}\right)$$
$$= I - D + (2x - x^2)D^2$$

or

$$T^{-1}(f) = f - f' + (2x - x^2)f''$$

One natural question arises concerning the matrix representation of a linear operator on a vector space V. What is the relationship between the matrix representation of a linear operator with respect to one basis and its matrix representation with respect to another basis? The answer is contained in the next theorem.

Theorem 3 Suppose $T: V \to V$ is a linear operator on a vector space V and $\mathfrak{B} = \{\boldsymbol{x}_1, \boldsymbol{x}_2, \ldots, \boldsymbol{x}_n\}$ and $\mathfrak{B}' = \{\boldsymbol{x}_1', \boldsymbol{x}_2', \ldots, \boldsymbol{x}_n'\}$ are two bases for V. Let Q be the matrix associated with the change of coordinates from the basis $\mathfrak{B}$ to the basis $\mathfrak{B}'$, i.e., $Q = [q_{ij}]_{(nn)}$, where

$$\boldsymbol{x}_j' = \sum_{i=1}^{n} q_{ij}\boldsymbol{x}_i$$

Let A_T be the matrix associated with T relative to the basis $\mathfrak{B}$, and A_T' be the matrix associated with T relative to the basis $\mathfrak{B}'$, then

$$Q^{-1}A_TQ = A_T'$$

PROOF Let $\boldsymbol{x}$ be a vector in V. Suppose

$$\begin{bmatrix} \alpha_1 \\ \alpha_2 \\ \vdots \\ \alpha_n \end{bmatrix} \underset{\mathfrak{B}}{\leftrightarrow} \boldsymbol{x} \underset{\mathfrak{B}'}{\leftrightarrow} \begin{bmatrix} \alpha_1' \\ \alpha_2' \\ \vdots \\ \alpha_n' \end{bmatrix}$$

are the coordinate n-tuples of the vector $\boldsymbol{x}$ in V with respect to the basis $\mathfrak{B}$

and $\mathcal{B}'$, respectively. We know, by definition of the matrices A_T and A'_T,

$$A_T\begin{bmatrix}\alpha_1\\ \alpha_2\\ \vdots\\ \alpha_n\end{bmatrix}\underset{\mathcal{B}}{\leftrightarrow}T(x)\underset{\mathcal{B}'}{\leftrightarrow}A'_T\begin{bmatrix}\alpha'_1\\ \alpha'_2\\ \vdots\\ \alpha'_n\end{bmatrix}$$

By §4.9, we also have

$$\begin{bmatrix}\alpha_1\\ \alpha_2\\ \vdots\\ \alpha_n\end{bmatrix}=Q\begin{bmatrix}\alpha'_1\\ \alpha'_2\\ \vdots\\ \alpha'_n\end{bmatrix}\qquad A_T\begin{bmatrix}\alpha_1\\ \alpha_2\\ \vdots\\ \alpha_n\end{bmatrix}=QA'_T\begin{bmatrix}\alpha'_1\\ \alpha'_2\\ \vdots\\ \alpha'_n\end{bmatrix}$$

Thus,

$$A_TQ\begin{bmatrix}\alpha'_1\\ \alpha'_2\\ \vdots\\ \alpha'_n\end{bmatrix}=QA'_T\begin{bmatrix}\alpha'_1\\ \alpha'_2\\ \vdots\\ \alpha'_n\end{bmatrix}$$

Since this holds for all n-tuples of coordinates with respect to $\mathcal{B}'$ we have by the lemma of §3.7

$$A_TQ = QA'_T$$

or

$$Q^{-1}A_TQ = A'_T \ ■$$

Two $n \times n$ matrices A and B are said to be **similar** if there is some invertible matrix Q, such that $A = QBQ^{-1}$. The foregoing theorem may be restated in the following manner: Two matrices that represent the same linear operator with respect to two (possibly different) bases are similar. Alternatively, two matrices are similar if and only if they represent the same linear operator with respect to two (possibly different) bases.

If A and B are similar matrices, it follows that $\det A = \det B$. Thus, we may define the determinant of a linear operator to be the determinant of its matrix representation with respect to some basis. By the previous remarks we see that the definition does not depend on the basis chosen.

Many problems involving matrices and linear operators become quite manageable if one makes a judicious choice of basis.

Example 4 In example 2 of §2.4, we considered a two-step operation on three beakers of water. In the first step, the water in the first two beakers is leveled, and in the second the water in the last two is leveled. Let v be the vector of water levels and $T(v)$ the vector that describes the water levels after the operation is performed. Then we saw that $T(v) = \begin{bmatrix} \frac{1}{2} & \frac{1}{2} & 0 \\ \frac{1}{4} & \frac{1}{4} & \frac{1}{2} \\ \frac{1}{4} & \frac{1}{4} & \frac{1}{2} \end{bmatrix} \begin{bmatrix} x_1 \\ x_2 \\ x_3 \end{bmatrix}$.

We call the matrix that occurs here B.

As in example 6 of §5.7, we wish to inquire what will happen if this operation is performed a large number of times. What we shall see is that the water levels in the three beakers rapidly become the same.

In this case, the problem is to compute B^n for large values of n. We will do this by computing the matrix representation of T relative to a suitable basis. In later chapters, we develop methods for choosing the basis. At this point, we simply write it down and use it to simplify computations.

The matrix B above represents T with respect to the standard basis $\boldsymbol{x}_1 = \boldsymbol{e}_1$, $\boldsymbol{x}_2 = \boldsymbol{e}_2$, $\boldsymbol{x}_3 = \boldsymbol{e}_3$. We take a new basis $\boldsymbol{x}_1' = \begin{bmatrix} 1 \\ 1 \\ 1 \end{bmatrix}$, $\boldsymbol{x}_2' = \begin{bmatrix} 1 \\ -1 \\ 0 \end{bmatrix}$, $\boldsymbol{x}_3' = \begin{bmatrix} -2 \\ 1 \\ 1 \end{bmatrix}$. Observe that $T(\boldsymbol{x}_1') = \boldsymbol{x}_1'$, $T(\boldsymbol{x}_2') = \boldsymbol{0}$, $T(\boldsymbol{x}_3') = \frac{1}{4}\boldsymbol{x}_3'$. Let Q be the matrix associated with the change of coordinates from the first to the second basis. Then $Q = \begin{bmatrix} 1 & 1 & -2 \\ 1 & -1 & 1 \\ 1 & 0 & 1 \end{bmatrix}$ and $Q^{-1} = \frac{1}{3}\begin{bmatrix} 1 & 1 & 1 \\ 0 & -3 & 3 \\ -1 & -1 & 2 \end{bmatrix}$.

By Theorem 3, $Q^{-1}BQ = \begin{bmatrix} 1 & 0 & 0 \\ 0 & 0 & 0 \\ 0 & 0 & \frac{1}{4} \end{bmatrix} = D$, since this matrix D represents T with respect to the basis $\boldsymbol{x}_1'$, $\boldsymbol{x}_2'$, $\boldsymbol{x}_3'$.

Now, D^n is easily computed. We use this fact to compute B^n. We have $Q^{-1}B^nQ = (Q^{-1}BQ)^n = D^n$. Thus,

$$B^n = QD^nQ^{-1} = Q\begin{bmatrix} 1 & 0 & 0 \\ 0 & 0 & 0 \\ 0 & 0 & \frac{1}{4^n} \end{bmatrix}Q^{-1} = \frac{1}{3}\begin{bmatrix} 1 & 1 & 1 \\ 1 & 1 & 1 \\ 1 & 1 & 1 \end{bmatrix} + \frac{1}{3 \cdot 4^n}\begin{bmatrix} 2 & 2 & -4 \\ -1 & -1 & 2 \\ -1 & -1 & 2 \end{bmatrix}$$

Thus, $T^n(v) = B^n v = \dfrac{1}{3}\begin{bmatrix} x_1 + x_2 + x_3 \\ x_1 + x_2 + x_3 \\ x_1 + x_2 + x_3 \end{bmatrix} + \dfrac{1}{3 \cdot 4^n}\begin{bmatrix} 2x_1 + 2x_2 - 4x_3 \\ -x_1 - x_2 + 2x_3 \\ -x_1 - x_2 + 2x_3 \end{bmatrix}$.

Now, we interpret this expression. Observe that in the formula for $T^n(v)$ as n becomes large, the quantity $\dfrac{1}{4^n}$ becomes small very rapidly, while all other quantities remain unchanged. This means that the system approaches a state in which a third of the water is in each of the beakers. For example, suppose that three quarts of water are placed in the first beaker, and no water in the other two. Suppose also that the whole operation is repeated six times.

Then in each beaker there will be approximately one quart of water. The exact amount of water in each beaker will differ from one quart by at most a teaspoon.

As in example 4 of §5.1 and example 6 of §5.7, this is an instance of a Markov process.

Example 5 Determine the function $\boldsymbol{R}_\theta$ from the plane into itself that reflects each point about the line l that makes an angle of θ degrees with the x axis.

Interpreting vector addition as given by the parallelogram law and scalar multiplication as signed magnification of length, $\boldsymbol{R}_\theta$ is easily seen to be linear.

Let $\boldsymbol{i}_\theta = (\cos\theta)\boldsymbol{i} + (\sin\theta)\boldsymbol{j}$ be the unit vector in the direction of the line l. Let $\boldsymbol{j}_\theta = -(\sin\theta)\boldsymbol{i} + (\cos\theta)\boldsymbol{j}$. (See Figure 5-13.) Of course, the vectors $\boldsymbol{i}_\theta$ and $\boldsymbol{j}_\theta$ are just the vectors obtained by rotating the standard basis vectors $\boldsymbol{i}$ and $\boldsymbol{j}$ by θ degrees. We calculate the matrix representation of the linear transformation R_θ in the $\boldsymbol{i}_\theta, \boldsymbol{j}_\theta$ coordinate system.

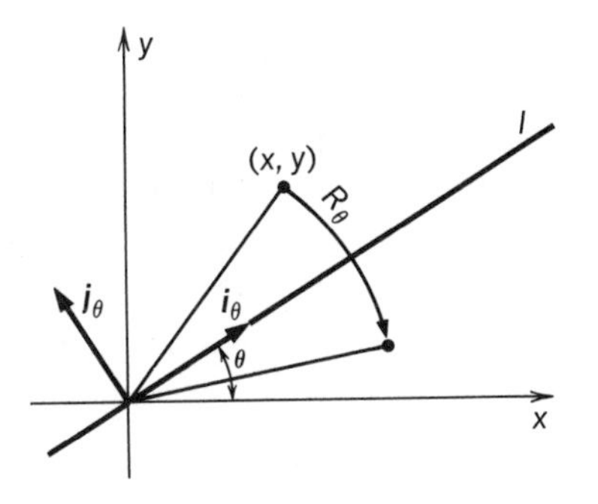

FIGURE 5-13

Clearly, we must have $R_\theta(\boldsymbol{i}_\theta) = \boldsymbol{i}_\theta$ and $R_\theta(\boldsymbol{j}_\theta) = -\boldsymbol{j}_\theta$. Thus, relative to the $\boldsymbol{i}_\theta, \boldsymbol{j}_\theta$ system, the matrix representation of R_θ is $\begin{bmatrix} 1 & 0 \\ 0 & -1 \end{bmatrix}$. Let B_θ be the matrix representation of R relative to the basis $\{\boldsymbol{i}, \boldsymbol{j}\}$. Let $Q = \begin{bmatrix} \cos\theta & -\sin\theta \\ \sin\theta & \cos\theta \end{bmatrix}$ be the matrix associated with the change of basis from the basis $\{\boldsymbol{i}, \boldsymbol{j}\}$ to the basis $\{\boldsymbol{i}_\theta, \boldsymbol{j}_\theta\}$.

By Theorem 3 it follows that $Q^{-1}BQ = \begin{bmatrix} 1 & 0 \\ 0 & -1 \end{bmatrix}$, or

$$B = Q\begin{bmatrix} 1 & 0 \\ 0 & -1 \end{bmatrix}Q^{-1} = \begin{bmatrix} \cos\theta & -\sin\theta \\ \sin\theta & \cos\theta \end{bmatrix}\begin{bmatrix} 1 & 0 \\ 0 & -1 \end{bmatrix}\begin{bmatrix} \cos\theta & \sin\theta \\ -\sin\theta & \cos\theta \end{bmatrix}$$

$$= \begin{bmatrix} \cos\theta & \sin\theta \\ \sin\theta & -\cos\theta \end{bmatrix}\begin{bmatrix} \cos\theta & \sin\theta \\ -\sin\theta & \cos\theta \end{bmatrix} = \begin{bmatrix} \cos^2\theta - \sin^2\theta & 2\sin\theta\cos\theta \\ 2\sin\theta\cos\theta & \sin^2\theta - \cos^2\theta \end{bmatrix}$$

$$= \begin{bmatrix} \cos 2\theta & \sin 2\theta \\ \sin 2\theta & -\cos 2\theta \end{bmatrix}$$

Thus, $R_\theta\left(\begin{bmatrix} x \\ y \end{bmatrix}\right) = \begin{bmatrix} \cos 2\theta & \sin 2\theta \\ \sin 2\theta & -\cos 2\theta \end{bmatrix}\begin{bmatrix} x \\ y \end{bmatrix}$.

A similar procedure may be used to derive the formula for the projection operator in §5.1.

EXERCISES

1. Let S and T be linear operators on $\boldsymbol{R}^3$ defined by

$$S(v) = \begin{bmatrix} 1 & 0 & 0 \\ 0 & 0 & 1 \\ 0 & 1 & 0 \end{bmatrix}\begin{bmatrix} x \\ y \\ z \end{bmatrix} \quad \text{and} \quad T(v) = \begin{bmatrix} 0 & 1 & 1 \\ 1 & 0 & 1 \\ 1 & 1 & 0 \end{bmatrix}\begin{bmatrix} x \\ y \\ z \end{bmatrix}.$$

Find the matrices associated with S and T relative to the following bases of $\boldsymbol{R}^3$.

(a) $\begin{bmatrix} 1 \\ 1 \\ 1 \end{bmatrix}, \begin{bmatrix} 0 \\ 1 \\ 1 \end{bmatrix}, \begin{bmatrix} 0 \\ 1 \\ -1 \end{bmatrix}$ (b) $\begin{bmatrix} 1 \\ 1 \\ 1 \end{bmatrix}, \begin{bmatrix} 1 \\ -1 \\ 0 \end{bmatrix}, \begin{bmatrix} 1 \\ 0 \\ -1 \end{bmatrix}$

2. Let T be the operator on M_{22} defined by $T(A) = A^T$. Find the matrix representation of T relative to the following bases for M_{22}.

(a) $\begin{bmatrix} 1 & 0 \\ 0 & 0 \end{bmatrix}, \begin{bmatrix} 0 & 1 \\ 0 & 0 \end{bmatrix}, \begin{bmatrix} 0 & 0 \\ 1 & 0 \end{bmatrix}, \begin{bmatrix} 0 & 0 \\ 0 & 1 \end{bmatrix}$

(b) $\begin{bmatrix} 1 & 0 \\ 0 & 0 \end{bmatrix}, \begin{bmatrix} 0 & 0 \\ 0 & 1 \end{bmatrix}, \begin{bmatrix} 0 & 1 \\ 1 & 0 \end{bmatrix}, \begin{bmatrix} 0 & 1 \\ -1 & 0 \end{bmatrix}$

3. Find the matrix representations of the following linear operators on P_3 relative to the basis $1, x, x^2, x^3$.
(a) $T(f) = f + f'$ (b) $T(f) = xf'$ (c) $T(f) = f - f(0)$
(d) $(T(f))(x) = f(x+1)$ (e) $T(f) = f(0) + f(1)x + f(2)x^2 + f(3)x^3$

4. Find the matrix representation relative to the basis $\{\boldsymbol{i}, \boldsymbol{j}\}$ for $\boldsymbol{R}^2$ of the operator T_θ where $T_\theta(\boldsymbol{i}_\theta) = \boldsymbol{i}_\theta$ and $T_\theta(\boldsymbol{j}_\theta) = 2\boldsymbol{j}_\theta$, with $\boldsymbol{i}_\theta = \cos\theta\boldsymbol{i} + \sin\theta\boldsymbol{j}$ and $\boldsymbol{j}_\theta = -\sin\theta\boldsymbol{i} + \cos\theta\boldsymbol{j}$.

5. Determine the matrix representation of the operator $T(f) = f - xf'$ on P_n relative to the basis $1, x, x^2, \ldots, x^n$.

6. Let

$$\begin{bmatrix} a \\ b \end{bmatrix}, \quad \begin{bmatrix} c \\ d \end{bmatrix}$$

be linearly independent vectors in $\boldsymbol{R}^2$. Let T be the linear transformation of $\boldsymbol{R}^2$ such that

$$T\left(\begin{bmatrix} a \\ b \end{bmatrix}\right) = \begin{bmatrix} c \\ d \end{bmatrix}, \quad T\left(\begin{bmatrix} c \\ d \end{bmatrix}\right) = \begin{bmatrix} a \\ b \end{bmatrix}$$

Show that

$$T\left(\begin{bmatrix} x \\ y \end{bmatrix}\right) = \frac{1}{ad - bc}\begin{bmatrix} cd - ab & a^2 - c^2 \\ d^2 - b^2 & ab - cd \end{bmatrix}\begin{bmatrix} x \\ y \end{bmatrix}$$

7. Show that all matrices of the form

$$\begin{bmatrix} \cos\theta & \sin\theta \\ \sin\theta & -\cos\theta \end{bmatrix}$$

are similar.

8. Let T_α be the linear operator on P_n defined by $T_\alpha(f(x)) = f(x+\alpha)$. Determine the matrix representation of T_α relative to the basis $1, x, x^2, \ldots, x^n$ for P_n. By showing that $(T_\alpha)^{-1} = T_{-\alpha}$, calculate the inverse of this matrix.

9. Let V be a vector space and $\boldsymbol{x}_1, \boldsymbol{x}_2, \boldsymbol{x}_3, \ldots, \boldsymbol{x}_n$ be a basis for V. Calculate the matrix representation of the following linear operators on V relative to the $\boldsymbol{x}_1, \boldsymbol{x}_2, \ldots, \boldsymbol{x}_n$ basis.
   (a) $T(\boldsymbol{x}_1) = \boldsymbol{x}_2,\ T(\boldsymbol{x}_2) = \boldsymbol{x}_3, \ldots, T(\boldsymbol{x}_{n-1}) = \boldsymbol{x}_n,\ T(\boldsymbol{x}_n) = \boldsymbol{0}$
   (b) $T(\boldsymbol{x}_1) = \boldsymbol{x}_2,\ T(\boldsymbol{x}_2) = \boldsymbol{x}_3, \ldots, T(\boldsymbol{x}_{n-1}) = \boldsymbol{x}_n,\ T(\boldsymbol{x}_n) = \alpha\boldsymbol{x}_1$
   (c) $T(\boldsymbol{x}_1) = \boldsymbol{x}_1,\ T(\boldsymbol{x}_2) = \boldsymbol{x}_1 + \boldsymbol{x}_2, \ldots, T(\boldsymbol{x}_{n-1}) = \boldsymbol{x}_1 + \boldsymbol{x}_2 + \cdots + \boldsymbol{x}_{n-1}$, $T(\boldsymbol{x}_n) = \boldsymbol{x}_1 + \boldsymbol{x}_2 + \cdots + \boldsymbol{x}_n$
   (d) $T(\boldsymbol{x}_1) = \lambda_1\boldsymbol{x}_1,\ T(\boldsymbol{x}_2) = \lambda_2\boldsymbol{x}_2, \ldots, T(\boldsymbol{x}_n) = \lambda_n\boldsymbol{x}_n$

10. In $\boldsymbol{R}^2$, let

$$\boldsymbol{i}_\theta = (\cos\theta)\boldsymbol{i} + (\sin\theta)\boldsymbol{j} \qquad \boldsymbol{j}_\theta = -(\sin\theta)\boldsymbol{i} + (\cos\theta)\boldsymbol{j}$$

   Let T_θ be the operator that carries $\boldsymbol{i}_\theta$ into $\boldsymbol{j}_\theta$ and $\boldsymbol{j}_\theta$ into $\boldsymbol{0}$.
   (a) Calculate the matrix representation of T relative to the basis $\{\boldsymbol{i}, \boldsymbol{j}\}$ for $\boldsymbol{R}^2$.
   (b) Show that $T^2 = 0$.

11. If A is a 2×2 nilpotent matrix with real entries, show that

$$A = \pm\begin{bmatrix} xy & -x^2 \\ y^2 & -xy \end{bmatrix}$$

   for some real numbers x and y. Show that any matrix of this sort is nilpotent.

12. Let S and T be linear operators on P_3 defined by $(S(f))(x) = f(x+1)$ and $T(f) = f + f' + \frac{1}{2}f'' + \frac{1}{6}f'''$. Show that $S = T$.

13. Find polynomial solutions to the following differential equations.
   (a) $x^2f'' + xf' + f = 1 + x + x^2$
   (b) $x^3f''' + f = x^3 + x^2 + 1$
   (c) $(1-x^2)f'' - 2xf' + f = x^2 + 1$

14. Find nonzero polynomial solutions to the following differential equations.
   (a) $xf' - f = 0$ (b) $(1-2x^2)f'' + xf' + 2f = 0$
   (c) $(1-x^2)f'' - 2xf' + 6f = 0$ (d) $x^3f''' + (1-x^2)f'' + xf' - 3f = 0$

15. Suppose that the $n \times n$ matrix A is similar to a diagonal matrix having diagonal entries $d_1, d_2, \ldots, d_n$. Show that $\det[xI + yA] = (x + yd_1)(x + yd_2) \cdots (x + yd_n)$.

16. Let V be a finite-dimensional vector space with basis B. Let S and T be linear operators on V and let A_S and A_T be their matrix representations relative to the basis B.
   (a) Show that S and T commute if and only if A_S and A_T commute.
   (b) Show that T is nilpotent if and only if A_T is nilpotent.
   (c) Show that T is idempotent if and only if A_T is idempotent.

17. Let T be a linear operator on $\boldsymbol{R}^2$ such that $T^2 = I$.
   (a) If $\boldsymbol{x}$ belongs to the range space of the operator $\frac{1}{2}(I+T)$, show that $T(\boldsymbol{x}) = \boldsymbol{x}$.
   (b) If $\boldsymbol{x}$ belongs to the range space of the operator $\frac{1}{2}(I-T)$, show that $T(\boldsymbol{x}) = -\boldsymbol{x}$.
   (c) If $T \neq \pm I$, show that there are nonzero vectors $\boldsymbol{x}_1$ and $\boldsymbol{x}_2$ such that $T(\boldsymbol{x}_1) = \boldsymbol{x}_1$ and $T(\boldsymbol{x}_2) = -\boldsymbol{x}_2$. Show that these vectors are linearly independent.

(d) If $A^2 = I_2$ and $A \neq \pm I_2$, where A is a 2×2 matrix, show that A is similar to the matrix

$$\begin{bmatrix} 1 & 0 \\ 0 & -1 \end{bmatrix}$$

18. As in example 4 of §5.1, three companies control the market for a certain product. In this case, however, the matrix A of loss and retention is $A = \begin{bmatrix} 0.7 & 0.3 & 0.2 \\ 0.2 & 0.6 & 0.1 \\ 0.1 & 0.1 & 0.7 \end{bmatrix}$.
If $\boldsymbol{v}$ is the state of the market, then $T(\boldsymbol{v}) = A\boldsymbol{v}$ is the state one year later.
(a) Find a nonzero vector $\boldsymbol{x}_1$ such that $T(\boldsymbol{x}_1) = \boldsymbol{x}_1$.
(b) Find a nonzero vector $\boldsymbol{x}_2$ such that $T(\boldsymbol{x}_2) = 0.4\boldsymbol{x}_2$.
(c) Find a nonzero vector $\boldsymbol{x}_3$ such that $T(\boldsymbol{x}_3) = 0.6\boldsymbol{x}_3$.
(d) Show that $\boldsymbol{x}_1$, $\boldsymbol{x}_2$, $\boldsymbol{x}_3$ is a basis, and compute the matrix representation of T relative to this basis.
(e) Show that the sequence of powers of A converges to the matrix $\frac{1}{24}\begin{bmatrix} 11 & 11 & 11 \\ 7 & 7 & 7 \\ 6 & 6 & 6 \end{bmatrix}$, and predict the long term prospects of each company.

19. Let $T : V \to V$ be a linear operator on an n-dimensional space V. If $r(T) = 1$, show that the matrix representation of T relative to any basis of V is of the form

$$\begin{bmatrix} a_1b_1 & a_2b_1 & \cdots & a_nb_1 \\ a_1b_2 & a_2b_2 & \cdots & a_nb_2 \\ & \vdots & & \\ a_1b_n & a_2b_n & \cdots & a_nb_n \end{bmatrix}$$

20. Prove that the differential equation

$$(a_0 + a_1x + a_2x^2)y'' + (b_0 + b_1x)y' + c_0y = 0$$

has a nonzero polynomial solution if and only if $n(n-1)a_2 + nb_1 + c_0 = 0$ for some nonnegative integer n.

21. If V and W are finite-dimensional vector spaces, show that $L(V, W)$ is finite-dimensional and $\dim L(V, W) = \dim V \dim W$.

22. Let A be a 2×2 matrix. Let C_A be the linear transformation of the space of 2×2 matrices into itself defined by

$$C_A(B) = AB - BA$$

(a) Show that $C_A = 0$ if and only if A is a scalar multiple of the identity.
(b) Relative to the basis

$$E_1 = \begin{bmatrix} 1 & 0 \\ 0 & 0 \end{bmatrix}, \quad E_2 = \begin{bmatrix} 0 & 0 \\ 1 & 0 \end{bmatrix}, \quad E_3 = \begin{bmatrix} 0 & 1 \\ 0 & 0 \end{bmatrix}, \quad E_4 = \begin{bmatrix} 0 & 0 \\ 0 & 1 \end{bmatrix}$$

show that the matrix associated to C_A is

$$\begin{bmatrix} 0 & c & -b & 0 \\ b & d-a & 0 & -b \\ -c & 0 & a-d & c \\ 0 & -c & b & 0 \end{bmatrix} \text{ if } A = \begin{bmatrix} a & b \\ c & d \end{bmatrix}$$

(c) If A is not a scalar multiple of the identity, show that the rank of the above matrix is 2.

(d) If A is not a scalar multiple of the identity, show that all matrices that commute with A can be written in the form $\alpha I_2 + \beta A$.

23. Show that any $n \times n$ matrix of rank r is similar to a matrix of the form

$$\begin{bmatrix} a_{11} & \cdots & a_{1r} & 0 & \cdots & 0 \\ a_{21} & \cdots & a_{2r} & 0 & \cdots & 0 \\ & & \vdots & & & \\ a_{n1} & \cdots & a_{nr} & 0 & \cdots & 0 \end{bmatrix}$$

24. Let W be a subspace of M_{nn}, the $n \times n$ matrices. If all nonzero elements of W are invertible, show that $\dim W \leqslant n$. (Hint: Take some n-vector v and consider the linear transformation carrying T in W into Tv in $\boldsymbol{R}^n$.)

products

1 INNER AND CROSS PRODUCTS IN R^3

So far in our study of vectors, we have neglected to ask if there are any circumstances in which it makes sense to speak of the product of two vectors. In this section, we correct this omission by discussing two sorts of products that are often very useful in physical applications. The first product we consider is called the **inner product**. The names **dot product** and **scalar product** are often used instead.

Let $\boldsymbol{a} = a_1\boldsymbol{i} + a_2\boldsymbol{j} + a_3\boldsymbol{k}$ and $\boldsymbol{b} = b_1\boldsymbol{i} + b_2\boldsymbol{j} + b_3\boldsymbol{k}$ be two vectors in $\boldsymbol{R}^3$. We define the **inner product** of $\boldsymbol{a}$ and $\boldsymbol{b}$, written $(\boldsymbol{a}, \boldsymbol{b})$, to be the real number $a_1b_1 + a_2b_2 + a_3b_3$. $(\boldsymbol{a}, \boldsymbol{b})$ is frequently denoted by $\boldsymbol{a}\cdot\boldsymbol{b}$, hence, the name dot product. Note that the inner product of two vectors is a scalar quantity.

If we agree to identify 1×1 matrices and scalars, the inner product may also be written in terms of matrix multiplication. Indeed, we have

$$\boldsymbol{a}^T\boldsymbol{b} = \begin{bmatrix} a_1 & a_2 & a_3 \end{bmatrix} \begin{bmatrix} b_1 \\ b_2 \\ b_3 \end{bmatrix} = a_1b_1 + a_2b_2 + a_3b_3$$

Certain properties of the inner product follow immediately from the definition. We suppose $\boldsymbol{a}$, $\boldsymbol{b}$, and $\boldsymbol{c}$ are vectors in $\boldsymbol{R}^3$, α and β are real scalars.

(1) $(\boldsymbol{a}, \boldsymbol{a}) \geqslant 0$
$(\boldsymbol{a}, \boldsymbol{a}) = 0$ if and only if $\boldsymbol{a} = \boldsymbol{0}$

(2) $(\alpha\boldsymbol{a}, \boldsymbol{b}) = \alpha(\boldsymbol{a}, \boldsymbol{b})$
$(\boldsymbol{a}, \beta\boldsymbol{b}) = \beta(\boldsymbol{a}, \boldsymbol{b})$

(3) $(\boldsymbol{a}, \boldsymbol{b} + \boldsymbol{c}) = (\boldsymbol{a}, \boldsymbol{b}) + (\boldsymbol{a}, \boldsymbol{c})$
$(\boldsymbol{a} + \boldsymbol{b}, \boldsymbol{c}) = (\boldsymbol{a}, \boldsymbol{c}) + (\boldsymbol{b}, \boldsymbol{c})$

(4) $(\boldsymbol{a}, \boldsymbol{b}) = (\boldsymbol{b}, \boldsymbol{a})$

To prove (1), observe that if $a = a_1 i + a_2 j + a_3 k$, $(a, a) = a_1^2 + a_2^2 + a_3^2$. Since a_1, a_2, and a_3 are real numbers $a_1^2 \geqslant 0$, $a_2^2 \geqslant 0$, $a_3^2 \geqslant 0$. Thus, $(a, a) \geqslant 0$. Moreover, if $a_1^2 + a_2^2 + a_3^2 = 0$, then $a_1 = a_2 = a_3 = 0$. Therefore, $a = 0$.

The proofs of the other properties of the inner product are easily obtained.

It follows from the Pythagorean theorem that the length of the vector $a = a_1 i + a_2 j + a_3 k$ is $\sqrt{a_1^2 + a_2^2 + a_3^2}$. (See Figure 6-1.) The length of the vector a is denoted by $|a|$. This quantity is often called the **norm** of a. Since $(a, a) = a_1^2 + a_2^2 + a_3^2$, it follows that $|a| = (a, a)^{1/2}$. Vectors of norm 1 are called unit vectors.

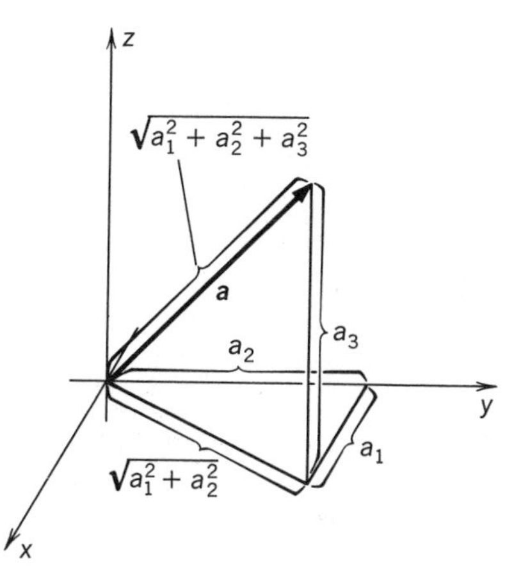

FIGURE 6-1

For example, in the plane the vector $i_\theta = (\cos \theta) i + (\sin \theta) j$ is the unit vector making an angle of θ degrees with the x axis. (See Figure 6-2.)

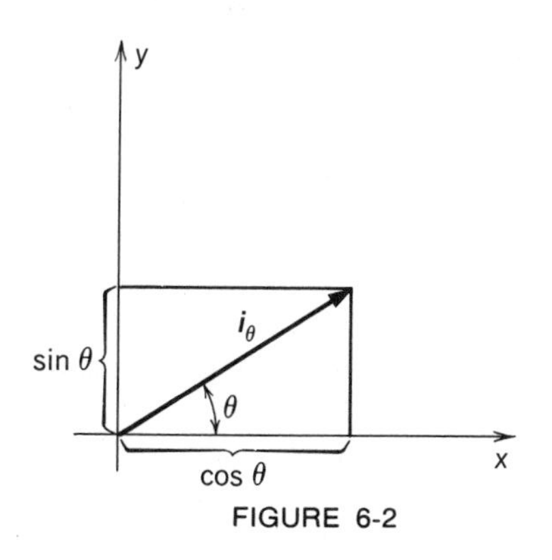

FIGURE 6-2

If a and b are vectors, we have seen that the vector $b - a$ is parallel to the directed line segment from the endpoint of a to the endpoint of b. It follows that the distance from the endpoint of a to the endpoint of b is $|b - a|$. (See Figure 6-3.) This distance is often written $d(a, b)$. For example, the distance from the endpoint of the vector i, that is, the point (1, 0, 0) to the

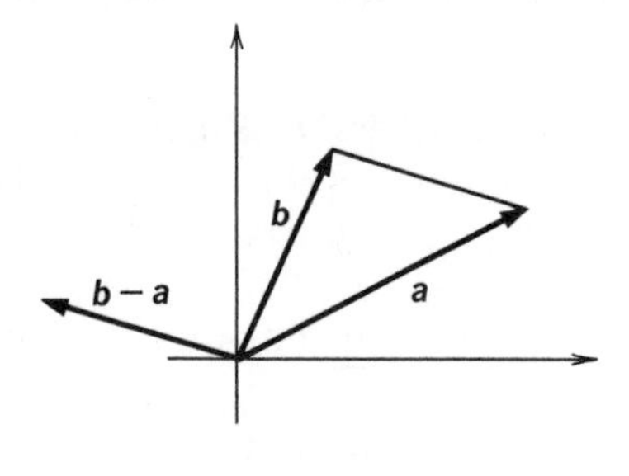

FIGURE 6-3

endpoint of the vector $\boldsymbol{j}$, (0, 1, 0), is

$$\sqrt{(1-0)^2+(0-1)^2+(0-0)^2} = \sqrt{2}$$

Based on our previous work with vectors, it is reasonable to expect that the inner product of two vectors has a geometric interpretation. Let $\boldsymbol{a}$ and $\boldsymbol{b}$ be two vectors in $\boldsymbol{R}^3$ and θ be the angle between them. (See Figure 6-4.) We will show that $(\boldsymbol{a}, \boldsymbol{b}) = |\boldsymbol{a}|\,|\boldsymbol{b}| \cos\theta$.

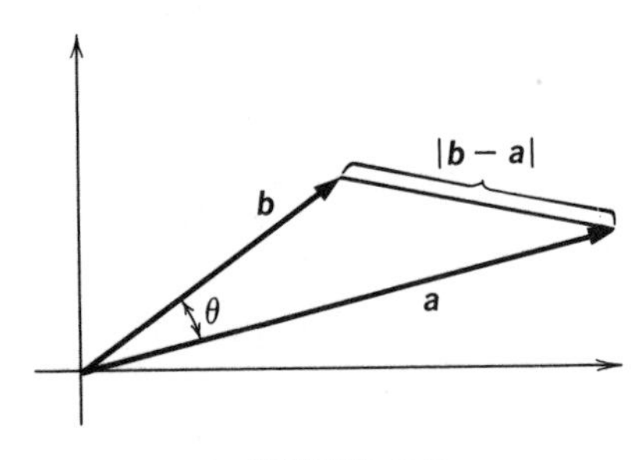

FIGURE 6-4

By the law of cosines, (§3.6), applied to the triangle with one vertex at the origin and adjacent sides determined by the vectors $\boldsymbol{a}$ and $\boldsymbol{b}$, it follows that

$$|\boldsymbol{b}-\boldsymbol{a}|^2 = |\boldsymbol{a}|^2 + |\boldsymbol{b}|^2 - 2|\boldsymbol{a}|\,|\boldsymbol{b}|\cos\theta$$

Since $|\boldsymbol{b}-\boldsymbol{a}|^2 = (\boldsymbol{b}-\boldsymbol{a}, \boldsymbol{b}-\boldsymbol{a})$, $|\boldsymbol{a}|^2 = (\boldsymbol{a}, \boldsymbol{a})$, $|\boldsymbol{b}|^2 = (\boldsymbol{b}, \boldsymbol{b})$, we can rewrite the last equation as

$$(\boldsymbol{b}-\boldsymbol{a}, \boldsymbol{b}-\boldsymbol{a}) = (\boldsymbol{a}, \boldsymbol{a}) + (\boldsymbol{b}, \boldsymbol{b}) - 2|\boldsymbol{a}|\,|\boldsymbol{b}|\cos\theta$$

Now,

$$\begin{aligned}(\boldsymbol{b}-\boldsymbol{a}, \boldsymbol{b}-\boldsymbol{a}) &= (\boldsymbol{b}, \boldsymbol{b}-\boldsymbol{a}) - (\boldsymbol{a}, \boldsymbol{b}-\boldsymbol{a}) \\ &= (\boldsymbol{b}, \boldsymbol{b}) - (\boldsymbol{b}, \boldsymbol{a}) - (\boldsymbol{a}, \boldsymbol{b}) + (\boldsymbol{a}, \boldsymbol{a}) \\ &= (\boldsymbol{a}, \boldsymbol{a}) + (\boldsymbol{b}, \boldsymbol{b}) - 2(\boldsymbol{a}, \boldsymbol{b})\end{aligned}$$

Thus, $(\boldsymbol{a}, \boldsymbol{a}) + (\boldsymbol{b}, \boldsymbol{b}) - 2(\boldsymbol{a}, \boldsymbol{b}) = (\boldsymbol{a}, \boldsymbol{a}) + (\boldsymbol{b}, \boldsymbol{b}) - 2|\boldsymbol{a}|\,|\boldsymbol{b}|\cos\theta$, or $(\boldsymbol{a}, \boldsymbol{b}) = |\boldsymbol{a}|\,|\boldsymbol{b}|\cos\theta$.

This formula means that the inner product of two vectors is the product of their lengths times the cosine of the angle between them. This relationship is often of value in problems of a geometric nature.

Example 1 Find the angle between the vectors $i + j + k$ and $i + j - k$, illustrated in Figure 6-5. By the above, $(i + j + k, i + j - k) = |i + j + k|\,|i + j - k| \cos \theta$. So $1 + 1 - 1 = (\sqrt{3})(\sqrt{3}) \cos \theta$, or $\cos \theta = \frac{1}{3}$. So $\theta = \cos^{-1}(\frac{1}{3})$.

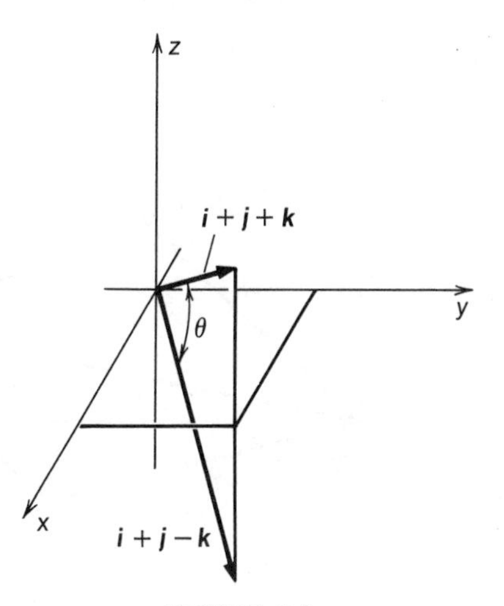

FIGURE 6-5

If a and b are nonzero vectors in R^3 and θ is the angle between them, we see that $(a, b) = 0$ if and only if $\cos \theta = 0$. From this it follows that the inner product of two nonzero vectors is zero if and only if the vectors are perpendicular. Often we say that perpendicular vectors are **orthogonal**. We adopt the convention that the zero vector is orthogonal to all vectors.

For example, the vectors $i_\theta = (\cos \theta)i + (\sin \theta)j$ and $j_\theta = -(\sin \theta)i + (\cos \theta)j$ are orthogonal since $(i_\theta, j_\theta) = -\cos \theta \sin \theta + \sin \theta \cos \theta = 0$. (See Figure 6-6.)

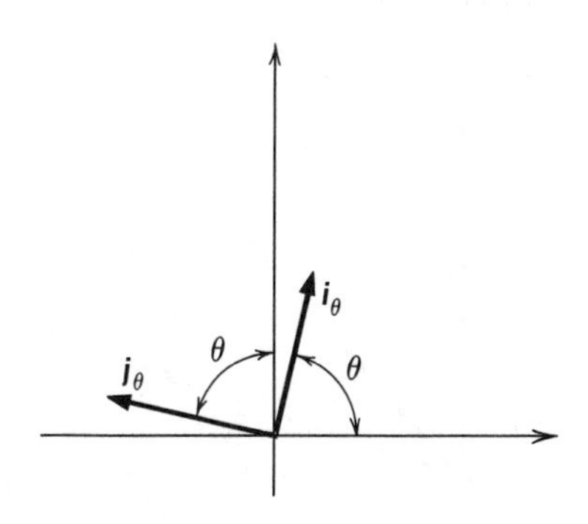

FIGURE 6-6

Example 2 Let $\boldsymbol{a}$ and $\boldsymbol{b}$ be two nonzero orthogonal vectors. Let $\boldsymbol{c}$ be another vector in the plane spanned by $\boldsymbol{a}$ and $\boldsymbol{b}$. As we have seen, there are scalars α and β such that $\boldsymbol{c} = \alpha\boldsymbol{a} + \beta\boldsymbol{b}$. We use the inner product to determine α and β. (See Figure 6-7.)

$$(\boldsymbol{a}, \boldsymbol{c}) = (\boldsymbol{a}, \alpha\boldsymbol{a} + \beta\boldsymbol{b}) = \alpha(\boldsymbol{a}, \boldsymbol{a}) + \beta(\boldsymbol{a}, \boldsymbol{b})$$

Since $\boldsymbol{a}$ and $\boldsymbol{b}$ are orthogonal, $(\boldsymbol{a}, \boldsymbol{b}) = 0$, and so $\alpha = (\boldsymbol{a}, \boldsymbol{c})/(\boldsymbol{a}, \boldsymbol{a}) = (\boldsymbol{a}, \boldsymbol{c})/|\boldsymbol{a}^2|$. Similarly, $\beta = (\boldsymbol{b}, \boldsymbol{c})/(\boldsymbol{b}, \boldsymbol{b}) = (\boldsymbol{b}, \boldsymbol{c})/|\boldsymbol{b}^2|$.

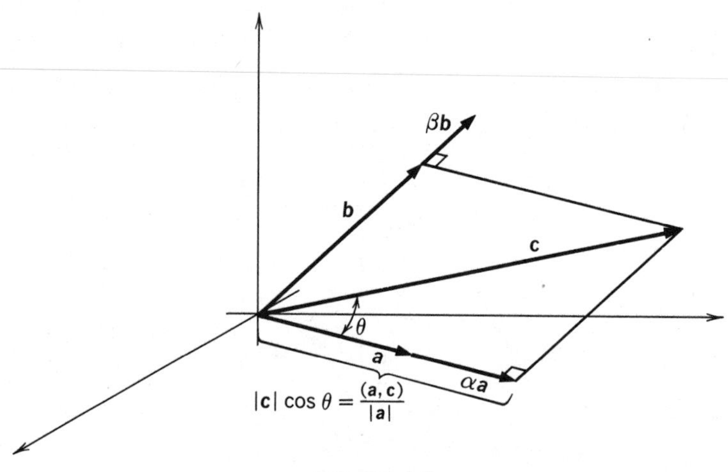

FIGURE 6-7

Example 3 Recall the projection operator P_θ of §5.2. If l is a line making an angle of θ degrees with the x axis, $P_\theta(\boldsymbol{v})$ is the vector along l, ending at the point where the perpendicular from the endpoint of $\boldsymbol{v}$ to the line l intersects l. (See Figure 6-8.) We now derive the formula for $P_\theta(\boldsymbol{v})$ using the inner product.

We know that $\boldsymbol{i}_\theta = (\cos\theta)\boldsymbol{i} + (\sin\theta)\boldsymbol{j}$ is the unit vector in the direction of l. If ψ is the angle between the vector $\boldsymbol{v}$ and the line l,

$$P_\theta(\boldsymbol{v}) = (|\boldsymbol{v}| \cos\psi)\boldsymbol{i}_\theta = (\boldsymbol{v}, \boldsymbol{i}_\theta)\boldsymbol{i}_\theta$$

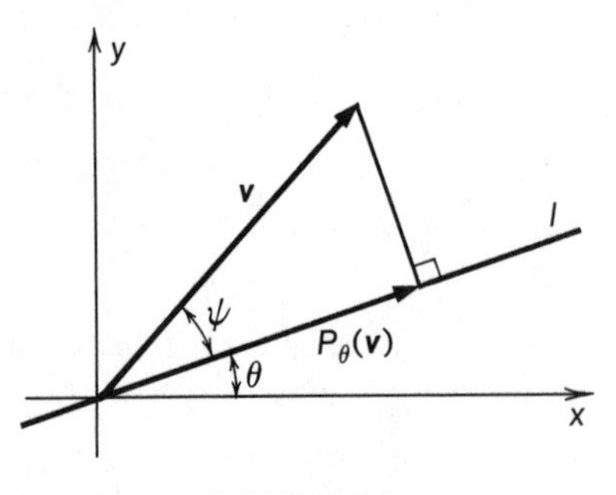

FIGURE 6-8

If $v = xi + yj$,

$$P_\theta(v) = P_\theta(xi + yj) = (xi + yj, i_\theta)i_\theta$$
$$= (x \cos\theta + y \sin\theta)((\cos\theta)i + (\sin\theta)j)$$

or

$$P_\theta\left(\begin{vmatrix} x \\ y \end{vmatrix}\right) = \begin{bmatrix} \cos^2\theta & \sin\theta\cos\theta \\ \sin\theta\cos\theta & \sin^2\theta \end{bmatrix}\begin{bmatrix} x \\ y \end{bmatrix}$$

Besides the inner product, there is another product of vectors that is sometimes useful. Let $a = a_1 i + a_2 j + a_3 k$ and $b = b_1 i + b_2 j + b_3 k$ be vectors in R^3. The **cross product** of a and b, denoted by $a \times b$, is defined to be the vector

$$a \times b = \begin{vmatrix} a_2 & a_3 \\ b_2 & b_3 \end{vmatrix} i - \begin{vmatrix} a_1 & a_3 \\ b_1 & b_3 \end{vmatrix} j + \begin{vmatrix} a_1 & a_2 \\ b_1 & b_2 \end{vmatrix} k$$

or symbolically,

$$a \times b = \begin{vmatrix} i & j & k \\ a_1 & a_2 & a_3 \\ b_1 & b_2 & b_3 \end{vmatrix}$$

Note that the cross product of two vectors is another vector. Just as the inner product is sometimes called the scalar product, the cross product is sometimes called the **vector product**.

Again, certain algebraic properties of the cross product follow immediately from the definition. Suppose a, b, and c are vectors and α, β, and γ are scalars.

(1) $a \times b = -(b \times a)$
(2) $a \times (\beta b + \gamma c) = \beta(a \times b) + \gamma(a \times c)$
$(\alpha a + \beta b) \times c = \alpha(a \times c) + \beta(b \times c)$

Note that $a \times a = -(a \times a)$ by (1). Thus, $a \times a = \mathbf{0}$.

For example,

$$(3i - j + k) \times (i + 2j - k) = \begin{vmatrix} i & j & k \\ 3 & -1 & 1 \\ 1 & 2 & -1 \end{vmatrix}$$
$$= -i + 4j + 7k$$

Our next goal is to present a geometric interpretation of the cross product analogous to that of the inner product.

First, we provide a formula for the product $\boldsymbol{a}\cdot(\boldsymbol{b}\times\boldsymbol{c})$. If $\boldsymbol{a}=a_1\boldsymbol{i}+a_2\boldsymbol{j}+a_3\boldsymbol{k}$, $\boldsymbol{b}=b_1\boldsymbol{i}+b_2\boldsymbol{j}+b_3\boldsymbol{k}$, and $\boldsymbol{c}=c_1\boldsymbol{i}+c_2\boldsymbol{j}+c_3\boldsymbol{k}$,

$$\boldsymbol{a}\cdot(\boldsymbol{b}\times\boldsymbol{c})=(a_1\boldsymbol{i}+a_2\boldsymbol{j}+a_3\boldsymbol{k})\cdot\left(\begin{vmatrix} b_2 & b_3\\ c_2 & c_3\end{vmatrix}\boldsymbol{i}-\begin{vmatrix} b_1 & b_3\\ c_1 & c_3\end{vmatrix}\boldsymbol{j}+\begin{vmatrix} b_1 & b_2\\ c_1 & c_2\end{vmatrix}\boldsymbol{k}\right)$$

$$=a_1\begin{vmatrix} b_2 & b_3\\ c_2 & c_3\end{vmatrix}-a_2\begin{vmatrix} b_1 & b_3\\ c_1 & c_3\end{vmatrix}+a_3\begin{vmatrix} b_1 & b_2\\ c_1 & c_2\end{vmatrix}$$

This can be written more concisely as

$$\boldsymbol{a}\cdot(\boldsymbol{b}\times\boldsymbol{c})=\begin{vmatrix} a_1 & a_2 & a_3\\ b_1 & b_2 & b_3\\ c_1 & c_2 & c_3\end{vmatrix}$$

Now suppose that $\boldsymbol{a}$ is a vector in the subspace of $\boldsymbol{R}^3$ spanned by the vectors $\boldsymbol{b}$ and $\boldsymbol{c}$. This means that the first row in the determinant expression for $\boldsymbol{a}\cdot(\boldsymbol{b}\times\boldsymbol{c})$ is a linear combination of the second and third rows, and therefore $\boldsymbol{a}\cdot(\boldsymbol{b}\times\boldsymbol{c})=0$. In other words, the vector $\boldsymbol{b}\times\boldsymbol{c}$ is orthogonal to any vector in the subspace spanned by $\boldsymbol{b}$ and $\boldsymbol{c}$, in particular to both $\boldsymbol{b}$ and $\boldsymbol{c}$.

Next, we calculate the magnitude of $\boldsymbol{b}\times\boldsymbol{c}$. Note that

$$|\boldsymbol{b}\times\boldsymbol{c}|^2=\begin{vmatrix} b_2 & b_2\\ c_3 & c_3\end{vmatrix}^2+\begin{vmatrix} b_1 & b_3\\ c_1 & c_3\end{vmatrix}^2+\begin{vmatrix} b_1 & b_2\\ c_1 & c_2\end{vmatrix}^2$$

$$=(b_1^2+b_2^2+b_3^2)(c_1^2+c_2^2+c_3^2)-(b_1c_1+b_2c_2+b_3c_3)^2$$

$$=|\boldsymbol{b}|^2\,|\boldsymbol{c}|^2-|\boldsymbol{b}|^2\,|\boldsymbol{c}|^2\cos^2\theta=|\boldsymbol{b}|^2\,|\boldsymbol{c}|^2\sin^2\theta$$

where θ is the angle between $\boldsymbol{b}$ and $\boldsymbol{c}$.

Combining our results, we see that $\boldsymbol{b}\times\boldsymbol{c}$ is a vector perpendicular to $\boldsymbol{b}$ and $\boldsymbol{c}$ with length $|\boldsymbol{b}|\,|\boldsymbol{c}|\,|\sin\theta|$.

If $\boldsymbol{b}$ and $\boldsymbol{c}$ are linearly dependent, $\theta=0$, and so $\boldsymbol{b}\times\boldsymbol{c}=\boldsymbol{0}$. If $\boldsymbol{b}$ and $\boldsymbol{c}$ are linearly independent, they span a plane and $\boldsymbol{b}\times\boldsymbol{c}$ is a vector perpendicular to this plane. The length of $\boldsymbol{b}\times\boldsymbol{c}$, $|\boldsymbol{b}|\,|\boldsymbol{c}|\,|\sin\theta|$, is just the area of the parallelogram with adjacent sides represented by the vectors and $\boldsymbol{b}$ and $\boldsymbol{c}$. (See Figure 6-9.)

Example 4 Find a unit vector orthogonal to the vectors $\boldsymbol{i}+\boldsymbol{j}$ and $\boldsymbol{j}+\boldsymbol{k}$.

A vector perpendicular to both $\boldsymbol{i}+\boldsymbol{j}$ and $\boldsymbol{j}+\boldsymbol{k}$ is the vector $(\boldsymbol{i}+\boldsymbol{j})\times(\boldsymbol{j}+\boldsymbol{k})=\begin{vmatrix} \boldsymbol{i} & \boldsymbol{j} & \boldsymbol{k}\\ 1 & 1 & 0\\ 0 & 1 & 1\end{vmatrix}=\boldsymbol{i}-\boldsymbol{j}+\boldsymbol{k}$. Since $|\boldsymbol{i}-\boldsymbol{j}+\boldsymbol{k}|=\sqrt{3}$, the vector $\dfrac{1}{\sqrt{3}}(\boldsymbol{i}-\boldsymbol{j}+\boldsymbol{k})$ is a unit vector perpendicular to $\boldsymbol{i}+\boldsymbol{j}$ and $\boldsymbol{j}+\boldsymbol{k}$.

Using the cross product, we may obtain an interesting geometric interpretation of determinants. Let $\boldsymbol{a}=a_1\boldsymbol{i}+a_2\boldsymbol{j}$ and $\boldsymbol{b}=b_1\boldsymbol{i}+b_2\boldsymbol{j}$ be two vectors

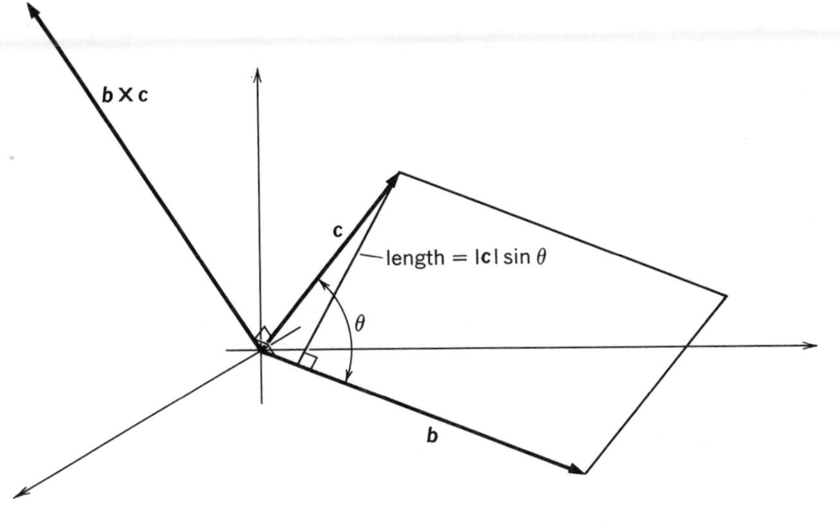

FIGURE 6-9

in the plane. If θ denotes the angle between $\boldsymbol{a}$ and $\boldsymbol{b}$, we have seen that $|\boldsymbol{a} \times \boldsymbol{b}| = |\boldsymbol{a}|\,|\boldsymbol{b}|\,|\sin\theta|$. As noted above, $|\boldsymbol{a}|\,|\boldsymbol{b}|\sin\theta$ is the area of the parallelogram with adjacent sides $\boldsymbol{a}$ and $\boldsymbol{b}$. (See Figure 6-10.) Using the definition of the cross product,

$$\boldsymbol{a} \times \boldsymbol{b} = \begin{vmatrix} \boldsymbol{i} & \boldsymbol{j} & \boldsymbol{k} \\ a_1 & a_2 & 0 \\ b_1 & b_2 & 0 \end{vmatrix} = \begin{vmatrix} a_1 & a_2 \\ b_1 & b_2 \end{vmatrix} \boldsymbol{k}$$

Thus, $|\boldsymbol{a} \times \boldsymbol{b}|$ is the absolute value of the determinant $\begin{vmatrix} a_1 & a_2 \\ b_1 & b_2 \end{vmatrix}$.

From this, it follows that the absolute value of the above determinant is the area of the parallelogram with adjacent sides $a_1\boldsymbol{i} + a_2\boldsymbol{j}$ and $b_1\boldsymbol{i} + b_2\boldsymbol{j}$.

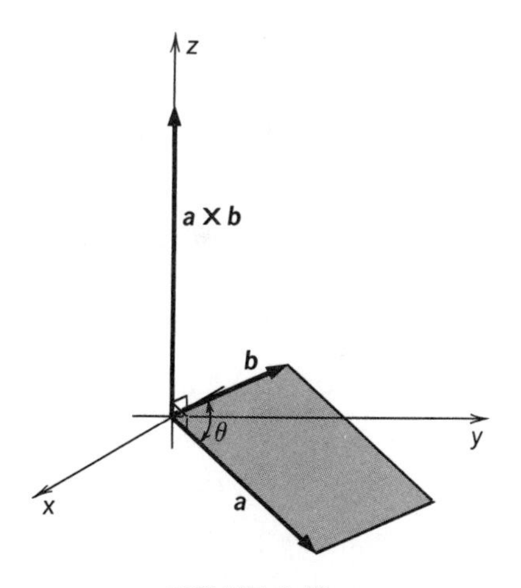

FIGURE 6-10

Example 5 Find the area of the triangle with vertices at the points (1, 1), (0, 2), and (3, 2). (See Figure 6-11.)

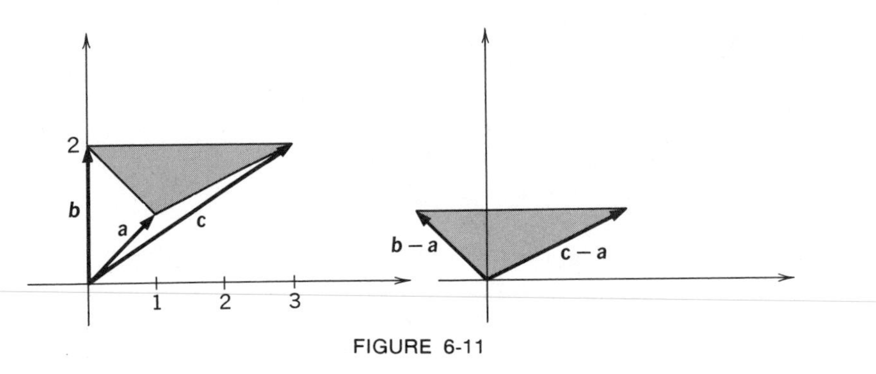

FIGURE 6-11

Let $\boldsymbol{a} = \boldsymbol{i} + \boldsymbol{j}$, $\boldsymbol{b} = 2\boldsymbol{j}$, and $\boldsymbol{c} = 3\boldsymbol{i} + 2\boldsymbol{j}$. It is clear that the triangle with vertices at the endpoints of the vectors $\boldsymbol{a}$, $\boldsymbol{b}$, and $\boldsymbol{c}$ has the same area as the triangle with vertices at $\boldsymbol{0}$, $\boldsymbol{b} - \boldsymbol{a}$, and $\boldsymbol{c} - \boldsymbol{a}$. Indeed, the latter is merely a translation of the former triangle. Since the area of this triangle is half the area of the parallelogram with adjacent sides $\boldsymbol{b} - \boldsymbol{a}$ and $\boldsymbol{c} - \boldsymbol{a}$, we find that the area of the triangle with vertices (1, 1), (0, 2), and (3, 2) is the absolute value of $\frac{1}{2}\begin{vmatrix} -1 & 1 \\ 2 & 1 \end{vmatrix} = -\frac{3}{2}$, i.e., $\frac{3}{2}$.

Analogous to the interpretation of 2×2 determinants as areas, there is an interpretation of 3×3 determinants as volumes. Let $\boldsymbol{a} = a_1\boldsymbol{i} + a_2\boldsymbol{j} + a_3\boldsymbol{k}$, $\boldsymbol{b} = b_1\boldsymbol{i} + b_2\boldsymbol{j} + b_3\boldsymbol{k}$, and $\boldsymbol{c} = c_1\boldsymbol{i} + c_2\boldsymbol{j} + c_3\boldsymbol{k}$ be vectors in $\boldsymbol{R}^3$. (See Figure 6-12.) We will show that the volume of the parallelepiped with adjacent sides $\boldsymbol{a}$, $\boldsymbol{b}$, and $\boldsymbol{c}$ is the absolute value of the determinant

$$\begin{vmatrix} a_1 & a_2 & a_3 \\ b_1 & b_2 & b_3 \\ c_1 & c_2 & c_3 \end{vmatrix}$$

As we now have seen, $|\boldsymbol{a} \times \boldsymbol{b}|$ is the area of the parallelogram with adjacent sides $\boldsymbol{a}$ and $\boldsymbol{b}$. Moreover, $(\boldsymbol{a} \times \boldsymbol{b}) \cdot \boldsymbol{c}$ is $|\boldsymbol{c}|\,|\boldsymbol{a} \times \boldsymbol{b}| \cos \psi$, where ψ is the angle that $\boldsymbol{c}$ makes with the normal to the plane spanned by $\boldsymbol{a}$ and $\boldsymbol{b}$. Since the volume of the parallelepiped with adjacent sides $\boldsymbol{a}$, $\boldsymbol{b}$, and $\boldsymbol{c}$ is the product of the area of the base $|\boldsymbol{a} \times \boldsymbol{b}|$ times the altitude $|\boldsymbol{c}| \cos \psi$, it follows that the volume is merely $|(\boldsymbol{a} \times \boldsymbol{b}) \cdot \boldsymbol{c}|$.

We saw earlier that $|(\boldsymbol{a} \times \boldsymbol{b}) \cdot \boldsymbol{c}|$ is the absolute value of the determinant

$$\begin{vmatrix} a_1 & a_2 & a_3 \\ b_1 & b_2 & b_3 \\ c_1 & c_2 & c_3 \end{vmatrix}$$

Therefore, the absolute value of the foregoing determinant is the volume of the parallelepiped with adjacent sides $\boldsymbol{a}$, $\boldsymbol{b}$, and $\boldsymbol{c}$.

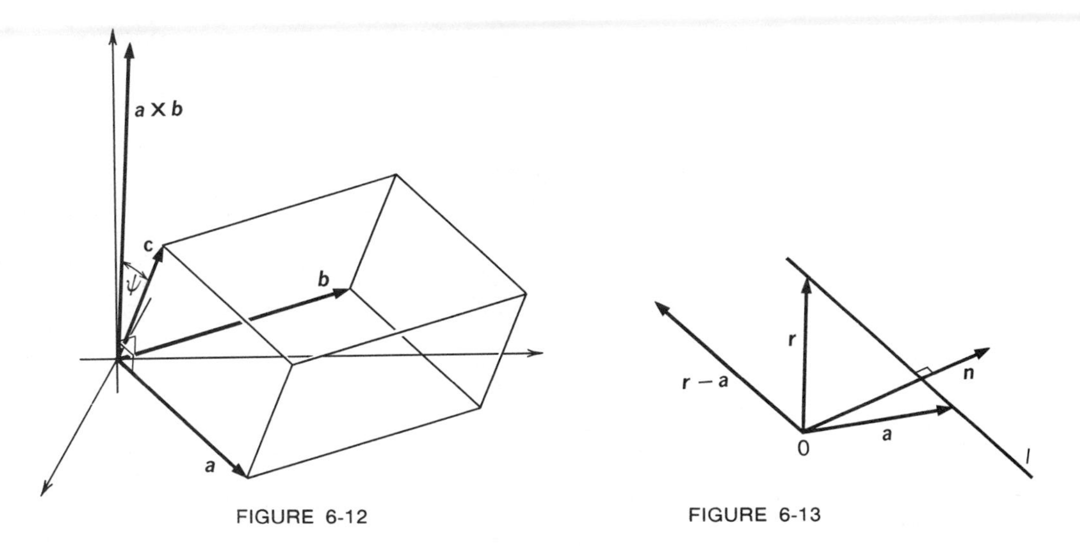

FIGURE 6-12

FIGURE 6-13

It is important to note that the cross product is neither commutative nor associative. Indeed, we saw earlier that $\boldsymbol{a} \times \boldsymbol{b} = -(\boldsymbol{b} \times \boldsymbol{a})$. To see that the cross product is not associative, consider $(\boldsymbol{i} \times \boldsymbol{j}) \times \boldsymbol{j}$ and $\boldsymbol{i} \times (\boldsymbol{j} \times \boldsymbol{j})$; now $(\boldsymbol{i} \times \boldsymbol{j}) \times \boldsymbol{j} = \boldsymbol{k} \times \boldsymbol{j} = -\boldsymbol{i}$ and $\boldsymbol{i} \times (\boldsymbol{j} \times \boldsymbol{j}) = \boldsymbol{i} \times \boldsymbol{0} = \boldsymbol{0}$. Thus, $(\boldsymbol{i} \times \boldsymbol{j}) \times \boldsymbol{j} \neq \boldsymbol{i} \times (\boldsymbol{j} \times \boldsymbol{j})$, and the cross product is not associative.

Using vector methods, we can derive the equation of a line in the plane. Let l be a line, $\boldsymbol{a}$ be a fixed vector ending somewhere on the line, and $\boldsymbol{n}$ be a nonzero vector perpendicular to the line. (See Figure 6-13.) If $\boldsymbol{r}$ is an arbitrary vector ending on the line, the vector $\boldsymbol{r} - \boldsymbol{a}$ is perpendicular to $\boldsymbol{n}$, and so $(\boldsymbol{r} - \boldsymbol{a}, \boldsymbol{n}) = 0$. If conversely, $(\boldsymbol{r} - \boldsymbol{a}, \boldsymbol{n}) = 0$, $\boldsymbol{r} - \boldsymbol{a}$ is perpendicular to $\boldsymbol{n}$, and so the vector $\boldsymbol{r} = (\boldsymbol{r} - \boldsymbol{a}) + \boldsymbol{a}$ lies on the line.

If $\boldsymbol{a} = a_1\boldsymbol{i} + a_2\boldsymbol{j}$, $\boldsymbol{n} = n_1\boldsymbol{i} + n_2\boldsymbol{j}$, and $\boldsymbol{r} = x\boldsymbol{i} + y\boldsymbol{j}$, it follows that the line consists of the locus of points (x, y) such that $(x - a_1)n_1 + (y - a_2)n_2 = 0$, or $n_1x + n_2y = a_1n_1 + a_2n_2$.

Example 6 Find a unit vector perpendicular to the line $2x + 3y = 1$.

By the preceding discussion it is clear that all vectors normal to the line $2x + 3y = 1$ are of the form $2\alpha\boldsymbol{i} + 3\alpha\boldsymbol{j}$. We choose α, so that $|2\alpha\boldsymbol{i} + 3\alpha\boldsymbol{j}| = 1$, or $\alpha = \pm 1/\sqrt{13}$. Thus, the vector $(2/\sqrt{13})\boldsymbol{i} + (3/\sqrt{13})\boldsymbol{j}$ is a unit vector normal to the given line.

In a manner similar to the above, it is possible to determine the equation of a plane in space. Let P be a plane in space, $\boldsymbol{a}$ be a vector ending on the plane, and $\boldsymbol{n}$ be a vector normal to the plane. (See Figure 6-14.)

If $\boldsymbol{r}$ is a vector in $\boldsymbol{R}^3$, $\boldsymbol{r}$ ends on the plane P if and only if $(\boldsymbol{r} - \boldsymbol{a}, \boldsymbol{n}) = 0$.

If $\boldsymbol{a} = a_1\boldsymbol{i} + a_2\boldsymbol{j} + a_3\boldsymbol{k}$, $\boldsymbol{n} = n_1\boldsymbol{i} + n_2\boldsymbol{j} + n_3\boldsymbol{k}$, and $\boldsymbol{r} = x\boldsymbol{i} + y\boldsymbol{j} + z\boldsymbol{k}$, it follows that the plane consists of the locus of points (x, y, z) such that $n_1x + n_2y + n_3z = n_1a_1 + n_2a_2 + n_3a_3$.

Thus, the locus of points in space satisfying an equation of the form $Ax + By + Cz + D = 0$, where A, B, and C are not all zero, is a plane.

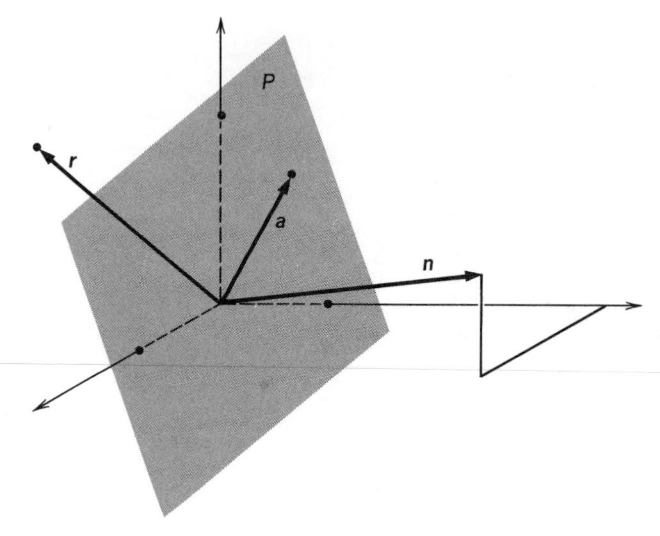

FIGURE 6-14

Example 7 Determine the equation of the plane perpendicular to the vector $2\boldsymbol{i}+\boldsymbol{j}-\boldsymbol{k}$ and containing the point (1, 2, 2).

From the above, it follows that the equation of the plane is of the form $2x+y-z=D$. Since (1, 2, 2) lies on the plane $2+2-2=D$. Thus, the equation of the plane is $2x+y-z=2$.

Example 8 Determine the equation of the plane containing the points (1, 1, 1), (2, 2, − 1), and (3, 1, − 1).

First method: The equation of the plane is of the form $Ax+By+Cz+D=0$. Since the points (1, 1, 1), (2, 2, − 1), and (3, 1, − 1) lie on the plane, we obtain the system

$$\begin{array}{l} A+B+C+D=0 \\ 2A+2B-C+D=0 \\ 3A+B-C+D=0 \end{array} \longrightarrow \begin{array}{l} -2A+2C=0 \\ -A+B=0 \\ 3A+B-C+D=0 \end{array}$$

that gives $A=B=C$ and $D=-3A$. Choose $A=1$, then $B=C=1$ and $D=-3$. Thus, the equation of the plane is $x+y+z=3$.

Second method: Let $\boldsymbol{a}=\boldsymbol{i}+\boldsymbol{j}+\boldsymbol{k}$, $\boldsymbol{b}=2\boldsymbol{i}+2\boldsymbol{j}-\boldsymbol{k}$, $\boldsymbol{c}=3\boldsymbol{i}+\boldsymbol{j}-\boldsymbol{k}$. Then, the vectors $\boldsymbol{b}-\boldsymbol{a}$ and $\boldsymbol{c}-\boldsymbol{a}$ are parallel to the plane. Thus, a normal to the plane is orthogonal to the vectors, $\boldsymbol{b}-\boldsymbol{a}$ and $\boldsymbol{c}-\boldsymbol{a}$. Thus, $\boldsymbol{n}=(\boldsymbol{b}-\boldsymbol{a})\times(\boldsymbol{c}-\boldsymbol{a})$ is normal to the plane. Since $\boldsymbol{n}=\begin{vmatrix} \boldsymbol{i} & \boldsymbol{j} & \boldsymbol{k} \\ 1 & 1 & -2 \\ 2 & 0 & -2 \end{vmatrix}=-2\boldsymbol{i}-2\boldsymbol{j}-2\boldsymbol{k}$.

Thus, the equation is of the form $-2x-2y-2z=D$. Since (1, 1, 1) lies on the plane, $D=-6$. Thus $-2x-2y-2z=-6$, which gives (on multiplication by $-\frac{1}{2}$), $x+y+z=3$.

In chapter 2, we mentioned that vectors in $\boldsymbol{R}^3$ appear as forces, velocities, and other quantities in physical problems. When a physical quantity is represented by a vector, the length of the vector gives the magnitude of the quantity. For example, if $\boldsymbol{F} = \boldsymbol{i} + 2\boldsymbol{j} + 2\boldsymbol{k}$ is a force, the magnitude of the force is 3 units in whatever measurement system is being used, for example, pounds or dynes. Similarly, $2\boldsymbol{i} + 3\boldsymbol{j}$ and $10\boldsymbol{i}$ represent forces of 5 and 10 units, respectively.

It is also possible to reverse this procedure. For example, find a force of 10 units in the direction of the vector $\boldsymbol{i} + \boldsymbol{j}$. First, a unit vector in that direction is $(\boldsymbol{i} + \boldsymbol{j})/\sqrt{2}$. Thus, $\boldsymbol{F} = 10(\boldsymbol{i} + \boldsymbol{j})/\sqrt{2}$ represents the force, since $\boldsymbol{F}$ has the right direction and magnitude.

When two forces act on the same point, the resultant force is just the vector sum of the forces. The same is true of velocities. For example, if a person walks across a moving train, his velocity vector with respect to the ground is the vector sum of the velocity vector of the train and his velocity vector with respect to the train. The following example illustrates this in another way.

Example 9 A tugboat captain wishes to proceed due east with all possible speed in a tugboat that is capable of 17 knots. There is a current of 8 knots heading due north.

What direction should he steer the boat to insure the easterly course, and what will his velocity be?

The velocity of the current is $8\boldsymbol{j}$. Let $\boldsymbol{v}$ be a vector of magnitude 17 in the direction that the boat should be headed. We wish to find $\boldsymbol{v}$ to insure the easterly course. The vector $\boldsymbol{v}$ is just the velocity of the boat with respect to the water. The true velocity of the boat is the sum of the velocity of the boat with respect to the water and the velocity of the water. Thus, the true velocity of the boat is $8\boldsymbol{j} + \boldsymbol{v}$.

Since the boat must head in an easterly course, $\boldsymbol{v}$ must be chosen so that $8\boldsymbol{j} + \boldsymbol{v}$ is some multiple $k\boldsymbol{i}$, of $\boldsymbol{i}$, the east vector. Thus, $8\boldsymbol{j} + \boldsymbol{v} = k\boldsymbol{i}$, or $\boldsymbol{v} = k\boldsymbol{i} - 8\boldsymbol{j}$. Since the length of $\boldsymbol{v}$ is 17, $17^2 = k^2 + 64$. And so $k^2 = 289 - 64 = 225$. Hence, $k = 15$. Thus, $\boldsymbol{v} = 15\boldsymbol{i} - 8\boldsymbol{j}$. If the boat is steered in this direction, its true velocity will be 15 knots due east.

EXERCISES

1. Find the angles between the vectors $\boldsymbol{i} + \boldsymbol{j}$, $\boldsymbol{j} + \boldsymbol{k}$, and $\boldsymbol{i} + \boldsymbol{j} + \boldsymbol{k}$.
2. Resolve the vector $\boldsymbol{i} + \boldsymbol{j} + \boldsymbol{k}$ into components that are parallel or perpendicular to $\boldsymbol{j} - \boldsymbol{k}$.
3. Find a unit vector in the direction of the vectors

   (a) $2\boldsymbol{i} + 3\boldsymbol{j}$ (b) $\boldsymbol{i} + 2\boldsymbol{j} + 2\boldsymbol{k}$ (c) $6\boldsymbol{i} + 3\boldsymbol{j} + 2\boldsymbol{k}$
4. Determine the equation of the line perpendicular to $x + y = 1$ and passing through $(2, -1)$.
5. Find the cosines of the interior angles of the triangle with vertices $(1, 0)$, $(0, 1)$, $(3, 2)$.

6. Find the equation of the plane perpendicular to the vector $2\boldsymbol{i} + 3\boldsymbol{j} - \boldsymbol{k}$ that contains the point (1, 1, 1).

7. Determine the angle between the planes defined by the equations $x + y + z = 1$ and $x + y - z = 3$.

8. Find the length of the diagonal of a cube whose side is of length s.

9. Find the area of a parallelogram having sides $\boldsymbol{i} + \boldsymbol{j}$ and $2\boldsymbol{i} + 3\boldsymbol{j}$.

10. A four-sided figure in the plane has successive vertices (1, 2), (5, 4), (6, 3), (2, 1). Show that the figure is a parallelogram and compute its area.

11. Find the equation of the plane containing the points (1, 2, 0), (1, 1, 1), and (3, − 1, 1).

12. Let A be a 2×2 real matrix with $\det A \neq 0$. Let T be the linear operator on $\boldsymbol{R}^2$ defined by $T(\boldsymbol{v}) = A\boldsymbol{v}$. Let P be a parallelogram having one vertex on the origin and area S. Show that $T(P)$ is a parallelogram having area $|\det A|S$.

13. If (x_1, y_1), (x_2, y_2), (x_3, y_3) are the vertices of a triangle, show that the area of the triangle is the absolute value of $\frac{1}{2}\begin{vmatrix} 1 & x_1 & y_1 \\ 1 & x_2 & y_2 \\ 1 & x_3 & y_3 \end{vmatrix}$.

14. Three particles leave the origin traveling in a straight line at constant velocities. Given that the velocities of the first two are $-\boldsymbol{i} - \boldsymbol{j} + 2\boldsymbol{k}$ and $\boldsymbol{i} + \boldsymbol{j} + \boldsymbol{k}$ and that the three particles always lie on a line. If the third particle lies in the xy plane, find its velocity.

15. An airplane leaves the origin flying in the direction of the vector $3\boldsymbol{i} + 4\boldsymbol{j}$ at the speed of 500 miles an hour. At the same time at a point 200 miles due east, another airplane leaves, flying in a straight line at a speed of 600 miles an hour. What path should the second plane take in order to intercept the first plane? When will the interception occur?

16. A quadrilateral has the four vertices (1, 1), (3, 2), (4, 4), (0, 2). Compute its area.

17. If $\boldsymbol{x}$ and $\boldsymbol{y}$ are vectors, show that $\frac{1}{2}(|\boldsymbol{x} + \boldsymbol{y}|^2 + |\boldsymbol{x} - \boldsymbol{y}|^2) = |\boldsymbol{x}|^2 + |\boldsymbol{y}|^2$. Interpret this geometrically.

18. Show that the points (x_1, y_1, z_1), (x_2, y_2, z_2), (x_3, y_3, z_3), and (x_4, y_4, z_4) are coplanar if and only if

$$\begin{vmatrix} x_1 & y_1 & z_1 & 1 \\ x_2 & y_2 & z_2 & 1 \\ x_3 & y_3 & z_3 & 1 \\ x_4 & y_4 & z_4 & 1 \end{vmatrix} = 0$$

19. If $\boldsymbol{a}$ and $\boldsymbol{b}$ are vectors in $\boldsymbol{R}^3$, show that the locus of points equidistant from the endpoints of $\boldsymbol{a}$ and $\boldsymbol{b}$ consists of the points $\boldsymbol{r}$ on the plane $(\boldsymbol{r}, \boldsymbol{b} - \boldsymbol{a}) = \frac{1}{2}((\boldsymbol{b}, \boldsymbol{b}) - (\boldsymbol{a}, \boldsymbol{a}))$.

20. If $\boldsymbol{x}$ is a unit vector and α is a scalar, under what circumstances is $\alpha\boldsymbol{x}$ a unit vector?

21. Determine the angle between the diagonal of a cube and one of its edges. Determine the angle between the diagonal of a cube and the diagonal of one of its faces.

22. Show that the endpoints of the vectors x, y, and z are collinear if and only if $(x - y) \times (z - y) = 0$.

23. If w, x, y, z are vectors, prove

$$(w \times x)\cdot(y \times z) = \begin{vmatrix} w\cdot y & w\cdot z \\ x\cdot y & x\cdot z \end{vmatrix}$$

24. Find a unit vector orthogonal to the vectors i and $j + tk$, where t is a real number.

25. Let n be a fixed vector. Describe the locus of points r such that $r\cdot n > 0$.

26. Prove that if x, y, and z are vectors, $x \times (y \times z) = (x, z)y - (x, y)z$.

27. Show that the cosine of the angle between the vector $ai + bj + ck$ and the x axis is $a/\sqrt{a^2 + b^2 + c^2}$. Find the cosine of the angle between the given vector and the y axis; the z axis.

28. Find a vector from the origin to the intersection of the medians of a triangle with vertices at (1, 0, 0), (0, 1, 0), and (1, 1, 1).

29. If x and y are vectors, show that $(|x|y + |y|x)/|x|\,|y|$ bisects the angle between x and y.

30. Let x be a vector in R^3. Let P be the function from R^3 to R^3, defined by $P_x(z) = x \times z$.
    (a) Show that P_x is a linear operator on R^3.
    (b) Show that N_{P_x} is spanned by x, if $x \neq 0$.
    (c) Show that the range of P_x consists precisely of the vectors orthogonal to x.
    (d) Show that relative to the standard basis i, j, and k the matrix representation of P_x is skew-symmetric.
    (e) Show that the mapping that carries x in R^3 into P_x determines an isomorphism from R^3 to the space of skew-symmetric matrices.

31. Let a and b be orthogonal unit vectors spanning a plane in R^3. Let z be another vector in R^3. Show that the closest vector to z in this plane is the vector $(z, a)a + (z, b)b$. Describe geometrically the function P from R^3 to R^3, where $P(z) = (z, a)a + (z, b)b$.

2 INNER PRODUCT IN R^n AND C^n

By analogy with the three-dimensional case, we define the inner product of two vectors

$$a = \begin{bmatrix} \alpha_1 \\ \alpha_2 \\ \vdots \\ \alpha_n \end{bmatrix} \quad \text{and} \quad b = \begin{bmatrix} \beta_1 \\ \beta_2 \\ \vdots \\ \beta_n \end{bmatrix}$$

in R^n, denoted by (a, b), to be the real number $(a, b) = \alpha_1\beta_1 + \alpha_2\beta_2 + \cdots + \alpha_n\beta_n$. The properties that were given in §6.1 for the inner product are again valid for inner products on an n-dimensional space.

In a similar manner, we define the norm of the vector $\boldsymbol{a}$, denoted by $|\boldsymbol{a}|$, by

$$|\boldsymbol{a}| = (\boldsymbol{a}, \boldsymbol{a})^{\frac{1}{2}} = \sqrt{\alpha_1^2 + \alpha_2^2 + \cdots + \alpha_n^2}$$

We note that

(1) $|\boldsymbol{a}| \geqslant 0$
   $|\boldsymbol{a}| = 0$ if and only if $\boldsymbol{a} = \boldsymbol{0}$
(2) $|\alpha \boldsymbol{a}| = |\alpha|\,|\boldsymbol{a}|$

For example, to prove (2),

$$\begin{aligned} |\alpha \boldsymbol{a}| = \sqrt{(\alpha\alpha_1)^2 + \cdots + (\alpha\alpha_n)^2} &= \sqrt{\alpha^2(\alpha_1^2 + \cdots + \alpha_n^2)} \\ &= |\alpha|\sqrt{\alpha_1^2 + \cdots + \alpha_n^2} \\ &= |\alpha|\,|\boldsymbol{a}| \end{aligned}$$

The inner product $(\boldsymbol{a}, \boldsymbol{b})$ may also be written $\boldsymbol{a}^T\boldsymbol{b}$. For we have

$$\boldsymbol{a}^T\boldsymbol{b} = \begin{bmatrix} \alpha_1 & \alpha_2 & \cdots & \alpha_n \end{bmatrix} \begin{bmatrix} \beta_1 \\ \beta_2 \\ \vdots \\ \beta_n \end{bmatrix} = \alpha_1\beta_1 + \cdots + \alpha_n\beta_n$$

There is an important inequality that relates the inner product of two vectors to the product of their norms.

Theorem 1 Cauchy–Schwarz Inequality If $\boldsymbol{x}$ and $\boldsymbol{y}$ belong to $\boldsymbol{R}^n$, $|(\boldsymbol{x}, \boldsymbol{y})| \leqslant |\boldsymbol{x}|\,|\boldsymbol{y}|$.

PROOF If $\boldsymbol{x} = \boldsymbol{0}$ or $\boldsymbol{y} = \boldsymbol{0}$, $(\boldsymbol{x}, \boldsymbol{y}) = 0$ and thus $|(\boldsymbol{x}, \boldsymbol{y})| = |\boldsymbol{x}|\,|\boldsymbol{y}|$.
If $\boldsymbol{x} \neq \boldsymbol{0}$ and $\boldsymbol{y} \neq \boldsymbol{0}$

$$\left(\frac{\boldsymbol{x}}{|\boldsymbol{x}|} - \frac{\boldsymbol{y}}{|\boldsymbol{y}|}, \frac{\boldsymbol{x}}{|\boldsymbol{x}|} - \frac{\boldsymbol{y}}{|\boldsymbol{y}|} \right) \geqslant 0$$

and

$$\left(\frac{\boldsymbol{x}}{|\boldsymbol{x}|} + \frac{\boldsymbol{y}}{|\boldsymbol{y}|}, \frac{\boldsymbol{x}}{|\boldsymbol{x}|} + \frac{\boldsymbol{y}}{|\boldsymbol{y}|} \right) \geqslant 0$$

Thus,

$$\frac{(x, x)}{|x|^2} - \frac{2(x, y)}{|x|\,|y|} + \frac{(y, y)}{|y|^2} \geqslant 0$$

$$\frac{(x, x)}{|x|^2} + \frac{2(x, y)}{|x|\,|y|} + \frac{(y, y)}{|y|^2} \geqslant 0$$

Since $(x, x) = |x|^2$ and $(y, y) = |y|^2$, we have

$$2 - \frac{2(x, y)}{|x|\,|y|} \geqslant 0$$

$$2 + \frac{2(x, y)}{|x|\,|y|} \geqslant 0$$

or

$$-|x|\,|y| \leqslant (x, y) \leqslant |x|\,|y|$$

It follows that $|(x, y)| \leqslant |x|\,|y|$. ■

In $\boldsymbol{R}^3$ we saw that $(x, y) = |x|\,|y| \cos\theta$, where θ is the angle between x and y. Since $|\cos\theta| \leqslant 1$, in $\boldsymbol{R}^3$ the Cauchy–Schwarz inequality is a consequence of our geometric interpretation of the inner product.

Immediately, from the Cauchy–Schwarz inequality, we obtain the following.

Theorem 2 If x and y belong to $\boldsymbol{R}^n$, $|x + y| \leqslant |x| + |y|$.

PROOF By definition,

$$\begin{aligned} |x + y|^2 &= (x + y, x + y) \\ &= (x, x) + 2(x, y) + (y, y) \\ &\leqslant |x|^2 + 2|x|\,|y| + |y|^2 \\ &= (|x| + |y|)^2 \end{aligned}$$

Since $|x + y|^2 \leqslant (|x| + |y|)^2$, it follows that $|x + y| \leqslant |x| + |y|$. ■

Theorem 2 can be interpreted geometrically. If the vectors x and y represent the sides of a parallelogram, then $x + y$ represents the diagonal of the parallelogram. Thus, Theorem 2 may be understood to mean that the sum of the lengths of two adjacent sides of a parallelogram is less than or equal to the length of the diagonal. (See Figure 6-15.)

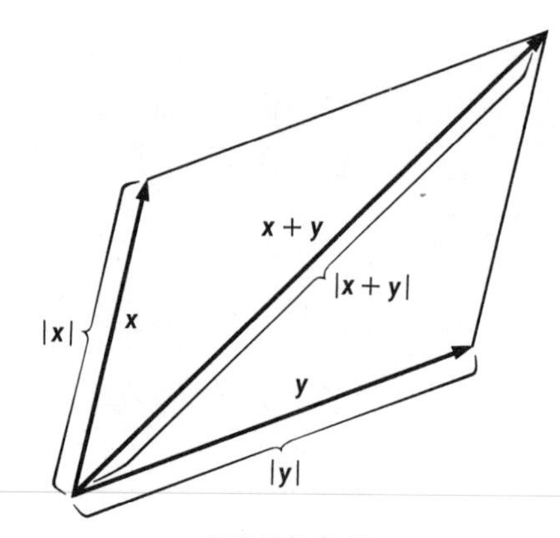

FIGURE 6-15

Letting $d(x, y) = |x - y|$, the distance between the endpoints of the vectors x and y, by Theorem 2, $|x - z| \leqslant |x - y| + |y - z|$, or $d(x, z) \leqslant d(x, y) + d(y, z)$. If we now consider the triangle with vertices the endpoints of x, y, and z, this last inequality may be reinterpreted as follows: The length of one side of a triangle is less than or equal to the sum of the lengths of the remaining two sides. (See Figure 6-16.)

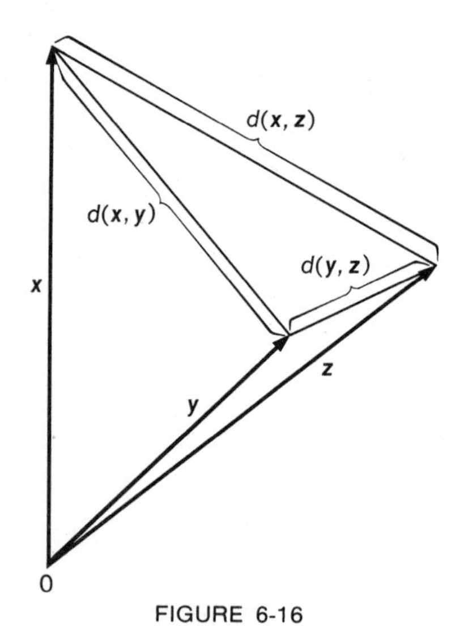

FIGURE 6-16

As in the three-dimensional case, vectors of norm 1 are said to be unit vectors. Two vectors whose inner product is zero are said to be orthogonal.

Under what circumstances does equality hold in the Cauchy–Schwarz inequality? If x and y are vectors in R^3 and $|(x, y)| = |x|\,|y|$, then the cosine of the angle between x and y is ± 1. In either case x and y are linearly dependent. (Why?) The following theorem indicates that this is always the case.

Theorem 3 Let x and y be vectors in R^n. Then, $|(x, y)| = |x|\,|y|$ if and only if x and y are linearly dependent.

PROOF First, suppose $|(x, y)| = |x|\,|y|$. If either $x = 0$ or $y = 0$, x and y are clearly linearly dependent. If not, we have either $(x, y) = |x|\,|y|$ or $(x, y) = -|x|\,|y|$. In the first case, it follows that

$$\left(\frac{x}{|x|} - \frac{y}{|y|}, \frac{x}{|x|} - \frac{y}{|y|}\right) = 0$$

In the second case,

$$\left(\frac{x}{|x|} + \frac{y}{|y|}, \frac{x}{|x|} + \frac{y}{|y|}\right) = 0$$

In either case,

$$\frac{x}{|x|} = \pm\frac{y}{|y|}$$

Thus, x and y are linearly dependent.

If, on the other hand, x and y are linearly dependent, one vector, say x, is a scalar multiple of the other. Thus, $x = \beta y$.

Then, $(x, y) = (\beta y, y) = \beta(y, y)$, while $|x| = |\beta|\,|y|$. Thus, $|(x, y)| = |\beta|\,|y|^2 = |x|\,|y|$. ■

Corollary If a and b are unit vectors and $(a, b) = 1$, then $a = b$.

A collection of vectors $x_1, x_2, \ldots, x_k$ in R^n is said to be an **orthonormal** system if any pair of distinct vectors is orthogonal and the norm of each vector is 1. In terms of the Kronecker delta,

$$(x_i, x_j) = \delta_{ij}$$

For example in R^4, the vectors

$$\begin{bmatrix} a \\ b \\ c \\ d \end{bmatrix}, \quad \begin{bmatrix} -b \\ a \\ d \\ -c \end{bmatrix}, \quad \begin{bmatrix} -c \\ -d \\ a \\ b \end{bmatrix}$$

constitute an orthonormal system if $a^2 + b^2 + c^2 + d^2 = 1$.

From the definition, it is clear that any subset of an orthonormal set of vectors is again orthonormal.

If an orthonormal system is a basis for R^n, it is called an orthonormal basis. For example, the standard basis $e_1, e_2, \ldots, e_n$ is an orthonormal basis for R^n. In R^2 and R^3, some examples of orthonormal bases are

$$\begin{bmatrix} \frac{2}{5} \\ \frac{3}{5} \end{bmatrix}, \quad \begin{bmatrix} \frac{3}{5} \\ -\frac{2}{5} \end{bmatrix} \quad \text{and} \quad \frac{1}{3}\begin{bmatrix} -1 \\ 2 \\ 2 \end{bmatrix}, \quad \frac{1}{3}\begin{bmatrix} 2 \\ -1 \\ 2 \end{bmatrix}, \quad \frac{1}{3}\begin{bmatrix} 2 \\ 2 \\ -1 \end{bmatrix}$$

Theorem 4 Let $x_1, x_2, \ldots, x_k$ be a collection of nonzero orthogonal vectors in R^n. Then $x_1, x_2, \ldots, x_k$ are linearly independent.

PROOF Suppose that $\alpha_1 x_1 + \alpha_2 x_2 + \cdots + \alpha_k x_k = \mathbf{0}$. If we take the inner product with x_i, it follows that $(\alpha_1 x_1 + \alpha_2 x_2 + \cdots + \alpha_k x_k, x_i) = 0$, or $\alpha_1(x_1, x_i) + \cdots + \alpha_k(x_k, x_i) = 0$. Since $(x_j, x_i) = 0$, if $j \neq i$, $\alpha_i(x_i, x_i) = 0$. Then as $x_i \neq \mathbf{0}$, $(x_i, x_i) \neq 0$. Thus, $\alpha_i = 0$. Thus, if $\alpha_1 x_1 + \cdots + \alpha_k x_k = \mathbf{0}$, $\alpha_1 = \cdots = \alpha_k = 0$. It follows that $x_1, \ldots, x_k$ are linearly independent. ■

From this theorem, it follows that the following collections of vectors in R^3 are linearly independent.

$$\begin{bmatrix} 1 \\ 1 \\ 1 \end{bmatrix}, \quad \begin{bmatrix} 1 \\ -1 \\ 0 \end{bmatrix}, \quad \begin{bmatrix} 1 \\ 1 \\ -2 \end{bmatrix} \quad \text{and} \quad \begin{bmatrix} 1 \\ -1 \\ 1 \end{bmatrix}, \quad \begin{bmatrix} 2 \\ 3 \\ 1 \end{bmatrix}, \quad \begin{bmatrix} -4 \\ 1 \\ 5 \end{bmatrix}$$

There is an inner product on C^n related to the inner product on R^n. If a and b are two vectors in C^n,

$$a = \begin{bmatrix} \alpha_1 \\ \alpha_2 \\ \vdots \\ \alpha_n \end{bmatrix}, \quad b = \begin{bmatrix} \beta_1 \\ \beta_2 \\ \vdots \\ \beta_n \end{bmatrix}$$

we define (a, b), the inner product of a and b, by

$$(a, b) = \alpha_1 \bar{\beta}_1 + \alpha_2 \bar{\beta}_2 + \cdots + \alpha_n \bar{\beta}_n$$

where $\bar{\beta}$ denotes the complex conjugate of the complex number β. If a and b are vectors in R^n, i.e., if all the components of a and b are real, the above definition of the inner product coincides with our earlier definition.

The properties of the complex inner product are similar to those of the real inner product. We assume a, b, c belong to C^n and α and β are complex scalars.

(1) $(a, a) \geqslant 0$
$(a, a) = 0$ if and only if $a = \mathbf{0}$
(2) $(a, b + c) = (a, b) + (a, c)$
$(a + b, c) = (a, c) + (b, c)$
(3) $(\alpha a, b) = \alpha(a, b)$
$(a, \beta b) = \bar{\beta}(a, b)$
(4) $(a, b) = \overline{(b, a)}$

For example, to prove (1), observe that $(a, a) = \alpha_1\bar{\alpha}_1 + \alpha_2\bar{\alpha}_2 + \cdots + \alpha_n\bar{\alpha}_n = |\alpha_1|^2 + \cdots + |\alpha_n|^2$. Since $|\alpha_1|^2 \geqslant 0, \ldots, |\alpha_n|^2 \geqslant 0$, we have $(a, a) \geqslant 0$. If $(a, a) = 0$, it follows that $|\alpha_1|^2 = |\alpha_2|^2 = \cdots = |\alpha_n|^2 = 0$, which implies that $a = \mathbf{0}$.

To prove (4),

$$(\boldsymbol{a}, \boldsymbol{b}) = \alpha_1 \bar{\beta}_1 + \cdots + \alpha_n \bar{\beta}_n \text{ and } (\boldsymbol{b}, \boldsymbol{a}) = \beta_1 \bar{\alpha}_1 + \cdots + \beta_n \bar{\alpha}_n.$$

Thus, $\overline{(\boldsymbol{b}, \boldsymbol{a})} = \bar{\beta}_1 \alpha_1 + \cdots + \bar{\beta}_n \alpha_n = (\boldsymbol{a}, \boldsymbol{b})$.

The proofs of the other properties are quite similar.

As before, we define the norm of the vector $\boldsymbol{a}$, denoted by $|\boldsymbol{a}|$, to be $|\boldsymbol{a}| = (\boldsymbol{a}, \boldsymbol{a})^{\frac{1}{2}}$. From the properties of the inner product, we again obtain

(1) $|\boldsymbol{a}| \geqslant 0$
$|\boldsymbol{a}| = 0$ if and only if $\boldsymbol{a} = \boldsymbol{0}$
(2) $|\alpha \boldsymbol{a}| = |\alpha|\, |\boldsymbol{a}|$

In the complex case, the Cauchy–Schwarz inequality is also valid.

Theorem 5 If $\boldsymbol{a}$ and $\boldsymbol{b}$ belong to $\boldsymbol{C}^n$, $|(\boldsymbol{a}, \boldsymbol{b})| \leqslant |\boldsymbol{a}|\, |\boldsymbol{b}|$.

PROOF The result is obvious if $\boldsymbol{a} = \boldsymbol{0}$, $\boldsymbol{b} = \boldsymbol{0}$, or $(\boldsymbol{a}, \boldsymbol{b}) = 0$. So assume $\boldsymbol{a} \neq \boldsymbol{0}$, $\boldsymbol{b} \neq \boldsymbol{0}$, and $(\boldsymbol{a}, \boldsymbol{b}) \neq 0$.

Since

$$\left(\frac{\boldsymbol{b}}{|\boldsymbol{b}|} - \frac{|(\boldsymbol{a}, \boldsymbol{b})|}{(\boldsymbol{a}, \boldsymbol{b})} \frac{\boldsymbol{a}}{|\boldsymbol{a}|}, \frac{\boldsymbol{b}}{|\boldsymbol{b}|} - \frac{|(\boldsymbol{a}, \boldsymbol{b})|}{(\boldsymbol{a}, \boldsymbol{b})} \frac{\boldsymbol{a}}{|\boldsymbol{a}|} \right) \geqslant 0$$

it follows that

$$\frac{(\boldsymbol{b}, \boldsymbol{b})}{|\boldsymbol{b}|^2} - \frac{|(\boldsymbol{a}, \boldsymbol{b})|}{|\boldsymbol{a}|\, |\boldsymbol{b}| \overline{(\boldsymbol{a}, \boldsymbol{b})}} (\boldsymbol{b}, \boldsymbol{a}) - \frac{|(\boldsymbol{a}, \boldsymbol{b})|}{|\boldsymbol{a}|\, |\boldsymbol{b}| (\boldsymbol{a}, \boldsymbol{b})} (\boldsymbol{a}, \boldsymbol{b}) + \frac{|(\boldsymbol{a}, \boldsymbol{b})|^2}{(\boldsymbol{a}, \boldsymbol{b}) \overline{(\boldsymbol{a}, \boldsymbol{b})}} \frac{(\boldsymbol{a}, \boldsymbol{a})}{|\boldsymbol{a}|^2} \geqslant 0$$

Since $\overline{(\boldsymbol{a}, \boldsymbol{b})} = (\boldsymbol{b}, \boldsymbol{a})$, $(\boldsymbol{a}, \boldsymbol{b})\overline{(\boldsymbol{a}, \boldsymbol{b})} = |(\boldsymbol{a}, \boldsymbol{b})|^2$, $(\boldsymbol{b}, \boldsymbol{b}) = |\boldsymbol{b}|^2$, and $(\boldsymbol{a}, \boldsymbol{a}) = |\boldsymbol{a}|^2$, it follows that

$$2 - 2 \frac{|(\boldsymbol{a}, \boldsymbol{b})|}{|\boldsymbol{a}|\, |\boldsymbol{b}|} \geqslant 0$$

or

$$|(\boldsymbol{a}, \boldsymbol{b})| \leqslant |\boldsymbol{a}|\, |\boldsymbol{b}| \qquad \blacksquare$$

As before the Cauchy–Schwarz inequality implies the norm inequality $|\boldsymbol{a} + \boldsymbol{b}| \leqslant |\boldsymbol{a}| + |\boldsymbol{b}|$.

The definitions of orthogonality, orthonormal system, and basis for complex vectors are formally the same as for real vectors. Moreover, theorems analogous to those holding for real vectors with the real inner product hold for complex vectors with the complex inner product.

EXERCISES

1. Prove that the following sets of vectors are orthogonal.

(a) $\begin{bmatrix} 3 \\ 4 \end{bmatrix}, \begin{bmatrix} -4 \\ 3 \end{bmatrix}$ (b) $\begin{bmatrix} 0 \\ 1 \\ 2 \end{bmatrix}, \begin{bmatrix} -1 \\ -2 \\ 1 \end{bmatrix}, \begin{bmatrix} 5 \\ -2 \\ 1 \end{bmatrix}$ (c) $\begin{bmatrix} 0 \\ 1 \\ 1 \end{bmatrix}, \begin{bmatrix} 1 \\ -1 \\ 1 \end{bmatrix}, \begin{bmatrix} 2 \\ 1 \\ -1 \end{bmatrix}$

2. Extend the orthonormal systems to orthonormal bases for $\boldsymbol{R}^4$.

(a) $\frac{1}{2}\begin{bmatrix} 1 \\ 1 \\ 1 \\ 1 \end{bmatrix}, \frac{1}{2}\begin{bmatrix} 1 \\ 1 \\ -1 \\ -1 \end{bmatrix}, \frac{1}{2}\begin{bmatrix} 1 \\ -1 \\ 1 \\ -1 \end{bmatrix}$ (b) $\frac{1}{5}\begin{bmatrix} 4 \\ 2 \\ 2 \\ 1 \end{bmatrix}, \frac{1}{5}\begin{bmatrix} 1 \\ -2 \\ 2 \\ -4 \end{bmatrix}, \frac{1}{5}\begin{bmatrix} -2 \\ 4 \\ 1 \\ -2 \end{bmatrix}$

3. If $\boldsymbol{a}$ and $\boldsymbol{b}$ are orthogonal unit vectors in $\boldsymbol{R}^3$, show that $\boldsymbol{a}, \boldsymbol{b}, \boldsymbol{a} \times \boldsymbol{b}$ is an orthonormal basis for $\boldsymbol{R}^3$.

4. Show that the vectors $\begin{bmatrix} a \\ b \end{bmatrix}$ and $\begin{bmatrix} -\bar{b} \\ \bar{a} \end{bmatrix}$ in $\boldsymbol{C}^2$ are orthogonal.

5. Let $\boldsymbol{u}_1, \ldots, \boldsymbol{u}_n$ be an orthonormal basis for $\boldsymbol{R}^n$. Let $\boldsymbol{x} = \alpha_1\boldsymbol{u}_1 + \cdots + \alpha_n\boldsymbol{u}_n$. Show that $\alpha_i = (\boldsymbol{x}, \boldsymbol{u}_i)$.

6. If $\boldsymbol{y}_1, \boldsymbol{y}_2, \ldots, \boldsymbol{y}_n$ are vectors orthogonal to the vector $\boldsymbol{x}$, show that any vector in $\text{sp}(\boldsymbol{y}_1, \boldsymbol{y}_2, \ldots, \boldsymbol{y}_n)$ is orthogonal to $\boldsymbol{x}$.

7. Let $\boldsymbol{x}_1, \boldsymbol{x}_2, \ldots, \boldsymbol{x}_n$ be vectors in $\boldsymbol{R}^n$. Let A be the matrix whose columns are $\boldsymbol{x}_1, \boldsymbol{x}_2, \ldots, \boldsymbol{x}_n$. Show that A^TA is diagonal if and only if every pair of distinct vectors in $\boldsymbol{x}_1, \boldsymbol{x}_2, \ldots, \boldsymbol{x}_n$ is orthogonal.

8. If $\boldsymbol{x}$ and $\boldsymbol{y}$ are vectors, show that $\frac{1}{2}(|\boldsymbol{x}+\boldsymbol{y}|^2 + |\boldsymbol{x}-\boldsymbol{y}|^2) = |\boldsymbol{x}|^2 + |\boldsymbol{y}|^2$.

9. Show that any orthonormal basis for $\boldsymbol{R}^2$ is of the form $\begin{bmatrix} a \\ b \end{bmatrix}, \begin{bmatrix} -b \\ a \end{bmatrix}$ or $\begin{bmatrix} a \\ b \end{bmatrix}, \begin{bmatrix} b \\ -a \end{bmatrix}$ where $a^2 + b^2 = 1$.

10. Let f be a linear transformation from $\boldsymbol{R}^n$ into $\boldsymbol{R}$. Show that there is a vector $\boldsymbol{y}$ in $\boldsymbol{R}^n$, such that $f(\boldsymbol{x}) = (\boldsymbol{x}, \boldsymbol{y})$.

11. Determine the length of the diagonal of an n-dimensional cube with edge of length 1.

12. Find an orthonormal basis for the set of solutions to the equations

$$x_1 - 2x_2 + 3x_3 - x_4 = 0$$
$$2x_1 + x_2 - x_3 + 2x_4 = 0$$

13. Let T be a linear operator on $\boldsymbol{R}^n$ having the property that $(T(\boldsymbol{x}), \boldsymbol{x}) = 0$ for all $\boldsymbol{x}$ in $\boldsymbol{R}^n$. Show that for all $\boldsymbol{x}$ and $\boldsymbol{y}$ in $\boldsymbol{R}^n$ that $(T(\boldsymbol{x}), \boldsymbol{y}) = -(T(\boldsymbol{y}), \boldsymbol{x})$. [Hint: Expand $(T(\boldsymbol{x}+\boldsymbol{y}), \boldsymbol{x}+\boldsymbol{y}) = 0$.]

14. Let A be an $n \times n$ matrix and T the linear operator on $\boldsymbol{R}^n$ defined by $T(\boldsymbol{x}) = A\boldsymbol{x}$. If $A = [a_{ij}]$, show that $(T(\boldsymbol{e}_i), \boldsymbol{e}_j) = a_{ji}$.

15. Let A be an $n \times n$ matrix and T the linear operator on $\boldsymbol{R}^n$ defined by $T(\boldsymbol{x}) = A\boldsymbol{x}$. Suppose that $(T(\boldsymbol{x}), \boldsymbol{x}) = 0$.
   (a) Show that A is skew-symmetric. [Hint: Use exercises 13 and 14.]
   (b) If n is odd, show that T is noninvertible.
   (c) If $n = 2$, show that T is a scalar multiple of R, where R is rotation by 90 degrees.

16. If $\boldsymbol{x}_1, \boldsymbol{x}_2, \ldots, \boldsymbol{x}_k$ is an orthogonal set of vectors, show that

$$|\boldsymbol{x}_1 + \boldsymbol{x}_2 + \cdots + \boldsymbol{x}_k|^2 = |\boldsymbol{x}_1|^2 + |\boldsymbol{x}_2|^2 + \cdots + |\boldsymbol{x}_k|^2$$

17. If $a^2 + b^2 + c^2 + d^2 = 1$, show that the vectors

$$\begin{bmatrix} a \\ b \\ c \\ d \end{bmatrix}, \quad \begin{bmatrix} -b \\ a \\ d \\ -c \end{bmatrix}, \quad \begin{bmatrix} -c \\ -d \\ a \\ b \end{bmatrix}, \quad \begin{bmatrix} -d \\ c \\ -b \\ a \end{bmatrix}$$

form an orthogonal basis for $\boldsymbol{R}^4$.

18. Find a nonzero linear transformation T from $\boldsymbol{R}^2$ into $\boldsymbol{R}^2$ such that for all $\boldsymbol{x}$, $(T(\boldsymbol{x}), \boldsymbol{x}) = 0$.

19. If $\boldsymbol{x}$ and $\boldsymbol{y}$ are orthogonal unit vectors, show that $|\boldsymbol{x} - \boldsymbol{y}| = \sqrt{2}$.

20. Let $\boldsymbol{x}$ be a unit vector. Consider the function f defined on the vectors of norm 1 by $f(\boldsymbol{y}) = |\boldsymbol{x} - \boldsymbol{y}|$.
    (a) Show that the maximum value of f is 2.
    (b) Show that the minimum value of f is 0.
    (c) Show that $f(\boldsymbol{y}) = \sqrt{2}$ if and only if $\boldsymbol{x}$ and $\boldsymbol{y}$ are orthogonal.

21. If $\boldsymbol{x}$ and $\boldsymbol{y}$ are vectors, show that $|\boldsymbol{x} + \boldsymbol{y}| = |\boldsymbol{x}| + |\boldsymbol{y}|$ only if $\boldsymbol{x}$ and $\boldsymbol{y}$ are linearly dependent.

22. If $\boldsymbol{x}$ and $\boldsymbol{y}$ are real vectors, determine the minimum value of the function $f(\lambda) = |\boldsymbol{x} + \lambda \boldsymbol{y}|^2$.

23. Show that the diagonals of a cube in $\boldsymbol{R}^3$ are not perpendicular.

24. If $\boldsymbol{a}$ and $\boldsymbol{b}$ are orthogonal unit vectors, show that the vectors $\alpha\boldsymbol{a} + \beta\boldsymbol{b}$ and $\gamma\boldsymbol{a} + \delta\boldsymbol{b}$ are orthogonal if and only if $\alpha\gamma + \beta\delta = 0$.

3 ORTHOGONAL COMPLEMENTS AND RELATED INEQUALITIES

If $\boldsymbol{x}$ is a unit vector in $\boldsymbol{R}^n$ (or $\boldsymbol{C}^n$) and $\boldsymbol{y}$ is another vector in the same space, the quantity $(\boldsymbol{y}, \boldsymbol{x})$ is called the **component** of $\boldsymbol{y}$ in the direction of $\boldsymbol{x}$. The vector $(\boldsymbol{y}, \boldsymbol{x})\boldsymbol{x}$ is called the **projection** of $\boldsymbol{y}$ onto $\boldsymbol{x}$. If $\boldsymbol{x}$ and $\boldsymbol{y}$ are vectors in $\boldsymbol{R}^3$, then $(\boldsymbol{y}, \boldsymbol{x})\boldsymbol{x}$ is the vector ending at the point where the perpendicular from the endpoint of $\boldsymbol{y}$ to the line determined by extending $\boldsymbol{x}$ intersects the line determined by extending $\boldsymbol{x}$. (See Figure 6-17.)

If $\boldsymbol{x}_1, \boldsymbol{x}_2, \ldots, \boldsymbol{x}_n$ is an orthonormal basis for $\boldsymbol{R}^n$ (or $\boldsymbol{C}^n$), there are scalars $\alpha_1, \alpha_2, \ldots, \alpha_n$ such that $\boldsymbol{x} = \alpha_1\boldsymbol{x}_1 + \alpha_2\boldsymbol{x}_2 + \cdots + \alpha_n\boldsymbol{x}_n$. It follows that $(\boldsymbol{x}, \boldsymbol{x}_i) = (\alpha_1\boldsymbol{x}_1 + \alpha_2\boldsymbol{x}_2 + \cdots + \alpha_n\boldsymbol{x}_n, \boldsymbol{x}_i) = \alpha_i(\boldsymbol{x}_i, \boldsymbol{x}_i) = \alpha_i$. Therefore, $\boldsymbol{x} = (\boldsymbol{x}, \boldsymbol{x}_1)\boldsymbol{x}_1 + (\boldsymbol{x}, \boldsymbol{x}_2)\boldsymbol{x}_2 + \cdots + (\boldsymbol{x}, \boldsymbol{x}_n)\boldsymbol{x}_n$.

We see in this case that the component of $\boldsymbol{x}$ in the direction of $\boldsymbol{x}_i$ is just the ith component of $\boldsymbol{x}$ relative to the basis $\boldsymbol{x}_1, \boldsymbol{x}_2, \ldots, \boldsymbol{x}_n$. In view of this fact, our choice of terminology is reasonable.

Theorem 1 Given a unit vector $\boldsymbol{x}$ in $\boldsymbol{R}^n$ (or $\boldsymbol{C}^n$) and another vector $\boldsymbol{y}$ in the same space, the vector $\boldsymbol{y}$ may be expressed as the sum of two vectors $\boldsymbol{u}$ and $\boldsymbol{v}$,

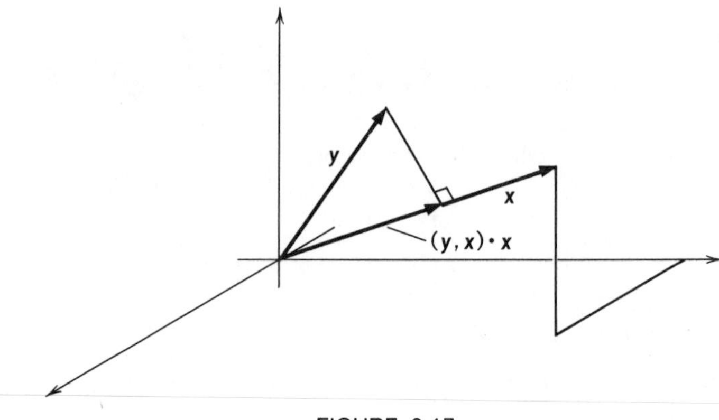

FIGURE 6-17

where $\boldsymbol{u}$ is parallel to $\boldsymbol{x}$ and $\boldsymbol{v}$ is perpendicular to $\boldsymbol{x}$. This decomposition is unique.

PROOF Let $\boldsymbol{u} = (\boldsymbol{y}, \boldsymbol{x})\boldsymbol{x}$ and $\boldsymbol{v} = \boldsymbol{y} - (\boldsymbol{y}, \boldsymbol{x})\boldsymbol{x}$. Clearly, $\boldsymbol{u}$ is parallel to $\boldsymbol{x}$ and $\boldsymbol{u} + \boldsymbol{v} = \boldsymbol{y}$. Since $(\boldsymbol{v}, \boldsymbol{x}) = (\boldsymbol{y} - (\boldsymbol{y}, \boldsymbol{x})\boldsymbol{x}, \boldsymbol{x}) = (\boldsymbol{y}, \boldsymbol{x}) - (\boldsymbol{y}, \boldsymbol{x})(\boldsymbol{x}, \boldsymbol{x}) = 0$, $\boldsymbol{v}$ is orthogonal to $\boldsymbol{x}$. This proves the existence of the decomposition asserted in the theorem.

We prove the uniqueness of the decomposition by showing that any decomposition is precisely the one exhibited above. To this end, suppose $\boldsymbol{y} = \boldsymbol{u} + \boldsymbol{v}$, where $\boldsymbol{u} = \alpha\boldsymbol{x}$ for some scalar α and $(\boldsymbol{v}, \boldsymbol{x}) = 0$. Then $(\boldsymbol{y}, \boldsymbol{x}) = (\boldsymbol{u} + \boldsymbol{v}, \boldsymbol{x}) = (\boldsymbol{u}, \boldsymbol{x}) + (\boldsymbol{v}, \boldsymbol{x})$. Since $(\boldsymbol{v}, \boldsymbol{x}) = 0$ and $\boldsymbol{u} = \alpha\boldsymbol{x}$, $(\boldsymbol{y}, \boldsymbol{x}) = \alpha(\boldsymbol{x}, \boldsymbol{x}) = \alpha$. Thus, $\boldsymbol{u} = (\boldsymbol{y}, \boldsymbol{x})\boldsymbol{x}$ and $\boldsymbol{v} = \boldsymbol{y} - (\boldsymbol{y}, \boldsymbol{x})\boldsymbol{x}$. ■

Example 1 Decompose the vector $5\boldsymbol{i} + 5\boldsymbol{j}$ into two parts, one parallel to and one perpendicular to $\frac{4}{5}\boldsymbol{i} + \frac{3}{5}\boldsymbol{j}$.

First, since $(5\boldsymbol{i} + 5\boldsymbol{j}, \frac{4}{5}\boldsymbol{i} + \frac{3}{5}\boldsymbol{j}) = 7$, the component of $5\boldsymbol{i} + 5\boldsymbol{j}$ in the direction of $\frac{4}{5}\boldsymbol{i} + \frac{3}{5}\boldsymbol{j}$ is $\frac{28}{5}\boldsymbol{i} + \frac{21}{5}\boldsymbol{j}$. Thus, by Theorem 1,

$$5\boldsymbol{i} + 5\boldsymbol{j} = \left(\tfrac{28}{5}\boldsymbol{i} + \tfrac{21}{5}\boldsymbol{j}\right) + \left(-\tfrac{3}{5}\boldsymbol{i} + \tfrac{4}{5}\boldsymbol{j}\right)$$

In this decomposition, the first vector is parallel to $\frac{4}{5}\boldsymbol{i} + \frac{3}{5}\boldsymbol{j}$ and the second is perpendicular to $\frac{4}{5}\boldsymbol{i} + \frac{3}{5}\boldsymbol{j}$. (See Figure 6-18.)

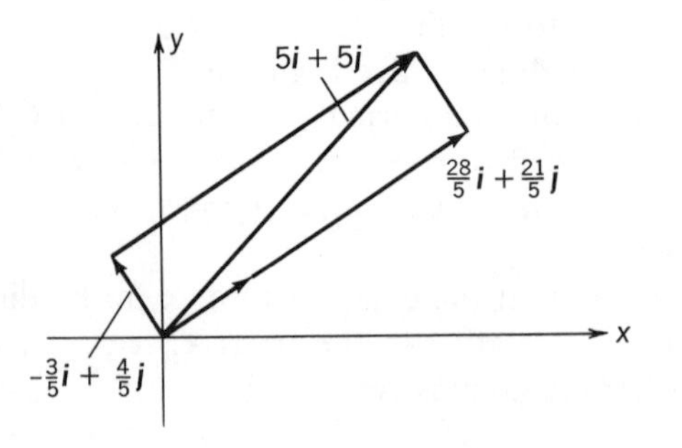

FIGURE 6-18

Next, we describe a procedure for obtaining an orthonormal set of vectors from an arbitrary collection of linearly independent vectors.

Let $\boldsymbol{x}_1, \boldsymbol{x}_2, \ldots, \boldsymbol{x}_k$ be a linearly independent collection of vectors in $\boldsymbol{R}^n$ (or $\boldsymbol{C}^n$). We define a new sequence of vectors $\boldsymbol{x}_1', \boldsymbol{x}_2', \ldots, \boldsymbol{x}_k'$ by

$$\begin{aligned}
\boldsymbol{x}_1' &= \boldsymbol{x}_1/|\boldsymbol{x}_1| \\
\boldsymbol{x}_2' &= (\boldsymbol{x}_2 - (\boldsymbol{x}_2, \boldsymbol{x}_1')\boldsymbol{x}_1')/|\boldsymbol{x}_2 - (\boldsymbol{x}_2, \boldsymbol{x}_1')\boldsymbol{x}_1'| \\
&\ \vdots \\
\boldsymbol{x}_i' &= (\boldsymbol{x}_i - (\boldsymbol{x}_i, \boldsymbol{x}_1')\boldsymbol{x}_1' - \cdots - (\boldsymbol{x}_i, \boldsymbol{x}_{i-1}')\boldsymbol{x}_{i-1}')/|\boldsymbol{x}_i - (\boldsymbol{x}_i, \boldsymbol{x}_1')\boldsymbol{x}_1' - \cdots (\boldsymbol{x}_i, \boldsymbol{x}_{i-1}')\boldsymbol{x}_{i-1}'| \\
\boldsymbol{x}_k' &= (\boldsymbol{x}_k - (\boldsymbol{x}_k, \boldsymbol{x}_1')\boldsymbol{x}_1' - \cdots - (\boldsymbol{x}_k, \boldsymbol{x}_{k-1}')\boldsymbol{x}_{k-1}')/|\boldsymbol{x}_k - (\boldsymbol{x}_k, \boldsymbol{x}_1')\boldsymbol{x}_1' - \cdots - (\boldsymbol{x}_k, \boldsymbol{x}_{k-1}')\boldsymbol{x}_{k-1}'|
\end{aligned}$$

Theorem 2 Gram–Schmidt Orthogonalization Process Let $\boldsymbol{x}_1, \boldsymbol{x}_2, \ldots, \boldsymbol{x}_k$ be a linearly independent collection of vectors in $\boldsymbol{R}^n$ (or $\boldsymbol{C}^n$). Then the vectors $\boldsymbol{x}_1', \boldsymbol{x}_2', \ldots, \boldsymbol{x}_k'$, defined above, have the properties

(1) $\boldsymbol{x}_1', \boldsymbol{x}_2', \ldots, \boldsymbol{x}_k'$ is an orthonormal set of vectors.
(2) $\text{sp}(\boldsymbol{x}_1', \ldots, \boldsymbol{x}_i') = \text{sp}(\boldsymbol{x}_1, \ldots, \boldsymbol{x}_i)$, $i = 1, 2, \ldots, k$.

PROOF It is clear by the linear independence of $\boldsymbol{x}_1, \boldsymbol{x}_2, \ldots, \boldsymbol{x}_k$, that $\boldsymbol{x}_1 \neq \boldsymbol{0}$, and, thus, the definition $\boldsymbol{x}_1' = \boldsymbol{x}_1/|\boldsymbol{x}_1|$ makes sense. It is also clear that $|\boldsymbol{x}_1'| = 1$ and $\text{sp}(\boldsymbol{x}_1') = \text{sp}(\boldsymbol{x}_1)$.

Assuming the truth of the theorem for $i-1$, we prove that it is valid for i. First, we show that $\boldsymbol{x}_i - (\boldsymbol{x}_i, \boldsymbol{x}_1')\boldsymbol{x}_1' - \cdots - (\boldsymbol{x}_i, \boldsymbol{x}_{i-1}')\boldsymbol{x}_{i-1}' \neq \boldsymbol{0}$, so that the definition of $\boldsymbol{x}_i'$ is meaningful. Indeed, if $\boldsymbol{x}_i - (\boldsymbol{x}_i, \boldsymbol{x}_1')\boldsymbol{x}_1' - \cdots - (\boldsymbol{x}_i, \boldsymbol{x}_{i-1}')\boldsymbol{x}_{i-1}' = \boldsymbol{0}$, then $\boldsymbol{x}_i$ belongs to $\text{sp}(\boldsymbol{x}_1', \ldots, \boldsymbol{x}_{i-1}') = \text{sp}(\boldsymbol{x}_1, \ldots, \boldsymbol{x}_{i-1})$. This implies that $\boldsymbol{x}_i$ is a linear combination of $\boldsymbol{x}_1, \boldsymbol{x}_2, \ldots, \boldsymbol{x}_{i-1}$, in contradiction to the linear independence of $\boldsymbol{x}_1, \boldsymbol{x}_2, \ldots, \boldsymbol{x}_i$. Hence, $\boldsymbol{x}_i - (\boldsymbol{x}_i, \boldsymbol{x}_1')\boldsymbol{x}_1' - \cdots - (\boldsymbol{x}_i, \boldsymbol{x}_{i-1}')\boldsymbol{x}_{i-1}' \neq \boldsymbol{0}$.

It is immediate from the definition of $\boldsymbol{x}_i'$ that $|\boldsymbol{x}_i'| = 1$. Now, if $j < i$,

$$\begin{aligned}
(\boldsymbol{x}_i', \boldsymbol{x}_j') &= \left(\frac{\boldsymbol{x}_i - (\boldsymbol{x}_i, \boldsymbol{x}_1')\boldsymbol{x}_1' - \cdots - (\boldsymbol{x}_i, \boldsymbol{x}_{i-1}')\boldsymbol{x}_{i-1}'}{|\boldsymbol{x}_i - (\boldsymbol{x}_i, \boldsymbol{x}_1')\boldsymbol{x}_1' - \cdots - (\boldsymbol{x}_i, \boldsymbol{x}_{i-1}')\boldsymbol{x}_{i-1}'|}, \boldsymbol{x}_j' \right) \\
&= \frac{1}{|\boldsymbol{x}_i - (\boldsymbol{x}_i, \boldsymbol{x}_1')\boldsymbol{x}_1' \cdots (\boldsymbol{x}_i, \boldsymbol{x}_{i-1}')\boldsymbol{x}_{i-1}'|}((\boldsymbol{x}_i, \boldsymbol{x}_j') - (\boldsymbol{x}_i, \boldsymbol{x}_j')(\boldsymbol{x}_j', \boldsymbol{x}_j'))
\end{aligned}$$

In the last step, we used the assumption that $\boldsymbol{x}_1', \ldots, \boldsymbol{x}_{i-1}'$ is an orthonormal system. Since $(\boldsymbol{x}_i, \boldsymbol{x}_j') - (\boldsymbol{x}_i, \boldsymbol{x}_j')(\boldsymbol{x}_j', \boldsymbol{x}_j') = 0$, $(\boldsymbol{x}_i', \boldsymbol{x}_j') = 0$. Thus, $\boldsymbol{x}_1', \ldots, \boldsymbol{x}_i'$ is an orthonormal system.

Since $\boldsymbol{x}_1', \boldsymbol{x}_2', \ldots, \boldsymbol{x}_i'$ belongs to $\text{sp}(\boldsymbol{x}_1, \boldsymbol{x}_2, \ldots, \boldsymbol{x}_i)$, $\text{sp}(\boldsymbol{x}_1', \ldots, \boldsymbol{x}_i') \subset \text{sp}(\boldsymbol{x}_1, \boldsymbol{x}_2, \ldots, \boldsymbol{x}_i)$. By Theorem 4, §6.2, $\boldsymbol{x}_1', \ldots, \boldsymbol{x}_i'$ are linearly independent. Since $\text{sp}(\boldsymbol{x}_1, \boldsymbol{x}_2, \ldots, \boldsymbol{x}_i)$ is a subspace of dimension i, by Theorem 2, §4.8, $\boldsymbol{x}_1', \ldots, \boldsymbol{x}_i'$ is a basis for $\text{sp}(\boldsymbol{x}_1, \boldsymbol{x}_2, \ldots, \boldsymbol{x}_i)$. Thus,

$$\text{sp}(\boldsymbol{x}_1', \ldots, \boldsymbol{x}_i') = \text{sp}(\boldsymbol{x}_1, \ldots, \boldsymbol{x}_i). \blacksquare$$

Example 2 Find an orthonormal basis for the subspace of $\boldsymbol{R}^4$ spanned by the vectors $\boldsymbol{x}_1, \boldsymbol{x}_2, \boldsymbol{x}_3$, if

$$\boldsymbol{x}_1 = \begin{bmatrix} -1 \\ 2 \\ 0 \\ 2 \end{bmatrix}, \quad \boldsymbol{x}_2 = \begin{bmatrix} 2 \\ -4 \\ 1 \\ -4 \end{bmatrix}, \quad \boldsymbol{x}_3 = \begin{bmatrix} -1 \\ 3 \\ 1 \\ 1 \end{bmatrix}$$

Now $|\boldsymbol{x}_1| = 3$, and so $\boldsymbol{x}'_1 = \frac{1}{3}\begin{bmatrix} -1 \\ \frac{2}{3} \\ 0 \\ \frac{2}{3} \end{bmatrix}$, and $\boldsymbol{x}_2 - (\boldsymbol{x}_2, \boldsymbol{x}'_1)\boldsymbol{x}'_1 = \begin{bmatrix} 0 \\ 0 \\ 1 \\ 0 \end{bmatrix}$. Thus, $\boldsymbol{x}'_2 = \begin{bmatrix} 0 \\ 0 \\ 1 \\ 0 \end{bmatrix}$. Then, $\boldsymbol{x}_3 - (\boldsymbol{x}_3, \boldsymbol{x}'_1)\boldsymbol{x}'_1 - (\boldsymbol{x}_3, \boldsymbol{x}'_2)\boldsymbol{x}'_2 = \boldsymbol{x}_3 - 3\boldsymbol{x}'_1 - \boldsymbol{x}'_2 = \begin{bmatrix} 0 \\ 1 \\ 0 \\ -1 \end{bmatrix}$. So, $\boldsymbol{x}'_3 = \frac{1}{\sqrt{2}}\begin{bmatrix} 0 \\ 1 \\ 0 \\ -1 \end{bmatrix}$. Thus, $\begin{bmatrix} -\frac{1}{3} \\ \frac{2}{3} \\ 0 \\ \frac{2}{3} \end{bmatrix}, \begin{bmatrix} 0 \\ 0 \\ 1 \\ 0 \end{bmatrix}, \frac{1}{\sqrt{2}}\begin{bmatrix} 0 \\ 1 \\ 0 \\ -1 \end{bmatrix}$ is an orthonormal basis for the subspace of $\boldsymbol{R}^4$ spanned by $\boldsymbol{x}_1, \boldsymbol{x}_2, \boldsymbol{x}_3$.

Theorem 3 Let H be a subspace of $\boldsymbol{R}^n$ (or $\boldsymbol{C}^n$); then H has an orthonormal basis.

PROOF Take any basis for H and apply the Gram–Schmidt orthogonalization process to obtain an orthonormal basis for H. ■

Let H be a subspace of $\boldsymbol{R}^n$ (or $\boldsymbol{C}^n$). The **orthogonal complement** of H, denoted by $H^\perp$, is the set of vectors $H^\perp = \{\, \boldsymbol{y} | (\boldsymbol{h}, \boldsymbol{y}) = 0 \text{ for all } \boldsymbol{h} \text{ in } H\}$. In words, $H^\perp$ consists of those vectors that are orthogonal to all vectors of H.

If $\boldsymbol{x}_1, \boldsymbol{x}_2, \ldots, \boldsymbol{x}_k$ is a basis for H, a vector $\boldsymbol{y}$ belongs to $H^\perp$ if and only if $\boldsymbol{y}$ is orthogonal to $\boldsymbol{x}_1, \boldsymbol{x}_2, \ldots, \boldsymbol{x}_k$. Indeed, if $\boldsymbol{y}$ belongs to $H^\perp$, $\boldsymbol{y}$ is orthogonal to all vectors in H, in particular to $\boldsymbol{x}_1, \boldsymbol{x}_2, \ldots, \boldsymbol{x}_k$. If, on the other hand, $\boldsymbol{y}$ is orthogonal to $\boldsymbol{x}_1, \boldsymbol{x}_2, \ldots, \boldsymbol{x}_k$, and if $\boldsymbol{x}$ belongs to H, there are scalars $\alpha_1, \alpha_2, \ldots, \alpha_k$ such that $\boldsymbol{x} = \alpha_1\boldsymbol{x}_1 + \alpha_2\boldsymbol{x}_2 + \cdots + \alpha_k\boldsymbol{x}_k$. Thus,

$$\begin{aligned} (\boldsymbol{x}, \boldsymbol{y}) &= (\alpha_1\boldsymbol{x}_1 + \alpha_2\boldsymbol{x}_2 + \cdots + \alpha_k\boldsymbol{x}_k, \boldsymbol{y}) \\ &= \alpha_1(\boldsymbol{x}_1, \boldsymbol{y}) + \alpha_2(\boldsymbol{x}_2, \boldsymbol{y}) + \cdots + \alpha_k(\boldsymbol{x}_k, \boldsymbol{y}) \\ &= 0 \end{aligned}$$

Thus, $\boldsymbol{y}$ is orthogonal to all vectors in H, and so $\boldsymbol{y}$ belongs to $H^\perp$.

Example 3 Find the orthogonal complement to the subspace of $\boldsymbol{R}^3$ spanned by the vectors $\boldsymbol{i} + \boldsymbol{j}$ and $\boldsymbol{j} + \boldsymbol{k}$.

To do this, we need only determine those vectors in $\boldsymbol{R}^3$ that are orthogonal to $\boldsymbol{i}+\boldsymbol{j}$ and $\boldsymbol{j}+\boldsymbol{k}$. Suppose that $x\boldsymbol{i}+y\boldsymbol{j}+z\boldsymbol{k}$ is such a vector. Then, it follows that

$$\begin{aligned} x+y \quad &= 0 \\ y+z &= 0 \end{aligned}$$

Solving this system, we see that the orthogonal complement consists of those vectors of the form $-c\boldsymbol{i}+c\boldsymbol{j}-c\boldsymbol{k}$. Thus, the orthogonal complement is the subspace of $\boldsymbol{R}^3$ spanned by the vector $-\boldsymbol{i}+\boldsymbol{j}-\boldsymbol{k}$.

In this example, the orthogonal complement is just the space of vectors normal to the plane spanned by $\boldsymbol{i}+\boldsymbol{j}$ and $\boldsymbol{j}+\boldsymbol{k}$. (See Figure 6-19.)

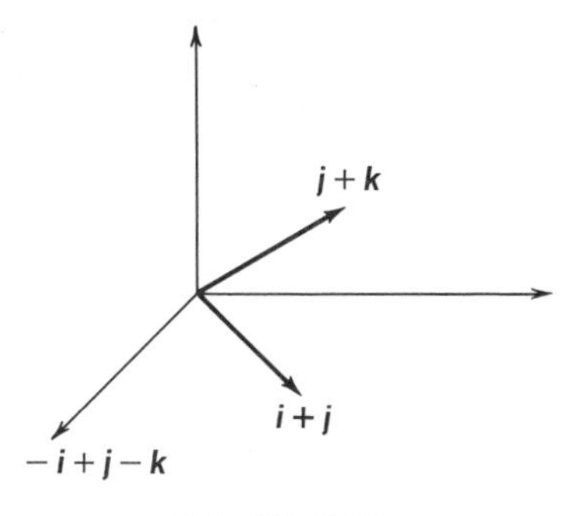

FIGURE 6-19

The following theorem tells us that the orthogonal complement of a given subspace is itself always a subspace.

Theorem 4 Let H be a subspace of $\boldsymbol{R}^n$ (or $\boldsymbol{C}^n$) and $H^\perp$ be its orthogonal complement. Then, $H^\perp$ is a subspace.

PROOF Suppose $\boldsymbol{y}_1$ and $\boldsymbol{y}_2$ belong to $H^\perp$. Then for all $\boldsymbol{x}$ in H, we have $(\boldsymbol{y}_1, \boldsymbol{x}) = 0$, $(\boldsymbol{y}_2, \boldsymbol{x}) = 0$. Then $(\boldsymbol{y}_1+\boldsymbol{y}_2, \boldsymbol{x}) = (\boldsymbol{y}_1, \boldsymbol{x}) + (\boldsymbol{y}_2, \boldsymbol{x}) = 0$. Thus, $\boldsymbol{y}_1 + \boldsymbol{y}_2$, being orthogonal to all vectors in H, belongs to $H^\perp$.

Similarly, if α is a scalar and $\boldsymbol{y}$ belongs to $H^\perp$, $(\alpha\boldsymbol{y}, \boldsymbol{x}) = \alpha(\boldsymbol{y}, \boldsymbol{x}) = 0$, for all $\boldsymbol{x}$ in H. Thus, $\alpha\boldsymbol{y}$ belongs to $H^\perp$. ■

Two subspaces H and K of a vector space V are said to be **complementary** if $H \cap K = 0$ and $H + K = V$. In other words, H and K have only the zero vector in common, and every vector in V can be expressed as the sum of a vector in H and a vector in K.

For example, if $\boldsymbol{a}$ and $\boldsymbol{b}$ are linearly independent vectors in $\boldsymbol{R}^2$, let H be the subspace spanned by $\boldsymbol{a}$ and let K be the subspace spanned by $\boldsymbol{b}$. Then H and K are complementary subspaces. Indeed, if $\boldsymbol{z}$ belongs to both H and K, then for suitable scalars α and β, $\boldsymbol{z} = \alpha\boldsymbol{a} = \beta\boldsymbol{b}$. Thus, $\alpha\boldsymbol{a} - \beta\boldsymbol{b} = \boldsymbol{0}$. However, since $\boldsymbol{a}$ and $\boldsymbol{b}$ are linearly independent, $\alpha = 0$ and $\beta = 0$. Thus, $\boldsymbol{z} = \boldsymbol{0}$ and $H \cap K = 0$. Likewise, since $\boldsymbol{a}$ and $\boldsymbol{b}$ are linearly independent, they span $\boldsymbol{R}^2$.

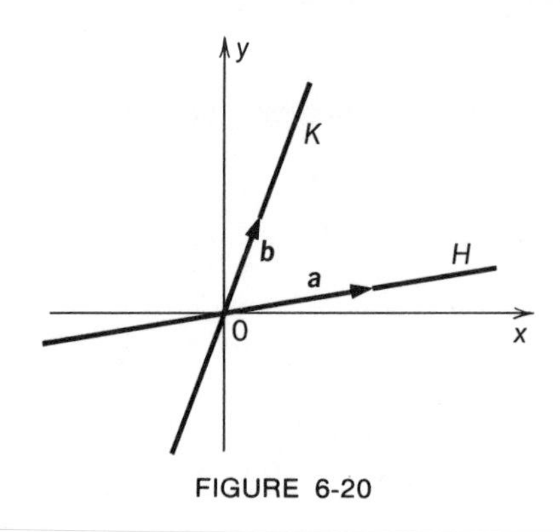

FIGURE 6-20

Thus, every vector in $\boldsymbol{R}^2$ is a linear combination of $\boldsymbol{a}$ and $\boldsymbol{b}$ and $H + K = \boldsymbol{R}^2$. (See Figure 6-20.)

Theorem 5 If H is a subspace of $\boldsymbol{R}^n$ (or $\boldsymbol{C}^n$), then H and $H^\perp$ are complementary subspaces, i.e., $H \cap H^\perp = 0$ and $H + H^\perp = \boldsymbol{R}^n$.

PROOF First, we show that H and $H^\perp$ have only the zero vector in common. Suppose $\boldsymbol{y}$ is a vector in both H and $H^\perp$. Then since $\boldsymbol{y}$ belongs to $H^\perp$ it is orthogonal to all vectors in H. Since $\boldsymbol{y}$ is also a vector in H, it follows that $(\boldsymbol{y}, \boldsymbol{y}) = 0$. Therefore, $\boldsymbol{y} = \boldsymbol{0}$. Hence, H and $H^\perp$ have only the zero vector in common.

Next we show that every vector may be expressed as the sum of a vector in H and $H^\perp$. Let $\boldsymbol{y}$ be an arbitrary vector in $\boldsymbol{R}^n$. Take $\boldsymbol{x}_1, \boldsymbol{x}_2, \ldots, \boldsymbol{x}_k$ to be an orthonormal basis for H. Let $\boldsymbol{z} = (\boldsymbol{y}, \boldsymbol{x}_1)\boldsymbol{x}_1 + (\boldsymbol{y}, \boldsymbol{x}_2)\boldsymbol{x}_2 + \cdots + (\boldsymbol{y}, \boldsymbol{x}_k)\boldsymbol{x}_k$. Clearly, $\boldsymbol{z}$ belongs to H. Then $\boldsymbol{y} = \boldsymbol{z} + (\boldsymbol{y} - \boldsymbol{z})$. To complete the proof, we must show that $\boldsymbol{y} - \boldsymbol{z}$ belongs to $H^\perp$. Note that

$$\begin{aligned}
(\boldsymbol{y} - \boldsymbol{z}, \boldsymbol{x}_i) &= (\boldsymbol{y}, \boldsymbol{x}_i) - (\boldsymbol{z}, \boldsymbol{x}_i) \\
&= (\boldsymbol{y}, \boldsymbol{x}_i) - ((\boldsymbol{y}, \boldsymbol{x}_1)\boldsymbol{x}_1 + \cdots + (\boldsymbol{y}, \boldsymbol{x}_k)\boldsymbol{x}_k, \boldsymbol{x}_i) \\
&= (\boldsymbol{y}, \boldsymbol{x}_i) - (\boldsymbol{y}, \boldsymbol{x}_i)(\boldsymbol{x}_i, \boldsymbol{x}_i) \\
&= 0
\end{aligned}$$

Since $\boldsymbol{y} - \boldsymbol{z}$ is orthogonal to all vectors in a basis for H, $\boldsymbol{y} - \boldsymbol{z}$ is orthogonal to all vectors in H and thus belongs to $H^\perp$. ■

For example, if H is the subspace of $\boldsymbol{R}^3$ consisting of the vectors on some plane through the origin, $H^\perp$ consists of all scalar multiples of some vector normal to the plane. Every vector in space may be expressed as the sum of a vector lying on the plane and some vector normal to the plane. Clearly, the plane and the line normal to the plane have only $\boldsymbol{0}$ in common.

If H and K are subspaces of $\boldsymbol{R}^n$ (or $\boldsymbol{C}^n$), other properties of the orthogonal complement are

(1) If $H \subset K$, then $K^\perp \subset H^\perp$.
(2) $(H^\perp)^\perp = H$.

To prove (1), observe that if $\boldsymbol{y}$ belongs to $K^{\perp}$, $\boldsymbol{y}$ is orthogonal to all vectors in K and thus to all vectors in H. Therefore, $\boldsymbol{y}$ belongs to $H^{\perp}$.

To prove (2), note first that if $\boldsymbol{x}$ belongs to H, $\boldsymbol{x}$ is orthogonal to all vectors in $H^{\perp}$, by definition of $H^{\perp}$, so $\boldsymbol{x}$ belongs to $(H^{\perp})^{\perp}$. Hence, $H \subset (H^{\perp})^{\perp}$. Next, suppose $\boldsymbol{y}$ belongs to $(H^{\perp})^{\perp}$. Then $\boldsymbol{y} = \boldsymbol{h} + \boldsymbol{k}$, where $\boldsymbol{h}$ belongs to H and $\boldsymbol{k}$ belongs to $H^{\perp}$. Then

$$\begin{aligned}(\boldsymbol{y}, \boldsymbol{y}) &= (\boldsymbol{h} + \boldsymbol{k}, \boldsymbol{h} + \boldsymbol{k}) \\ &= (\boldsymbol{h}, \boldsymbol{h}) + (\boldsymbol{h}, \boldsymbol{k}) + (\boldsymbol{k}, \boldsymbol{h}) + (\boldsymbol{k}, \boldsymbol{h}) \\ &= (\boldsymbol{h}, \boldsymbol{h}) + (\boldsymbol{k}, \boldsymbol{k})\end{aligned}$$

Thus, $|\boldsymbol{y}|^2 = |\boldsymbol{h}|^2 + |\boldsymbol{k}|^2$. Also, $(\boldsymbol{y}, \boldsymbol{y}) = (\boldsymbol{y}, \boldsymbol{h} + \boldsymbol{k}) = (\boldsymbol{y}, \boldsymbol{h}) + (\boldsymbol{y}, \boldsymbol{k})$.

By hypothesis $\boldsymbol{y}$ belongs to $(H^{\perp})^{\perp}$. Thus, $(\boldsymbol{y}, \boldsymbol{k}) = 0$ and $(\boldsymbol{y}, \boldsymbol{y}) = (\boldsymbol{y}, \boldsymbol{h}) = (\boldsymbol{h} + \boldsymbol{k}, \boldsymbol{h}) = (\boldsymbol{h}, \boldsymbol{h})$. Thus, $|\boldsymbol{y}|^2 = |\boldsymbol{h}|^2$. Therefore, $|\boldsymbol{k}|^2 = 0$, and so $\boldsymbol{k} = \boldsymbol{0}$, and $\boldsymbol{y} = \boldsymbol{h}$ belongs to H. Combining this with the result above, we see that $(H^{\perp})^{\perp} = H$.

There is a useful inequality relating the norm of a vector to its components relative to some orthonormal system. It is known as **Bessel's inequality**.

Theorem 6 Let $\boldsymbol{x}_1, \boldsymbol{x}_2, \ldots, \boldsymbol{x}_k$ be an orthonormal set of vectors in $\boldsymbol{R}^n$ (or $\boldsymbol{C}^n$). Let $\boldsymbol{x}$ be another vector in the same space. Then

$$|\boldsymbol{x}|^2 \geqslant \sum_{i=1}^{k} |(\boldsymbol{x}, \boldsymbol{x}_i)|^2$$

Moreover, if $\boldsymbol{x}_1, \ldots, \boldsymbol{x}_k$ is an orthonormal basis, equality holds.

PROOF Let H be the subspace spanned by $\boldsymbol{x}_1, \ldots, \boldsymbol{x}_k$. By Theorem 5, $\boldsymbol{x} = \boldsymbol{h} + \boldsymbol{l}$, where $\boldsymbol{h}$ belongs to H and $\boldsymbol{l}$ belongs to $H^{\perp}$. Thus, $\boldsymbol{h} = \alpha_1 \boldsymbol{x}_1 + \alpha_2 \boldsymbol{x}_2 + \cdots + \alpha_k \boldsymbol{x}_k$, since $\boldsymbol{x}_1, \ldots, \boldsymbol{x}_k$ span H, and $\boldsymbol{x} = \alpha_1 \boldsymbol{x}_1 + \alpha_2 \boldsymbol{x}_2 + \cdots + \alpha_k \boldsymbol{x}_k + \boldsymbol{l}$. Since $\boldsymbol{l}$ belongs to $H^{\perp}$,

$$\begin{aligned}(\boldsymbol{x}, \boldsymbol{x}_i) &= (\alpha_1 \boldsymbol{x}_1 + \cdots + \alpha_k \boldsymbol{x}_k + \boldsymbol{l}, \boldsymbol{x}_i) \\ &= \alpha_i + (\boldsymbol{l}, \boldsymbol{x}_i) \\ &= \alpha_i\end{aligned}$$

Hence we can rewrite $\boldsymbol{x}$ in the form $\boldsymbol{x} = (\boldsymbol{x}, \boldsymbol{x}_1)\boldsymbol{x}_1 + \cdots + (\boldsymbol{x}, \boldsymbol{x}_k)\boldsymbol{x}_k + \boldsymbol{l}$. Therefore,

$$\begin{aligned}(\boldsymbol{x}, \boldsymbol{x}) &= ((\boldsymbol{x}, \boldsymbol{x}_1)\boldsymbol{x}_1 + \cdots + (\boldsymbol{x}, \boldsymbol{x}_k)\boldsymbol{x}_k + \boldsymbol{l}, (\boldsymbol{x}, \boldsymbol{x}_1)\boldsymbol{x}_1 + \cdots + (\boldsymbol{x}, \boldsymbol{x}_k)\boldsymbol{x}_k + \boldsymbol{l}) \\ &= |(\boldsymbol{x}, \boldsymbol{x}_1)|^2 + \cdots + |(\boldsymbol{x}, \boldsymbol{x}_k)|^2 + (\boldsymbol{l}, \boldsymbol{l})\end{aligned}$$

Thus, $|\boldsymbol{x}|^2 \geqslant |(\boldsymbol{x}, \boldsymbol{x}_1)|^2 + \cdots + |(\boldsymbol{x}, \boldsymbol{x}_k)|^2$.

If $\boldsymbol{x}_1, \ldots, \boldsymbol{x}_k$ is an orthonormal basis, $H^{\perp} = \boldsymbol{0}$. Thus, $\boldsymbol{l} = \boldsymbol{0}$, and $|\boldsymbol{x}|^2 = |(\boldsymbol{x}, \boldsymbol{x}_1)|^2 + \cdots + |(\boldsymbol{x}, \boldsymbol{x}_k)|^2$. ■

EXERCISES

1. Show that the orthogonal complement of the subspace of $\boldsymbol{R}^2$ spanned by $\boldsymbol{i}_\theta = (\cos\theta)\boldsymbol{i} + (\sin\theta)\boldsymbol{j}$ is the subspace spanned by $\boldsymbol{j}_\theta = -(\sin\theta)\boldsymbol{i} + \cos\theta\boldsymbol{j}$.

2. For the following vectors $\boldsymbol{u}$ and $\boldsymbol{x}$, show that $\boldsymbol{u}$ is a unit vector, and find the component of $\boldsymbol{x}$ in the direction of $\boldsymbol{u}$.

$$\text{(a)}\quad \boldsymbol{u} = \tfrac{1}{3}\begin{bmatrix} 2 \\ 1 \\ 2 \end{bmatrix}, \quad \boldsymbol{x} = \begin{bmatrix} 1 \\ 0 \\ 2 \end{bmatrix} \qquad \text{(b)}\quad \boldsymbol{u} = \tfrac{1}{7}\begin{bmatrix} 6 \\ 2 \\ -3 \end{bmatrix}, \quad \boldsymbol{x} = \begin{bmatrix} 7 \\ 1 \\ 3 \end{bmatrix}$$

3. Determine an orthonormal basis for the subspace of $\boldsymbol{R}^n$ spanned by the following sets of vectors.

$$\text{(a)}\ \begin{bmatrix} 1 \\ 1 \\ 1 \end{bmatrix}, \begin{bmatrix} -1 \\ 1 \\ 1 \end{bmatrix} \qquad \text{(b)}\ \begin{bmatrix} 2 \\ 2 \\ -1 \\ 0 \end{bmatrix}, \begin{bmatrix} 2 \\ 3 \\ 1 \\ -2 \end{bmatrix}, \begin{bmatrix} 3 \\ 4 \\ 5 \\ -2 \end{bmatrix} \qquad \text{(c)}\ \begin{bmatrix} 6 \\ 2 \\ -3 \end{bmatrix}, \begin{bmatrix} 8 \\ -1 \\ -1 \end{bmatrix}$$

4. Let $\boldsymbol{x}_1, \ldots, \boldsymbol{x}_k$ be an orthonormal system in $\boldsymbol{R}^n$. Show that $\boldsymbol{x}_1, \ldots, \boldsymbol{x}_k$ is a subset of some orthonormal basis for $\boldsymbol{R}^n$.

5. Each of the following sets of vectors spans a subspace V of $\boldsymbol{R}^n$. Determine an orthonormal basis for the orthogonal complement of V.

$$\text{(a)}\ \begin{bmatrix} 1 \\ 1 \\ 1 \\ 1 \end{bmatrix}, \begin{bmatrix} 1 \\ 1 \\ -1 \\ -1 \end{bmatrix} \qquad \text{(b)}\ \begin{bmatrix} 1 \\ 1 \\ 1 \\ 0 \end{bmatrix}, \begin{bmatrix} 0 \\ 0 \\ 1 \\ 1 \end{bmatrix} \qquad \text{(c)}\ \begin{bmatrix} 1 \\ 2 \\ 3 \end{bmatrix}, \begin{bmatrix} 2 \\ 0 \\ 1 \end{bmatrix}$$

6. Let $\boldsymbol{u}_1, \boldsymbol{u}_2, \boldsymbol{u}_3$ be an orthonormal basis for $\boldsymbol{R}^3$. Let $\boldsymbol{v}$ be a unit vector orthogonal to $\boldsymbol{u}_3$. Show that $\boldsymbol{v} = a\boldsymbol{u}_1 + b\boldsymbol{u}_2$ where $a^2 + b^2 = 1$.

7. Let $\boldsymbol{a}$ and $\boldsymbol{b}$ be linearly independent vectors in $\boldsymbol{R}^3$. Let H be the subspace spanned by $\boldsymbol{a}$ and $\boldsymbol{b}$. Show that $\boldsymbol{a} \times \boldsymbol{b}$ spans the orthogonal complement of H.

8. If H is a subspace of $\boldsymbol{R}^n$ and $H^\perp$ is its orthogonal complement, we saw that any vector in $\boldsymbol{R}^n$ may be expressed as the sum of a vector in H and a vector in $H^\perp$. Show that this expression is unique.

9. If $\boldsymbol{x}_1, \boldsymbol{x}_2, \ldots, \boldsymbol{x}_k$ is an orthonormal set of vectors in $\boldsymbol{R}^n$, show that $\boldsymbol{x}_1, \boldsymbol{x}_2, \ldots, \boldsymbol{x}_k$ is an orthonormal basis if and only if for all $\boldsymbol{x}$ in $\boldsymbol{R}^n$, $|\boldsymbol{x}|^2 = |(\boldsymbol{x}, \boldsymbol{x}_1)|^2 + |(\boldsymbol{x}, \boldsymbol{x}_2)|^2 + \cdots + |(\boldsymbol{x}, \boldsymbol{x}_n)|^2$.

10. Show that the subspaces of $\boldsymbol{R}^4$ spanned by the following pairs of vectors are complementary.

$$\text{(a)}\ \begin{bmatrix} 1 \\ 2 \\ 0 \\ 1 \end{bmatrix}, \begin{bmatrix} 1 \\ 0 \\ -1 \\ 0 \end{bmatrix} \qquad \text{(b)}\ \begin{bmatrix} 0 \\ 2 \\ 1 \\ 0 \end{bmatrix}, \begin{bmatrix} 3 \\ 1 \\ 2 \\ 1 \end{bmatrix} \qquad \text{(c)}\ \begin{bmatrix} 2 \\ 3 \\ 1 \\ 0 \end{bmatrix}, \begin{bmatrix} -1 \\ 1 \\ 0 \\ 1 \end{bmatrix}$$

11. Let H be a subspace of $\boldsymbol{R}^n$ and $\boldsymbol{x}_1, \boldsymbol{x}_2, \ldots, \boldsymbol{x}_k$ be an orthonormal basis for H. If $\boldsymbol{x}$ belongs to $\boldsymbol{R}^n$, let $P(\boldsymbol{x}) = (\boldsymbol{x}, \boldsymbol{x}_1)\boldsymbol{x}_1 + (\boldsymbol{x}, \boldsymbol{x}_2)\boldsymbol{x}_2 + \cdots + (\boldsymbol{x}, \boldsymbol{x}_k)\boldsymbol{x}_k$.
    (a) Show that P is a linear transformation.
    (b) Show that $P^2 = P$.
    (c) Show that the range space of P is H and the nullspace of P is $H^\perp$.
    (d) Show that the range space of $(I - P)$ is $H^\perp$ and the nullspace of $I - P$ is H.
    (e) If $\boldsymbol{x}$ is a vector in $\boldsymbol{R}^n$ and $\boldsymbol{y}$ is a vector in H, show that $|\boldsymbol{x} - \boldsymbol{y}| \geqslant |\boldsymbol{x} - P(\boldsymbol{x})|$, i.e., $P(\boldsymbol{x})$ is the closest vector in H to $\boldsymbol{x}$.

12. If $\boldsymbol{x}_1, \boldsymbol{x}_2, \ldots, \boldsymbol{x}_n$ is an orthonormal basis for $\boldsymbol{R}^n$ and $\boldsymbol{y}$ and $\boldsymbol{z}$ are vectors in $\boldsymbol{R}^n$, show that

$$(\boldsymbol{y}, \boldsymbol{z}) = (\boldsymbol{y}, \boldsymbol{x}_1)(\boldsymbol{z}, \boldsymbol{x}_1) + (\boldsymbol{y}, \boldsymbol{x}_2)(\boldsymbol{z}, \boldsymbol{x}_2) + \cdots + (\boldsymbol{y}, \boldsymbol{x}_n)(\boldsymbol{z}, \boldsymbol{x}_n)$$

13. If H is an h-dimensional subspace of $\boldsymbol{R}^n$, and if K is a complementary subspace of H, show that K is of dimension $n - h$.

14. If H and K are subspaces of $\boldsymbol{R}^n$ with the same orthogonal complement, show that $H = K$.

15. If $\boldsymbol{x}_1, \boldsymbol{x}_2, \ldots, \boldsymbol{x}_n$ is an orthonormal basis for $\boldsymbol{R}^n$, show that

$$(\text{sp}(\boldsymbol{x}_1, \boldsymbol{x}_2, \ldots, \boldsymbol{x}_i))^{\perp} = \text{sp}(\boldsymbol{x}_{i+1}, \ldots, \boldsymbol{x}_n)$$

16. If $\boldsymbol{x}_1, \boldsymbol{x}_2, \ldots, \boldsymbol{x}_k$ is an orthogonal set of vectors in H, a subspace of $\boldsymbol{R}^n$, and $\boldsymbol{y}_1, \boldsymbol{y}_2, \ldots, \boldsymbol{y}_k$ is an orthogonal set of vectors in $H^{\perp}$, show that $\boldsymbol{x}_1 + \boldsymbol{y}_1, \boldsymbol{x}_2 + \boldsymbol{y}_2, \ldots, \boldsymbol{x}_k + \boldsymbol{y}_k$ is an orthogonal set of vectors.

4 ORTHOGONAL AND UNITARY MATRICES AND OPERATORS

If A is a real $n \times n$ matrix and $A^TA = I_n$, A is said to be an **orthogonal matrix**.

For example, if

$$A = \begin{bmatrix} \frac{4}{5} & 0 & -\frac{3}{5} \\ -\frac{9}{25} & \frac{4}{5} & -\frac{12}{25} \\ \frac{12}{25} & \frac{3}{5} & \frac{16}{25} \end{bmatrix}$$

$$A^TA = \begin{bmatrix} \frac{4}{5} & -\frac{9}{25} & \frac{12}{25} \\ 0 & \frac{4}{5} & \frac{3}{5} \\ -\frac{3}{5} & -\frac{12}{25} & \frac{16}{25} \end{bmatrix} \begin{bmatrix} \frac{4}{5} & 0 & -\frac{3}{5} \\ -\frac{9}{25} & \frac{4}{5} & -\frac{12}{25} \\ \frac{12}{25} & \frac{3}{5} & \frac{16}{25} \end{bmatrix} = \begin{bmatrix} 1 & 0 & 0 \\ 0 & 1 & 0 \\ 0 & 0 & 1 \end{bmatrix}$$

Thus, A is an orthogonal matrix. Also,

$$\begin{bmatrix} \cos\theta & \sin\theta \\ -\sin\theta & \cos\theta \end{bmatrix} \begin{bmatrix} \cos\theta & -\sin\theta \\ \sin\theta & \cos\theta \end{bmatrix} = \begin{bmatrix} 1 & 0 \\ 0 & 1 \end{bmatrix}$$

Thus,

$$\begin{bmatrix} \cos\theta & -\sin\theta \\ \sin\theta & \cos\theta \end{bmatrix}$$

is an orthogonal matrix. Clearly, any orthogonal matrix is invertible and $A^{-1} = A^T$.

If A is an $n \times n$ matrix and $A_1, A_2, \ldots, A_n$ are its columns, then

$$A^T A = \begin{bmatrix} (A_1, A_1) & (A_1, A_2) & \cdots & (A_1, A_n) \\ (A_2, A_1) & (A_2, A_2) & \cdots & (A_2, A_n) \\ & & \vdots & \\ (A_n, A_1) & (A_n, A_2) & \cdots & (A_n, A_n) \end{bmatrix}$$

Thus, the matrix A is orthogonal if and only if its columns form an orthonormal basis for $\boldsymbol{R}^n$.

A complex $n \times n$ matrix is said to be **unitary** if $U^* U = I_n$. Just as in the real case, a matrix U is unitary if and only if its columns form an orthonormal basis for $\boldsymbol{C}^n$. For example, if $|a|^2 + |b|^2 = 1$, the matrix

$$\begin{bmatrix} a & b \\ -\bar{b} & \bar{a} \end{bmatrix}$$

is unitary. Indeed,

$$\begin{bmatrix} \bar{a} & -b \\ \bar{b} & a \end{bmatrix} \begin{bmatrix} a & b \\ -\bar{b} & \bar{a} \end{bmatrix} = \begin{bmatrix} |a|^2 + |b|^2 & 0 \\ 0 & |a|^2 + |b|^2 \end{bmatrix} = \begin{bmatrix} 1 & 0 \\ 0 & 1 \end{bmatrix}$$

If A is an orthogonal matrix, from $A^T A = I_n$, it follows that $\det A^T \det A = 1$, or $\det A = \pm 1$.

If A is an $n \times n$ orthogonal matrix, consider the linear operator on $\boldsymbol{R}^n$, $T(\boldsymbol{x}) = A\boldsymbol{x}$. Then

$$\begin{aligned} (T(\boldsymbol{x}), T(\boldsymbol{y})) &= (T(\boldsymbol{x}))^T (T(\boldsymbol{y})) \\ &= (A\boldsymbol{x})^T (A\boldsymbol{y}) \\ &= (\boldsymbol{x}^T A^T)(A\boldsymbol{y}) \\ &= \boldsymbol{x}^T (A^T A)\boldsymbol{y} = \boldsymbol{x}^T \boldsymbol{y} = (\boldsymbol{x}, \boldsymbol{y}) \end{aligned}$$

In other words, the linear operator induced by an orthogonal matrix leaves the inner product invariant. Likewise, the linear operator on $\boldsymbol{C}^n$ induced by multiplication by a unitary operator leaves the complex inner product invariant.

If $T : \boldsymbol{R}^n \rightarrow \boldsymbol{R}^n$ (or $\boldsymbol{C}^n \rightarrow \boldsymbol{C}^n$) is a linear operator with the property that for all vectors $\boldsymbol{x}$ and $\boldsymbol{y}$, $(T(\boldsymbol{x}), T(\boldsymbol{y})) = (\boldsymbol{x}, \boldsymbol{y})$, T is said to be an **isometry**. As we have seen above, any linear operator induced by an orthogonal or unitary matrix is an isometry.

The following theorem summarizes some important properties of isometries.

Theorem 1 Let $T : \boldsymbol{R}^n \to \boldsymbol{R}^n$ (or $\boldsymbol{C}^n \to \boldsymbol{C}^n$) be an isometry. Then

(i) $|T(\boldsymbol{x})| = |\boldsymbol{x}|$.
(ii) T is an isomorphism.
(iii) If $\boldsymbol{x}$ and $\boldsymbol{y}$ are orthogonal vectors, $T(\boldsymbol{x})$ and $T(\boldsymbol{y})$ are also orthogonal.
(iv) If $\boldsymbol{x}_1, \boldsymbol{x}_2, \ldots, \boldsymbol{x}_n$ is an orthonormal basis, $T(\boldsymbol{x}_1), T(\boldsymbol{x}_2), \ldots, T(\boldsymbol{x}_n)$ is an orthonormal basis.

PROOF

(i) $|T(\boldsymbol{x})| = (T(\boldsymbol{x}), (T(\boldsymbol{x}))^{\frac{1}{2}} = (\boldsymbol{x}, \boldsymbol{x})^{\frac{1}{2}}$, since T is an isometry. Thus, $|T(\boldsymbol{x})| = |\boldsymbol{x}|$.
(ii) Since T is a linear operator on a finite-dimensional space, if T is one-one, T is an isomorphism. Since $|T(\boldsymbol{x})| = |\boldsymbol{x}|$, if $T(\boldsymbol{x}) = \boldsymbol{0}$, then $\boldsymbol{x} = \boldsymbol{0}$. Thus, T is one-one.
(iii) If $(\boldsymbol{x}, \boldsymbol{y}) = 0$, then $(T(\boldsymbol{x}), T(\boldsymbol{y})) = (\boldsymbol{x}, \boldsymbol{y}) = 0$.
(iv) If $\boldsymbol{x}_1, \boldsymbol{x}_2, \ldots, \boldsymbol{x}_n$ is an orthonormal basis, by (i), $T(\boldsymbol{x}_1), T(\boldsymbol{x}_2), \ldots, T(\boldsymbol{x}_n)$ are each unit vectors. By (iii), if $i \neq j$, $T(\boldsymbol{x}_i)$ and $T(\boldsymbol{x}_j)$ are orthogonal, since $\boldsymbol{x}_i$ and $\boldsymbol{x}_j$ are orthogonal. By (ii), if $\boldsymbol{x}_1, \boldsymbol{x}_2, \ldots, \boldsymbol{x}_n$ form a basis, so do $T(\boldsymbol{x}_1), T(\boldsymbol{x}_2), \ldots, T(\boldsymbol{x}_n)$. Thus, $T(\boldsymbol{x}_1), T(\boldsymbol{x}_2), \ldots, T(\boldsymbol{x}_n)$ form an orthogonal basis. ■

As we may have expected from earlier results, the matrix representation of an isometry with respect to an orthonomal basis has a special form.

Theorem 2 Let $T : \boldsymbol{R}^n \to \boldsymbol{R}^n$ be a linear operator. Let $\boldsymbol{x}_1, \boldsymbol{x}_2, \ldots, \boldsymbol{x}_n$ be an orthonormal basis for $\boldsymbol{R}^n$. Suppose A_T is the matrix representation of T relative to the basis $\boldsymbol{x}_1, \boldsymbol{x}_2, \ldots, \boldsymbol{x}_n$. Then T is an isometry if and only if A_T is an orthogonal matrix.

PROOF If

$$T(\boldsymbol{x}_j) = \Sigma a_{ij} \boldsymbol{x}_i$$

we have that $A_T = [a_{ij}]_{(nn)}$.

Since $\boldsymbol{x}_1, \boldsymbol{x}_2, \ldots, \boldsymbol{x}_n$ is an orthonormal basis,

$$(T(\boldsymbol{x}_j), T(\boldsymbol{x}_k)) = \left(\sum_{i=1}^{n} a_{ij} \boldsymbol{x}_i, \sum_{i=1}^{n} a_{ik} \boldsymbol{x}_i \right)$$

$$= \sum_{i=1}^{n} a_{ij} a_{ik}$$

If T is an isometry, $T(\boldsymbol{x}_1), T(\boldsymbol{x}_2), \ldots, T(\boldsymbol{x}_n)$ is an orthonormal basis, and thus

$$\sum_{i=1}^{n} a_{ij} a_{ij} = 1 \quad \text{and} \quad \sum_{i=1}^{n} a_{ij} a_{ik} = 0 \text{ if } j \neq k$$

Therefore, the columns of A_T form an orthonormal basis and A_T is orthogonal.

If, on the other hand, A_T is orthogonal,

$$\sum_{i=1}^{n} a_{ij}a_{ij} = 1 \quad \text{and} \quad \sum_{i=1}^{n} a_{ij}a_{ik} = 0 \text{ if } j \neq k$$

Thus $(T(\boldsymbol{x}_j), T(\boldsymbol{x}_j)) = 1$ and $(T(\boldsymbol{x}_j), T(\boldsymbol{x}_k)) = 0$ if $j \neq k$. Thus, the vectors $T(\boldsymbol{x}_1), T(\boldsymbol{x}_2), \ldots, T(\boldsymbol{x}_n)$ form an orthonormal basis for $\boldsymbol{R}^n$. If $\boldsymbol{y} = \sum_{i=1}^{n} \beta_i \boldsymbol{x}_i$ and $\boldsymbol{z} = \sum_{i=1}^{n} \gamma_i \boldsymbol{x}_i$ are two vectors in $\boldsymbol{R}^n$, it follows that

$$\begin{aligned}(T(\boldsymbol{y}), T(\boldsymbol{z})) &= (\beta_1 T(\boldsymbol{x}_1) + \cdots + \beta_n T(\boldsymbol{x}_n), \gamma_1 T(\boldsymbol{x}_1) + \cdots + \gamma_n T(\boldsymbol{x}_n)) \\ &= \beta_1\gamma_1 + \beta_2\gamma_2 + \cdots + \beta_n\gamma_n \\ &= (\boldsymbol{y}, \boldsymbol{z})\end{aligned}$$

Therefore, T is an isometry of $\boldsymbol{R}^n$. ■

As an example, consider the orthonormal basis for $\boldsymbol{R}^3$,

$$\begin{aligned}\boldsymbol{i}_{\theta\phi} &= (\sin\theta\cos\phi)\boldsymbol{i} + (\sin\theta\sin\phi)\boldsymbol{j} + (\cos\theta)\boldsymbol{k} \\ \boldsymbol{j}_{\theta\phi} &= -(\sin\phi)\boldsymbol{i} + (\cos\phi)\boldsymbol{j} \\ \boldsymbol{k}_{\theta\phi} &= (\cos\theta\cos\phi)\boldsymbol{i} + (\cos\theta\sin\phi)\boldsymbol{j} - (\sin\theta)\boldsymbol{k}\end{aligned}$$

(See Figure 6-21.)

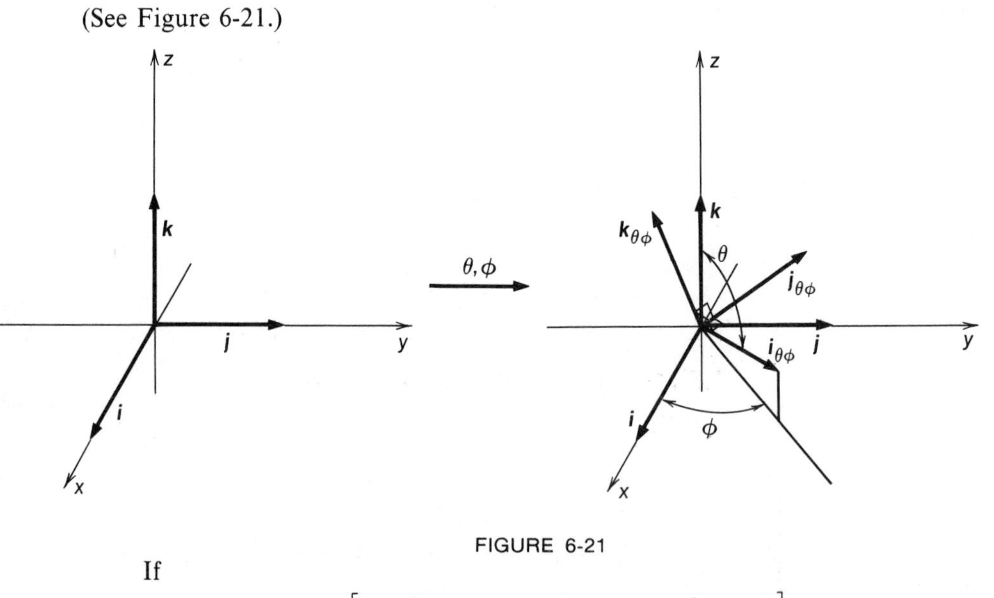

FIGURE 6-21

If

$$A_{\theta\phi} = \begin{bmatrix} \sin\theta\cos\phi & -\sin\phi & \cos\theta\cos\phi \\ \sin\theta\sin\phi & \cos\phi & \cos\theta\sin\phi \\ \cos\theta & 0 & -\sin\theta \end{bmatrix}$$

since the columns of $A_{\theta\phi}$ form an orthonormal system, $A_{\theta\phi}$ is an orthogonal matrix. Thus, the linear operator $T(\boldsymbol{x}) = A_{\theta\phi}\boldsymbol{x}$ is an isometry of $\boldsymbol{R}^3$.

As another result, we determine all 2×2 orthogonal matrices.

Example 1 Let A be a 2×2 orthogonal matrix. Then for some real number θ, we have

$$A = \begin{bmatrix} \cos\theta & -\sin\theta \\ \sin\theta & \cos \end{bmatrix} \quad \text{or} \quad A = \begin{bmatrix} \cos\theta & \sin\theta \\ \sin\theta & -\cos\theta \end{bmatrix}$$

To prove this, first note that if A is a 2×2 orthogonal matrix, its columns, say $\boldsymbol{a}$ and $\boldsymbol{b}$, form an orthogonal basis for $\boldsymbol{R}^2$. Let θ be the angle that $\boldsymbol{a}$ makes with the $\boldsymbol{x}$ axis. Since $\boldsymbol{a}$ is a unit vector, $\boldsymbol{a} = \boldsymbol{i}_\theta = (\cos\theta)\boldsymbol{i} + (\sin\theta)\boldsymbol{j}$. Since $\boldsymbol{b}$ is a unit vector orthogonal to $\boldsymbol{a}$, $\boldsymbol{b} = \pm\boldsymbol{j}_\theta = \pm(-(\sin\theta)\boldsymbol{i} + (\cos\theta)\boldsymbol{j})$. Thus,

$$A = \begin{bmatrix} \cos\theta & -\sin\theta \\ \sin\theta & \cos\theta \end{bmatrix} \quad \text{or} \quad \begin{bmatrix} \cos\theta & \sin\theta \\ \sin\theta & -\cos\theta \end{bmatrix}$$

In the previous chapter, we saw that a matrix of the first type induces a rotation of the plane by θ degrees, while matrices of the second type induce a reflection of the plane about the line making an angle of $\frac{1}{2}\theta$ degrees with the x axis. Thus, any 2×2 orthogonal matrix induces a rotation or reflection of the plane.

EXERCISES

1. Which of the following matrices are orthogonal?

(a) $\frac{1}{3}\begin{bmatrix} -1 & 2 & 2 \\ 2 & -1 & 2 \\ 2 & 2 & -1 \end{bmatrix}$ (b) $\begin{bmatrix} 1 & -1 & 1 \\ 1 & 1 & 1 \\ 1 & 0 & -2 \end{bmatrix}$

(c) $\frac{1}{7}\begin{bmatrix} 6 & 3 & -2 \\ -2 & 6 & 3 \\ 3 & -2 & 6 \end{bmatrix}$ (d) $\begin{bmatrix} 1 & 0 & 1 \\ 0 & 1 & 0 \\ 1 & 0 & -1 \end{bmatrix}$

2. Show that an isometry of $\boldsymbol{R}^3$ preserves the angle between any two vectors.

3. Show that in $\boldsymbol{R}^n$, $(\boldsymbol{x}, \boldsymbol{y}) = \frac{1}{2}(|\boldsymbol{x} + \boldsymbol{y}|^2 - |\boldsymbol{x}|^2 - |\boldsymbol{y}|^2)$.

4. If $A : \boldsymbol{R}^n \to \boldsymbol{R}^n$ is a linear operator such that $|A\boldsymbol{x}| = |\boldsymbol{x}|$ for all $\boldsymbol{x}$ in $\boldsymbol{R}^n$, show that A is an isometry. [Hint: Use exercise 3.]

5. If S and T are isometries of $\boldsymbol{R}^n$ (or $\boldsymbol{C}^n$), show that $S \circ T$ and T^{-1} are also isometries.

6. If U and V are orthogonal matrices, show that UV and U^{-1} are also orthogonal matrices. Prove the same if U and V are unitary.

7. Let $\boldsymbol{x}$ be a given unit vector in $\boldsymbol{R}^n$. Show that there is an orthogonal matrix whose first column is $\boldsymbol{x}$.

8. Let $\boldsymbol{u}$ and $\boldsymbol{v}$ be orthogonal vectors. If $\boldsymbol{u} + \boldsymbol{v}$ and $\boldsymbol{u} - \boldsymbol{v}$ are orthogonal, show that $|\boldsymbol{u}| = |\boldsymbol{v}|$.

9. Let T be a linear operator on $\boldsymbol{R}^2$. Suppose that T has the following property: Whenever $\boldsymbol{a}$ and $\boldsymbol{b}$ are orthogonal, then $T(\boldsymbol{a})$ and $T(\boldsymbol{b})$ are orthogonal. Show that T is a scalar multiple of an isometry. (Hint: Let $\boldsymbol{u} = T(\boldsymbol{i})$ and $\boldsymbol{v} = T(\boldsymbol{j})$. Use exercise 8 to show that $|\boldsymbol{u}| = |\boldsymbol{v}|$.)

10. Let T be a linear operator on $\boldsymbol{R}^3$ that preserves the angle between any two vectors. Show that T is a nonzero scalar multiple of an isometry and conversely.

11. If A is an orthogonal matrix, $\det A = 1$, and a_{ij} and A_{ij} are the entry and cofactor in position (i, j), respectively, show that $a_{ij} = A_{ij}$. What happens if $\det A = -1$?

12. If U is a unitary matrix, show that $|\det U| = 1$.

13. If $\boldsymbol{x}_1, \boldsymbol{x}_2, \ldots, \boldsymbol{x}_n$ are vectors in $\boldsymbol{R}^n$, and $[\boldsymbol{x}_1, \boldsymbol{x}_2, \ldots, \boldsymbol{x}_n]$ is the $n \times n$ matrix whose ith column is the vector $\boldsymbol{x}_i$, show that $|\det[\boldsymbol{x}_1, \boldsymbol{x}_2, \ldots, \boldsymbol{x}_n]| \leqslant |\boldsymbol{x}_1|\,|\boldsymbol{x}_2| \ldots |\boldsymbol{x}_n|$. Interpret this geometrically in $\boldsymbol{R}^2$ and $\boldsymbol{R}^3$.

14. If A is an orthogonal matrix and $I_n + A$ is nonsingular, show that $(I_n - A)(I_n + A)^{-1}$ is skew-symmetric.

15. Calculate the orthogonal matrix associated with a rotation of $\boldsymbol{R}^3$ of θ degrees about the z axis.

16. Calculate the orthogonal matrix associated with a rotation of $\boldsymbol{R}^3$ by θ degrees about the z axis, followed by a rotation of θ degrees about the y axis.

17. Show that any 2×2 unitary matrix of determinant 1 is of the form

$$\begin{bmatrix} a & b \\ -\bar{b} & \bar{a} \end{bmatrix}$$

where $|a|^2 + |b|^2 = 1$.

18. Let T be a linear operator on $\boldsymbol{R}^n$. Let $\boldsymbol{x}_1, \boldsymbol{x}_2, \ldots, \boldsymbol{x}_n$ be an orthonormal basis for $\boldsymbol{R}^n$. If $T(\boldsymbol{x}_1), T(\boldsymbol{x}_2), \ldots, T(\boldsymbol{x}_n)$ is an orthonormal basis for $\boldsymbol{R}^n$, show that T is an isometry.

19. Let T be a linear transformation of $\boldsymbol{R}^2$ that carries the unit circle into itself. Show that T is an isometry.

20. Let U be an orthogonal matrix. Show that the following are equivalent: (a) U is symmetric and (b) $U^2 = I$.

21. Let C be an $n \times n$ matrix having all diagonal entries, a, and all off diagonal entries, b. If C is orthogonal and $b \neq 0$, show that $a = -\dfrac{n-2}{n}d$ and $b = \dfrac{2}{n}d$, with $d = \pm 1$.

22. If Q is a real $n \times n$ matrix and $Q^T Q$ is a scalar multiple of the identity, show that Q is a scalar multiple of an orthogonal matrix.

23. Show that the matrix

$$\begin{bmatrix} \frac{1}{2}(1+i) & i/\sqrt{3} & (3+i)/2\sqrt{15} \\ -\frac{1}{2} & 1/\sqrt{3} & (4+3i)/2\sqrt{15} \\ \frac{1}{2} & -i/\sqrt{3} & 5i/2\sqrt{15} \end{bmatrix}$$

is unitary.

24. Let $\lambda_1, \lambda_2, \ldots, \lambda_n$ be complex numbers of absolute value 1. If the columns of a unitary matrix are multiplied by $\lambda_1, \lambda_2, \ldots, \lambda_n$, respectively, show that the resulting matrix is unitary.

25. Prove that if an orthogonal matrix is triangular, it is diagonal and all its diagonal entries are ± 1.

26. Let A be a real symmetric matrix and S be a real skew-symmetric matrix. Suppose that $AS = SA$ and $\det[A - S] \neq 0$. Show that $(A + S)(A - S)^{-1}$ is orthogonal.

27. Let $\boldsymbol{a}$ be the vector $\boldsymbol{a} = \boldsymbol{i} + \boldsymbol{j} + \boldsymbol{k}$ in $\boldsymbol{R}^3$. Find the orthogonal matrix representing a rotation of θ degrees about the line determined by the vector $\boldsymbol{a}$.

28. Let A be a real $n \times n$ matrix. Let $S(A)$ be the function that assigns to A the sum of the squares of the entries of A. If Q is an orthogonal matrix, show that $S(QAQ^{-1}) = S(A)$.

29. Let $\boldsymbol{x}_1, \boldsymbol{x}_2, \ldots, \boldsymbol{x}_n$ and $\boldsymbol{y}_1, \boldsymbol{y}_2, \ldots, \boldsymbol{y}_n$ be orthonormal bases for $\boldsymbol{R}^n$. Show that there is an isometry such that $T(\boldsymbol{x}_1) = \boldsymbol{y}_1$, $T(\boldsymbol{x}_2) = \boldsymbol{y}_2, \ldots, T(\boldsymbol{x}_n) = \boldsymbol{y}_n$.

5 GENERAL INNER PRODUCTS

Just as the concept of an abstract vector space was formulated by studying the principal properties of vector addition and scalar multiplication on $\boldsymbol{R}^n$, so too it is possible to consider abstract inner product spaces by singling out the most important properties of the dot product on $\boldsymbol{R}^n$. If V is a real vector space, a function that assigns to every pair of vectors $\boldsymbol{x}$ and $\boldsymbol{y}$ in V a real number $(\boldsymbol{x}, \boldsymbol{y})$ is said to be an inner product on V, if it has the following properties.

(1) $(\boldsymbol{x}, \boldsymbol{x}) \geqslant 0$
$(\boldsymbol{x}, \boldsymbol{x}) = 0$ if and only if $\boldsymbol{x} = \boldsymbol{0}$
(2) $(\boldsymbol{x}, \boldsymbol{y} + \boldsymbol{z}) = (\boldsymbol{x}, \boldsymbol{y}) + (\boldsymbol{x}, \boldsymbol{z})$
$(\boldsymbol{x} + \boldsymbol{y}, \boldsymbol{z}) = (\boldsymbol{x}, \boldsymbol{z}) + (\boldsymbol{y}, \boldsymbol{z})$
(3) $(\alpha\boldsymbol{x}, \boldsymbol{y}) = \alpha(\boldsymbol{x}, \boldsymbol{y})$
$(\boldsymbol{x}, \beta\boldsymbol{y}) = \beta(\boldsymbol{x}, \boldsymbol{y})$
(4) $(\boldsymbol{x}, \boldsymbol{y}) = (\boldsymbol{y}, \boldsymbol{x})$

The vector space V, together with its inner product, is said to constitute an **inner product space**.

As we noted earlier, if $V = \boldsymbol{R}^n$ and we define $(\boldsymbol{x}, \boldsymbol{y}) = \boldsymbol{x}^T\boldsymbol{y}$ for $\boldsymbol{x}$ and $\boldsymbol{y}$ in $\boldsymbol{R}^n$, then we obtain an inner product on $\boldsymbol{R}^n$. In addition to this, there are many other natural inner products.

Example 1 If $V = \boldsymbol{R}^n$ and $c_1, c_2, \ldots, c_n$ are n strictly positive real numbers, if we define

$$(\boldsymbol{x}, \boldsymbol{y}) = c_1\alpha_1\beta_1 + c_2\alpha_2\beta_2 + \cdots + c_n\alpha_n\beta_n$$

when

$$x = \begin{bmatrix} \alpha_1 \\ \alpha_2 \\ \vdots \\ \alpha_n \end{bmatrix} \quad \text{and} \quad y = \begin{bmatrix} \beta_1 \\ \beta_2 \\ \vdots \\ \beta_n \end{bmatrix}$$

then (,) is an inner product on $\boldsymbol{R}^n$.

Example 2 If $V = P_n$, the space of polynomials with real coefficients of degree less than or equal to n in a variable x, and if we define

$$(f, g) = \int_a^b f(x)\,g(x)\,dx$$

where a and b are real numbers with $a < b$, we obtain an inner product on P_n.

We verify some of the properties in this case.

(1) If f is a real polynomial, then $f^2(x) \geqslant 0$. Thus, $(f, f) = \int_a^b f^2(x)\,dx \geqslant 0$. If $(f, f) = 0$, then $\int_a^b f^2(x)\,dx = 0$. Since $f^2(x) \geqslant 0$, this can happen only if f is zero everywhere in the interval $[a, b]$. But any polynomial of degree less than or equal to n that has more than n roots must be identically 0. Thus, $f = 0$.

(2) $(f, g + h) = \int_a^b f(x)(g(x) + h(x))\,dx = \int_a^b f(x)\,g(x)\,dx + \int_a^b f(x)h(x)\,dx = (f, g) + (f, h)$

(3) $(f, g) = \int_a^b f(x)\,g(x)\,dx = \int_a^b g(x)f(x)\,dx = (g, f)$

If A is an $n \times n$ matrix, $A = [a_{ij}]_{(nn)}$, the trace of A, denoted by tr A, has already been defined as

$$\text{tr } A = a_{11} + a_{22} + \cdots + a_{nn}$$

Notice that the trace of A is the sum of the diagonal entries of A.

Immediate consequences of the definition are

$$\text{tr } \alpha A = \alpha \text{tr } A$$

$$\text{tr}(A + B) = \text{tr } A + \text{tr } B$$

$$\text{tr } A = \text{tr } A^T$$

The first two properties imply that the trace function is a linear transformation from the space of matrices into the reals (or complexes if we are dealing with complex matrices).

Example 3 Using the trace function, we define an inner product on the space of $n \times n$ matrices as follows: If A and B are $n \times n$ matrices

$$(A, B) = \operatorname{tr} AB^T$$

To prove (1), we note that

$$\operatorname{tr} AA^T = \sum_{i,j=1}^{n} a_{ij}^2, \quad \text{if } A = [a_{ij}]_{(nn)}$$

Thus, the trace of AA^T is the sum of the squares of the entries of A. Therefore, $\operatorname{tr} AA^T \geqslant 0$. If $\operatorname{tr} AA^T = 0$, all entries of A must be zero, and thus $A = 0$.

That the other properties of the inner product are satisfied follows easily from the corresponding properties of the trace.

The definitions and theorems concerning abstract inner products are formally the same as those of the special inner product considered in the first five sections of this chapter. For example, the norm of the vector $\boldsymbol{x}$, denoted as before by $|\boldsymbol{x}|$, is defined to be $(\boldsymbol{x}, \boldsymbol{x})^{\frac{1}{2}}$. Two vectors $\boldsymbol{x}$ and $\boldsymbol{y}$ are orthogonal if $(\boldsymbol{x}, \boldsymbol{y}) = 0$. A vector of norm 1 is said to be a unit vector. Orthonormal sets of vectors and bases are defined as before.

All theorems proved for the special inner product on $\boldsymbol{R}^n$ are still valid for general inner products. Indeed, a cursory perusal of the proofs of these theorems offered in earlier sections shows that only properties (1)–(4) of the inner product were used during the course of the demonstration. Thus, the theorems hold for all functions, defined on pairs of vectors, that satisfy properties (1)–(4), that is, general inner products.

Thus, we have the Cauchy–Schwarz inequality, $|(\boldsymbol{x}, \boldsymbol{y})| \leqslant |\boldsymbol{x}|\,|\boldsymbol{y}|$, for vectors $\boldsymbol{x}$ and $\boldsymbol{y}$ in V, and the norm inequality, $|\boldsymbol{x} + \boldsymbol{y}| \leqslant |\boldsymbol{x}| + |\boldsymbol{y}|$. In the case of example 2 above, the Cauchy–Schwarz inequality becomes

$$\left|\int_a^b f(x)g(x)\,dx\right| \leqslant \left(\int_a^b f^2(x)\,dx\right)^{\frac{1}{2}} \left(\int_a^b g^2(x)\,dx\right)^{\frac{1}{2}}$$

In the case of example 3, it becomes

$$|\operatorname{tr} AB^T| \leqslant (\operatorname{tr} AA^T)^{\frac{1}{2}} (\operatorname{tr} BB^T)^{\frac{1}{2}}$$

In a similar vein, we see that any set of nonzero orthogonal vectors is linearly independent, that the Gram–Schmidt orthogonalization procedure is valid, and that any finite-dimensional real vector space with an inner product admits an orthonormal basis with respect to this inner product. We illustrate with examples.

Example 4 In P_3, with inner product

$$(f, g) = \int_{-1}^{+1} f(x) g(x)\, dx$$

the polynomials $1, x, x^2 - \frac{1}{3}, x^3 - \frac{3}{5}x$ form an orthogonal collection of vectors. Thus, they are linearly independent and form a basis. By multiplying each polynomial by a suitable constant, we may obtain an orthonormal basis.

Example 5 Consider the space of 2×2 matrices with real entries, with the inner product of example 3 above.

An easy verification shows that the matrices

$$\begin{bmatrix} 1 & 0 \\ 0 & 0 \end{bmatrix}, \quad \begin{bmatrix} 0 & 1 \\ 0 & 0 \end{bmatrix}, \quad \begin{bmatrix} 0 & 0 \\ 1 & 0 \end{bmatrix}, \quad \begin{bmatrix} 0 & 0 \\ 0 & 1 \end{bmatrix}$$

form an orthonormal basis.

If V_1 and V_2 are real vector spaces with inner products $(\ ,\ )_1$ and $(\ ,\ )_2$, respectively, an invertible linear transformation $T : V_1 \to V_2$ is said to be an isometry if

$$(T(\boldsymbol{x}), T(\boldsymbol{y}))_2 = (\boldsymbol{x}, \boldsymbol{y})_1,$$

for all vectors $\boldsymbol{x}$ and $\boldsymbol{y}$ in V_1.

Just as $\boldsymbol{R}^n$ is the model real vector space, $\boldsymbol{R}^n$ with the standard inner product is the model real inner product space. We have the following theorem.

Theorem Let V_1 and V_2 be two real n-dimensional inner product spaces with inner product $(\ ,\ )_1$ and $(\ ,\ )_2$, respectively. Then there is an isometry from V_1 onto V_2.

PROOF Take $\boldsymbol{x}_1, \boldsymbol{x}_2, \ldots, \boldsymbol{x}_n$ as an orthonormal basis for V_1 and $\boldsymbol{y}_1, \boldsymbol{y}_2, \ldots, \boldsymbol{y}_n$ as an orthonormal basis for V_2. By §5.2, there is a linear transformation T from V_1 to V_2 such that $T(\boldsymbol{x}_1) = \boldsymbol{y}_1, T(\boldsymbol{x}_2) = \boldsymbol{y}_2, \ldots, T(\boldsymbol{x}_n) = \boldsymbol{y}_n$. T is clearly an isomorphism. To see that T is an isometry let $\boldsymbol{x} = \alpha_1\boldsymbol{x}_1 + \alpha_2\boldsymbol{x}_2 + \cdots + \alpha_n\boldsymbol{x}_n$ and $\boldsymbol{y} = \beta_1\boldsymbol{x}_1 + \beta_2\boldsymbol{x}_2 + \cdots + \beta_n\boldsymbol{x}_n$ be vectors in V_1. Then

$$\begin{aligned}(\boldsymbol{x}, \boldsymbol{y})_1 &= (\alpha_1\boldsymbol{x}_1 + \alpha_2\boldsymbol{x}_2 + \cdots + \alpha_n\boldsymbol{x}_n, \beta_1\boldsymbol{x}_1 + \beta_2\boldsymbol{x}_2 + \cdots + \beta_n\boldsymbol{x}_n) \\ &= \alpha_1\beta_1 + \alpha_2\beta_2 + \cdots + \alpha_n\beta_n\end{aligned}$$

since $\boldsymbol{x}_1, \boldsymbol{x}_2, \ldots, \boldsymbol{x}_n$ is an orthonormal basis, and

$$\begin{aligned}(T(\boldsymbol{x}), T(\boldsymbol{y}))_2 &= (\alpha_1\boldsymbol{y}_1 + \alpha_2\boldsymbol{y}_2 + \cdots + \alpha_n\boldsymbol{y}_n, \beta_1\boldsymbol{y}_1 + \beta_2\boldsymbol{y}_2 + \cdots + \beta_n\boldsymbol{y}_n) \\ &= \alpha_1\beta_1 + \alpha_2\beta_2 + \cdots + \alpha_n\beta_n\end{aligned}$$

since $y_1, y_2, \ldots, y_n$ is an orthonormal basis. Thus, $(T(x), T(y))_2 = (x, y)_1$. Therefore, T is an isometry from V_1 to V_2. ■

This theorem enables us to reduce many results about arbitrary inner product spaces to results about R^n with the familiar dot product as inner product.

It is also possible to define general complex inner products by abstracting the properties of the standard complex inner product on C^n. All properties, definitions, theorems, and proofs are the same as before.

EXERCISES

1. Let $x = \alpha_1 i + \alpha_2 j$ and $y = \beta_1 i + \beta_2 j$. Show that the following are inner products on R^2.
   (a) $(x, y) = 2\alpha_1\beta_1 + 3\alpha_2\beta_2$.
   (b) $(x, y) = 2\alpha_1\beta_1 + \alpha_1\beta_2 + \alpha_2\beta_1 + \alpha_2\beta_2$.

2. Let m be a continuous strictly positive function on the interval $[a, b]$. If f and g are polynomials in P_n, show that
$$(f, g) = \int_a^b f(x)g(x)m(x)\,dx$$
is an inner product on P_n.

3. On P_3, consider the inner product $(f, g) = \int_{-1}^{1} f(x)g(x)\,dx$.
   (a) Compute the inner products: $(1, x)$, $(1, x^2)$, $(x + 1, x)$, (x^2, x^3), $(1 + x^2, x)$.
   (b) Determine an orthonormal basis for P_3.
   (c) Determine an orthonormal basis for the orthogonal complement of the subspace spanned by 1 and x.

4. Consider the inner product on the space of $n \times n$ matrices of example 3 in the text. Show that the norm of the matrix A is the square root of the sum of the squares of the entries of A.

5. In P_n, let the inner product of f and g be given by
$$(f, g) = \int_{-1}^{+1} f(x)g(x)\,dx$$
   (a) Show that the orthogonal complement of the subspace of even polynomials in P_n is the subspace of odd polynomials in P_n.
   (b) Show that the linear transformation that carries $f(x)$ into $f(-x)$ is an isometry.

6. Let V be a real vector space. Show that the sum of two inner products on V is again an inner product on V.

7. Let V be a real finite-dimensional vector space and $(\ ,\ )$ be an inner product on V. Let $f: V \to R$ be a linear transformation from V into R. Show that there is a unique vector y in V such that for all x in V
$$f(x) = (x, y)$$

8. Let x_1, x_2, x_3, x_4 be four distinct points on the real line. If f and g lie in P_3, define $(f, g) = f(x_1)g(x_1) + f(x_2)g(x_2) + f(x_3)g(x_3) + f(x_4)g(x_4)$. Show that this is an inner product. If three points had been used instead of four, would it still be an inner product?

9. Let $\boldsymbol{x}_1, \boldsymbol{x}_2, \ldots, \boldsymbol{x}_n$ be a basis for $\boldsymbol{R}^n$. Show that there is an inner product on $\boldsymbol{R}^n$ with respect to which $\boldsymbol{x}_1, \boldsymbol{x}_2, \ldots, \boldsymbol{x}_n$ is an orthonormal basis.

10. Regarding the complex numbers as a vector space over the reals, define $(z_1, z_2) = \frac{1}{2}(z_1\bar{z}_2 + z_2\bar{z}_1)$.
    (a) Show that $(\ ,\ )$ is an inner product.
    (b) If $z_1 = \alpha_1 + \alpha_2 \boldsymbol{i}$ and $z_2 = \beta_1 + \beta_2 \boldsymbol{i}$, show that

$$(z_1, z_2) = \alpha_1\beta_1 + \alpha_2\beta_2$$

    Show that (z, z) is the square of the absolute value of the complex number z in the usual sense.
    (c) Let M_a be the linear transformation of $\boldsymbol{C}^n$ into itself, defined by $M_a(z) = az$. Show that $(M_a z_1, M_a z_2) = a\bar{a}(z_1, z_2)$.
    (d) With M_a defined as in (c) show that M_a is an isometry if and only if $|a| = 1$.
    (e) Letting $T(z) = \bar{z}$, show that T is an isometry.

11. In the space of $n \times n$ matrices with the inner product of example 3 in the text, let E_{ij} be the $n \times n$ matrix with 0 in all entries except (i, j) and 1 in entry (i, j). Show that the matrices E_{ij}, $1 \leqslant i \leqslant n$ and $1 \leqslant j \leqslant n$, form an orthonormal basis for M_{nn}.

12. In the space of $n \times n$ matrices with the inner product of example 3 in the text, let T be the linear operator defined by $T(B) = AB$, where A is an $n \times n$ orthogonal matrix. Show that T is an isometry.

13. Let $\boldsymbol{x}_1, \boldsymbol{x}_2, \ldots, \boldsymbol{x}_n$ be an orthonormal basis for $\boldsymbol{R}^n$ with the standard inner product. Show that the matrices $A_{ij} = \boldsymbol{x}_i\boldsymbol{x}_j^T$ form an orthonormal basis for the space of $n \times n$ matrices with the inner product of example 3 in the text.

14. In the space P_n of polynomials of degree not exceeding n with real coefficients in a variable x, consider the inner products

$$(f, g)_1 = \int_a^b f(x)\,g(x)\,dx$$

$$(f, g)_2 = \int_c^d f(x)\,g(x)\,dx$$

    where $a < b$ and $c < d$. Let T be the function from P_n (regarded as an inner product space with the first inner product) into P_n (regarded as an inner product space with the second inner product), defined by

$$(T(f))(x) = \sqrt{\frac{b-a}{d-c}}\; f\!\left(\frac{b-a}{d-c}\,x + \frac{ad-bc}{d-c}\right)$$

    Show that T is an isometry.

15. Consider the space of $n \times n$ matrices with the inner product of example 3 in the text.
    (a) Show that the orthogonal complement of the subspace of diagonal matrices is the subspaces of matrices all of whose diagonal entries are 0.
    (b) Show that the orthogonal complement of the subspace of symmetric matrices is the subspace of skew-symmetric matrices.

eigenvalues and canonical forms

1 EIGENVALUES AND EIGENVECTORS

In many of the applications of the theory of linear transformations, the following problem arises. Given a linear operator T on a vector space V, determine those scalars λ and those vectors x in V that satisfy the equation $T(x) = \lambda x$. This is known as the eigenvalue problem.

If T is a linear operator on a vector space V and $T(x) = \lambda x$ for some nonzero x in V and some scalar λ, λ is said to be an **eigenvalue** of T. The vector x is said to be an **eigenvector** of T corresponding to the eigenvalue λ. Often eigenvalues are called **characteristic values** and eigenvectors are called **characteristic vectors**. If A is an $n \times n$ matrix with real entries, a real number λ for which there is some nonzero real column n-vector x such that $Ax = \lambda x$ is said to be an eigenvalue of A, and x is said to be an eigenvector of A. Complex eigenvalues and eigenvectors are defined similarly.

Intuitively, eigenvectors correspond to the fixed directions of the linear operator T.

Example 1 Let T be the linear operator on $\boldsymbol{R}^2$ defined by $T(x) = \begin{bmatrix} 4 & -1 \\ 2 & 1 \end{bmatrix} x$. Since $T\left(\begin{bmatrix} 1 \\ 1 \end{bmatrix}\right) = \begin{bmatrix} 4 & -1 \\ 2 & 1 \end{bmatrix}\begin{bmatrix} 1 \\ 1 \end{bmatrix} = 3\begin{bmatrix} 1 \\ 1 \end{bmatrix}$ and $T\left(\begin{bmatrix} 1 \\ 2 \end{bmatrix}\right) = \begin{bmatrix} 4 & -1 \\ 2 & 1 \end{bmatrix}\begin{bmatrix} 1 \\ 2 \end{bmatrix} = 2\begin{bmatrix} 1 \\ 2 \end{bmatrix}$, it follows that 3 and 2 are eigenvalues of T with corresponding eigenvectors $\begin{bmatrix} 1 \\ 1 \end{bmatrix}$ and $\begin{bmatrix} 1 \\ 2 \end{bmatrix}$, respectively. In this case, the linear operator T fixes the lines through the origin determined by scalar multiples of the eigenvectors. (See Figure 7-1.)

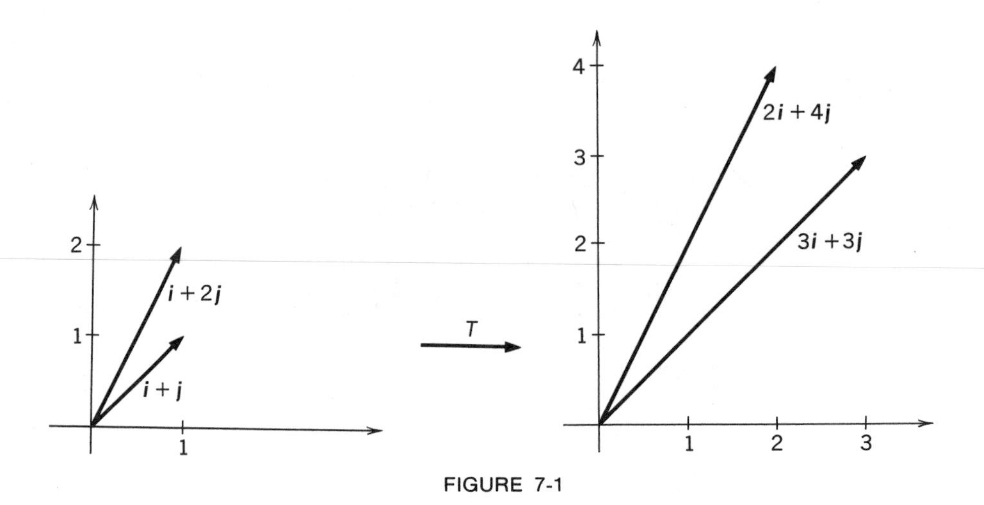

FIGURE 7-1

Earlier in the text, we encountered several examples of eigenvalues and eigenvectors, although at the time we did not regard them as such. For instance, in example 3 of §2.5 we considered a model for population growth in which the 4-vector $\boldsymbol{p}$ exhibits the population of each age group. Multiplying $\boldsymbol{p}$ by the matrix T gives the population distribution $T\boldsymbol{p}$ one year later. At the end of that example, we noted that there was a vector $\boldsymbol{p}_0$ such that $T\boldsymbol{p}_0 = \boldsymbol{p}_0$. Thus, T has eigenvalues 1 and $\boldsymbol{p}_0$ is the corresponding eigenvector. In this example, determining all the eigenvectors corresponding to the eigenvalue 1 will give the population distributions that are stable from year to year.

Example 2 In example 4 of §5.1, we considered a situation in which three companies controlled the market for a certain product. A matrix A was given that indicated retention and loss of customers. If the distribution of the market is given by the 3-vector $\boldsymbol{v}$, then the distribution one year later is given by $A\boldsymbol{v}$.

What are the stable distributions of the market?

Here the problem is to determine those vectors $\boldsymbol{v}$ such that $A\boldsymbol{v} = \boldsymbol{v}$, i.e., all eigenvectors corresponding to the eigenvalue 1. Equivalently, we wish to determine those vectors $\boldsymbol{v}$ for which $(A - I_3)\boldsymbol{v} = \boldsymbol{0}$. In other words, we must solve the linear equations

$$-\tfrac{1}{2}x + \tfrac{1}{3}y + \tfrac{1}{6}z = 0$$

$$\tfrac{1}{4}x - \tfrac{1}{2}y + \tfrac{1}{6}z = 0$$

$$\tfrac{1}{4}x + \tfrac{1}{6}y - \tfrac{1}{3}z = 0$$

for x, y, and z. Solving as usual, we find that $x = 10c$, $y = 9c$, and $z = 12c$, with c arbitrary. In this case, it is natural to choose c so that $x + y + z = 1$, or $c = \frac{1}{31}$. Thus, in a stable market the shares of the companies are $\frac{10}{31}$, $\frac{9}{31}$, and $\frac{13}{31}$, respectively. The vector $\boldsymbol{v}_0$ with these components is an eigenvector corresponding to the eigenvalue 1.

Since $T(\boldsymbol{x}) = \lambda\boldsymbol{x}$ is equivalent to $(T - \lambda I)(\boldsymbol{x}) = \mathbf{0}$, it follows that $\boldsymbol{x}$ is an eigenvector of T corresponding to the eigenvalue λ if and only if $\boldsymbol{x}$ belongs to the nullspace of $\lambda I - T$. We call the collection of all eigenvectors of T corresponding to the eigenvalue λ the **eigenspace** of T corresponding to the eigenvalue λ. It follows that the eigenspace of T corresponding to the eigenvalue λ is just the nullspace of $\lambda I - T$. In particular, it is clear that an eigenspace is a subspace of the vector space upon which T operates.

Example 3 Let T be the linear operator on P_n defined by $T(f)(x) = f(-x)$. In other words, T carries the polynomial $f(x)$ into the polynomial $f(-x)$. Since $T(1) = 1$, 1 is an eigenvalue of T. Since $T(x) = -x$, -1 is an eigenvalue of T. We determine the eigenspaces of T corresponding to the eigenvalues 1 and -1.

The eigenspace corresponding to the eigenvalue 1 consists of those polynomials f that satisfy $f(x) = f(-x)$, i.e., the even polynomials. The polynomials $1, x^2, x^4, \ldots$ form a basis for this space. The eigenspace corresponding to the eigenvalue -1 consists of those polynomials f that satisfy $f(-x) = -f(x)$, i.e., the odd polynomials. The polynomials $x, x^3, x^5, \ldots$ form a basis for this space.

In example 2, we saw that determining the eigenvectors corresponding to a given eigenvalue is simply a question of solving simultaneous linear equations. Next, we shall develop a method for determining which scalars are eigenvalues of a given linear operator.

Theorem 1 Let T be a linear operator on a real vector space V. Let λ be a real number. There is a nonzero vector $\boldsymbol{x}$ in V such that $T(\boldsymbol{x}) = \lambda\boldsymbol{x}$ if and only if $\det[T - \lambda I] = 0$.

PROOF Note that the nullspace of $T - \lambda I$ is nonzero if and only if $\det[T - \lambda I] = 0$. ■

An analogous result holds if the word "real" in the above theorem is replaced throughout by "complex."

The polynomial $f(x) = \det[xI - T]$ is called the **characteristic polynomial** of T and is denoted by C_T. It is calculated by choosing a basis for the vector space V and determining the matrix representation of the operator T relative to this basis. If A represents T relative to the chosen basis, $C_T(x) = \det[xI_n - A]$. If A' represents T with respect to another basis, in §5.8 we saw that there is an invertible matrix B such that $A' = BAB^{-1}$.

Then

$$\begin{aligned}\det[xI_n - A'] &= \det[xI_n - BAB^{-1}]\\ &= \det[B(xI_n - A)B^{-1}]\\ &= \det B \det[xI_n - A]\det B^{-1}\\ &= \det[xI_n - A]\end{aligned}$$

Consequently, the characteristic polynomial of T is independent of the basis used to compute it.

In a similar vein, we denote the characteristic polynomial of the matrix A by C_A. The previous paragraph implies that similar matrices have the same characteristic polynomial.

From our definition of the characteristic polynomial and from Theorem 1, it follows that the roots of the characteristic polynomial are precisely the eigenvalues of the linear operator.

If $A = [a_{ij}]_{(nn)}$ is an $n \times n$ matrix,

$$C_A(x) = \begin{vmatrix} x - a_{11} & -a_{12} & -a_{13} & \cdots & -a_{1n} \\ -a_{21} & x - a_{22} & -a_{23} & \cdots & -a_{2n} \\ -a_{31} & -a_{32} & x - a_{33} & \cdots & -a_{3n} \\ & & & \vdots & \\ -a_{n1} & -a_{n2} & -a_{n3} & \cdots & x - a_{nn} \end{vmatrix}$$

It is clear that $C_A(x)$ is a polynomial of degree n with leading coefficient 1:

$$C_A(x) = x^n + a_{n-1}x^{n-1} + \cdots + a_1x + a_0$$

Expanding the above determinant for $C_A(x)$ yields that the coefficient of x^{n-1} is

$$-(a_{11} + a_{22} + \cdots + a_{nn}) = -\operatorname{tr} A$$

The constant term a_0 in the above polynomial may be calculated by setting $x = 0$. Then $a_0 = C_A(0) = \det(-A)$, or $a_0 = (-1)^n \det A$.

Example 4 Let $A = \begin{bmatrix} a & b \\ c & d \end{bmatrix}$ be a 2×2 matrix. Then

$$C_A(x) = \begin{vmatrix} x - a & -b \\ -c & x - d \end{vmatrix} = x^2 - (a + d)x + ad - bc = x^2 - (\operatorname{tr} A)x + \det A$$

Thus, in the two-dimensional case, the characteristic polynomial of A is completely determined by the determinant and trace of A. The eigenvalues of A may be found by solving the equation $x^2 - (\operatorname{tr} A)x + \det A = 0$.

For example, if $A = \begin{bmatrix} 2 & 3 \\ -1 & -2 \end{bmatrix}$, $\operatorname{tr} A = 0$, $\det A = -1$, and $C_A(x) = x^2$

-1. Since the roots of the equation $x^2 - 1 = 0$ are $x = 1, -1$, the eigenvalues of A are 1 and -1. By solving linear equations, we find that corresponding eigenvectors are $\begin{bmatrix} 3 \\ -1 \end{bmatrix}$ and $\begin{bmatrix} 1 \\ -1 \end{bmatrix}$.

Example 5 Let $A = \begin{bmatrix} 0 & 1 & 1 \\ 1 & 0 & -1 \\ 1 & 5 & 3 \end{bmatrix}$. Find the eigenvalues of A. $C_A(x) = \begin{vmatrix} x & -1 & -1 \\ -1 & x & 1 \\ -1 & -5 & x-3 \end{vmatrix} = x(x^2 - 3x + 5) + (-x + 3 + 1) - (5 + x) = x^3 - 3x^2 + 3x - 1 = (x-1)^3$. From this, it follows that the only eigenvalue of A is 1. The eigenvectors of A corresponding to the eigenvalue 1 can be found by determining the nullspace of $I_3 - A$.

The characteristic polynomials of certain matrices are quite easy to compute. For example, the characteristic polynomial of I_n is simply $(x-1)^n$. More generally, if D is an $n \times n$ diagonal matrix with diagonal entries $d_1, d_2, \ldots, d_n$, the characteristic polynomial of D is $(x - d_1)(x - d_2) \ldots (x - d_n)$.

We list a few additional properties of the characteristic polynomial.

Theorem 2 Let A be an $n \times n$ matrix. Then, $C_{A^T}(x) = C_A(x)$. Moreover, if A is invertible, $C_{A^{-1}}(x) = (\det A)^{-1}(-x)^n C_A(\frac{1}{x})$.

PROOF We prove the first statement. We have $C_{A^T}(x) = \det[xI_n - A^T] = \det[(xI_n - A)^T] = \det[xI_n - A] = C_A(x)$.

Next, suppose that A is invertible. Then $C_{A^{-1}}(x) = \det[xI_n - A^{-1}] = \det[A^{-1}(xA - I_n)] = (\det A)^{-1} \det[xA - I_n] = (\det A)^{-1}(-x)^n \det[\frac{1}{x} I_n - A] = (\det A)^{-1}(-x)^n C_A(\frac{1}{x})$. ■

Example 6 A 3×3 orthogonal matrix U is said to be a rotation matrix if $\det U = 1$. We will show that 1 is an eigenvalue of a rotation matrix. Geometrically this means that every rotation of $\boldsymbol{R}^3$ fixes a vector, namely the eigenvector corresponding to the eigenvalue 1. In other words, every rotation of $\boldsymbol{R}^3$ has an axis.

We wish to show that $C_U(1) = 0$. By Theorem 2, $C_U(1) = C_{U^T}(1)$ and $C_{U^{-1}}(1) = 1(-1)^3 C_U(1) = -C_U(1)$. Since $UU^T = I_3$, $U^T = U^{-1}$. Thus, $C_U(1) = -C_U(1)$. It follows that $C_U(1) = 0$.

A similar argument shows that a 3×3 orthogonal matrix with determinant -1 has -1 as an eigenvalue.

Example 7 Recall that a square matrix A is stochastic if all its entries are nonnegative and its column sums are 1. In example 4 of both §5.1 and §5.8, we saw instances of such matrices. In many applications of such matrices, we are concerned with those vectors $\boldsymbol{v}$ such that $A\boldsymbol{v} = \boldsymbol{v}$, i.e., the eigenvectors corresponding to the eigenvalue 1. In example 2 above, we found such eigenvectors. We will show that a stochastic matrix always has 1 as an eigenvalue. Thus, there always exist nonzero vectors $\boldsymbol{v}$ such that $A\boldsymbol{v} = \boldsymbol{v}$.

Let A be an $n \times n$ matrix all of whose column sums are 1. Let $\boldsymbol{u}$ be the column n-vector having all entries 1. Note that in the matrix A^T all row sums are 1. It follows from this that $A^T\boldsymbol{u} = \boldsymbol{u}$. Since $\boldsymbol{u} \neq 0$, 1 is an eigenvalue of A^T. Thus, $C_{A^T}(1) = 0$. Since $C_{A^T}(x) = C_A(x)$, it follows that $C_A(1) = 0$. Thus, 1 is an eigenvalue of A.

It is important to note that a real matrix may have no real eigenvalues. For example, consider the matrix

$$B_\theta = \begin{bmatrix} \cos\theta & -\sin\theta \\ \sin\theta & \cos\theta \end{bmatrix}$$

We have seen that B_θ induces a rotation of the plane by θ degrees. If $0 < \theta < \pi$, it is clear that no vector in $\boldsymbol{R}^2$ is carried into a scalar multiple of itself. Thus, B_θ has no real eigenvalues.

In this case, the characteristic polynomial is $x^2 - (2\cos\theta)x + 1$. Its roots are $\cos\theta + i\sin\theta$ and $\cos\theta - i\sin\theta$. If $0 < \theta < \pi$, $\sin\theta \neq 0$, and neither root is real, a conclusion that was to be expected from the above geometric considerations.

It is known that any polynomial with complex coefficients has a complex root. Thus, any $n \times n$ matrix with complex entries has some complex eigenvalue. Consequently, any real matrix has some complex eigenvalue. Sometimes this fact, coupled with other information about a given matrix, can be used to show that the matrix has a real eigenvalue.

There is an important relationship between eigenvectors and linear independence that is brought forth in the following theorem.

Theorem 3 Let T be a linear operator on a vector space V. Let $\lambda_1, \lambda_2, \ldots, \lambda_n$ be distinct eigenvalues of T and let $\boldsymbol{x}_1, \boldsymbol{x}_2, \ldots, \boldsymbol{x}_n$ be eigenvectors (necessarily nonzero vectors) corresponding to the eigenvalues $\lambda_1, \lambda_2, \ldots, \lambda_n$, respectively. Then $\boldsymbol{x}_1, \boldsymbol{x}_2, \ldots, \boldsymbol{x}_n$ are linearly independent.

PROOF Suppose $\alpha_1\boldsymbol{x}_1 + \alpha_2\boldsymbol{x}_2 + \cdots + \alpha_n\boldsymbol{x}_n = \boldsymbol{0}$. Applying T to this equation and using $T(\boldsymbol{x}_i) = \lambda_i\boldsymbol{x}_i$, we get

$$\alpha_1\lambda_1\boldsymbol{x}_1 + \alpha_2\lambda_2\boldsymbol{x}_2 + \cdots + \alpha_n\lambda_n\boldsymbol{x}_n = \boldsymbol{0}$$

Multiplying the first of the above equations by λ_1 and subtracting from the second, we obtain

$$\alpha_2(\lambda_2 - \lambda_1)\boldsymbol{x}_2 + \alpha_3(\lambda_3 - \lambda_1)\boldsymbol{x}_3 + \cdots + \alpha_n(\lambda_n - \lambda_1)\boldsymbol{x}_n = \boldsymbol{0}$$

Applying T to this equation,

$$\alpha_2(\lambda_2 - \lambda_1)\lambda_2\boldsymbol{x}_2 + \alpha_3(\lambda_3 - \lambda_1)\lambda_3\boldsymbol{x}_3 + \cdots + \alpha_n(\lambda_n - \lambda_1)\lambda_n\boldsymbol{x}_n = \boldsymbol{0}$$

Multiplying the first of the preceding two equations by λ_2 and subtracting from the second, we obtain

$$\alpha_3(\lambda_3 - \lambda_1)(\lambda_3 - \lambda_2)\boldsymbol{x}_3 + \cdots + \alpha_n(\lambda_n - \lambda_1)(\lambda_n - \lambda_2)\boldsymbol{x}_n = \boldsymbol{0}$$

By repeating this process, we eventually obtain

$$\alpha_n(\lambda_n - \lambda_1)(\lambda_n - \lambda_2) \ldots (\lambda_n - \lambda_{n-1})\boldsymbol{x}_n = \boldsymbol{0}$$

Since $\lambda_1, \lambda_2, \ldots, \lambda_n$ are distinct, $(\lambda_n - \lambda_1)(\lambda_n - \lambda_2) \ldots (\lambda_n - \lambda_{n-1}) \neq 0$, and since it is also true that $\boldsymbol{x}_n \neq \boldsymbol{0}$, it follows that $\alpha_n = 0$.

Thus, $\alpha_1\boldsymbol{x}_1 + \cdots + \alpha_{n-1}\boldsymbol{x}_{n-1} = \boldsymbol{0}$. Using the same procedure on this equation repeatedly yields

$$\alpha_1 = \alpha_2 = \cdots = \alpha_{n-1} = 0$$

Thus, we have shown that whenever $\alpha_1\boldsymbol{x}_1 + \alpha_2\boldsymbol{x}_2 + \cdots + \alpha_n\boldsymbol{x}_n = \boldsymbol{0}$, $\alpha_1 = \alpha_2 = \cdots \alpha_n = 0$, and so $\boldsymbol{x}_1, \boldsymbol{x}_2, \ldots, \boldsymbol{x}_n$ are linearly independent. ∎

Recall that two real matrices C and D are said to be similar if there is a real invertible matrix B such that $B^{-1}CB = D$. There is an analogous definition for similarity of complex matrices. We say that a real matrix is **diagonalizable** over the reals if it is similar over the reals to some real diagonal matrix. There is an analogous definition for diagonalizability over the complexes.

A linear operator T on a vector space V is said to be diagonalizable if there is some basis for V such that the matrix representation of T, with respect to some basis for V, is diagonal. Since the matrix representations of T, with respect to the various bases of V, are all similar, T is diagonalizable if and only if one (and hence all) of its matrix representations are diagonalizable.

Suppose that the linear operator T is represented, with respect to the basis $\boldsymbol{x}_1, \ldots, \boldsymbol{x}_n$, by the diagonal matrix D having diagonal entries $d_1, \ldots, d_n$. Then $T(\boldsymbol{x}_i) = d_i\boldsymbol{x}_i$. Thus, the basis vectors $\boldsymbol{x}_1, \ldots, \boldsymbol{x}_n$ are eigenvectors of T. Thus, if T is diagonalizable, there is a basis consisting entirely of eigenvectors of T. Conversely, if there is a basis consisting entirely of eigenvectors of T, then the matrix representation of T, with respect to this basis, is diagonal, and so T is diagonalizable.

In terms of matrices this says: An $n \times n$ matrix A is diagonalizable over the reals if and only if $\boldsymbol{R}^n$ has a basis consisting of eigenvectors of A.

An analogous statement holds for the diagonalization of complex matrices.

We give a condition on the characteristic polynomial that guarantees that a matrix is diagonalizable.

Theorem 4 Let A be an $n \times n$ matrix with real entries and suppose $C_A(x) = (x - \lambda_1)(x - \lambda_2) \ldots (x - \lambda_n)$, where $\lambda_1, \ldots, \lambda_n$ are distinct real numbers. Then A is diagonalizable over the reals.

PROOF Take $\boldsymbol{x}_1, \ldots, \boldsymbol{x}_n$ eigenvectors of A corresponding to the eigenvalues $\lambda_1, \lambda_2, \ldots, \lambda_n$, respectively. By Theorem 3 above and by virtue of the fact that $\lambda_1, \lambda_2, \ldots, \lambda_n$ are distinct numbers, $\boldsymbol{x}_1, \boldsymbol{x}_2, \ldots, \boldsymbol{x}_n$ are linearly independent. But by Theorem 2, §4.8, $\boldsymbol{x}_1, \boldsymbol{x}_2, \ldots, \boldsymbol{x}_n$ also span $\boldsymbol{R}^n$, and thus $\boldsymbol{R}^n$ has a basis of eigenvectors of A. Consequently, A is diagonalizable. ∎

In this case we may also provide an explicit means for calculating the matrix B such that $B^{-1}AB = D$ is diagonal. Indeed, let $\boldsymbol{x}_1, \ldots, \boldsymbol{x}_n$ be the eigenvectors corresponding to $\lambda_1, \ldots, \lambda_n$. These vectors may be found by solving the system of linear equations $(A - \lambda_i)\boldsymbol{x} = \boldsymbol{0}$.

Let $B = [\boldsymbol{x}_1, \boldsymbol{x}_2, \ldots, \boldsymbol{x}_n]$, that is, B is the matrix whose ith column is $\boldsymbol{x}_i$. Then $B\boldsymbol{e}_i = \boldsymbol{x}_i$ and $B^{-1}\boldsymbol{x}_i = \boldsymbol{e}_i$. Consequently,

$$\begin{aligned}(B^{-1}AB)\boldsymbol{e}_i &= B^{-1}(A\boldsymbol{x}_i)\\ &= B^{-1}(\lambda_i\boldsymbol{x}_i)\\ &= \lambda_i\boldsymbol{e}_i\end{aligned}$$

Thus, $B^{-1}AB$ is a diagonal matrix with successive diagonal entries $\lambda_1, \lambda_2, \ldots, \lambda_n$.

Example 8 Let

$$A = \begin{bmatrix} -2 & 3 & 1 \\ 0 & 1 & 1 \\ -3 & 4 & 1 \end{bmatrix}$$

Then

$$C_A(x) = \begin{vmatrix} x+2 & -3 & -1 \\ 0 & x-1 & -1 \\ 3 & -4 & x-1 \end{vmatrix} = x^3 - 4x = x(x-2)(x+2)$$

Thus, the eigenvalues of A are 0, 2, and -2, which are real and distinct. It follows that there is a real invertible 3×3 matrix B such that

$$B^{-1}AB = \begin{bmatrix} 0 & 0 & 0 \\ 0 & 2 & 0 \\ 0 & 0 & -2 \end{bmatrix}$$

B may be determined explicitly by finding the eigenvectors corresponding to 0, 2, and -2 and writing these as successive columns of the matrix B.

Example 9 Let

$$B_\theta = \begin{bmatrix} \cos\theta & -\sin\theta \\ \sin\theta & \cos\theta \end{bmatrix}$$

Earlier, we saw that B_θ has the complex eigenvalues

$$\cos\theta + i\sin\theta, \quad \cos\theta - i\sin\theta$$

Thus, by Theorem 4, there is a complex matrix C such that

$$CB_\theta C^{-1} = \begin{bmatrix} \cos\theta + i\sin\theta & 0 \\ 0 & \cos\theta - i\sin\theta \end{bmatrix}$$

Note that though B_θ has no real eigenvalues, it is diagonalizable over the complexes.

Many calculations involving diagonal matrices are easily performed. It is, for example, quite simple to compute high powers of diagonal matrices. Such calculations can also be performed with diagonalizable matrices. For example, if C is diagonalizable and $B^{-1}CB = D$ is diagonal, to compute C^n, we need only compute D^n. Then, $C^n = BD^nB^{-1}$. In example 4 of §5.8, this procedure was used to compute high powers of a stochastic matrix. We present another example of this type.

Example 10 Compute C^n if $C = \begin{bmatrix} 4 & -2 \\ 1 & 1 \end{bmatrix}$.

In this case the characteristic polynomial of C is $x^2 - 5x + 6 = (x-2)(x-3)$. Thus, C has eigenvalues 2 and 3, and by Theorem 4, C is diagonalizable. The eigenvectors corresponding to the eigenvalues 3 and 2, respectively, can be obtained by solving the systems

$$\begin{aligned} -x + 2y &= 0 \\ -x + 2y &= 0 \end{aligned} \quad \text{and} \quad \begin{aligned} -2x + 2y &= 0 \\ x - y &= 0 \end{aligned}$$

We find that $\begin{bmatrix} 2 \\ 1 \end{bmatrix}$ and $\begin{bmatrix} 1 \\ 1 \end{bmatrix}$ are eigenvectors corresponding to the eigenvalues 3 and 2, respectively. We set $B = \begin{bmatrix} 2 & 1 \\ 1 & 1 \end{bmatrix}$. It follows that $B^{-1} = \begin{bmatrix} 1 & -1 \\ -1 & 2 \end{bmatrix}$ and that $B^{-1}CB = \begin{bmatrix} 3 & 0 \\ 0 & 2 \end{bmatrix}$. Therefore,

$$\begin{aligned} C^n = B\begin{bmatrix} 3^n & 0 \\ 0 & 2^n \end{bmatrix} B^{-1} &= \begin{bmatrix} 2 & 1 \\ 1 & 1 \end{bmatrix}\begin{bmatrix} 3^n & 0 \\ 0 & 2^n \end{bmatrix}\begin{bmatrix} 1 & -1 \\ -1 & 2 \end{bmatrix} \\ &= \begin{bmatrix} 2 \cdot 3^n - 2^n & -2 \cdot 3^n + 2^{n+1} \\ 3^n - 2^n & -3^n + 2^{n+1} \end{bmatrix} \end{aligned}$$

A similar procedure shows that if f is any polynomial

$$f(C) = \begin{bmatrix} 2f(3) - f(2) & -2f(3) + 2f(2) \\ f(3) - f(2) & -f(3) + 2f(2) \end{bmatrix}$$

Diagonalizable matrices are easy to compute with. However, it is not true that all matrices are diagonalizable.

Example 11 Let $A = \begin{bmatrix} 0 & 1 \\ 0 & 0 \end{bmatrix}$. Then $C_A(x) = x^2$.

Suppose that A is diagonalizable. Then there is an invertible matrix C such that $C^{-1}AC = \begin{bmatrix} d_1 & 0 \\ 0 & d_2 \end{bmatrix}$. Since similar matrices have the same characteristic polynomial, $(x - d_1)(x - d_2) = x^2$. It follows that $d_1 = d_2 = 0$. Thus, $C^{-1}AC = 0$. If we multiply this equation by C on the left and C^{-1} on the right, it follows that $A = 0$, which is a contradiction. Therefore, A is not diagonalizable.

EXERCISES

1. Determine the characteristic polynomial and eigenvalues of the following matrices.

(a) $\begin{bmatrix} 2 & 3 \\ -1 & -2 \end{bmatrix}$ (b) $\begin{bmatrix} 3 & 2 \\ -2 & -1 \end{bmatrix}$ (c) $\begin{bmatrix} 3 & 2 \\ -2 & -2 \end{bmatrix}$ (d) $\begin{bmatrix} 2 & 1 \\ 0 & -1 \end{bmatrix}$

(e) $\begin{bmatrix} 7 & 8 & 8 \\ -10 & -11 & -10 \\ 4 & 4 & 3 \end{bmatrix}$ (f) $\begin{bmatrix} 1 & 0 & 1 \\ 0 & 1 & 0 \\ 1 & 0 & 1 \end{bmatrix}$ (g) $\begin{bmatrix} 0 & 1 & 1 \\ 1 & 0 & 1 \\ 1 & 1 & 0 \end{bmatrix}$

2. Determine the eigenvalues and corresponding eigenspaces of the following matrices.

(a) $\begin{bmatrix} 1 & 1 & 1 \\ 0 & 1 & 1 \\ 0 & 0 & 1 \end{bmatrix}$ (b) $\begin{bmatrix} -5 & 3 & 0 \\ -6 & 4 & 2 \\ 2 & -1 & 1 \end{bmatrix}$ (c) $\begin{bmatrix} 0 & 1 & 0 \\ -1 & 2 & 0 \\ 1 & 1 & -1 \end{bmatrix}$

(d) $\begin{bmatrix} 1 & 1 & 1 \\ 1 & 1 & 1 \\ 1 & 1 & 1 \end{bmatrix}$

3. Show that the following matrices are diagonalizable, and find the diagonal matrix to which each is similar.

(a) $\begin{bmatrix} 1 & 1 & 1 \\ 0 & 2 & 1 \\ 0 & 0 & 3 \end{bmatrix}$ (b) $\begin{bmatrix} 2 & 2 & 0 \\ 1 & 1 & 2 \\ 1 & 1 & 2 \end{bmatrix}$ (c) $\begin{bmatrix} -2 & -4 & -5 \\ 1 & 3 & 1 \\ 2 & 2 & 5 \end{bmatrix}$

4. Determine the axis of rotation of the following rotation matrices.

(a) $\frac{1}{7}\begin{bmatrix} 3 & -2 & 6 \\ -6 & -3 & 2 \\ 2 & -6 & -3 \end{bmatrix}$ (b) $\frac{1}{3}\begin{bmatrix} 2 & -1 & 2 \\ -2 & -2 & 1 \\ 1 & -2 & -2 \end{bmatrix}$ (c) $\begin{bmatrix} 0 & 0 & 1 \\ 1 & 0 & 0 \\ 0 & 1 & 0 \end{bmatrix}$

5. Determine the characteristic polynomial and eigenvalues of the following operator on P_3.

(a) $D(f) = f'$ (b) $(S(f))(x) = f(x+1)$ (c) $T(f) = (1 - x^2)f'' - 2xf'$

6. If $A = [a_{ij}]$ is an $n \times n$ upper triangular matrix, show that $C_A(x) = (x - a_{11})(x - a_{22}) \ldots (x - a_{nn})$.

7. Let A be an $n \times n$ matrix having characteristic polynomial $(x - a)^n$. If A is diagonalizable, show that $A = aI_n$.

8. Let A be an $n \times n$ matrix. If the sum of the entries on all columns of A is equal to s, show that s is an eigenvalue of A.

9. As in example 4 of §5.1, three companies control the market for a certain product. The following stochastic matrices give the retention and loss of customers, as in

example 4. Compute the stable states of the market.

(a) $\begin{bmatrix} \frac{3}{4} & \frac{1}{8} & \frac{1}{4} \\ \frac{1}{8} & \frac{3}{4} & \frac{1}{8} \\ \frac{1}{8} & \frac{1}{8} & \frac{5}{8} \end{bmatrix}$ (b) $\begin{bmatrix} \frac{1}{2} & \frac{1}{4} & \frac{1}{2} \\ \frac{1}{4} & \frac{1}{2} & \frac{1}{4} \\ \frac{1}{4} & \frac{1}{4} & \frac{1}{4} \end{bmatrix}$ (c) $\begin{bmatrix} \frac{1}{3} & \frac{1}{2} & \frac{4}{9} \\ \frac{1}{3} & \frac{1}{4} & \frac{4}{9} \\ \frac{1}{3} & \frac{1}{4} & \frac{1}{9} \end{bmatrix}$

10. For each of the following matrices A, find a matrix B such that $B^{-1}AB$ is diagonal.

(a) $\begin{bmatrix} 6 & 1 & -3 \\ -5 & 0 & 3 \\ 7 & 1 & -4 \end{bmatrix}$ (b) $\begin{bmatrix} 1 & 1 & -1 \\ 2 & 1 & 2 \\ 2 & -1 & 4 \end{bmatrix}$ (c) $\begin{bmatrix} 3 & 2 & -2 \\ 4 & 1 & -2 \\ 8 & 4 & -5 \end{bmatrix}$

11. Show that two diagonalizable matrices are similar if and only if they have the same characteristic polynomial.

12. (a) If A is an $n \times n$ matrix and $A^T = \lambda A$, show that $\lambda = \pm 1$, if $A \neq 0$.
(b) Show that the only eigenvalues of the linear transformation of the space of $n \times n$ matrices into itself defined by $T(A) = A^T$ are ± 1.
(c) Determine the eigenspaces of each eigenvalue.

13. Find a general formula for the nth power of the following matrices.

(a) $\begin{bmatrix} 2 & -2 \\ 2 & -3 \end{bmatrix}$ (b) $\begin{bmatrix} 3 & 1 \\ 1 & 3 \end{bmatrix}$ (c) $\begin{bmatrix} 2 & 1 \\ 1 & 2 \end{bmatrix}$

14. Let T be a linear operator on a finite-dimensional vector space V. Show that T is invertible if and only if 0 is not an eigenvalue of T.

15. Let $A = \begin{bmatrix} a & b \\ c & d \end{bmatrix}$ be a 2×2 real matrix, and suppose that all of the entries of A are positive numbers.
(a) Show that the eigenvalues of A are given by the formula

$$\tfrac{1}{2}\left((a+d) \pm \sqrt{(a-d)^2 + 4bc}\right)$$

(b) Show that the eigenvalues of A are real and distinct.
(c) Show that at least one eigenvalue of A is positive.
(d) Show that there is an eigenvector corresponding to the largest eigenvalue such that both components of the eigenvector are positive.

16. As in example 3 of §2.5, we study the population changes of a certain animal species. Here we suppose that there are only three age groups. Let $\mathbf{v}$ be the 3-vector whose components give the number of members of each age group. Let $T(\mathbf{v})$ give the population distribution after a year has elapsed. We suppose that $T(\mathbf{v})$ is given by

$$T(\mathbf{v}) = A\mathbf{v} = \begin{bmatrix} \frac{1}{4} & \frac{5}{6} & \frac{1}{4} \\ \frac{3}{4} & 0 & 0 \\ 0 & \frac{2}{3} & 0 \end{bmatrix} \mathbf{v}$$

(a) Interpret the entries of the matrix A.
(b) Find the characteristic polynomial and eigenvalues of A.
(c) Find a matrix B such that $B^{-1}AB$ is diagonal.

(d) Prove that

$$A^n = \frac{1}{30}\begin{bmatrix} 16 & 16 & 4 \\ 12 & 12 & 3 \\ 8 & 8 & 2 \end{bmatrix} + \frac{1}{30}\left(-\frac{1}{2}\right)^n \begin{bmatrix} 20 & -20 & -10 \\ -30 & 30 & 15 \\ 40 & -40 & -20 \end{bmatrix} + \frac{1}{30}\left(-\frac{1}{4}\right)^n \begin{bmatrix} -6 & 4 & 6 \\ 18 & -12 & -18 \\ -48 & 32 & 48 \end{bmatrix}$$

(e) If the initial population is $\begin{bmatrix} x_1 \\ x_2 \\ x_3 \end{bmatrix}$, show that the population distribution approaches $\dfrac{(4x_1 + 4x_2 + x_3)}{30}\begin{bmatrix} 4 \\ 3 \\ 2 \end{bmatrix}$ after enough time has elapsed. (In fact, after 10 years the distribution is correct to within 1% of the total population.)

17. Let $A_1, A_2, A_3, \ldots$ be a sequence of real $n \times n$ matrices. The sequence is said to approach the limit L if for all i and j, the sequence of entries in position (i, j) approaches the entry of L in position (i, j). This is written as $\lim A_n = L$.
   (a) If $\lim A_n = L$ and B is an invertible $n \times n$ matrix, show that $\lim B^{-1}A_nB = B^{-1}LB$.
   (b) Let A be a 2×2 real matrix having positive entries. Show that the sequence $A, A^2, A^3, \ldots$ of powers of A has a limit if and only if the eigenvalues λ_1 and λ_2 of A satisfy $-1 < \lambda_1 \leqslant 1$ and $-1 < \lambda_2 \leqslant 1$. [Hint: Use exercise 15.]
   (c) If in (b) neither eigenvalue is 1, show that the limit of the sequence is 0.

18. On the vector space P_n, show that the linear operator $T(f) = xf'$ is diagonalizable.

19. For each of the following matrices A find a complex matrix B such that $B^{-1}AB$ is diagonal.

$$\text{(a)} \begin{bmatrix} 3 & -4 \\ 4 & 3 \end{bmatrix} \qquad \text{(b)} \begin{bmatrix} 0 & 1 \\ -1 & 0 \end{bmatrix} \qquad \text{(c)} \begin{bmatrix} 0 & 0 & 1 \\ 1 & 0 & 0 \\ 0 & 1 & 0 \end{bmatrix}$$

20. Let a, b, and c be arbitrary numbers. A sequence is defined according to the following rule: $x_1 = a$, $x_2 = b$, $x_3 = c$, and for all $k > 3$, x_k is the average of the three elements of the sequence that precedes it. Thus, $x_4 = \frac{1}{3}(a + b + c)$, $x_5 = \frac{1}{9}(a + 4b + 4c)$, and so on. Show that the limit of the sequence is $\frac{1}{6}(a + 2b + 3c)$. [Hint: Let v_k be the vector with first component x_{k+2}, second component x_{k+1}, and third component x_k. Find a matrix A such that $v_{k+1} = Av_k$ for all k, and compute the powers of A.]

21. Let A and B be two $n \times n$ matrices (real or complex), and let x be a fixed scalar.
   (a) Show that if λ is sufficiently large,

$$\det\left[xI_n - (A - \lambda I_n)B\right] = \det\left[xI_n - B(A - \lambda I_n)\right]$$

   [Hint: If λ is big enough, $A - \lambda I_n$ is invertible.]
   (b) Using part (a) show that for all λ

$$\det\left[xI_n - (A - \lambda I_n)B\right] = \det\left[xI_n - B(A - \lambda I_n)\right]$$

   [Hint: Two polynomials that are equal at infinitely many points are equal everywhere.]
   (c) Show that $C_{AB}(x) = C_{BA}(x)$.

22. Let A be an $n \times n$ real matrix with n real distinct eigenvalues. If B is real and commutes with A, show that B is diagonalizable.

23. If

$$A = \begin{bmatrix} a_{11} & a_{12} & a_{13} \\ a_{21} & a_{22} & a_{23} \\ a_{31} & a_{32} & a_{33} \end{bmatrix}$$

show that the coefficient of x in $C_A(x)$ is

$$\begin{vmatrix} a_{11} & a_{12} \\ a_{21} & a_{22} \end{vmatrix} + \begin{vmatrix} a_{11} & a_{13} \\ a_{31} & a_{33} \end{vmatrix} + \begin{vmatrix} a_{22} & a_{23} \\ a_{32} & a_{33} \end{vmatrix}$$

2 SYMMETRIC MATRICES

If A and B are two $n \times n$ matrices and there is an orthogonal matrix U such that $A = U^{-1}BU$, we say that A and B are **orthogonally similar**. Since U is orthogonal, $U^{-1} = U^T$. Thus, $A = U^TBU$. If U is unitary and $A = U^{-1}BU$, A and B are said to be **unitarily similar**. In this section, we show that any symmetric matrix is orthogonally similar to a diagonal matrix. In particular, it follows that any symmetric matrix is diagonalizable.

In this section, we often use the standard inner product on $\boldsymbol{R}^n$ defined by $(\boldsymbol{x}, \boldsymbol{y}) = \boldsymbol{x}^T\boldsymbol{y}$. On $\boldsymbol{C}^n$ we use the inner product $(\boldsymbol{x}, \boldsymbol{y}) = \boldsymbol{x}^*\boldsymbol{y}$ of §6.2.

Suppose that A is a real symmetric matrix and $T(\boldsymbol{x}) = A\boldsymbol{x}$ is the linear operator induced by A. We observe that

$$(T(\boldsymbol{x}), \boldsymbol{y}) = (T(\boldsymbol{x}))^T\boldsymbol{y} = (A\boldsymbol{x})^T\boldsymbol{y} = \boldsymbol{x}^TA^T\boldsymbol{y} = \boldsymbol{x}^TA\boldsymbol{y} = (\boldsymbol{x}, T(\boldsymbol{y}))$$

Thus, T has the property that $(T(\boldsymbol{x}), \boldsymbol{y}) = (\boldsymbol{x}, T(\boldsymbol{y}))$. Any linear operator on an inner product space that has this property is said to be **symmetric**. It is not hard to show (using exercise 14 of §6.2) that the matrix representation of a symmetric operator, with respect to the standard basis for $\boldsymbol{R}^n$ (or any other orthonormal basis) is symmetric. In proving things about symmetric matrices, it is generally useful to keep the property $(T(\boldsymbol{x}), \boldsymbol{y}) = (\boldsymbol{x}, T(\boldsymbol{y}))$ in mind.

Our first objective is to prove that all eigenvalues of a symmetric matrix are real numbers.

Theorem 1 Let A be an $n \times n$ real symmetric matrix. Then any eigenvalue of A is real.

PROOF Let T be the linear operator on $\boldsymbol{C}^n$ defined by $T(\boldsymbol{x}) = A\boldsymbol{x}$. Let $(\ , \ )$ be the standard inner product on $\boldsymbol{C}^n$, i.e., $(\boldsymbol{x}, \boldsymbol{y}) = \boldsymbol{x}^*\boldsymbol{y}$, for all $\boldsymbol{x}$ and $\boldsymbol{y}$ in $\boldsymbol{C}^n$. As above, $(T(\boldsymbol{x}), \boldsymbol{y}) = (\boldsymbol{x}, T(\boldsymbol{y}))$.

If $T(\boldsymbol{x}) = \lambda\boldsymbol{x}$,

$$(T(\boldsymbol{x}), \boldsymbol{x}) = (\lambda\boldsymbol{x}, \boldsymbol{x}) = \lambda(\boldsymbol{x}, \boldsymbol{x})$$

and

$$(\boldsymbol{x}, T(\boldsymbol{x})) = (\boldsymbol{x}, \lambda\boldsymbol{x}) = \bar{\lambda}(\boldsymbol{x}, \boldsymbol{x})$$

Since

$$(T(x), x) = (x, T(x)), \text{ as } T \text{ is symmetric}$$

$$\lambda(x, x) = \bar{\lambda}(x, x)$$

Since $x \neq \mathbf{0}$, $(x, x) \neq 0$, and $\lambda = \bar{\lambda}$. Thus, λ is real. ■

Note that the same result holds if A is an $n \times n$ Hermitian matrix, that is, $A = A^*$. Indeed,

$$(Ax, y) = (Ax)^* y = x^* A^* y = x^* A y = (x, Ay)$$

If $Ax = \lambda x$, the proof above yields $\lambda = \bar{\lambda}$.

If T is a symmetric operator on an inner product space, by observing that the matrix representation of T with respect to an orthonormal basis is Hermitian, it follows that all eigenvalues of T are real.

We say that a matrix is **orthogonally diagonalizable** if it is orthogonally similar to a diagonal matrix. In §7.1 we saw that a matrix A is diagonalizable over the reals if and only if $\mathbf{R}^n$ admits a basis of eigenvectors of A. In the same way, it may be shown that a matrix A is orthogonally diagonalizable if and only if $\mathbf{R}^n$ admits an orthonormal basis of eigenvectors of A. With the same idea in mind we say that a linear operator T on an inner product space V is orthogonally diagonalizable if V admits an orthonormal basis of T eigenvectors.

We now prove the main theorem of this section.

Theorem 2 Let T be a symmetric linear operator on a finite-dimensional inner product space V. Then V has an orthonormal basis of eigenvectors of T.

PROOF The proof is by induction on dim V. The theorem is obvious if dim $V = 1$.

We know that T has a (possibly complex) eigenvalue. (Take any root of the characteristic polynomial.) By Theorem 1, this root is real. Thus, T has a real eigenvalue. Let x be an eigenvector corresponding to this eigenvalue. So we have $T(x) = \lambda x$, with λ real. We may assume $|x| = 1$.

Let H be the orthogonal complement of x in V. We claim: If h belongs to H, so does $T(h)$.

Indeed,

$$\begin{aligned} (T(h), x) &= (h, T(x)) && \text{(by symmetry of } T\text{)} \\ &= (h, \lambda x) \\ &= \lambda(h, x) \\ &= 0 && \text{(since } h \text{ belongs to } H\text{)} \end{aligned}$$

Thus, $T(h)$ belongs to H, as well.

It follows that if T is restricted to H, T is a linear operator on H. Since $\dim H = \dim V - 1$, by induction hypothesis, H has an orthonormal basis of T eigenvectors, say $x_2, x_3, \ldots, x_n$. Thus, $x_1, x_2, \ldots, x_n$ is an orthonormal basis for V of eigenvectors of T. ■

As a consequence of this theorem, we have

(1) If A is a real symmetric matrix, A is orthogonally similar to a real diagonal matrix.
(2) If A is a Hermitian matrix, A is unitarily similar to a real diagonal matrix.

It is interesting to note that the only matrices that are orthogonally similar to diagonal matrices are symmetric. Indeed, let A be a matrix that is orthogonally similar to the diagonal matrix D. Then, there is an orthogonal matrix U such that $U^{-1}AU = D$. It follows that $A = UDU^{-1}$. Since U is orthogonal, $U^{-1} = U^T$. Thus, $A = UDU^T$. Therefore, $A^T = (UDU^T)^T = (U^T)^T D^T U^T = UD^T U^T$. Now since D is diagonal, $D^T = D$. Thus, $A^T = UDU^T$. It follows that $A^T = A$, and so A is symmetric. Thus, a real matrix is orthogonally diagonalizable if and only if it is symmetric.

Example 1 If $A = \begin{bmatrix} 2 & 1 \\ 1 & 2 \end{bmatrix}$, find an orthogonal matrix U such that $U^{-1}AU$ is diagonal.

In this case, $C_A(x) = x^2 - 4x + 3 = (x-1)(x-3)$. Thus, the eigenvalues of A are 1 and 3. To find the corresponding eigenvectors, we must solve the systems

$$\begin{aligned} 2x + y &= x \\ x + 2y &= y \end{aligned} \quad \text{and} \quad \begin{aligned} 2x + y &= 3x \\ x + 2y &= 3y \end{aligned}$$

Doing this, we obtain the eigenvectors $\begin{bmatrix} c \\ -c \end{bmatrix}$ and $\begin{bmatrix} d \\ d \end{bmatrix}$, with c and d arbitrary. Next, we choose c and d so that the eigenvectors have length 1. Thus, take $c = d = \frac{1}{\sqrt{2}}$. The matrix U can be found by writing the eigenvectors in its successive columns. We find that

$$U = \frac{1}{\sqrt{2}} \begin{bmatrix} 1 & 1 \\ -1 & 1 \end{bmatrix}$$

Note that U is an orthogonal matrix. By the way U has been chosen, it follows that $U^{-1}AU$ is the diagonal matrix with diagonal entries 1 and 3. We verify this although it is not really necessary. Note

$$U^{-1}AU = U^TAU = \frac{1}{\sqrt{2}} \begin{bmatrix} 1 & -1 \\ 1 & 1 \end{bmatrix} \begin{bmatrix} 2 & 1 \\ 1 & 2 \end{bmatrix} \cdot \frac{1}{\sqrt{2}} \begin{bmatrix} 1 & 1 \\ -1 & 1 \end{bmatrix}$$

$$= \frac{1}{\sqrt{2}} \begin{bmatrix} 1 & -1 \\ 1 & 1 \end{bmatrix} \cdot \frac{1}{\sqrt{2}} \begin{bmatrix} 1 & 3 \\ -1 & 3 \end{bmatrix} = \begin{bmatrix} 1 & 0 \\ 0 & 3 \end{bmatrix}$$

Example 2 To what diagonal matrix is the following matrix A similar?

$$A = \begin{bmatrix} 3 & 2 & 2 \\ 2 & 2 & 0 \\ 2 & 0 & 4 \end{bmatrix}$$

Since A is symmetric, we know that A is similar to a diagonal matrix. To find which diagonal matrix A is similar to, we need only compute its characteristic polynomial.

$$C_A(x) = \begin{vmatrix} x-3 & -2 & -2 \\ -2 & x-2 & 0 \\ -2 & 0 & x-4 \end{vmatrix} = x^3 - 9x + 18x = x(x-3)(x-6)$$

Thus, A is similar to a diagonal matrix having diagonal entries 0, 3, and 6.

Example 3 For the following matrix A, find a matrix U such that $U^{-1}AU$ is diagonal.

$$A = \begin{bmatrix} -1 & 2 & 2 \\ 2 & -1 & 2 \\ 2 & 2 & -1 \end{bmatrix}$$

First, we compute the characteristic polynomial

$$C_A(x) = \begin{vmatrix} x+1 & -2 & -2 \\ -2 & x+1 & -2 \\ -2 & -2 & x+1 \end{vmatrix} = x^3 + 3x^2 - 9x - 27 = (x-3)(x+3)^2$$

Next, the eigenspaces corresponding to the eigenvalues 3 and -3 can be found by solving the equations

$$\begin{aligned} -4x + 2y + 2z &= 0 \\ 2x - 4y + 2z &= 0 \\ 2x + 2y - 4z &= 0 \end{aligned} \quad \text{and} \quad \begin{aligned} 2x + 2y + 2z &= 0 \\ 2x + 2y + 2z &= 0 \\ 2x + 2y + 2z &= 0 \end{aligned}$$

The solution to the first system is $x = y = z = c$, with c arbitrary. For the second system, any vector satisfying $x + y + z = 0$ is a solution. Thus, a unit eigenvector corresponding to the eigenvalue 3 is

$$\begin{bmatrix} \frac{1}{\sqrt{3}} \\ \frac{1}{\sqrt{3}} \\ \frac{1}{\sqrt{3}} \end{bmatrix}$$

For the eigenvalue -3, we must choose two orthogonal unit vectors satisfying $x + y + z = 0$. We write one unit vector satisfying this equation. A second is then determined by the equation and the fact that it is orthogonal to the first.

We take as our first vector

$$\begin{bmatrix} \frac{1}{\sqrt{2}} \\ -\frac{1}{\sqrt{2}} \\ 0 \end{bmatrix}$$

Then the second satisfies $x + y + z = 0$ and $x - y = 0$. It thus has the form $x = c, y = c, z = -2c$. To insure that the vector is a unit vector, we take $c = \frac{1}{\sqrt{6}}$. Thus,

$$U = \begin{bmatrix} \frac{1}{\sqrt{3}} & \frac{1}{\sqrt{2}} & \frac{1}{\sqrt{6}} \\ \frac{1}{\sqrt{3}} & -\frac{1}{\sqrt{2}} & \frac{1}{\sqrt{6}} \\ \frac{1}{\sqrt{3}} & 0 & -\frac{2}{\sqrt{6}} \end{bmatrix}$$

EXERCISES

1. For each of the following matrices A determine an orthogonal matrix U such that $U^{-1}AU$ is diagonal.

$$\text{(a)} \begin{bmatrix} 2 & 3 \\ 3 & 2 \end{bmatrix} \quad \text{(b)} \begin{bmatrix} 5 & 2 \\ 2 & 2 \end{bmatrix} \quad \text{(c)} \begin{bmatrix} 7 & 6 \\ 6 & 2 \end{bmatrix} \quad \text{(d)} \begin{bmatrix} 0 & 1 \\ 1 & 0 \end{bmatrix}$$

$$\text{(e)} \begin{bmatrix} 2 & -4 & 2 \\ -4 & 2 & -2 \\ 2 & -2 & -1 \end{bmatrix} \quad \text{(f)} \begin{bmatrix} -3 & -6 & 0 \\ -6 & 0 & 6 \\ 0 & 6 & 3 \end{bmatrix} \quad \text{(g)} \begin{bmatrix} -1 & -2 & 1 \\ -2 & 2 & -2 \\ 1 & -2 & -1 \end{bmatrix}$$

2. Find a 3×3 matrix A having eigenvalues $1, 0, -1$ and corresponding eigenvectors

$$\tfrac{1}{3}\begin{bmatrix} -1 \\ 2 \\ 2 \end{bmatrix}, \quad \tfrac{1}{3}\begin{bmatrix} 2 \\ -1 \\ 2 \end{bmatrix}, \quad \tfrac{1}{3}\begin{bmatrix} 2 \\ 2 \\ -1 \end{bmatrix}$$

3. For each of the following matrices A determine a unitary matrix U such that $U^{-1}AU$ is diagonal.

$$\text{(a)} \begin{bmatrix} 0 & i \\ -i & 0 \end{bmatrix} \quad \text{(b)} \begin{bmatrix} 1 & 1+i \\ 1-i & 0 \end{bmatrix}$$

4. Let B be a real 3×3 matrix that is similar to a diagonal matrix having diagonal entries x_1, x_2, x_3. Show that $aI_3 + bB$ is similar to a diagonal matrix having diagonal entries $a + x_1b, a + x_2b, a + x_3b$. To which diagonal matrices are the following matrices similar?

$$\text{(a)} \begin{bmatrix} a & b \\ b & a \end{bmatrix} \quad \text{(b)} \begin{bmatrix} a & b & b \\ b & a & b \\ b & b & a \end{bmatrix} \quad \text{(c)} \begin{bmatrix} a & 0 & b \\ 0 & a & 0 \\ b & 0 & a \end{bmatrix}$$

5. To what diagonal matrix is the matrix $\begin{bmatrix} a & c \\ c & b \end{bmatrix}$ similar?

6. Let A be an $n \times n$ complex matrix such that $A^* = -A$. Show that A is unitarily similar to a diagonal matrix whose diagonal entries are purely imaginary. [Hint: Show that iA is Hermitian.]

7. Let A be an $n \times n$ real skew-symmetric matrix. Show that $I + A$ is invertible and $(I - A)(I + A)^{-1}$ is orthogonal.

8. Show that two real symmetric matrices are orthogonally similar if and only if they have the same characteristic polynomial.

9. Show that two real symmetric matrices that are similar are orthogonally similar.

10. Let A be an $n \times n$ real symmetric matrix having eigenvalues $\lambda_1, \lambda_2, \ldots, \lambda_n$. Show that $\det A = \lambda_1\lambda_2 \ldots \lambda_n$.

11. Show that the only real eigenvalues of an orthogonal matrix are ± 1.

12. Let A be an $n \times n$ real symmetric matrix. Show that the following are equivalent:
    (a) A is orthogonal.
    (b) $A^2 = I_n$.
    (c) All eigenvalues of A are ± 1.

13. Show that eigenvectors belonging to different eigenvalues of a symmetric matrix are orthogonal.

3 UPPER TRIANGULAR FORM

In the last section, we saw that every real symmetric matrix is similar to a diagonal matrix. In this section we shall show that every $n \times n$ complex matrix is similar to an upper triangular matrix of a special type.

Often in this section we consider block matrices of the type

$$\begin{bmatrix} B_1 & 0 & 0 & \cdots & 0 \\ 0 & B_2 & 0 & \cdots & 0 \\ 0 & 0 & B_3 & \cdots & 0 \\ & \vdots & & & \\ 0 & 0 & 0 & \cdots & B_k \end{bmatrix}$$

where B_1 is an $n_1 \times n_1$ matrix, B_2 an $n_2 \times n_2$ matrix, and so on. For example, we might write

$$\begin{bmatrix} b_{11} & b_{12} & 0 & 0 \\ b_{21} & b_{22} & 0 & 0 \\ 0 & 0 & c_{11} & c_{12} \\ 0 & 0 & c_{21} & c_{22} \end{bmatrix} = \begin{bmatrix} B & 0 \\ 0 & C \end{bmatrix}$$

with B and C the 2×2 matrices $[b_{ij}]$ and $[c_{ij}]$.

Manipulation of block matrices is quite simple. For example, if B_1 and B_2 are 2×2 matrices and C_1 and C_2 are 3×3 matrices, it is not hard to show that

$$\begin{bmatrix} B_1 & 0 \\ 0 & C_1 \end{bmatrix} + \begin{bmatrix} B_2 & 0 \\ 0 & C_2 \end{bmatrix} = \begin{bmatrix} B_1 + B_2 & 0 \\ 0 & C_1 + C_2 \end{bmatrix}$$

and

$$\begin{bmatrix} B_1 & 0 \\ 0 & C_1 \end{bmatrix} \begin{bmatrix} B_2 & 0 \\ 0 & C_2 \end{bmatrix} = \begin{bmatrix} B_1 B_2 & 0 \\ 0 & C_1 C_2 \end{bmatrix}$$

The main result of this section is:

Theorem 1 Let A be an $n \times n$ matrix having characteristic polynomial $C_A(x) = (x - \lambda_1)^{n_1} \ldots (x - \lambda_k)^{n_k}$, where $\lambda_1, \lambda_2, \ldots, \lambda_k$ are distinct. Then A is similar to a matrix of the form

$$\begin{bmatrix} B_1 & 0 & \ldots & 0 \\ 0 & B_2 & \ldots & 0 \\ & \vdots & & \\ 0 & 0 & \ldots & B_k \end{bmatrix}$$

where B_i is an $n_i \times n_i$ upper triangular matrix all of whose diagonal entries are λ_i.

For example, if $C_A(x) = (x - \lambda_1)^2(x - \lambda_2)^2$, with $\lambda_1 \neq \lambda_2$, then A is similar to a matrix of the form

$$\begin{bmatrix} \lambda_1 & s & 0 & 0 \\ 0 & \lambda_1 & 0 & 0 \\ 0 & 0 & \lambda_2 & t \\ 0 & 0 & 0 & \lambda_2 \end{bmatrix}$$

PROOF The proof is by induction on n. Let $T(\boldsymbol{x}) = A\boldsymbol{x}$ be the linear operator on $\boldsymbol{R}^n$ induced by A. A similar proof is valid if $\lambda_1, \ldots, \lambda_n$ are complex, replacing $\boldsymbol{R}^n$ by $\boldsymbol{C}^n$.

Since λ_1 is an eigenvalue of A, there is a vector $\boldsymbol{x}_1 \neq \boldsymbol{0}$, such that $T(\boldsymbol{x}_1) = \lambda_1 \boldsymbol{x}_1$. Take a basis $\boldsymbol{x}_1, \boldsymbol{x}_2, \ldots, \boldsymbol{x}_n$ for $\boldsymbol{R}^n$ containing the vector $\boldsymbol{x}_1$. The matrix representation of T relative to this basis is of the form

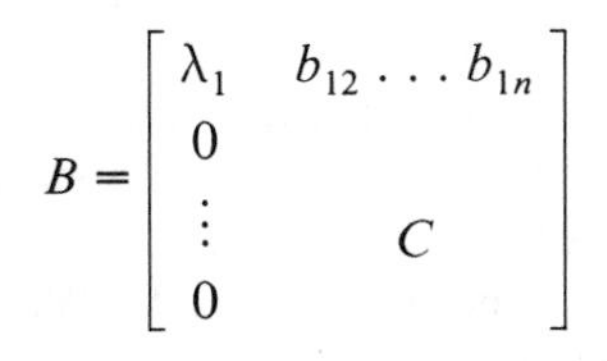

$$B = \begin{bmatrix} \lambda_1 & b_{12} \ldots b_{1n} \\ 0 & \\ \vdots & C \\ 0 & \end{bmatrix}$$

where C is some $(n-1) \times (n-1)$ matrix. Since A and B represent T with respect to different bases, A and B are similar. Thus, $C_B(x) = C_A(x)$. Note also that

$$xI - B = \begin{bmatrix} x - \lambda_1 & -b_{12} \dots -b_{1n} \\ 0 & \\ & xI - C \\ 0 & \end{bmatrix}$$

It follows that $C_A(x) = C_B(x) = (x - \lambda_1)C_C(x)$. Consequently, $C_C(x) = (x - \lambda_1)^{n_1 - 1}(x - \lambda_2)^{n_2} \dots (x - \lambda_k)^{n_k}$.

Now we apply induction to the matrix C. By induction $R^{-1}CR = D$, where D is a matrix of the form

$$D = \begin{bmatrix} D_1 & 0 & \dots & 0 \\ 0 & D_2 & \dots & 0 \\ & \vdots & & \\ 0 & 0 & \dots & D_k \end{bmatrix}$$

where D_1 is an $(n_1 - 1) \times (n_1 - 1)$ upper triangular matrix having all diagonal entries λ_1 and for $i > 1$, D_i is an $n_i \times n_i$ upper triangular matrix with diagonal entries λ_i. Now, we have that

$$\begin{bmatrix} 1 & 0 & \dots & 0 \\ 0 & & & \\ \vdots & & R^{-1} & \\ 0 & & & \end{bmatrix} \begin{bmatrix} \lambda_1 & b_{12} & \dots & b_{1n} \\ 0 & & & \\ \vdots & & C & \\ 0 & & & \end{bmatrix} \begin{bmatrix} 1 & 0 & \dots & 0 \\ 0 & & & \\ \vdots & & R & \\ 0 & & & \end{bmatrix}$$

$$= \begin{bmatrix} \lambda_1 & s_2 & \dots & s_n \\ 0 & & & \\ \vdots & & R^{-1}CR & \\ 0 & & & \end{bmatrix} = \left[\begin{array}{cccc|cccc} \lambda_1 & s_2 & \dots & s_{n_1} & s_{n_1+1} & & \dots & s_n \\ 0 & & & & 0 & & & 0 \\ \vdots & & D_1 & & \vdots & & & \vdots \\ 0 & & & & 0 & & \dots & 0 \\ \hline 0 & \dots & & 0 & D_2 & 0 & \dots & 0 \\ 0 & \dots & & 0 & 0 & D_3 & \dots & 0 \\ \vdots & & & \vdots & \vdots & & & \\ 0 & \dots & & 0 & 0 & 0 & \dots & D_k \end{array}\right]$$

Now, the last matrix is similar to B and hence to A. It is the matrix representation of T with respect to some basis $\boldsymbol{y}_1, \dots, \boldsymbol{y}_n$ for $\boldsymbol{R}^n$. Note that the first block of this matrix is an $n_1 \times n_1$ upper triangular matrix having diagonal entries λ_1. The remaining blocks D_i are also upper triangular matrices of order n_i having diagonal entries λ_i. We modify the basis vectors $\boldsymbol{y}_1, \dots, \boldsymbol{y}_n$ to replace the entries $s_{n_1+1}, \dots, s_n$ by zeros.

First, in the basis $\boldsymbol{y}_1, \dots, \boldsymbol{y}_n$ we will replace the basis element $\boldsymbol{y}_{n_1+1}$ by $\boldsymbol{y}_{n_1+1} + t\boldsymbol{y}_1$, where t will be chosen so that the matrix representation of T has the same form as before with $s_{n_1+1} = 0$.

Now, $T(y_{n_1+1}) = \lambda_2 y_{n_1+1} + s_{n_1+1} y_1$. Thus,

$$T(y_{n_1+1} + ty_1) = \lambda_2 y_{n_1+1} + s_{n_1+1} y_1 + t\lambda_1 y_1$$

$$= \lambda_2(y_{n_1+1} + ty_1) + (s_{n_1+1} + t(\lambda_1 - \lambda_2))\, y_1$$

Since $\lambda_1 - \lambda_2 \neq 0$, t can be chosen so that $s_{n_1+1} + t(\lambda_1 - \lambda_2) = 0$. It follows that $T(y_{n_1+1} + ty_1) = \lambda_2(y_{n_1+1} + ty_1)$. Thus, the matrix representation of T with respect to the basis $y_1, \ldots, y_{n_1}, y_{n_1+1} + ty_1, y_{n_1+2}, \ldots, y_n$ has the same form as before, except that $s_{n_1+1} = 0$.

By repeating the process of adding suitable multiples of y_1 to the remaining vectors of the basis, we eventually obtain a basis in which the matrix representation of T has the same form with $s_{n_1+1} = \cdots = s_n = 0$. ■

Now, if A is any $n \times n$ complex matrix, $C_A(x)$ can be factored in the form $(x - \lambda_1)^{n_1} \ldots (x - \lambda_k)^{n_k}$. Thus, A is similar to an upper triangular matrix. If A is an $n \times n$ real matrix and $C_A(x) = (x - \lambda_1)^{n_1} \ldots (x - \lambda_k)^{n_k}$ with $\lambda_1, \ldots, \lambda_k$ all real, then A is similar to a real upper triangular matrix. However, if A has a complex eigenvalue that is not real, A will not be similar to an upper triangular real matrix. (See exercise 6 of §7.1.)

Our next goal is to give a method for determining an upper triangular matrix to which a given matrix is similar. First, we note a property of upper triangular matrices having the same diagonal entry.

Lemma Let D be an $n \times n$ upper triangular matrix having all diagonal entries λ. Then, $(D - \lambda I)^n = 0$.

PROOF It is not hard to prove this in general, but we write it out explicitly for $n = 3$. Suppose that $B = \begin{bmatrix} \lambda & a & b \\ 0 & \lambda & c \\ 0 & 0 & \lambda \end{bmatrix}$. Then $B - \lambda I = \begin{bmatrix} 0 & a & b \\ 0 & 0 & c \\ 0 & 0 & 0 \end{bmatrix}$ and

$$(B - \lambda I)^2 = \begin{bmatrix} 0 & 0 & ac \\ 0 & 0 & 0 \\ 0 & 0 & 0 \end{bmatrix}, \quad (B - \lambda I)^3 = \begin{bmatrix} 0 & 0 & 0 \\ 0 & 0 & 0 \\ 0 & 0 & 0 \end{bmatrix} \quad ■$$

We now introduce a new concept. If T is a linear operator on a vector space V and $C_T(x) = (x - \lambda_1)^{n_1} \ldots (x - \lambda_k)^{n_k}$, then the **generalized eigenspace** of T corresponding to the eigenvalue λ_i is the nullspace of the operator $(T - \lambda_i I)^{n_i}$. For example, if D is an $n \times n$ upper triangular matrix with all diagonal entries λ, then the last lemma shows that the generalized eigenspace is the entire vector space.

Example 1 Suppose that T is the linear operator on $\boldsymbol{R}^3$ induced by the matrix $A = \begin{bmatrix} -1 & 1 & 1 \\ -2 & 2 & 1 \\ -1 & 1 & 1 \end{bmatrix}$. Then,

$$C_A(x) = \begin{vmatrix} x+1 & -1 & -1 \\ 2 & x-2 & -1 \\ 1 & -1 & x-1 \end{vmatrix} = x^3 - 2x^2 + x = x(x-1)^2$$

We find the generalized eigenspace determined by the eigenvalue 1. Observe that

$$A - I = \begin{bmatrix} -2 & 1 & 1 \\ -2 & 1 & 1 \\ -1 & 1 & 0 \end{bmatrix} \quad \text{and} \quad (A-I)^2 = \begin{bmatrix} 1 & 0 & -1 \\ 1 & 0 & -1 \\ 0 & 0 & 0 \end{bmatrix}$$

The nullspace of the last matrix is spanned by the vectors $\boldsymbol{i} + \boldsymbol{k}$ and $\boldsymbol{j}$. This is the generalized eigenspace determined by 1. In this example, the generalized eigenspace corresponding to 0 is simply the nullspace of A.

We now have:

Theorem 2 Let T be a linear operator on a vector space V having $C_T(x) = (x - \lambda_1)^{n_1} \ldots (x - \lambda_k)^{n_k}$ with $\lambda_1, \ldots, \lambda_k$ distinct. Then

(1) The generalized eigenspace of T corresponding to λ_i is of dimension n_i.
(2) If T has the block representation of Theorem 1 with respect to the basis $\boldsymbol{y}_1, \ldots, \boldsymbol{y}_n$, then $\boldsymbol{y}_1, \ldots, \boldsymbol{y}_{n_1}$ is a basis for the generalized eigenspace corresponding to λ_1, and so on.

PROOF By Theorem 1, there is a basis $\boldsymbol{y}_1, \ldots, \boldsymbol{y}_n$ for V such that the matrix representation of T relative to this basis is

$$\begin{bmatrix} B_1 & 0 & \ldots & 0 \\ 0 & B_2 & \ldots & 0 \\ & & \vdots & \\ 0 & 0 & \ldots & B_k \end{bmatrix}$$

where B_i is an $n_i \times n_i$ upper triangular matrix with diagonal entries λ_i. Now the matrix representation of $(T - \lambda_1 I)^{n_1}$ relative to this basis is

$$\begin{bmatrix} (B_1 - \lambda_1 I)^{n_1} & 0 & \ldots & 0 \\ 0 & (B_2 - \lambda_1 I)^{n_1} & \ldots & 0 \\ & \vdots & & \\ 0 & 0 & (B_k - \lambda_1 I)^{n_1} & \end{bmatrix}$$

Now, B_1 is an $n_1 \times n_1$ matrix with diagonal entries λ_1. By the lemma, $(B_1 - \lambda_1 I)^{n_1} = 0$. All of the other matrices are upper triangular having diagonal entries $(\lambda_i - \lambda_1)^{n_1} \neq 0$. Thus, the nullspace of this matrix is of dimension n_1 and spanned by the vectors $\boldsymbol{y}_1, \ldots, \boldsymbol{y}_{n_1}$. ■

Corollary (Cayley–Hamilton) $C_T(T) = 0$.

PROOF $C_T(T) = (T - \lambda_1 I)^{n_1} \ldots (T - \lambda_k I)^{n_k}$. Now, we compute $C_T(T)$ relative to the basis $\boldsymbol{y}_1, \ldots, \boldsymbol{y}_n$ of the theorem. $(T - \lambda_1)^{n_1}$ is 0 on the first block, as we saw in the proof. Likewise, $(T - \lambda_2 I)^{n_2}$ is 0 on the second block, and so on. When the product of all of the matrices is taken, we find $C_T(T) = 0$. ■

What the corollary says, in effect, is that any matrix satisfies its characteristic polynomial.

Example 2 Let $A = \begin{bmatrix} 3 & 2 \\ 1 & 4 \end{bmatrix}$. Then $C_A(x) = x^2 - 7x + 10$. Then, $A^2 - 7A + 10I = \begin{bmatrix} 11 & 14 \\ 7 & 18 \end{bmatrix} - 7\begin{bmatrix} 3 & 2 \\ 1 & 4 \end{bmatrix} + \begin{bmatrix} 10 & 0 \\ 0 & 10 \end{bmatrix} = \begin{bmatrix} 0 & 0 \\ 0 & 0 \end{bmatrix}$.

We now present an algorithm for computing a basis that yields the matrix representation of Theorem 1. Theorem 2 tells us, in effect, that we must compute bases for each of the generalized eigenspaces. By being a little more careful, we may insure that the resulting matrix is in upper triangular form. Note that we have the following nested inclusion of subspaces:

$$N_{T-\lambda_1 I} \subset N_{(T-\lambda_1 I)^2} \subset N_{(T-\lambda_1 I)^3} \subset \cdots$$

Indeed, it is clear that if $(T - \lambda_1 I)^k \boldsymbol{x} = \boldsymbol{0}$, then $(T - \lambda_1 I)^{k+1}\boldsymbol{x} = \boldsymbol{0}$.

To find an upper triangular basis, first find a basis $\boldsymbol{x}_1, \ldots, \boldsymbol{x}_p$ for $N_{T-\lambda_1 I}$. Next, extend this to a basis $\boldsymbol{x}_1, \ldots, \boldsymbol{x}_p, \boldsymbol{y}_1, \ldots, \boldsymbol{y}_q$ for $N_{(T-\lambda_1 I)^2}$. Then by adding vectors $\boldsymbol{z}_1, \ldots, \boldsymbol{z}_r$, find a basis for $N_{(T-\lambda_1 I)^3}$. Continue the process until a basis of n_1 vectors for the generalized eigenspace are found. Then repeat the whole procedure for the other eigenvalues. Note that since $(T - \lambda_1 I)^2 \boldsymbol{y}_i = \boldsymbol{0}$, $(T - \lambda_1 I)\boldsymbol{y}_i$ is a linear combination of $\boldsymbol{x}_1, \ldots, \boldsymbol{x}_p$. Thus, $T(\boldsymbol{y}_i) = \lambda_1 \boldsymbol{y}_i + c_{1i}\boldsymbol{x}_i + \cdots + c_{pi}\boldsymbol{x}_p$. Repeating the argument for the $\boldsymbol{z}$'s and so on, it follows that the representation is upper triangular.

In the next section, we shall refine this process to achieve a sharper upper triangular form.

Example 3 Find a basis so that the linear transformation induced by the following matrix A is upper triangular.

$$A = \begin{bmatrix} -3 & 2 & 1 & 1 \\ -6 & 3 & 3 & 1 \\ -3 & 2 & 0 & 2 \\ -2 & 2 & 1 & 0 \end{bmatrix}$$

Then

$$C_A(x) = \begin{vmatrix} x+3 & -2 & -1 & -1 \\ 6 & x-3 & -3 & -1 \\ 3 & -2 & x & -2 \\ 2 & -2 & -1 & x \end{vmatrix} = \begin{vmatrix} x+1 & 0 & -1 & -x-1 \\ 0 & x+3 & -3 & -3x-1 \\ 2x+3 & -2x-2 & x & x^2-2 \\ 0 & 0 & -1 & 0 \end{vmatrix}$$

$$= \begin{vmatrix} x+1 & 0 & -x-1 \\ 0 & x+3 & -3x-1 \\ 2x+3 & -2x-2 & x^2-2 \end{vmatrix} = \begin{vmatrix} x+1 & 0 & 0 \\ 0 & x+3 & -3x-1 \\ 2x+3 & -2x-2 & x^2+2x+1 \end{vmatrix}$$

$$= (x+1)\big((x+3)(x^2+2x+1) - (2x+2)(3x+1)\big)$$

$$= (x+1)^2((x+3)(x+1) - 2(3x+1))$$

$$= (x+1)^2(x^2 - 2x + 1) = (x+1)^2(x-1)^2$$

Next we compute the nullspaces of the matrices $A + I$, $(A + I)^2$, $A - I$, $(A - I)^2$. We have

$$A + I = \begin{bmatrix} -2 & 2 & 1 & 1 \\ -6 & 4 & 3 & 1 \\ -3 & 2 & 1 & 2 \\ -2 & 2 & 1 & 1 \end{bmatrix}, \quad (A + I)^2 = \begin{bmatrix} -13 & 8 & 6 & 3 \\ -23 & 12 & 10 & 5 \\ -13 & 8 & 6 & 3 \\ -13 & 8 & 6 & 3 \end{bmatrix}$$

By solving linear equations, the nullspaces of the first and second matrices are spanned by the vectors

$$\boldsymbol{x}_1 = \begin{bmatrix} 1 \\ -1 \\ 3 \\ 1 \end{bmatrix} \quad \text{and} \quad \boldsymbol{x}_1 = \begin{bmatrix} 1 \\ -1 \\ 3 \\ 1 \end{bmatrix}, \quad \boldsymbol{x}_2 = \begin{bmatrix} 0 \\ 0 \\ 1 \\ -2 \end{bmatrix}$$

Similar calculations show that the nullspace of $A - I$ and $(A - I)^2$ is spanned by

$$\boldsymbol{x}_3 = \begin{bmatrix} 1 \\ 1 \\ 1 \\ 1 \end{bmatrix} \quad \text{and} \quad \boldsymbol{x}_3 = \begin{bmatrix} 1 \\ 1 \\ 1 \\ 1 \end{bmatrix}, \quad \boldsymbol{x}_4 = \begin{bmatrix} 0 \\ 1 \\ 0 \\ 0 \end{bmatrix}$$

Now, we have

$$A\boldsymbol{x}_1 = -\boldsymbol{x}_1,\ A\boldsymbol{x}_2 = -\boldsymbol{x}_1 - \boldsymbol{x}_2,\ A\boldsymbol{x}_3 = \boldsymbol{x}_3,\ A\boldsymbol{x}_4 = \boldsymbol{x}_4 + 2\boldsymbol{x}_3$$

Take $B = [\boldsymbol{x}_1, \boldsymbol{x}_2, \boldsymbol{x}_3, \boldsymbol{x}_4]$. Then it follows that

$$B^{-1}AB = \begin{bmatrix} -1 & -1 & 0 & 0 \\ 0 & -1 & 0 & 0 \\ 0 & 0 & 1 & 2 \\ 0 & 0 & 0 & 1 \end{bmatrix}$$

EXERCISES

1. Find the characteristic polynomial and generalized eigenspaces of the following matrices.

   (a) $\begin{bmatrix} -1 & 1 & 1 \\ -3 & 2 & 2 \\ -2 & 1 & 2 \end{bmatrix}$ (b) $\begin{bmatrix} -2 & 1 & 1 \\ -6 & 3 & 2 \\ -1 & 1 & 0 \end{bmatrix}$ (c) $\begin{bmatrix} -1 & 1 & 1 \\ -3 & 1 & 3 \\ 0 & 1 & 0 \end{bmatrix}$

2. For each of the following matrices A find a matrix B so that $B^{-1}AB$ is in the upper triangular form of Theorem 1.

   (a) $\begin{bmatrix} 2 & 1 & -1 \\ 0 & -1 & 4 \\ -3 & -1 & 0 \end{bmatrix}$ (b) $\begin{bmatrix} -1 & 1 & 2 \\ 0 & -1 & -1 \\ -1 & 1 & 2 \end{bmatrix}$ (c) $\begin{bmatrix} 3 & 1 & 2 & -1 \\ 3 & 2 & 3 & -1 \\ -1 & -1 & -1 & 0 \\ 4 & 1 & 2 & -2 \end{bmatrix}$

3. Let A be an $n \times n$ matrix whose distinct eigenvalues are $\lambda_1, \ldots, \lambda_k$. Show that A is diagonalizable if and only if the eigenspace corresponding to λ_i is the generalized eigenspace corresponding to λ_i for all i.

4. If A is an $n \times n$ nilpotent matrix, show that the only eigenvalue of A is 0.

5. If A is an $n \times n$ nilpotent matrix, show that $A^n = 0$.

6. Show that there is no 2×2 matrix A such that

$$A^2 = \begin{bmatrix} 0 & 1 \\ 0 & 0 \end{bmatrix}$$

7. Let A be a real 3×3 matrix such that $\det A > 0$. Show that A has a positive eigenvalue.

8. Let A be a real 2×2 matrix and $\boldsymbol{x}$ a 2-vector such that $\boldsymbol{x}$ and $A\boldsymbol{x}$ are linearly independent. Show that the linear operator T which A induces on $\boldsymbol{R}^2$ is represented relative to the basis $\boldsymbol{x}, A\boldsymbol{x}$ by the matrix $\begin{bmatrix} 0 & -\det A \\ 1 & \operatorname{tr} A \end{bmatrix}$.

9. Let T be a linear operator on $\boldsymbol{R}^2$ such that for all vectors $\boldsymbol{x}$ in $\boldsymbol{R}^2$, $\boldsymbol{x}$ and $T(\boldsymbol{x})$ are linearly dependent. Show that T is a scalar multiple of the identity.

10. Let A and B be two real 2×2 matrices neither of which is a scalar multiple of the identity. Suppose that $C_A(x) = C_B(x)$. Show that A and B are similar. (Hint: Use exercises 8 and 9.)

11. Show that the characteristic polynomial of the matrix

$$\begin{bmatrix} 0 & 0 & -a_0 \\ 1 & 0 & -a_1 \\ 0 & 1 & -a_2 \end{bmatrix}$$

is $x^3 + a_2x^2 + a_1x + a_0$.

12. Let A be a 2×2 matrix having distinct eigenvalues s and t. Show that
(a) $A^2 = (s + t)A - stI$.
(b) Using induction, prove the formula

$$A^n = \left(\frac{s^n - t^n}{s - t} \right) A + \left(\frac{st^n - ts^n}{s - t} \right) I_2$$

13. Let A be the 2×2 stochastic matrix $A = \begin{bmatrix} 1-a & b \\ a & 1-b \end{bmatrix}$, where $0 < a < 1$ and $0 < b < 1$. Prove that

$$\lim_{n \to \infty} A^n = \frac{1}{a+b} A + \frac{a+b-1}{a+b} I_2 = \frac{1}{a+b} \begin{bmatrix} b & b \\ a & a \end{bmatrix}$$

[Hint: Use exercise 12.]

14. Let T be a linear operator on V whose characteristic polynomial is $(x - \lambda_1)^{n_1} \ldots (x - \lambda_k)^{n_k}$ with $\lambda_1, \ldots, \lambda_k$ distinct. If $j \geqslant n_i$, show that $r((T - \lambda_i)^j) = n - n_i$, where $n = \dim V$. [Hint: Take a basis so that T has the matrix representation of Theorem 1 and proceed as in Theorem 2.]

15. Let N be an $n \times n$ nilpotent matrix. Show that $I + N$ has a square root, i.e., there is a matrix B such that $B^2 = I + N$. The following steps are suggested.

(a) Let $(1+x)^{\frac{1}{2}} = a_0 + a_1x + a_2x^2 + a_3x^3 + \cdots$ be the McLaurin expansion of the function $(1+x)^{\frac{1}{2}}$. Show that $a_0^2 = 1$, $2a_0a_1 = 1$, and $\sum_{i=0}^{m} a_i a_{m-i} = 0$ if $m \geqslant 2$.

(b) Let $B = a_0I + a_1N + a_2N^2 + \cdots$, and show that $B^2 = I + N$.

16. Let A be an $n \times n$ complex matrix. Show that A has a square root if $\det A \neq 0$. [Hint: Use Theorem 1 and exercise 15.]

4 JORDAN NORMAL FORM

In the last section, we proved that every $n \times n$ complex matrix is similar to a matrix of the form

$$A = \begin{bmatrix} B_1 & 0 & \cdots & 0 \\ 0 & B_2 & \cdots & 0 \\ & & \vdots & \\ 0 & 0 & \cdots & B_k \end{bmatrix}$$

where B_i is an $n_i \times n_i$ upper triangular matrix having all diagonal entries the same number λ_i.

In this section, we study linear operators having characteristic polynomial $(x-\lambda)^n$ for some λ, i.e., linear operators having only one distinct eigenvalue λ. By Theorem 1 of §7.3, the matrix representation of such an operator with respect to a suitable basis is an upper triangular matrix with diagonal entries all equal to λ. Our goal is to formulate a more precise upper triangular form. We shall see that a matrix B with characteristic polynomial $(x-\lambda)^n$ is similar to a matrix

$$J = \begin{bmatrix} J_1 & 0 & \cdots & 0 \\ 0 & J_2 & \cdots & 0 \\ & & \vdots & \\ 0 & 0 & \cdots & J_r \end{bmatrix}$$

in which J_i is an $m_i \times m_i$ matrix of the form

$$\begin{bmatrix} \lambda & 1 & 0 & \cdots & 0 \\ 0 & \lambda & 1 & \cdots & 0 \\ 0 & 0 & \lambda & \cdots & 0 \\ & & & \vdots & \\ 0 & 0 & & \cdots & \lambda \end{bmatrix}$$

In words, J_i has all diagonal entries λ, all entries immediately above the diagonal 1, and all other entries 0. J_i is called a **Jordan block** of size m_i. The matrix J is called the **Jordan normal form** of B.

We list all of the possible Jordan normal forms in the 2×2 and 3×3 cases for matrices with characteristic polynomial $(x - \lambda)^n$.

$$\begin{bmatrix} \lambda & 1 \\ 0 & \lambda \end{bmatrix}, \begin{bmatrix} \lambda & 0 \\ 0 & \lambda \end{bmatrix}, \begin{bmatrix} \lambda & 1 & 0 \\ 0 & \lambda & 1 \\ 0 & 0 & \lambda \end{bmatrix}, \begin{bmatrix} \lambda & 1 & 0 \\ 0 & \lambda & 0 \\ 0 & 0 & \lambda \end{bmatrix}, \begin{bmatrix} \lambda & 0 & 0 \\ 0 & \lambda & 0 \\ 0 & 0 & \lambda \end{bmatrix}$$

The first matrix has one Jordan block of size 2. The second has two Jordan blocks of size 1. Any 2×2 matrix having characteristic polynomial $(x - \lambda)^2$ is similar to the first or second matrix. The first 3×3 matrix has one Jordan block of size 3. The second has one Jordan block of size 2 and one of size 1. The third has three Jordan blocks of size 1. Any 3×3 matrix having characteristic polynomial $(x - \lambda)^3$ is similar to one of these three matrices. Similarly, there are five Jordan normal forms for 4×4 matrices with characteristic polynomial $(x - \lambda)^4$.

The Jordan normal form of an arbitrary $n \times n$ matrix C is defined as follows. First, find a matrix to which C is similar and that is of the upper triangular form of the matrix A above. Then, determine the Jordan normal form of each of the blocks B_i. It follows that C is similar to a matrix

$$\begin{bmatrix} J_1 & 0 & \dots & 0 \\ 0 & J_2 & \dots & 0 \\ & & \vdots & \\ 0 & 0 & \dots & J_m \end{bmatrix}$$

where each J_i is a Jordan block having diagonal entry λ_i. Different Jordan blocks may have different diagonal entries. In the 3×3 case, this means that there are three additional Jordan normal forms

$$\begin{bmatrix} \lambda & 1 & 0 \\ 0 & \lambda & 0 \\ 0 & 0 & \mu \end{bmatrix}, \begin{bmatrix} \lambda & 0 & 0 \\ 0 & \lambda & 0 \\ 0 & 0 & \mu \end{bmatrix}, \begin{bmatrix} \lambda & 0 & 0 \\ 0 & \mu & 0 \\ 0 & 0 & \nu \end{bmatrix}$$

where we suppose that λ, μ, and ν are all distinct. Any matrix with characteristic polynomial $(x - \lambda)^2(x - \mu)$ is similar to the first or second matrix. We have already seen (Theorem 4 of §7.1) that any matrix with characteristic polynomial $(x - \lambda)(x - \mu)(x - \nu)$ is similar to the third matrix.

Thus, in proving the existence of a Jordan normal form, we can confine ourselves to the case in which the matrix has a single eigenvalue.

Suppose that T is a linear operator on a vector space V with the characteristic polynomial $(x - \lambda)^n$. By the Cayley–Hamilton theorem, $(T - \lambda I)^n = 0$. Thus, if we set $D = T - \lambda I$, it follows that $D^n = 0$, i.e., D is nilpotent. We shall show that every nilpotent operator has a Jordan normal form. Then, since $T = \lambda I + D$, it follows that T has a Jordan normal form. Accordingly, in what follows we concentrate on nilpotent operators.

Thus, we suppose that D is a nilpotent operator on a finite-dimensional vector space V. We prove that D has a Jordan normal form by giving an algorithm for computing the Jordan normal form. We confine our proof to the case in which $D^3 = 0$. The proof in the general case is essentially the same. It simply requires more steps. Before we begin the proof itself, we note two lemmas.

Lemma 1 Let $\boldsymbol{x}_1, \ldots, \boldsymbol{x}_j, \boldsymbol{y}_1, \ldots, \boldsymbol{y}_k$ be a basis for a vector space V. Let $\boldsymbol{v}_1, \ldots, \boldsymbol{v}_k$ be vectors that are linear combinations of the vectors $\boldsymbol{x}_1, \ldots, \boldsymbol{x}_j$. Then $\boldsymbol{x}_1, \ldots, \boldsymbol{x}_j, \boldsymbol{y}_1 + \boldsymbol{v}_1, \ldots, \boldsymbol{y}_k + \boldsymbol{v}_k$ is still a basis for V.

PROOF By Theorem 2 of §4.8, it suffices to show that the given collection of vectors is linearly independent. So, we suppose that $\alpha_1\boldsymbol{x}_1 + \cdots + \alpha_j\boldsymbol{x}_j + \beta_1(\boldsymbol{y}_1 + \boldsymbol{v}_1) + \cdots + \beta_k(\boldsymbol{y}_k + \boldsymbol{v}_k) = \mathbf{0}$. Since the vectors $\boldsymbol{v}_1, \ldots, \boldsymbol{v}_k$ are linear combinations of the vectors $\boldsymbol{x}_1, \ldots, \boldsymbol{x}_j$, it follows that there are scalars $\gamma_1, \ldots, \gamma_j$ such that $\gamma_1\boldsymbol{x}_1 + \cdots + \gamma_j\boldsymbol{x}_j + \beta_1\boldsymbol{y}_1 + \cdots + \beta_k\boldsymbol{y}_k = \mathbf{0}$. Since $\boldsymbol{x}_1, \ldots, \boldsymbol{x}_j, \boldsymbol{y}_1, \ldots, \boldsymbol{y}_k$ are linearly independent, it follows that $\gamma_1 = \cdots = \gamma_j = \beta_1 = \cdots = \beta_k = 0$. Thus, it follows that $\alpha_1\boldsymbol{x}_1 + \cdots + \alpha_j\boldsymbol{x}_j = \mathbf{0}$. Hence, $\alpha_1 = \cdots = \alpha_j = 0$, since $\boldsymbol{x}_1, \ldots, \boldsymbol{x}_j$ are linearly independent. ■

Lemma 2 Let T be a linear operator on a vector space V. Suppose that $\boldsymbol{u}_1, \ldots, \boldsymbol{u}_m$ is a basis for N_T. Suppose that $\boldsymbol{v}_1, \ldots, \boldsymbol{v}_n$ are vectors in V such that $T(\boldsymbol{v}_1), \ldots, T(\boldsymbol{v}_n)$ is a basis for R_T. Then $\boldsymbol{u}_1, \ldots, \boldsymbol{u}_m, \boldsymbol{v}_1, \ldots, \boldsymbol{v}_n$ is a basis for V.

This was proved in the course of proving Theorem 3 of §5.4.

Now, we take a linear operator D on a finite-dimensional vector space V and suppose that $D^3 = 0$. We will give a method for finding a basis for V such that the matrix representation of D relative to this basis is in Jordan normal form. The procedure for finding such a basis is divided into three steps.

(1) Choose vectors $\boldsymbol{x}_1, \ldots, \boldsymbol{x}_r$ in V such that $D^2(\boldsymbol{x}_1), \ldots, D^2(\boldsymbol{x}_r)$ is a basis for R_{D^2}.

It is clear that such a choice is always possible: Take $\boldsymbol{u}_1, \ldots, \boldsymbol{u}_r$ as a basis for R_{D^2} and choose $\boldsymbol{x}_1, \ldots, \boldsymbol{x}_r$ so that $D^2(\boldsymbol{x}_i) = \boldsymbol{u}_i$.

Claim 1: $D^2(\boldsymbol{x}_1), D(\boldsymbol{x}_1), \boldsymbol{x}_1, \ldots, D^2(\boldsymbol{x}_r), D(\boldsymbol{x}_r), \boldsymbol{x}_r$ are linearly independent vectors.

Proof of claim: Suppose that

$$\mathbf{0} = \alpha_1 D^2(\boldsymbol{x}_1) + \beta_1 D(\boldsymbol{x}_1) + \gamma_1\boldsymbol{x}_1 + \cdots + \alpha_r D^2(\boldsymbol{x}_r) + \beta_r D(\boldsymbol{x}_r) + \gamma_r\boldsymbol{x}_r$$

Apply the operator D^2 to this equation. Since $D^4(\boldsymbol{x}_i) = D^3(\boldsymbol{x}_i) = \mathbf{0}$, it follows that $\gamma_1 D^2(\boldsymbol{x}_1) + \cdots + \gamma_r D^2(\boldsymbol{x}_r) = \mathbf{0}$. Then, since $D^2(\boldsymbol{x}_1), \ldots, D^2(\boldsymbol{x}_r)$ is a basis for R_{D^2}, it follows that $\gamma_1 = \cdots = \gamma_r = 0$. Thus,

$$\mathbf{0} = \alpha_1 D^2(\boldsymbol{x}_1) + \beta_1 D(\boldsymbol{x}_1) + \cdots + \alpha_r D^2(\boldsymbol{x}_r) + \beta_r D(\boldsymbol{x}_r)$$

Now, apply the operator D to this equation and use the fact that $D^3(x_i) = \mathbf{0}$. It follows that $\beta_1 D^2(x_1) + \cdots + \beta_r D^2(x_r) = \mathbf{0}$. Thus, as before, $\beta_1 = \cdots = \beta_r = 0$. This leaves that $\mathbf{0} = \alpha_1 D^2(x_1) + \cdots + \alpha_r D^2(x_r)$. Again, it follows that $\alpha_1 = \cdots = \alpha_r = 0$, and the claim is proved.

It follows from this claim that $D^2(x_1), D(x_1), \ldots, D^2(x_r), D(x_r)$ are linearly independent vectors in R_D.

(2) Choose vectors $y_1, \ldots, y_s$ in V such that $D^2(x_1), D(x_1), \ldots, D^2(x_r), D(x_r), D(y_1), \ldots, D(y_s)$ is a basis for R_D and so that $D^2(y_1) = \cdots = D^2(y_s) = \mathbf{0}$.

First, we show how it is possible to choose such vectors $y_1, \ldots, y_s$. We begin by choosing vectors $t_1, \ldots, t_s$ in R_D so that $D^2(x_1), D(x_1), \ldots, D^2(x_r), D(x_r), t_1, \ldots, t_s$ is a basis for R_D.

Since $t_1, \ldots, t_s$ lie in R_D, $D(t_1), \ldots, D(t_s)$ lie in R_{D^2}. Since $D^2(x_1), \ldots, D^2(x_r)$ is a basis for R_{D^2}, it follows that there are scalars $\eta_1, \ldots, \eta_r$ so that $D(t_i) = \eta_1 D^2(x_1) + \cdots + \eta_r D^2(x_r) = D(v_i)$, where $v_i = \eta_1 D(x_1) + \cdots + \eta_r D(x_r)$. Thus, $D(t_i - v_i) = \mathbf{0}$ and v_i is a linear combination of $D(x_1), \ldots, D(x_r)$. By Lemma 1, it follows that $D^2(x_1), D(x_1), \ldots, D^2(x_r), D(x_r), t_1 - v_1, \ldots, t_s - v_s$ is still a basis for R_D. Thus, we may choose $y_1, \ldots, y_s$ in V so that $D(y_i) = t_i - v_i$. Since $D(t_i - v_i) = \mathbf{0}$, $D^2(y_i) = \mathbf{0}$. Thus, step 2 can always be performed.

Claim 2: The following vectors are linearly independent: $D^2(x_1), D(x_1), x_1, \ldots, D^2(x_r), D(x_r), x_r, D(y_1), y_1, \ldots, D(y_s), y_s$.

Proof of claim: Suppose that

$$\mathbf{0} = \alpha_1 D^2(x_1) + \beta_1 D(x_1) + \gamma_1 x_1 + \cdots + \delta_1 D(y_1) + \epsilon_1 y_1 + \cdots$$

Apply the operator D to this equation. Then since $D^3(x_i) = \mathbf{0}$ and $D^2(y_i) = \mathbf{0}$, it follows that

$$\mathbf{0} = \beta_1 D^2(x_1) + \gamma_1 D(x_1) + \cdots + \epsilon_1 D(y_1) + \cdots$$

Since $D^2(x_1), D(x_1), \ldots, D(y_1), \ldots$ is a basis for R_D, it follows that $\beta_1 = \cdots = \gamma_1 = \cdots = \epsilon_1 = \cdots = 0$. Therefore, $\mathbf{0} = \alpha_1 D^2(x_1) + \cdots + \delta_1 D(y_1) + \cdots$. By the reasoning just used, it follows that $\alpha_1 = \cdots = \delta_1 = \cdots = 0$, proving the claim.

Now, note that it follows from the claim that $D^2(x_1), \ldots, D^2(x_r), D(y_1), \ldots, D(y_s)$ is a linearly independent set of vectors in N_D.

(3) Choose vectors $z_1, \ldots, z_t$ in N_D so that $D^2(x_1), \ldots, D^2(x_r), D(y_1), \ldots, D(y_s), z_1, \ldots, z_t$ is a basis for N_D.

Claim 3: $D^2(x_1), D(x_1), x_1, \ldots, D(y_1), y_1, \ldots, z_1, \ldots$ is a basis for V.

Proof of claim: We apply Lemma 2 with the vectors $u_1, \ldots, u_m$ of Lemma 2 being the vectors $D^2(x_1), \ldots, D(y_1), \ldots, z_1, \ldots$. Our choice of the vectors $z_1, \ldots, z_t$ in step 3 guarantees that the vectors $u_1, \ldots, u_m$ are a basis for

N_D. The vectors $v_1, \ldots, v_n$ are taken as the vectors $D(\boldsymbol{x}_1), \boldsymbol{x}_1, \ldots, \boldsymbol{y}_1, \ldots$. Our choice in step 2 guarantees that the vectors $D^2(\boldsymbol{x}_1)$, $D(\boldsymbol{x}_1), \ldots, D(\boldsymbol{y}_1), \ldots$ form a basis for R_D. Thus, the claim follows.

Next, we compute the matrix representation of D with respect to the basis

$$D^2(\boldsymbol{x}_1), D(\boldsymbol{x}_1), \boldsymbol{x}_1, \ldots, D(\boldsymbol{y}_1), \boldsymbol{y}_1, \ldots, \boldsymbol{z}_1, \ldots$$

Observe that D sends $D^2(\boldsymbol{x}_1)$ to $\mathbf{0}$, D sends $D(\boldsymbol{x}_1)$ to $D^2(\boldsymbol{x}_1)$, and D sends $\boldsymbol{x}_1$ to $D(\boldsymbol{x}_1)$. Thus, on the subspace spanned by $D^2(\boldsymbol{x}_1), D(\boldsymbol{x}_1), \boldsymbol{x}_1$, D is represented by the matrix

$$\begin{bmatrix} 0 & 1 & 0 \\ 0 & 0 & 1 \\ 0 & 0 & 0 \end{bmatrix}$$

This is, of course, a Jordan block of size 3. On each of the subspaces having bases $D^2(\boldsymbol{x}_i), D(\boldsymbol{x}_i), \boldsymbol{x}_i$, D is represented as a Jordan block of size 3 relative to this basis.

Similarly, D sends $D(\boldsymbol{y}_i)$ to 0 (as $D^2(\boldsymbol{y}_i) = 0$, by step 2) and D sends $\boldsymbol{y}_i$ to $D(\boldsymbol{y}_i)$. Thus, on the subspace of V with basis $D(\boldsymbol{y}_i), \boldsymbol{y}_i$, D is represented relative to this basis by the matrix

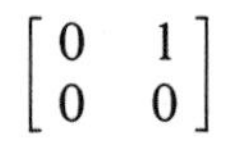

This is a Jordan block of size 2. Finally, as $D(\boldsymbol{z}_i) = \mathbf{0}$, each basis vector $\boldsymbol{z}_i$ contributes a block of size 1.

Thus, relative to the basis

$$D^2(\boldsymbol{x}_1), D(\boldsymbol{x}_1), \boldsymbol{x}_1, \ldots, D(\boldsymbol{y}_1), \boldsymbol{y}_1, \ldots, \boldsymbol{z}_1, \ldots$$

D is represented by a matrix of the form

$$\begin{bmatrix}
\begin{matrix} 0 & 1 & 0 \\ 0 & 0 & 1 \\ 0 & 0 & 0 \end{matrix} & & & & & & & \\
& \vdots \quad \vdots & & & & & & \\
& & \begin{matrix} 0 & 1 & 0 \\ 0 & 0 & 1 \\ 0 & 0 & 0 \end{matrix} & & & & & \\
& & & \begin{matrix} 0 & 1 \\ 0 & 0 \end{matrix} & & & & \\
& & & & \vdots \quad \vdots & & & \\
& & & & & \begin{matrix} 0 & 1 \\ 0 & 0 \end{matrix} & & \\
& & & & & & 0 & \\
& & & & & & \vdots \quad \vdots & \\
& & & & & & & 0
\end{bmatrix}$$

Thus, with a suitable choice of basis, the matrix representation of D is in Jordan normal form. With a little more effort, we can describe the number of Jordan blocks of a given size. The Jordan blocks of size 3 come from the basis vectors $D^2(\boldsymbol{x}_1), D(\boldsymbol{x}_1), \boldsymbol{x}_1, \ldots, D^2(\boldsymbol{x}_r), D(\boldsymbol{x}_r), \boldsymbol{x}_r$. Thus, there are r Jordan blocks of size 3. From step 1 it follows that r is the dimension of R_{D^2}. Thus, $r = r(D^2)$, the rank of D^2. From the basis vectors, $D(\boldsymbol{y}_1), \boldsymbol{y}_1, \ldots, D(\boldsymbol{y}_s), \boldsymbol{y}_s$ there arise s Jordan blocks of size 2. By step 2, it follows that $2r + s = r(D)$. Therefore, $s = r(D) - 2r(D^2)$. The number of Jordan blocks of size 1 is t. By step 3, $r + s + t = n(D) = \dim V - r(D)$. It follows that $t = \dim V - 2r(D) + r(D^2)$. With the convention, $D^0 = I$, this may be re-written as $t = r(D^0) - 2r(D) + r(D^2)$.

Extending the methods above gives the general result.

Theorem If D is a nilpotent operator on a finite-dimensional vector space V, the matrix representation of D with respect to a suitable basis of V is in Jordan normal form. The number of Jordan blocks of size i is given by the formula $r(D^{i-1}) - 2r(D^i) + r(D^{i+1})$.

We indicate in the exercises how to derive the general formula for the number of Jordan blocks. Instead, we find the Jordan normal form of a specific nilpotent matrix.

Example 1 For the following nilpotent matrix D, find a matrix B such that $B^{-1}DB$ is in Jordan normal form.

$$D = \begin{bmatrix} 2 & 1 & 2 & 0 & -1 \\ 3 & 1 & 3 & -1 & 1 \\ -3 & -1 & -2 & 0 & 2 \\ 3 & 2 & 4 & 0 & -1 \\ 2 & 1 & 2 & 0 & -1 \end{bmatrix} \quad D^2 = \begin{bmatrix} -1 & 0 & 1 & -1 & 4 \\ -1 & 0 & 1 & -1 & 4 \\ 1 & 0 & -1 & 1 & -4 \\ -2 & 0 & 2 & -2 & 8 \\ -1 & 0 & 1 & -1 & 4 \end{bmatrix} \quad D^3 = 0$$

As usual, $\boldsymbol{e}_1, \ldots, \boldsymbol{e}_5$ is the standard basis for $\boldsymbol{R}^5$. Note that $D\boldsymbol{e}_i$ is just the ith column of D. Let $\boldsymbol{v}$ be the vector that is the first column of D^2. First, we compute bases for R_{D^2} and R_D.

Clearly, $\boldsymbol{v}$ is a basis for R_{D^2} and $r(D^2) = 1$. Next we observe that $D\boldsymbol{e}_2, D\boldsymbol{e}_4, D\boldsymbol{e}_5$ is a basis for R_D. It is not hard to check that these vectors are linearly independent. Since $D\boldsymbol{e}_1 = D\boldsymbol{e}_2 - 3D\boldsymbol{e}_4 - D\boldsymbol{e}_5$ and $D\boldsymbol{e}_3 = 2D\boldsymbol{e}_2 - D\boldsymbol{e}_4$, the vectors span R_D. Thus, $r(D) = 3$. Already from the formulas presented earlier, it follows that the Jordan normal form of D has $r(D^2) = 1$ Jordan block of size 3, $r(D) - 2r(D^2) = 1$ Jordan block of size 2, and no Jordan block of size 1. In order to find the matrix B, we follow the procedure of the text.

First, we must compute $\boldsymbol{x}$ so that $D^2(\boldsymbol{x})$ spans R_{D^2}. By inspecting D^2, we find there are a large number of possibilities. We take $\boldsymbol{x} = \boldsymbol{e}_4$. Next, we must compute $\boldsymbol{y}$ so that $D^2\boldsymbol{y} = \boldsymbol{0}$ and so that $D^2\boldsymbol{x}, D\boldsymbol{x}, D\boldsymbol{y}$ is a basis for R_D. Thus, we must find $\boldsymbol{t} = D\boldsymbol{y}$ in R_D so that $D\boldsymbol{t} = \boldsymbol{0}$ and such that $D^2\boldsymbol{x}, D\boldsymbol{x}, \boldsymbol{t}$ is a basis for R_D. Now, $D\boldsymbol{x} = D\boldsymbol{e}_4$ and $D^2\boldsymbol{x} = DD\boldsymbol{e}_4 = D(-\boldsymbol{e}_2) = -D\boldsymbol{e}_2$. Thus, $\operatorname{sp}(D^2\boldsymbol{x}, D\boldsymbol{x}) = \operatorname{sp}(D\boldsymbol{e}_4, D\boldsymbol{e}_2)$. Thus, $\boldsymbol{t}$ must be chosen as some element of $\operatorname{sp}(D\boldsymbol{e}_4, D\boldsymbol{e}_2, D\boldsymbol{e}_5)$ that does not lie in $\operatorname{sp}(D\boldsymbol{e}_4, D\boldsymbol{e}_2)$ and such that $D\boldsymbol{t} = \boldsymbol{0}$.

Observe that $D(De_4) = D^2e_4 = v$, $D(De_2) = \mathbf{0}$, and $D(De_5) = -4v$. Thus, if we set $t = De_5 + 4De_4$, we have $Dt = \mathbf{0}$ and t does not lie in sp(De_4, De_2). Now choose y so that $Dy = t$. A suitable choice for y is $e_5 + 4e_4$.

Thus, the matrix representation of D with respect to the basis D^2x, Dx, x, Dy, y is in Jordan normal form. Rewriting the basis in terms of $x = e_4$ and $y = e_5 + 4e_4$, we have $D^2e_4, De_4, e_4, De_5 + 4De_4, e_5 + 4e_4$. Taking B to be the matrix whose successive columns are these vectors, we see that $B^{-1}DB$ is in Jordan normal form. Thus,

$$B = \begin{bmatrix} -1 & 0 & 0 & -1 & 0 \\ -1 & -1 & 0 & -3 & 0 \\ 1 & 0 & 0 & 2 & 0 \\ -2 & 0 & 1 & -1 & 4 \\ -1 & 0 & 0 & -1 & 1 \end{bmatrix}$$

It is now a simple matter to determine the Jordan normal form of an operator having a single eigenvalue. Indeed, let T be a linear operator on a vector space V having characteristic polynomial $(x - \lambda)^n$. Set $D = T - \lambda I$. Then by the Cayley–Hamilton theorem, $D^n = 0$. Thus, there is a basis $x_1, \ldots, x_n$ such that the matrix representation of D relative to this basis is in Jordan normal form. The matrix representations of D and $T = \lambda I + D$ with respect to this basis are

$$\begin{bmatrix} J_1 & 0 & \cdots & 0 \\ 0 & J_2 & \cdots & 0 \\ & & \vdots & \\ 0 & 0 & \cdots & J_k \end{bmatrix} \quad \text{and} \quad \begin{bmatrix} \lambda I + J_1 & 0 & \cdots & 0 \\ 0 & \lambda I + J_2 & \cdots & 0 \\ & & \vdots & \\ 0 & 0 & \cdots & \lambda I + J_k \end{bmatrix}$$

with $J_1, \ldots, J_k$ Jordan blocks having diagonal entries 0. Thus, $\lambda I + J_1, \ldots, \lambda I + J_k$ are Jordan blocks having diagonal entries λ. Since the number of Jordan blocks of size k for D is $r(D^{k-1}) - 2r(D^k) + r(D^{k+1})$, the number of Jordan blocks of size k for T is $r((T - \lambda I)^{k-1}) - 2r((T - \lambda I)^k) + r((T - \lambda I)^{k+1})$.

Example 2 Find the Jordan normal form of the matrix

$$A = \begin{bmatrix} -2 & -1 & 2 \\ 2 & 2 & -1 \\ -3 & -1 & 3 \end{bmatrix}$$

We begin by finding the characteristic polynomial of A.

$$C_A(x) = \begin{vmatrix} x+2 & 1 & -2 \\ -2 & x-2 & 1 \\ 3 & 1 & x-3 \end{vmatrix} = x^3 - 3x^2 + 3x - 1 = (x-1)^3$$

Next, calculate $A - I$, $(A - I)^2$, and the ranks of these matrices.

$$A - I = \begin{bmatrix} -3 & -1 & 2 \\ 2 & 1 & -1 \\ -3 & -1 & 2 \end{bmatrix}, \quad (A - I)^2 = \begin{bmatrix} 1 & 0 & -1 \\ -1 & 0 & 1 \\ 1 & 0 & -1 \end{bmatrix}, \quad (A - I)^3 = 0$$

Clearly, the first and second columns of $A - I$ are linearly independent. Thus, $r(A - I) \geqslant 2$. Since $A - I$ is singular, $r(A - I) = 2$. Also, $r(A - I)^2 = 1$. It follows that in the Jordan normal form of A there is one Jordan block of size 3. Thus, the Jordan normal form of A is

$$\begin{bmatrix} 1 & 1 & 0 \\ 0 & 1 & 1 \\ 0 & 0 & 1 \end{bmatrix}$$

It is also possible to describe the number of Jordan blocks of a given type when the linear transformation has several distinct eigenvalues. First we note:

Lemma 3 Let C be the block matrix

$$C = \begin{bmatrix} A & 0 \\ 0 & B \end{bmatrix}$$

where A is an $m \times m$ matrix and B is an $n \times n$ matrix. Then $r(C) = r(A) + r(B)$.

PROOF Suppose that A has rank s and B has rank t. In §5.5, we saw that there are invertible matrices G_1 and G_2 of order m and H_1 and H_2 of order n such that

$$G_1 A G_2 = \begin{bmatrix} I_s & 0 \\ 0 & 0 \end{bmatrix} \quad \text{and} \quad H_1 B H_2 = \begin{bmatrix} I_t & 0 \\ 0 & 0 \end{bmatrix}$$

It follows that

$$\begin{bmatrix} G_1 & 0 \\ 0 & H_1 \end{bmatrix} \begin{bmatrix} A & 0 \\ 0 & B \end{bmatrix} \begin{bmatrix} G_2 & 0 \\ 0 & H_2 \end{bmatrix} = \begin{bmatrix} I_s & 0 & 0 & 0 \\ 0 & 0 & 0 & 0 \\ 0 & 0 & I_t & 0 \\ 0 & 0 & 0 & 0 \end{bmatrix}$$

and

$$\begin{bmatrix} G_i & 0 \\ 0 & H_i \end{bmatrix}^{-1} = \begin{bmatrix} G_i^{-1} & 0 \\ 0 & H_i^{-1} \end{bmatrix}$$

Together, these imply C has rank $s + t$. ■

Corollary Let T be a linear operator on a finite-dimensional vector space V and λ an eigenvalue of T. Then in the Jordan normal form of T, the number of Jordan blocks having size k and diagonal entry λ is

$$r\big((T - \lambda I)^{k-1}\big) - 2r\big((T - \lambda I)^{k}\big) + r\big((T - \lambda I)^{k+1}\big)$$

PROOF Suppose that the characteristic polynomial of T is $(x - \lambda_1)^{n_1} \ldots (x - \lambda_m)^{n_m}$ with $\lambda_1, \ldots, \lambda_m$ distinct and $\lambda = \lambda_1$. Let $\boldsymbol{x}_1, \ldots, \boldsymbol{x}_n$ be a basis for V such that the matrix representation of T relative to this basis has the block form of Theorem 1 of the last section. With respect to this basis, the matrix representations of T and $(T-\lambda_1 I)^p$ are, respectively,

$$\begin{bmatrix} B_1 & 0 & \dots & 0 \\ 0 & B_2 & \dots & 0 \\ & & \vdots & \\ 0 & 0 & \dots & B_m \end{bmatrix} \quad \text{and} \quad \begin{bmatrix} (B_1-\lambda_1 I)^p & 0 & \dots & 0 \\ 0 & (B_2-\lambda_1 I)^p & \dots & 0 \\ & \vdots & & \\ 0 & 0 & \dots & (B_m-\lambda_1 I)^p \end{bmatrix}$$

where B_i is an $n_i \times n_i$ upper triangular matrix having diagonal entries λ_i. The rank of the matrix $(B_1-\lambda_1 I)^p$ is simply $r((B_1-\lambda_1 I)^p)$. The matrices $(B_2-\lambda_1 I)^p, \dots, (B_m-\lambda_1 I)^p$ are all upper triangular having the nonzero diagonal entries $(\lambda_2-\lambda_1)^p, \dots, (\lambda_m-\lambda_1)^p$, respectively. Consequently, the rank of the lower right-hand block is $n_2 + \cdots + n_m = n - n_1$. By Lemma 3, it follows that the rank of $(T-\lambda_1 I)^p$ is $r((B_1-\lambda_1 I)^p) + n - n_1$. Now, the number of Jordan blocks of T of size k having diagonal entry λ_1 is simply the number of Jordan blocks of B_1 of size k, i.e., $r((B_1-\lambda_1 I)^{k-1}) - 2r((B_1-\lambda_1 I)^k) + r(B_1 - \lambda_1 I)^{k+1}$. When we substitute into this expression $r((B_1-\lambda_1 I)^p) = r((T-\lambda_1 I)^p) - (n-n_1)$ for $p = k-1, k, k+1$, the corollary follows. ■

Thus, to determine the Jordan normal form of a matrix A, it is first necessary to find $C_A(x)$. Then factor $C_A(x) = (x-\lambda_1)^{n_1} \dots (x-\lambda_k)^{n_k}$. Last, for each i compute the decreasing sequence of numbers

$$r(A-\lambda_i I) \geqslant r\left((A-\lambda_i I)^2\right) \geqslant r\left((A-\lambda_i I)^3\right) \geqslant \dots$$

The proof of the corollary shows that $r((A-\lambda_i I)^p) = n - n_i$ if $p \geqslant n_i$. Thus, it is only necessary to compute the terms of this sequence until the number $n - n_i$ is reached. After that the sequence is constant. Then from the corollary the number of Jordan blocks of each size can be read off.

Example 3 Find the Jordan normal form of the matrix

$$A = \begin{bmatrix} -2 & 1 & 3 & -1 \\ 3 & 0 & -2 & 2 \\ 1 & 1 & 2 & 1 \\ 1 & -1 & -3 & 0 \end{bmatrix} \quad \text{Then } C_A(x) = \begin{vmatrix} x+2 & -1 & -3 & 1 \\ -3 & x & 2 & -2 \\ -1 & -1 & x-2 & -1 \\ -1 & 1 & 3 & x \end{vmatrix}$$

$$= \begin{vmatrix} x+2 & x+1 & 3x+3 & x^2+2x+1 \\ -3 & x-3 & -7 & -3x-2 \\ -1 & -2 & x-5 & -x-1 \\ -1 & 0 & 0 & 0 \end{vmatrix} = (x+1)\begin{vmatrix} 1 & 3 & x+1 \\ x-3 & -7 & -3x-2 \\ -2 & x-5 & -x-1 \end{vmatrix}$$

$$= (x+1)\begin{vmatrix} 1 & 0 & 0 \\ x-3 & -3x+2 & -x^2-x+1 \\ -2 & x+1 & x+1 \end{vmatrix} = (x+1)^2(x-1)^2$$

$$\text{Now, } A+I = \begin{bmatrix} -1 & 1 & 3 & -1 \\ 3 & 1 & -2 & 2 \\ 1 & 1 & 3 & 1 \\ 1 & -1 & -3 & 1 \end{bmatrix} \quad \text{and} \quad A-I = \begin{bmatrix} -3 & 1 & 3 & -1 \\ 3 & -1 & -2 & 2 \\ 1 & 1 & 1 & 1 \\ 1 & -1 & -3 & -1 \end{bmatrix}$$

Now it is not difficult to check that the first three columns of both of these matrices are linearly independent. It follows that $r(A + I) = r(A - I) = 3$. By the remark above if $p > 2$, $r((A + I)^p) = r((A - I)^p) = 2$. Thus, the number of Jordan 2-blocks corresponding to both 1 and -1 is $3 - 2 \cdot 2 + 2 = 1$. It follows that the Jordan normal form of the matrix is

$$\begin{bmatrix} 1 & 1 & 0 & 0 \\ 0 & 1 & 0 & 0 \\ 0 & 0 & -1 & 1 \\ 0 & 0 & 0 & -1 \end{bmatrix}$$

EXERCISES

1. Find the Jordan normal form of the following matrices.

(a) $\begin{bmatrix} 2 & -1 \\ 1 & 0 \end{bmatrix}$ (b) $\begin{bmatrix} 2 & -3 \\ 3 & -4 \end{bmatrix}$ (c) $\begin{bmatrix} 4 & 2 \\ -1 & 1 \end{bmatrix}$ (d) $\begin{bmatrix} 1 & -2 \\ \frac{1}{2} & -1 \end{bmatrix}$

(e) $\begin{bmatrix} 0 & 2 & -1 \\ 1 & 2 & -2 \\ 0 & 2 & -1 \end{bmatrix}$ (f) $\begin{bmatrix} -2 & 2 & 1 \\ -9 & 5 & -1 \\ 3 & -2 & 0 \end{bmatrix}$ (g) $\begin{bmatrix} 4 & 1 & -1 \\ -3 & 1 & 2 \\ 3 & 1 & 0 \end{bmatrix}$

2. For each of the following nilpotent matrices A, find a matrix B so that $B^{-1}AB$ is in Jordan normal form.

(a) $\begin{bmatrix} -1 & -1 & 2 \\ -2 & -1 & 3 \\ -1 & -1 & 2 \end{bmatrix}$ (b) $\begin{bmatrix} 4 & 1 & -1 & 2 \\ -4 & -1 & 2 & -1 \\ 4 & 1 & -1 & 2 \\ -4 & -1 & 1 & -2 \end{bmatrix}$

3. Show that the matrix A has only one Jordan block of size 3 if and only if $ab \neq 0$, where

$$A = \begin{bmatrix} 1 & a & c \\ 0 & 1 & b \\ 0 & 0 & 1 \end{bmatrix}$$

4. Let D be the linear operator on P_n defined by $D(f) = f'$. Show that the matrix representation of T relative to the basis $1, x, \frac{1}{2}x^2, \frac{1}{6}x^3, \ldots, \frac{1}{n!}x^n$ is in Jordan normal form.

5. Let A be an $n \times n$ complex matrix. Show that A is similar over the complexes to A^T.

6. Let J be the $n \times n$ nilpotent matrix that is a single Jordan block of size n. Show that $r(J^k) = n - k$.

7. Let D be an $n \times n$ nilpotent matrix having j_1 Jordan blocks of size 1, j_2 of size $2, \ldots, j_k$ of size k. Let $d_i = r(D^i)$. Show that

$$\begin{aligned} j_1 + 2j_2 + 3j_3 + \cdots + kj_k &= d_0 (= n) \\ j_2 + 2j_3 + \cdots + (k-1)j_k &= d_1 \\ j_3 + \cdots + (k-2)j_k &= d_2 \\ &\vdots \\ j_k &= d_k \end{aligned}$$

[Hint: Use exercise 6 and Lemma 3.]

8. By proving

$$\begin{bmatrix} 1 & 2 & 3 & \cdots & k \\ 0 & 1 & 2 & \cdots & k-1 \\ 0 & 0 & 1 & \cdots & k-2 \\ & & & \vdots & \\ 0 & 0 & 0 & \cdots & 1 \end{bmatrix}^{-1} = \begin{bmatrix} 1 & -2 & 1 & 0 & \cdots & 0 \\ 0 & 1 & -2 & 1 & \cdots & 0 \\ 0 & 0 & 1 & -2 & \cdots & 0 \\ & & & & \vdots & \\ 0 & 0 & 0 & 0 & \cdots & 1 \end{bmatrix}$$

and using exercise 7, obtain the formula of the theorem of this section (§7.4).

9. Show that every operator on $\boldsymbol{R}^n$ of rank 1 is diagonalizable or nilpotent.

10. Let A be an $n \times n$ complex matrix all of whose eigenvalues are real. Show that A is similar to a real matrix.

11. Let M be an $n \times n$ matrix. Show that the following are equivalent: (a) $M^2 = 0$, and (b) $M = AB$ where A and B are $n \times n$ matrices such that $BA = 0$.

12. Let D be a nilpotent matrix. Let j_i be the number of Jordan blocks of D of size i. Show that there is a matrix B such that $B^2 = D$ if and only if $j_1 \geqslant j_2 + \cdots + j_k$, with k the size of the largest Jordan block.

13. Let A be an $n \times n$ complex matrix. Show that all of the entries of A^m go to 0 as m approaches infinity if and only if all eigenvalues of A are of absolute value less than 1.

14. Let N be a nilpotent matrix and c a nonzero number. Show that N and cN are similar.

15. Let D be an $n \times n$ nilpotent matrix such that $D^{n-1} \neq 0$. Show that D has one Jordan block.

16. Let A be a 2×2 matrix having characteristic polynomial $(x - \lambda)^2$. Show that $A^n = (1 - n)\lambda^n I_2 + n\lambda^{n-1}A$.

17. Let A be a complex $n \times n$ matrix. Show that A is the sum of two matrices B and N where B is diagonalizable, N is nilpotent, and B and N commute.

5 BILINEAR FORMS

In an earlier section of this book, we discussed the inner product and the cross product. We noted that

$$(\alpha \boldsymbol{a} + \beta \boldsymbol{b}) \times \boldsymbol{c} = \alpha(\boldsymbol{a} \times \boldsymbol{b}) + \beta(\boldsymbol{b} \times \boldsymbol{c})$$

$$\boldsymbol{a} \times (\beta \boldsymbol{b} + \gamma \boldsymbol{c}) = \beta(\boldsymbol{a} \times \boldsymbol{b}) + \gamma(\boldsymbol{a} \times \boldsymbol{c})$$

$$(\alpha \boldsymbol{a} + \beta \boldsymbol{b}, \boldsymbol{c}) = \alpha(\boldsymbol{a}, \boldsymbol{c}) + \beta(\boldsymbol{b}, \boldsymbol{c})$$

$$(\boldsymbol{a}, \beta \boldsymbol{b} + \gamma \boldsymbol{c}) = \beta(\boldsymbol{a}, \boldsymbol{b}) + \gamma(\boldsymbol{a}, \boldsymbol{c})$$

We wish to study, in general, functions of this type. Accordingly, we define a bilinear function b on a vector space V to a vector space W to be a function defined on the ordered pairs of elements of V such that

$$b(\alpha \boldsymbol{x} + \beta \boldsymbol{y}, \boldsymbol{z}) = \alpha b(\boldsymbol{x}, \boldsymbol{z}) + \beta b(\boldsymbol{y}, \boldsymbol{z})$$

$$b(\boldsymbol{x}, \beta \boldsymbol{y} + \gamma \boldsymbol{z}) = \beta b(\boldsymbol{x}, \boldsymbol{y}) + \gamma b(\boldsymbol{x}, \boldsymbol{z})$$

with $\boldsymbol{x}, \boldsymbol{y}, \boldsymbol{z}$ in V and α, β, γ scalars.

In this section, all vector spaces are real. If $W = \boldsymbol{R}$ in the above definition, the bilinear function is said to be a **bilinear form**. Thus, the inner product is a bilinear form. The cross product is not.

A bilinear function is symmetric if $b(\boldsymbol{x}, \boldsymbol{y}) = b(\boldsymbol{y}, \boldsymbol{x})$ for all $\boldsymbol{x}$ and $\boldsymbol{y}$ in V. It is alternate or skew-symmetric if $b(\boldsymbol{x}, \boldsymbol{y}) = -b(\boldsymbol{y}, \boldsymbol{x})$, for all $\boldsymbol{x}$ and $\boldsymbol{y}$ in V. The cross product is alternate, the inner product symmetric.

If b is a bilinear form and there is a nonzero vector $\boldsymbol{x}$ such that $b(\boldsymbol{x}, \boldsymbol{y}) = 0$ for all $\boldsymbol{y}$ in V, the bilinear form is said to be **degenerate**. If a bilinear form is not degenerate, it is called **nondegenerate**. The inner product is a nondegenerate bilinear form. Indeed, if $\boldsymbol{x}$ is a vector in $\boldsymbol{R}^n$ such that $b(\boldsymbol{x}, \boldsymbol{y}) = 0$ for all $\boldsymbol{y}$ in $\boldsymbol{R}^n$, then $b(\boldsymbol{x}, \boldsymbol{x}) = 0$. Thus, $\boldsymbol{x} = 0$.

A bilinear form is said to be **positive** if for all vectors $\boldsymbol{x}$, $b(\boldsymbol{x}, \boldsymbol{x}) \geqslant 0$. If for all $\boldsymbol{x}$, $b(\boldsymbol{x}, \boldsymbol{x}) \leqslant 0$, b is said to be **negative**. As we have seen, the inner product is a positive bilinear form.

If A is an $n \times n$ matrix, the function $b(\boldsymbol{x}, \boldsymbol{y}) = \boldsymbol{x}^T A \boldsymbol{y}$, defined for all pairs of vectors $\boldsymbol{x}, \boldsymbol{y}$ in $\boldsymbol{R}^n$, is clearly bilinear. If $A = I_n$ this bilinear form is just the inner product. We shall show that all bilinear forms on $\boldsymbol{R}^n$ are of this type.

Theorem 1 If b is a bilinear form on $\boldsymbol{R}^n$, there is a unique real $n \times n$ matrix A such that $b(\boldsymbol{x}, \boldsymbol{y}) = \boldsymbol{x}^T A \boldsymbol{y}$, for all $\boldsymbol{x}$ and $\boldsymbol{y}$ in $\boldsymbol{R}^n$.

PROOF Define $A = [b(\boldsymbol{e}_i, \boldsymbol{e}_j)]_{(nn)}$, with $\boldsymbol{e}_1, \ldots, \boldsymbol{e}_n$ the standard basis for $\boldsymbol{R}^n$. Note that $\boldsymbol{e}_i{}^T A \boldsymbol{e}_j = b(\boldsymbol{e}_i, \boldsymbol{e}_j)$. Let c be the bilinear form $c(\boldsymbol{x}, \boldsymbol{y}) = \boldsymbol{x}^T A \boldsymbol{y}$. Then the bilinear forms b and c agree on all basis vectors. We show that they agree everywhere.

Now

$$b\left(\sum_i x_i \boldsymbol{e}_i, \sum_j y_j \boldsymbol{e}_j\right) = \sum_{i,j} x_i y_j b(\boldsymbol{e}_i, \boldsymbol{e}_j)$$

$$= \sum_{i,j} x_i y_j c(\boldsymbol{e}_i, \boldsymbol{e}_j)$$

$$= c\left(\sum_i x_i \boldsymbol{e}_i, \sum_j y_j \boldsymbol{e}_j\right)$$

Thus, $b(\boldsymbol{x}, \boldsymbol{y}) = \boldsymbol{x}^T A \boldsymbol{y}$.

A is unique, for if $b(\boldsymbol{x}, \boldsymbol{y}) = \boldsymbol{x}^T A' \boldsymbol{y}$, $b(\boldsymbol{e}_i, \boldsymbol{e}_j) = \boldsymbol{e}_i^T A' \boldsymbol{e}_j$. If $A' = [a'_{ij}]_{(nn)}$, $b(\boldsymbol{e}_i, \boldsymbol{e}_j) = a'_{ij}$. Since we saw earlier that $b(\boldsymbol{e}_i, \boldsymbol{e}_j) = a_{ij}$, $a_{ij} = a'_{ij}$. Thus, $A = A'$. ■

A bilinear form $b_A(\boldsymbol{x}, \boldsymbol{y}) = \boldsymbol{x}^T A \boldsymbol{y}$ is symmetric if and only if the matrix A is symmetric. Now, since $\boldsymbol{x}^T A \boldsymbol{y}$ is a real number, $(\boldsymbol{x}^T A \boldsymbol{y})^T = \boldsymbol{x}^T A \boldsymbol{y}$. Thus,

$y^TA^Tx = x^TAy$. Therefore, $x^TAy = y^TAx$ if and only if $y^TA^Tx = y^TAx$. By the uniqueness part of the previous theorem, this is equivalent to $A^T = A$.

Likewise, it can be shown that b_A is skew-symmetric if and only if A is skew-symmetric.

A function Q from $\boldsymbol{R}^n$ to R is said to be a **quadratic form** if there is a bilinear b on $\boldsymbol{R}^n$ such that $Q(x) = b(x, x)$. By Theorem 1 any quadratic form can be written as x^TAx for some $n \times n$ matrix $A = [a_{ij}]_{(nn)}$. Thus, if

$$x = x_1e_1 + \cdots + x_ne_n, \qquad Q(x) = \sum_{i,j=1}^{n} a_{ij}x_ix_j$$

Thus, for example, every quadratic form on R^2 can be written in the form $a_{11}x_1^2 + a_{12}x_1x_2 + a_{21}x_2x_1 + a_{22}x_2^2$. This expression, of course, equals $a_{11}x_1^2 + (a_{12} + a_{21})x_1x_2 + a_{22}x_2^2$. Thus, the quadratic form associated with the matrix

$$A' = \begin{bmatrix} a_{11} & a_{12} \\ a_{21} & a_{22} \end{bmatrix}$$

is the same as that associated with the symmetric matrix

$$A' = \begin{bmatrix} a_{11} & \frac{1}{2}(a_{12} + a_{21}) \\ \frac{1}{2}(a_{12} + a_{21}) & a_{12} \end{bmatrix}$$

That every quadratic form is the quadratic form associated with a symmetric bilinear form is the content of the following theorem.

Theorem 2 Let Q be a quadratic form on $\boldsymbol{R}^n$. Then there is a unique symmetric bilinear form b on $\boldsymbol{R}^n$ such that $Q(x) = b(x, x)$.

PROOF If Q is a quadratic form, there is some bilinear form c (not necessarily symmetric) such that $Q(x) = c(x, x)$. Define $b(x, y) = \frac{1}{2}(c(x, y) + c(y, x))$.

Clearly, b is a bilinear form. It is likewise clear that $b(x, y) = b(y, x)$. Thus, b is symmetric. Moreover, $b(x, x) = c(x, x) = Q(x)$.

Next, we show that b is uniquely determined by Q. For suppose b_1 and b_2 are symmetric bilinear forms and $Q(x) = b_1(x, x) = b_2(x, x)$. Then

$$\begin{aligned} b_1(x, y) &= \tfrac{1}{4}(b_1(x + y, x + y) - b_1(x - y, x - y)) \\ &= \tfrac{1}{4}(b_2(x + y, x + y) - b_2(x - y, x - y)) \\ &= b_2(x, y) \end{aligned}$$

Thus, b is uniquely determined. ■

If C is an $n \times n$ nonsingular matrix and b is a bilinear form on $\boldsymbol{R}^n$, we may define a new bilinear form by $b'(x, y) = b(Cx, Cy)$. We say that b' is **congruent** to b.

We interpret congruence in terms of matrices. If $b(\mathbf{x}, \mathbf{y}) = \mathbf{x}^T A\mathbf{y}$ and $b'(\mathbf{x}, \mathbf{y}) = \mathbf{x}^T B\mathbf{y}$ and b' is congruent to b, there is a nonsingular $n \times n$ matrix C such that $\mathbf{x}^T B\mathbf{y} = \mathbf{x}^T C^T A C\mathbf{y}$. Since this holds for all $\mathbf{x}$ and $\mathbf{y}$, by Theorem 1, $B = C^T A C$.

If conversely, $B = C^T A C$, then $b'(\mathbf{x}, \mathbf{y}) = b(C\mathbf{x}, C\mathbf{y})$ and b' is congruent to b. If $B = C^T A C$, then $(C^{-1})^T B C^{-1} = A$. Thus, b is congruent to b', if b' is congruent to b.

If b is a bilinear form we say that the **rank** of b is the rank of the associated matrix A. If b and b' are congruent bilinear forms, and b is associated with the matrix A, then b' is associated with $C^T A C$, for some nonsingular matrix C. Thus, it follows that the two bilinear forms b and b' have the same rank.

Just as we defined equivalence and orthogonal equivalence of matrices, we can define congruence and orthogonal congruence of bilinear forms and quadratic forms. Thus, we say that two bilinear forms b and b' are **orthogonally congruent** if there is an orthogonal matrix A such that $b(\mathbf{x}, \mathbf{y}) = b(A\mathbf{x}, A\mathbf{y})$. Likewise, two quadratic forms Q and Q' are orthogonally congruent if $Q'(\mathbf{x}) = Q(A\mathbf{x})$, for some orthogonal A. This is equivalent to the statement that their associated symmetric bilinear forms are orthogonally congruent.

Clearly, the simplest class of quadratic forms are those of the form

$$\lambda_1 x_1^2 + \lambda_2 x_2^2 + \cdots + \lambda_n x_n^2$$

These correspond to diagonal matrices. Accordingly, the following theorem is of interest.

Theorem 3 Any quadratic form Q is orthogonally congruent to one of the type

$$\lambda_1 x_1^2 + \cdots + \lambda_n x_n^2$$

PROOF Now by Theorem 2 of this section, $Q(\mathbf{x}) = \mathbf{x}^T A\mathbf{x}$, for some symmetric matrix A. By Theorem 2 of section 7-2 there is an orthogonal matrix C such that $C^T A C$ is diagonal. ■

Example 1 Show that the curve

$$ax^2 + bxy + cy^2 = d, \qquad d \neq 0$$

is an ellipse if $4ac - b^2 > 0$, and an hyperbola if $4ac - b^2 < 0$.

Let

$$Q(\mathbf{x}) = \begin{bmatrix} x \\ y \end{bmatrix}^T \begin{bmatrix} a & \frac{1}{2}b \\ \frac{1}{2}b & c \end{bmatrix} \begin{bmatrix} x \\ y \end{bmatrix}$$

Clearly, $Q(\mathbf{x}) = d$ represents the above curve. By Theorem 3, there is an orthogonal matrix A such that $Q(A\boldsymbol{x}) = \lambda_1 x_1^2 + \lambda_2 x_2^2$ and

$$A^T \begin{bmatrix} a & \frac{1}{2}b \\ \frac{1}{2}b & c \end{bmatrix} A = \begin{bmatrix} \lambda_1 & 0 \\ 0 & \lambda_2 \end{bmatrix}$$

The curve $\lambda_1 x_1^2 + \lambda_2^2 x_2^2 = d$ is an ellipse if $\lambda_1\lambda_2 > 0$, an hyperbola if $\lambda_1\lambda_2 < 0$. Since A is orthogonal, $A^T = A^{-1}$, and

$$\det\left(A^T \begin{bmatrix} a & \frac{1}{2}b \\ \frac{1}{2}b & c \end{bmatrix} A \right) = ac - \tfrac{1}{4}b^2 = \lambda_1\lambda_2$$

Since the curve $\lambda_1 x_1^2 + \lambda_2 x_2^2 = d$ was obtained from $ax^2 + bxy + cy^2 = d$ by rotation of axes, $ax^2 + bxy + cy^2 = d$ is an ellipse if $4ac - b^2 > 0$, an hyperbola if $4ac - b^2 < 0$.

The next result follows immediately from Theorem 3.

Theorem 4 Every quadratic form is congruent to one of the type

$$x_1^2 + \cdots + x_p^2 - x_{p+1}^2 - \cdots - x_r^2$$

PROOF One replaces x_i by $x_i/\sqrt{\lambda_i}$ if $\lambda_i > 0$ and $x_i/\sqrt{-\lambda_i}$ if $\lambda_i < 0$. ∎

Thus, every symmetric bilinear form b is congruent to one of the form $x_1y_1 + \cdots + x_ky_k - x_{k+1}y_{k+1} - \cdots - x_{k+r}y_{k+r}$. The rank of this form is clearly $k + r$. The number $k - r$ is called the **signature**. We shall show that the signature is uniquely determined. In other words, we wish to show that if b is also congruent to $x_1y_1 + \cdots + x_{k'}y_{k'} - x_{k'+1}y_{k'+1} - \cdots - x_{k'+r'}y_{k'+r'}$ then $k' - r' = k - r$. Since congruent forms have the same rank, $k + r = k' + r'$. Thus, to show $k' - r' = k - r$, it is sufficient to show $k = k'$.

Lemma Let W be a subspace of $\boldsymbol{R}^n$ for which $b(\boldsymbol{x}, \boldsymbol{x}) \leqslant 0$ for all $\boldsymbol{x}$ in W with b the form $x_1y_1 + \cdots + x_ky_k - x_{k+1}y_{k+1} - \cdots - x_{k+r}y_{k+r}$. Then $\dim W \leqslant n - k$.

PROOF Let V be the subspace of $\boldsymbol{R}^n$ spanned by $\boldsymbol{e}_{k+1}, \ldots, \boldsymbol{e}_n$. Clearly, $\dim V = n - k$. Let U be the subspace of $\boldsymbol{R}^n$ spanned by $\boldsymbol{e}_1, \ldots, \boldsymbol{e}_k$. Let $\boldsymbol{y}_1, \ldots, \boldsymbol{y}_m$ be a basis for W. Then each $\boldsymbol{y}_i$ can be expressed uniquely in the form $\boldsymbol{u}_i + \boldsymbol{v}_i$ where $\boldsymbol{u}_i$ belongs to U and $\boldsymbol{v}_i$ belongs to V. If $m > n - k$, there are real scalars $\alpha_1, \ldots, \alpha_m$, not all 0, such that

$$\sum_{i=1}^{m} \alpha_i \boldsymbol{v}_i = 0$$

Then

$$\boldsymbol{z} = \sum_{i=1}^{m} \alpha_i \boldsymbol{y}_i = \sum_{i=1}^{m} \alpha_i(\boldsymbol{u}_i + \boldsymbol{v}_i) = \sum_{i=1}^{m} \alpha_i \boldsymbol{u}_i$$

belongs to W. Thus, $b(z, z) \leqslant 0$.

But z also belongs to U, and the bilinear form b on U is just the inner product of §6.2. Thus, $b(z, z) \geqslant 0$. Consequently, $b(z, z) = 0$. Again, since b on U is the inner product of §6.2, $z = \mathbf{0}$. Thus, $\Sigma_{i=1}^{n} \alpha_i y_i = 0$, in contradiction to the fact that $y_1, \ldots, y_m$ is a basis for W and not all the α_i's are 0. Thus, we must have $m \leqslant n - k$. ■

Theorem 5 Sylvester's Law of Inertia The signature of a symmetric bilinear form is uniquely determined.

PROOF In the notation preceding the lemma, we must show that $k = k'$. By the lemma $n - k \leqslant n - k'$ and $n - k \leqslant n - k$. Thus, $k = k'$. ■

EXERCISES

1. Show that the bilinear forms on $\boldsymbol{R}^n$ form a vector space using the usual addition and scalar multiplication for real functions. Show that this space is isomorphic to the space of real $n \times n$ matrices.
2. Determine the rank and signature of the following quadratic forms:

   (a) $x^2 + xy + y^2$ (b) $x^2 + xy + 3y^2$
   (c) $x^2 + 2xy + y^2$ (d) $x^2 - y^2$

3. Let $b_A(\boldsymbol{x}, \boldsymbol{y}) = \boldsymbol{x}^T A \boldsymbol{y}$ be a symmetric bilinear form. Show that b_A is positive if and only if the eigenvalues of A are nonnegative. Show that b_A is nondegenerate if and only if the eigenvalues of A are all nonzero.
4. Let b be a symmetric bilinear form on $\boldsymbol{R}^n$. Show that b is positive if and only if its rank and signature are equal.
5. Let b be a bilinear form on a vector space V. Let W be the collection of all vectors $\boldsymbol{x}$ in V such that $b(\boldsymbol{x}, \boldsymbol{y}) = 0$ for all $\boldsymbol{y}$ in V.
   (a) Show that W is a subspace of V.
   (b) Let r be the rank of b. Show that $\dim W + r = \dim V$.
6. Let A be a nonzero symmetric 2×2 matrix of rank 1. Interpret geometrically the sets $b_A(\boldsymbol{x}, \boldsymbol{x}) = c$ for all values of c.
7. Show that two symmetric bilinear forms are congruent if and only if they have the same rank and signature.
8. Determine the rank and signature of the quadratic form $xy + yz + xz$.
9. Show that every bilinear form can be expressed in a unique way as the sum of a symmetric and skew-symmetric bilinear form. Interpret this in terms of matrices.

ANSWERS TO SELECTED EXERCISES

Section 1.1

1. (a) $x = \frac{23}{7}, y = \frac{3}{7}$. (c) No solution. (e) $x + 2y = 3$.
3. (a) $x = 4, y = 3, z = -1$.
5. 5 units of A, 20 of B.

Section 1.2

1. (a) Not equivalent; first system has solution $(\frac{1}{3}, \frac{1}{3})$, second has solution (1, 2).
   (c) Not equivalent; first system has solution (0, 0), second system has solution $x + y = 0$.
   (e) Not equivalent; first system has solution (3, 1), second has solution (0, 4).

Section 1.3

1. (a) ix $= 1, y = 3, z = 7$. (c) No solution. (e) No solution.
   (g) $x_1 = a, x_2 = b, x_3 = \frac{1}{2}a + b, x_4 = -\frac{3}{2}a$.
3. $x = -t, y = 1 - t$.
4. (a) $a - 3b + c = 0$.
7. Yes, by using exercise 2(a).
9. Let a be the weight of the first. Then, the second, third, and fourth are: $\frac{3}{2}a$, $\frac{1}{2}a$, $\frac{5}{2}a$.
11. $2KMnO_4 + 8H_2SO_4 + 10KBr \rightarrow 6K_2SO_4 + 5Br_2 + 2MnSO_4 + 8H_2O$.
13. 3 of type 1, 4 of type 2, and 2 of type 3.

Section 1.4

1. (a) $x = 3c, y = -2c, z = -7c$. (c) $x_1 = -3a, x_2 = 5a, x_3 = a, x_4 = 5a$.

Section 2.1

1. (a) $\begin{bmatrix} 4 \\ 2 \\ 12 \end{bmatrix}$ (c) $\begin{bmatrix} 22 \\ 39 \\ -15 \end{bmatrix}$ (e) $\begin{bmatrix} 13 \\ 1 \\ 28 \\ 21 \end{bmatrix}$

11. $x = 9, y = -2$.

13. $\begin{bmatrix} 12 \\ 14 \\ 13 \end{bmatrix}$ 15. 700, 450, 400.

Section 2.2

3. $\boldsymbol{r}(t) = \boldsymbol{i} + t(\boldsymbol{i} + \boldsymbol{j} + \boldsymbol{k})$, (0, − 1, − 1), (3, 2, 2).
7. $(-\frac{1}{9}, -\frac{1}{9}, \frac{4}{9})$.
13. (a) $(1, \frac{1}{2}, \frac{1}{2})$ (b) (1, 1, 0) (c) (0, 0, 1).
17. $(\frac{7}{10}, \frac{5}{10})$.

Section 2.3

1. (a) $\begin{bmatrix} 2 & 5 \\ 3 & 6 \end{bmatrix}$ (c) $\begin{bmatrix} 2 & 5 & 10 \\ 5 & 8 & 13 \end{bmatrix}$ 2(b) $\begin{bmatrix} 9 & 21 \\ -4 & 2 \end{bmatrix}$
3. $X = -\frac{3}{2}A + \frac{1}{2}B - \frac{1}{2}C$.
5. $X = 2A - 3B$, $Y = -A + 2B$.
7. 64.

Section 2.4

1. (a) $\begin{bmatrix} 13 & 18 \\ 13 & 2 \end{bmatrix}$ (c) $\begin{bmatrix} 3 & 1 \\ 14 & 11 \end{bmatrix}$ (e) $\begin{bmatrix} 12 & 9 \\ 8 & 5 \\ 13 & 4 \end{bmatrix}$

3. [6]

Section 3.1

1. (a) 25 (b) −9 (c) 2 (d) 1 (e) 1 (f) −1.

Section 3.2

1. (a) − 14 (c) 1 (e) 340 (g) −3.
7. $a_{11}a_{22} \ldots a_{nn}$.

Section 3.4

1. (a) − 18 (c) −6 (e) 9 (g) 104.

Section 3.5

1. (a) $\frac{1}{5}\begin{bmatrix} -6 & -5 & 4 \\ 3 & 5 & -2 \\ 7 & 5 & -3 \end{bmatrix}$ (c) $\begin{bmatrix} 1 & -4 & 1 \\ 1 & 1 & -2 \\ -1 & 1 & 1 \end{bmatrix}$

(e) $\begin{bmatrix} 1 & -2 & 1 & 0 \\ 1 & -2 & 2 & -3 \\ 0 & 1 & -1 & 1 \\ -2 & 3 & -2 & 3 \end{bmatrix}$ (g) $\frac{1}{40}\begin{bmatrix} -9 & 1 & 1 & 11 \\ 1 & 1 & 11 & -9 \\ 1 & 11 & -9 & 1 \\ 11 & -9 & 1 & 1 \end{bmatrix}$

13. (a) $\dfrac{1}{1+x^2}\begin{bmatrix} 1 & 0 & -x \\ 0 & 1+x^2 & 0 \\ x & 0 & 1 \end{bmatrix}$ (b) $\dfrac{1}{x^3-2x}\begin{bmatrix} x^2-1 & x & 1 \\ x & x^2 & x \\ 1 & x & x^2-1 \end{bmatrix}$, if $x \neq 0, \pm\sqrt{2}$.

(c) $\frac{1}{x^3+1}\begin{bmatrix} -x & 1 & x^2 \\ x^2 & -x & 1 \\ 1 & x^2 & -x \end{bmatrix}$, if $x \neq -1, \frac{1 \pm i\sqrt{3}}{2}$.

15. $\frac{1}{a^2+b^2+c^2+1}\begin{bmatrix} 1+c^2 & -a-bc & -b+ac \\ a-bc & 1+b^2 & -c-ab \\ b+ac & c-ab & 1+a^2 \end{bmatrix}$.

23. 6.

Section 3.6

1. (a) $x = -\frac{11}{13}, y = \frac{45}{13}, z = \frac{23}{13}$.
   (c) $x = 7, y = -1, z = 4$.

2. (a) $\frac{1}{14}\begin{bmatrix} -7 & -5 & 11 \\ -7 & -1 & 5 \\ 7 & 7 & -7 \end{bmatrix}$ (c) $\begin{bmatrix} 2 & -1 & -5 & 2 \\ -2 & 1 & 3 & -1 \\ 1 & -1 & 8 & -3 \\ 1 & -1 & 3 & -1 \end{bmatrix}$

5. $\frac{1}{6}, \frac{1}{2}, \frac{1}{3}$.

Section 3.7

1. (a) $\frac{1}{5}\begin{bmatrix} -10 & 5 & 5 \\ 0 & -1 & 2 \\ 5 & -1 & -3 \end{bmatrix}$ (c) $\begin{bmatrix} 7 & -3 & -3 \\ -1 & 1 & 0 \\ -1 & 0 & 1 \end{bmatrix}$

   (e) $\frac{1}{4}\begin{bmatrix} -30 & 3 & 1 & 6 \\ 24 & -10 & -2 & -4 \\ -4 & 6 & 2 & 0 \\ -10 & 3 & 1 & 2 \end{bmatrix}$ (g) $\frac{1}{4}\begin{bmatrix} 1 & 1 & 1 & 1 \\ 1 & 1 & -1 & -1 \\ 1 & -1 & 1 & -1 \\ 1 & -1 & -1 & 1 \end{bmatrix}$

3. $$\begin{bmatrix} 1 + a_2\lambda_2 + \cdots + a_n\lambda_n & -a_2 & -a_3 & \cdots & -a_n \\ -\lambda_2 & 1 & 0 & \cdots & 0 \\ -\lambda_3 & 0 & 1 & \cdots & 0 \\ & \vdots & & & \\ -\lambda_n & 0 & 0 & \cdots & 1 \end{bmatrix}$$

5. $\begin{bmatrix} 0 & \frac{5}{4} & 0 & 0 \\ 0 & 0 & \frac{4}{3} & 0 \\ 0 & 0 & 0 & 3 \\ 4 & -\frac{1}{2} & -\frac{8}{3} & -9 \end{bmatrix}$, $\begin{bmatrix} 1250 \\ 800 \\ 900 \\ 400 \end{bmatrix}$

7. $$\frac{1}{2}\begin{bmatrix} 1 & -1 & 0 & 0 & \cdots & 0 \\ 1 & 0 & -1 & 0 & \cdots & 0 \\ 1 & 0 & 0 & -1 & \cdots & 0 \\ \cdots & \cdots & \cdots & \cdots & \cdots & \cdots \\ 1 & 0 & 0 & 0 & & -1 \\ 3-n & 1 & 1 & 1 & & 1 \end{bmatrix}$$

Section 4.9

1. (a) $\begin{bmatrix} x-y \\ y \end{bmatrix}$ (b) $\begin{bmatrix} \frac{1}{2}(x+y) \\ \frac{1}{2}(x-y) \end{bmatrix}$ (c) $\begin{bmatrix} -x+y \\ -x \end{bmatrix}$ (d) $\begin{bmatrix} 3x-5y \\ -x+2y \end{bmatrix}$
5. (a) $\lambda = -2$ or 1. (b) All values of λ.

Section 5.1

5. $\begin{bmatrix} \frac{3}{4} & -\frac{1}{4} \\ -\frac{1}{8} & \frac{3}{4} \end{bmatrix}$
7. $\begin{bmatrix} \frac{2}{3} & \frac{1}{2} \\ \frac{1}{3} & \frac{1}{2} \end{bmatrix}$

Section 5.2

1. All are induced by matrices of the form $\begin{bmatrix} a & 0 \\ 0 & b \end{bmatrix}$.
7. All are induced by matrices of the form:
(a) $\begin{bmatrix} a & b \\ 0 & d \end{bmatrix}$ (b) $\begin{bmatrix} a & 0 \\ c & d \end{bmatrix}$ (c) $\begin{bmatrix} a & b \\ c & a+b-c \end{bmatrix}$

Section 5.5

1. (a) 2 (c) 3 (e) 3.

Section 5.8

1. (a) $\begin{bmatrix} 1 & 0 & 0 \\ 0 & 1 & 0 \\ 0 & 0 & -1 \end{bmatrix}$, $\begin{bmatrix} 2 & 2 & 0 \\ 0 & -1 & 0 \\ 0 & 0 & -1 \end{bmatrix}$
13. (a) $1 + \frac{1}{2}x + \frac{1}{5}x^2$ (b) $\frac{1}{7}x^3 + x^2 + 1$ (c) $-\frac{1}{5}x^2 + \frac{7}{5}$.

Section 6.1

1. $\cos^{-1}(\frac{1}{2})$, $\cos^{-1}(\sqrt{\frac{2}{3}})$, $\cos^{-1}(\sqrt{\frac{2}{3}})$.
3. (a) $\frac{1}{\sqrt{13}}(2\boldsymbol{i} + 3\boldsymbol{j})$ (b) $\frac{1}{3}(\boldsymbol{i} + 2\boldsymbol{j} + 2\boldsymbol{k})$ (c) $\frac{1}{7}(6\boldsymbol{i} + 3\boldsymbol{j} + 2\boldsymbol{k})$.
5. 90°, $\cos^{-1}(\frac{1}{\sqrt{5}})$, $\cos^{-1}(\frac{2}{\sqrt{5}})$.
7. $\cos^{-1}(\frac{1}{3})$.
9. 1.
11. $x + y + z = 3$.
13. The route is given by $\boldsymbol{r}(t) = (200 - 200\sqrt{5}t)\boldsymbol{i} + 400t\boldsymbol{j}$. Interception will occur at $t = \dfrac{2}{3 + 2\sqrt{5}}$.
21. $\cos^{-1}(\frac{1}{\sqrt{3}})$, $\cos^{-1}(\sqrt{\frac{2}{3}})$.

Section 7.1

1. (a) $x^2 - 1$; 1, -1 (c) $x^2 - x - 2$; 2, -1 (e) $x^3 + x^2 - x - 1$; 1, -1.
(g) $x^3 - 3x - 2$; 2, -1.

2. (a) 1, sp($\boldsymbol{i}$) (c) 1, sp($\boldsymbol{i}+\boldsymbol{j}+\boldsymbol{k}$); -1, sp($\boldsymbol{k}$).

3. (b) $\begin{bmatrix} 0 & 0 & 0 \\ 0 & 1 & 0 \\ 0 & 0 & 4 \end{bmatrix}$

4. (a) $\begin{bmatrix} 2 \\ -1 \\ 1 \end{bmatrix}$ (c) $\begin{bmatrix} 1 \\ 1 \\ 1 \end{bmatrix}$

9. (a) $\begin{bmatrix} \frac{5}{12} \\ \frac{1}{3} \\ \frac{1}{4} \end{bmatrix}$ (c) $\begin{bmatrix} \frac{5}{12} \\ \frac{1}{3} \\ \frac{1}{4} \end{bmatrix}$

13. (a) $\frac{1}{3}\begin{bmatrix} 4-(-2)^n & -2+2(-2)^n \\ 2-2(-2)^n & -1+4(-2)^n \end{bmatrix}$ (c) $\frac{1}{2}\begin{bmatrix} 1+3^n & 3^n-1 \\ 3^n-1 & 1+3^n \end{bmatrix}$

Section 7.2

1. (a) $\frac{1}{\sqrt{2}}\begin{bmatrix} 1 & 1 \\ -1 & 1 \end{bmatrix}$ (c) $\frac{1}{\sqrt{13}}\begin{bmatrix} 2 & 3 \\ -3 & 2 \end{bmatrix}$ (e) $\frac{1}{3}\begin{bmatrix} 1 & 2 & 2 \\ 2 & 1 & -2 \\ 2 & -2 & 1 \end{bmatrix}$

3. (a) $\frac{1}{\sqrt{2}}\begin{bmatrix} i & -i \\ 1 & 1 \end{bmatrix}$

5. $\frac{1}{2}\begin{bmatrix} a+b+\sqrt{(a-b)^2+4c^2} & 0 \\ 0 & a+b-\sqrt{(a-b)^2+4c^2} \end{bmatrix}$

Section 7.3

1. (a) $(x-1)^3$; sp($\boldsymbol{i}, \boldsymbol{j}, \boldsymbol{k}$) (c) $x(x-1)^2$; eigenvalues and generalized eigenspaces are: 0, sp($\boldsymbol{i}+\boldsymbol{k}$); 1, sp($\boldsymbol{i}+\boldsymbol{j}+\boldsymbol{k}$); -1, sp($\boldsymbol{i}-3\boldsymbol{j}+3\boldsymbol{k}$).

Section 7.4

1. (a) $\begin{bmatrix} 1 & 1 \\ 0 & 1 \end{bmatrix}$ (c) $\begin{bmatrix} 2 & 0 \\ 0 & 3 \end{bmatrix}$ (e) $\begin{bmatrix} 1 & 0 & 0 \\ 0 & 0 & 1 \\ 0 & 0 & 0 \end{bmatrix}$

   (g) $\begin{bmatrix} 1 & 0 & 0 \\ 0 & 2 & 1 \\ 0 & 0 & 2 \end{bmatrix}$

index

A 6
B 7
C 8
D 9
E 0
F 1
G 2
H 3
I 4
J 5